Electroporation Protocols

METHODS IN MOLECULAR BIOLOGY™

John M. Walker, SERIES EDITOR

Electroporation Protocols

Preclinical and Clinical Gene Medicine

Shulin Li

Editor

Department of Comparative Biomedical Sciences,
School of Veterinary Medicine, Louisiana State University,
Baton Rouge, LA, USA

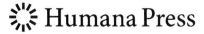 Humana Press

© 2008 Humana Press
999 Riverview Drive, Suite 208
Totowa, New Jersey 07512

humanapress.com

Due diligence has been taken by the publishers, editors, and authors of this book to assure the accuracy of the information published and to describe generally accepted practices. The contributors herein have carefully checked to ensure that the drug selections and dosages set forth in this text are accurate and in accord with the standards accepted at the time of publication. Notwithstanding, as new research, changes in government regulations, and knowledge from clinical experience relating to drug therapy and drug reactions constantly occurs, the reader is advised to check the product information provided by the manufacturer of each drug for any change in dosages or for additional warnings and contraindications. This is of utmost importance when the recommended drug herein is a new or infrequently used drug. It is the responsibility of the treating physician to determine dosages and treatment strategies for individual patients. Further it is the responsibility of the health care provider to ascertain the Food and Drug Administration status of each drug or device used in their clinical practice. The publisher, editors, and authors are not responsible for errors or omissions or for any consequences from the application of the information presented in this book and make no warranty, express or implied, with respect to the contents in this publication.

This publication is printed on acid-free paper.

ANSI Z39.48-1984 (American Standards Institute) Permanence of Paper for Printed Library Materials.

Production Editor: Michele Seugling

Cover illustration: Courtesy of Shulin Li

Cover design by Nancy Fallatt

For additional copies, pricing for bulk purchases, and/or information about other Humana titles, contact Humana at the above address or at any of the following numbers: Tel.: 973-256-1699; Fax: 973-256-8341; E-mail: orders@humanapr.com; or visit our website at www.humanapress.com.

Printed in the United States of America. 10 9 8 7 6 5 4 3 2 1

ISBN: 978-1-59745-194-9 (e-book)

Library of Congress Control Number: 2007934681

Preface

Delivery of genetic materials into the target tissues or cells via electric pulses for treating or preventing diseases is referred to as electroporation gene therapy. The focus of this book is to provide hands-on protocols for, and in-depth knowledge on, nonviral delivery of genetic materials, including DNA, oligo-DNA, and RNA, into the targeted immune cells, tissues, and model animals for treating diseases.

When the light bulb was invented, no one, including those with the most imaginative minds, thought of the future applications of electric fields for delivering genes to prevent or treat diseases. No one thought about this type of application even after electric shock was used for restarting the suddenly rested heart in the clinical setting, after genes were discovered, after the gene was successfully delivered into the mammalian cells via electroporation by the pioneer Pr. Eberhard Neumann in 1982, or after electroporation was routinely used in many laboratories for transfecting the genes of interest into cell lines to study the gene function. As detailed in the first review chapter by Dr. Lluis Mir in Part I of this book, there are many reasons for the lack of initial interest in using this technique for gene delivery in a clinical setting, one of the extreme reasons being that electric pulses are associated with torture and the death penalty. This electroporation technology was eventually applied for gene delivery in animals and in humans because of the success in using electroporation for delivering the antineoplastic drug bleomycin to treat highly malignant tumors. Therefore, a review of electroporation chemotherapy is presented in Part I to provide the bridging information between electroporation, electroporation chemotherapy, and electroporation gene therapy, although the latter is the central theme of this book.

To help investigators who are new to this exciting technology, reviews are provided in Part I of this book to cover the theory on how DNA is delivered into cells and tissues via electroporation and the common devices that are used for performing electroporation in vitro and in vivo. To help those who are familiar with this field but anxious to translate this approach to larger animals and clinical study, this book provides not only the basic methods for delivering genes into target cells (Part II) and targeted tissues (Part III), and for the treatment of diseases in rodent models, but also the protocols or generic protocols for delivering genes into dogs, pigs, horses, nonhuman primates, and humans (Parts IV–VI).

The methods for delivering genes into immune cells, such as dendritic cells, are described in Part II because high transfection efficiency of mRNA or DNA into immune cells is achievable with the current approach, and immune-cell-based vaccinations seem very effective in the clinical setting for treating malignancy. Moreover, the development of in situ electroporation, as discussed in Part II, and the discovery of electroporation buffers make transfection of cells more effective via electroporation without the introduction of viral DNA vectors. Therefore, it is our vision that transfection of therapeutic genes or vaccines into immune cells and other types of cells via electroporation for treating diseases may eventually have the same impact as in vivo electroporation gene therapy. This part will help investigators interested in cell-based vaccinations consider electroporation as an alternative approach. To address the clinical demand, a chapter describing large-volume cell transfection via electroporation is also included. In addition, a unique protocol to discuss how to use electroporation for eliminating adipose cells has been included to illustrate a novel application of this technique.

In Part III, the focus is on tissue-targeted gene delivery via electroporation. What is unique about this part is the inclusion of not only the accessible tissues but also the internal organs such as lung and fragile tissues such as eyes for electroporation-mediated gene delivery. In addition, delivery of DNA into other model species such as zebra fishes and delivery of oligo-DNA into the targeted tissues via electroporation are also included in this section. This coverage aims to give the investigator another look into the versatility of this technique and stimulate some novel thinking in applying this technique in other unexplored tissues.

Parts IV through VI deal with the applications of electroporation-based gene delivery, which include DNA vaccinations and treatment of cancer and various other diseases. Three unique aspects are included in these sections: applications in larger animals such as dogs, applications in humans, and applications in a wide spectrum of disease models. This wide-spectrum coverage of different models may help expand the vision of investigators on this technology in future applications. This encyclopedia approach aims to build sufficient information for investigators who want to get to know the electroporation gene therapy field in one stop and for investigators who want to translate this technique from bench to bed. These sections also provide enough protocols and knowledge for veterinarians to use this approach for treating diseases in pets.

Although a wide spectrum of disease models are described in the chapters of Parts IV through VI, a high degree of focus is given to cancer electroporation gene therapy and DNA vaccination because our view is that these two areas are closer than any other areas to undergoing human trials and for eventually treating human diseases using this technology. The electroporation-based DNA vaccination will greatly reduce the cost, speed up the development, and increase the applications of new vaccines. A few chapters in Part V are also provided to describe systemically or locally delivering both genes and antineoplastic drugs for treating malignancy because the treatment of tumors eventually requires combination therapy, and electroporation provides a unique platform for administering both. These early

discoveries and applications will hopefully serve as seeds for the future blossoming of this field. In summary, this book will benefit investigators, readers, and the public in this subject.

The editor thanks all authors who worked so hard to get this book to press on time. He extends his special thanks to his graduate assistant Jeffry Cutrera and undergraduate student Robin Barret for helping to format the chapters, and post-doctoral researcher Boyu Zhang for drawing figures for several chapters to fit the format of this book. Finally, the editor is very thankful to his Department Head, Gary Wise, PhD, for providing space and time to accomplish this work.

Shulin Li

Contents

Contributors

Cornell Allen
MaxCyte, Inc., Gaithersburg, MD, USA

Aikaterini Anagnostopoulou
Ask Science Products, Inc., Kingston, Ontario, Canada

Khursheed Anwer
Expression Genetics, Inc., Mock Road, Huntsville, AL, USA

Rozanne Arulanandam
Department of Microbiology and Immunology and Pathology,
Queen's University, Kingston, Ontario, Canada

Pascal Bigey
Unité de Pharmacologie Chimique et Génétique, René Descartes Paris 5
University, Faculté de Pharmacie, Paris, France

Aude Bonehill
Department of Physiology and Immunology, Medical School of the Vrije
Universiteit Brussel (VUB), Brussels, Belgium

Patricia A. Brown
VGX Animal Health, The Woodlands, TX, USA

Heather L. Brownell
Department of Microbiology and Immunology and Pathology, Queen's University,
Kingston, Ontario, Canada

Oliver Brüstle
Institute of Reconstructive Neurobiology, Life and Brain Center, University of
Bonn and Hertie Foundation, Bonn, Germany

Jun Cao
Department of Microbiology and Immunology and Pathology, Queen's University,
Kingston, Ontario, Canada

Constance L. Cepko
Department of Genetics and Howard Hughes Medical Institute, Harvard Medical
School, Boston, MA, USA

Nge Cheong
Department of Paediatrics, National University of Singapore, Singapore

Jurgen Corthals
Department of Physiology and Immunology, Medical School
of the Vrije Universiteit Brussel (VUB), Brussels, Belgium

Abie Craiu
Science Research Laboratory, Inc., Somerville, MA, USA
and
Kendle International, Inc., Thousand Oaks, CA, USA

Jeffry Cutrera
Department of Comparative Biomedical Sciences, Louisiana State University,
Baton Rouge, LA, USA

David A. Dean
Division of Pulmonary and Critical Care Medicine, Feinberg School of Medicine,
Northwestern University, Chicago, IL, USA

Ruxandra Draghia-Akli
VGX Pharmaceuticals, Immune Therapeutics Division, The Woodlands, TX, USA

Igor R. Efimov
Department of Biomedical Engineering, Washington University in St. Louis,
St. Louis, MO, USA

Vadim V. Fedorov
Department of Biomedical Engineering, Washington University in St. Louis,
St. Louis, MO, USA

Aurora Ferrer
Department of Cellular and Molecular Neuroscience, Division of Neuroscience
and Mental Health, Imperial College London, Charing Cross Campus, London,
United Kingdom

Jason G. Fewell
Expression Genetics, Inc., Huntsville, AL, USA

Kevin L. Firth
Ask Science Products, Inc., Kingston, Ontario, Canada

Helen Foster
Department of Cellular and Molecular Neuroscience, Division of Neuroscience
and Mental Health, Imperial College London, Charing Cross Campus,
London, United Kingdom

Joseph C. Fratantoni
MaxCyte, Inc., Gaithersburg, MD, USA

Julie Gehl
Department of Oncology, Copenhagen University Hospital at Herlev,
Herlev Ringvej, Denmark

Muriel Golzio
IPBS-CNRS, Toulouse, France

James G. Granneman
Department of Psychiatry, Wayne State University School of Medicine,
Detroit, MI, USA

John W. Harmon
Department of Surgery, Johns Hopkins Bayview Medical Center, Baltimore,
MD, USA

Carlo Heirman
Department of Physiology and Immunology, Medical School of the Vrije
Universiteit Brussel (VUB), Brussels, Belgium

Masatsugu Hori
Department of Cardiovascular Dynamics, Research Institute, National
Cardiovascular Center, Suita, Japan

Holly M. Horton
Vical, Inc., San Diego, CA, USA

Sek-Wen Hui
Cancer Biology Department, Roswell Park Cancer Institute, and Department
of Molecular and Cellular Biophysics and Biochemistry, University at Buffalo,
Buffalo, NY, USA

Enyu Imai
Department of Cardiovascular Dynamics, Research Institute, National
Cardiovascular Center, Suita, Japan

Akito Inadome
Department of Urology, Graduate School of Medical Sciences, Kumamoto
University, Kumamoto, Japan

Hitoshi Iwashita
Department of Urology, Graduate School of Medical Sciences, Kumamoto
University, Kumamoto, Japan

Robert M.I. Kapsa
Howard Florey Institute and St. Vincent's Hospital, (Melbourne) Parkville,
Victoria, Australia
and
The Bionic Ear Institute, East Melbourne, Victoria, Australia

Amir S. Khan
VGX Pharmaceuticals, Immune Therapeutics Division, The Woodlands,
Texas, USA

Alan King
Cyto Pulse Sciences, Glen Burnie, MD, USA

Rune Kjeken
Inovio Biomedical Corp., San Diego, CA, USA

Philipp Koch
Institute of Reconstructive Neurobiology, Life and Brain Center, University
of Bonn and Hertie Foundation, Bonn, Germany

Kazuo Komamura
Department of Cardiovascular Dynamics, Research Institute, National
Cardiovascular Center, Suita, Japan

Peggy A. Lalor
Vical, Inc., San Diego, CA, USA

Shulin Li
Department of Comparative Biomedical Sciences, School of Veterinary Medicine,
Louisiana State University, Baton Rouge, LA, USA

Linda N. Liu
MaxCyte, Inc., Gaithersburg, MD, USA

Lixin Liu
Department of Surgery, Johns Hopkins Bayview Medical Center,
Baltimore, MD, USA

Jill McMahon
Department of Cellular and Molecular Neuroscience, Division of Neuroscience
and Mental Health, Imperial College London, Charing
Cross Campus, London, United Kingdom

Guy P. Marti
Department of Surgery, Johns Hopkins Bayview Medical Center, Baltimore,
MD, USA

Koichi Masunaga
Department of Urology, Graduate School of Medical Sciences, Kumamoto
University, Kumamoto, Japan

Iacob Mathiesen
Inovio AS, Oslo, Norway

Takahiko Matsuda
Department of Genetics and Howard Hughes Medical Institute, Harvard Medical
School, Boston, MA, USA

Kunio Matsumoto
Department of Cardiovascular Dynamics, Research Institute, National
Cardiovascular Center, Suita, Japan

Neal Mauldin
Veterinary Clinical Sciences, Louisiana State University,
Baton Rouge, LA, USA

Babu M. Medi
DelSite Biotechnologies, Inc., Irving, TX, USA

Annelies Michiels
Department of Physiology and Immunology, Medical School of the Vrije
Universiteit Brussel (VUB), Brussels, Belgium

Lluis M. Mir
CNRS UMR 8121, Institut Gustave-Roussy, Villejuif, France
Univ Paris-Sud, UMR 8121

Tomoharu Miyashita
Department of Surgery, Johns Hopkins Bayview Medical Center, Baltimore,
MD, USA

Jun-ichi Miyazaki
Department of Cardiovascular Dynamics, Research Institute, National
Cardiovascular Center, Suita, Japan

Parsa Mohebi
Department of Surgery, Johns Hopkins Bayview Medical Center, Baltimore,
MD, USA

Toshikazu Nakamura
Department of Cardiovascular Dynamics, Research Institute, National
Cardiovascular Center, Suita, Japan

Vladimir P. Nikolski
Department of Biological Engineering, Washington University in St. Louis,
St. Louis, MO, USA

Lars Nolden
Institute of Reconstructive Neurobiology, Life and Brain Center, University
of Bonn and Hertie Foundation, Bonn, Germany

James E. Norton
Division of Pulmonary and Critical Care Medicine, Feinberg School of Medicine,
Northwestern University, Chicago, IL, USA

Masayuki Otani
Department of Urology, Graduate School of Medical Sciences, Kumamoto
University, Kumamoto, Japan

Christian Ottensmeier
Southampton University Hospital, Southampton, United Kingdom

Francesca Papalini
Laboratory of Tumor Immunology, Immunology Center, INRCA Research
Department, Ancona, Italy

Sara Pierpaoli
Laboratory of Tumor Immunology, Immunology Center, INRCA Research
Department, Ancona, Italy

Pavel Pisa
Department of Oncology and Pathology, Cancer Center Karolinska R8:01,
Karolinska Hospital, Stockholm, Sweden

Mauro Provinciali
Laboratory of Tumor Immunology, Immunology Center, INRCA Gerontology
Research Department, Ancona, Italy

Anita F. Quigley
Howard Florey Institute and St. Vincent's Hospital, (Melbourne) Parkville,
Victoria, Australia
and
The Bionic Ear Institute, East Melbourne, Victoria, Australia

Dietmar Rabussay
Inovio Biomedical Corp., San Diego, CA, USA

K. Murali Rambabu
Center for Cellular and Molecular Biology, Hyderabad, India

John D.A. Ramos
Department of Biological Sciences, College of Science, University of Santo
Tomas, Espana, Manila, Philippines

N. Madhusudhana Rao
Center for Cellular and Molecular Biology, Hyderabad, India

S. Harinarayana Rao
Reliance Life Sciences Pvt. Ltd, Navi Mumbai, India

Leda Raptis
Department of Microbiology and Immunology, Queen's University, Kingston,
Ontario, Canada

Alain P. Rolland
Vical, Inc., San Diego, CA, USA

Marie-Pierre Rols
Institut de Pharmacologie et de Biologie Structurale, Toulouse, France

Anna-Karin Roos
Department of Oncology and Pathology, Cancer Center Karolinska R8:01,
Karolinska Institute, Stockholm, Sweden

David Scadden
Center for Regenerative Medicine, Massachusetts General Hospital,
Harvard Medical School, Boston, MA, USA

Daniel Scherman
Unité de Pharmacologie Chimique et Génétique, Faculté de Pharmacie,
René Descartes Paris 5 University, Paris, France

Kathrin Schoenberg
University Clinic of Düsseldorf, Institute for Transplantation Diagnostics
and Cell Therapeutics, Düsseldorf, Germany

Henrike Siemen
Institute of Reconstructive Neurobiology, Life and Brain Center, University of
Bonn and Hertie Foundation, Bonn, Germany
and
Department of Obstetrics and Gynecology, Stanford University, Stanford, CA, USA

Keijiro Shiomitsu
Veterinary Clinical Sciences, Louisiana State University, Baton Rouge, LA, USA

Rama Shivakumar
MaxCyte, Inc., Gaithersburg, MD, USA

Jagdish Singh
Department of Pharmaceutical Sciences, North Dakota State University, Fargo,
ND, USA

Arianna Smorlesi
Laboratory of Tumor Immunology, Immunology Center, INRCA Research
Department, Ancona, Italy

Raji Sundararajan
ECET Department, Knoy Hall of Technology, 401 Grant St.,
West Lafayette, IN, USA

Justin Teissié
IPBS-CNRS, Toulouse, France

Stefanie Terstegge
Institute of Reconstructive Neurobiology, Life and Brain Center, University
of Bonn and Hertie Foundation, Bonn, Germany

Kris Thielemans
Department of Physiology and Immunology, Medical School of the Vrije
Universiteit Brussel (VUB), Brussels, Belgium

Torunn Elisabeth Tjelle
Inovio AS, Oslo, Norway

Evangelia Tomai
Ask Science Products, Inc., Kingston, Ontario, Canada

Marina Torrero
Department of Comparative Biomedical Sciences, Louisiana State University,
Baton Rouge, LA, USA

Capucine Trollet
Unité de Pharmacologie Chimique et Génétique, Faculté de Pharmacie,
René Descartes Paris 5 University, Paris, France

Hans-Ingo Trompeter
University Clinic of Düsseldorf, Institute for Transplantation Diagnostics
and Cell Therapeutics, Düsseldorf, Germany

Sandra Tuyaerts
Department of Physiology and Immunology, Medical School of the Vrije
Universiteit Brussel (VUB), Brussels, Belgium

Markus Uhrberg
University Clinic of Düsseldorf, Institute for Transplantation Diagnostics
and Cell Therapeutics, Düsseldorf, Germany

Adina Vultur
Department of Microbiology and Immunology and Pathology, Queen's University,
Kingston, Ontario, Canada

Jiaai Wang
Department of Surgery, Johns Hopkins Bayview Medical Center,
Baltimore, MD, USA

Dominic J. Wells
Department of Cellular and Molecular Neuroscience, Division of Neuroscience
and Mental Health, Imperial College London, Charing Cross Campus,
London, United Kingdom

Kim E. Wells
Department of Cellular and Molecular Neuroscience, Division of Neuroscience
and Mental Health, Imperial College London, Charing Cross Campus,
London, United Kingdom

Sharon H.A. Wong
Howard Florey Institute and St. Vincent's Hospital, (Melbourne) Parkville,
Victoria, Australia
and
The Bionic Ear Institute, East Melbourne, Victoria, Australia

Yuhong Xu
School of Pharmacy, Shanghai Jiao Tong University, Shanghai, People's Republic
of China

Kaw Yan Chua
Department of Paediatrics, National University of Singapore, Singapore
and
A member of Immunology Programme, National University of Singapore,
Singapore

Masaki Yoshida
Department of Urology, Graduate School of Medical Sciences, Kumamoto
University, Kumamoto, Japan

Yong-Gang Zhao
Institute Pasteur of Shanghai, Chinese Academy of Sciences, Shanghai, China

Rui Zhou
Division of Pulmonary and Critical Care Medicine, Feinberg
School of Medicine, Northwestern University, Chicago, IL, USA

Shiguo Zhu
Department of Comparative Biomedical Sciences, Louisiana State University,
Baton Rouge, LA, USA

Part I
Basic Concepts of DNA Transfer
via Electroporation

Chapter 1
Application of Electroporation Gene Therapy: Past, Current, and Future

Lluis M. Mir

Abstract Twenty-five years after the publication of the first report on gene transfer in vitro in cultured cells by the means of electric pulse delivery, reversible cell electroporation for gene transfer and gene therapy (DNA electrotransfer) is at a crossroad in its development. Present knowledge on the effects of cell exposure to appropriate electric field pulses, particularly at the level of the cell membrane, is reported here as an introduction to the large range of applications described in this book. The importance of the models of electric field distribution in tissues and of the correct choice of electrodes and applied voltages is highlighted. The mechanisms involved in DNA electrotransfer, which include cell electropermeabilization and DNA electrophoresis, are also surveyed. The feasibility of electric pulse for gene transfer in humans is discussed taking into account that electric pulse delivery is already regularly used for localized drug delivery in the treatment of cutaneous and subcutaneous solid tumors by electrochemotherapy. Because recent technological developments have made DNA electrotransfer more efficient and safer, this nonviral gene therapy approach is now ready to reach the clinical stage. A good understanding of DNA electrotransfer principles and a respect for safe procedures will be key elements for the successful future transition of DNA electrotransfer to the clinics.

Keywords: electroporation, electropermeabilization, DNA electrotransfer, electric pulses, gene delivery, nonviral gene therapy

1. Introduction

Electroporation for gene transfer and gene therapy is one of the biotechnological and biomedical applications of cell electroporation. Reversible cell electroporation and DNA electrotransfer emerged in 1982 (1), have been developing since then, and now are arriving at a crossroad in their development. Who could imagine, 25 years ago, that this laboratory technique, which often killed a large part of the cells exposed to the electric pulses, could be used for biomedical purposes and directly

applied to humans? It was not evident that one day physicians could safely deliver "electroporating" high-voltage pulses to patients not only because of the gap the technique had to cross, but also because of the fears that all societies ineluctably express when somebody addresses the question of the delivery of intense electric pulses to humans. Since Mary Shelley's book, intense electric pulses are associated with malefic creatures, Frankenstein being of course the name that everybody knows. Dramatically, electric pulses are also associated with worse cases, namely, instruments of torture and death penalty.

Now, researchers and physicians, together, have to demonstrate to society that short and intense electric pulses can be beneficial to mankind. Indeed, the demonstration in animals of the potential of these pulses is complete, as testified by the present book *Electroporation Protocols: Experimental and Clinical Gene Medicine*. Thus, the next step will be the treatment of patients with this nonviral gene therapy method.

Gene therapy has slowed down and needs more than one great success to make a new start. Will the emergence of electroporation gene therapy be one of these successes that will help put the entire field in movement again? Probably all the authors contributing to this book are working in this direction and are already building this success, which is on the horizon, as the first clinical trials have been (or will soon be) activated, as reported recently by R. Heller and M. Fons at the 2006 annual meeting of the American Society for Gene Therapy.

2. What Happens to a Cell or a Tissue Exposed to an Electric Field?

The inside of the cell, where all the biochemical reactions necessary for life occur, is rich in ions and highly conductive. At least for animal cells, the external medium, either in vitro (culture medium) or in vivo (biological fluids), is also highly conductive. Contrarily, the plasma membrane, which insulates the inside of the cell from the outside, is nonconductive. In an electric field, all the charges inside and outside the cell will be electrophoretically displaced, but their movement will be stopped at the plasma membrane. Thus, charges of opposite sign will accumulate at the two sides of the plasma membrane and this accumulation will result in the induction of a transmembrane potential difference, ΔV_M, at point M of the cell surface, which, under valid simplifying hypothesis, follows the equation

$$\Delta V_M = 1.5 \times r \times E_{ext} \times \cos\theta$$

where r is the radius of the cell, E_{ext} is the external electric field strength, and θ is the polar angle with respect to the electric field direction (2–4). Once the transmembrane potential difference is established (in less than 1 µs), membrane

structural changes occur in the area of the cell surface where ΔV_M surpasses a threshold value. All the experimental arguments demonstrate that ΔV_M has to be maintained for at least 30–40 µs to make these structural changes complete, which explains why, with square wave pulses, 100 µs has been a very common pulse duration for more than 20 years (5, 6). The conductivity of tissues exposed to electric pulses changes rapidly during this period of time, and these changes in tissue conductivity are one of the measurable signatures of cell electroporation. But—what is cell electroporation?

3. Cell Electroporation

It is important to be precise in this introductory review about what we do and do not know about cell electroporation after 25 years of research. First of all, electroporation is a commonly used term. It is so common that the title of this book is *Electroporation Protocols: Experimental and Clinical Gene Medicine*. However, what consistently defines an "electroporated" cell? It is its increased permeability to otherwise low permeant or nonpermeant molecules (nonpermeant molecules are all those molecules, usually hydrophilic and of relatively larger size, that do not diffuse across the plasma membrane and for which there are no active transporters at the plasma membrane (6); low permeant molecules are molecules that may diffuse through the plasma membrane even though their diffusion is limited or hindered by the normal membrane structure). Thus, the functional demonstration of achieving cell "electroporation" requires the experimental demonstration of the "electropermeabilization" of the cells, and this last term should have been the exact term used anywhere. Indeed, it is easy to understand what (electro)permeabilization means. There are several theories to explain cell electropermeabilization, and the most common one is referred to as the electroporation theory (1).

The electroporation theory was developed by Pr. Eberhard Neumann, in a paper that was published in 1982 (1). That work is the basis of the entire content of the present book. Indeed, Pr. Neumann, a pioneer as well as a remarkable visionary researcher, reported the first transfer of genes in living cells by means of electric pulses, and the present book *Electroporation Protocols: Experimental and Clinical Gene Medicine* appears 25 years after the seminal paper, as a sort of homage rendered to Pr. Neumann to celebrate this anniversary. In that article, Pr. Neumann developed a theory of electroporation, which states that normal fluctuations of the membrane are enlarged by the electric-pulse-induced transmembrane voltage difference, resulting in large hydrophobic pores. These pores then become hydrophilic by a rotation of the lipids at the limits between the lipid bilayer and the aqueous medium that tries to fill the hydrophobic "pore" (conduit). It is, however, necessary to recognize that, in spite of the active basic research of many groups, there is still a large lack of knowledge on the mechanism of cell electropermeabilization

(or more precisely, on the mechanisms of the electric-pulse-provoked permeabilization of the plasma membrane). Cell electroporation, the duration of which is much longer than the pulse duration, can also be described as the modification of the membrane impermeability to ions and hydrophilic and/or charged molecules. The changes in the diffusion coefficient of these molecules through the membrane can be the result of the hydration and partial loss of the ordered membrane structure (see later) or, as described earlier, the generation of "metastable" hydrophilic pores. In conditions where cells remain alive after their exposure to the electric pulses, no one has been able to see or find evidence of these pores by any means. (The freeze-etching electron microscopy images of pores reported (7) were artifacts due to the choice of a hypoosmotic electroporation medium.) It is normal to use the term reversible electroporation when the consequences of the exposure of the cells to the electric pulses are transient, and thus, reversible. The term irreversible electroporation is reserved for cases where these consequences are permanent, leading to the impossibility of restoring the membrane structure and barrier function, and thus, the impossibility of maintaining cell homeostasis (as permanent cell (electro)permeabilization results in continuous inflow of Na^+ ions and continuous outflow of K^+ ions). In irreversible electroporation, the permeabilization structures could actually be permanent hydrophilic pores in the cell membrane. The natural consequence of irreversible electroporation is cell death (8, 9).

The mechanisms of cell membrane electropermeabilization have recently been discussed by Dr. J. Teissie and colleagues (10) in a paper subtitled "A Minireview of Our Present (Lack of?) Knowledge." Changes that may contribute to making the cell membrane reversibly permeable without "pores" can result from the following:

1. The electrocompression forces generated by electric field associated with the transmembrane potential difference ΔV_M. For example, a ΔV_M value of about 500 mV across a 5-nm cell membrane is associated to an electric field strength of 1×10^8 V/m, or 1,000,000 V/cm, not far from the field strength in nightlights. These electrocompressive forces bring the two lipid layers closer than the distance that allows the two lipid layers standing one parallel to the other (thus provoking the disruption of the ordered stacking of the lipids of the two lipid layers).
2. Changes in the lipids' polar head orientation, detected by ^{31}P magnetic resonance changes (11).
3. The penetration of water in the lipid layer (hydration of the membrane) resulting from the two previous structural changes of the lipids. The membrane hydration clearly appears in recent simulations of the membranes exposed to high transmembrane potential differences, using validated molecular dynamics programs (12, 13).
4. Possible changes in the structure of transmembrane proteins (10), etc.

In all, these changes sustain a theory of reversible electropermeabilization that does not require the presence of holes (pores) in the cell membrane with the stabilized structure proposed in the usual electroporation theory. In summary, electroporation is a theory, electropermeabilization is an alternate theory, and electropermeabilization is the real consequence of the exposure of cells to appropriate electric pulses.

4. Gene Therapy by Electroporation

Gene therapy by electroporation is another statement that needs to be clarified. In vivo, after the delivery of a train of eight square wave pulses of 100-μs duration each, the membrane of muscle cells or tumor cells remains permeabilized for several minutes (14, 15). However, if DNA is injected in the tissue after the delivery of the electric pulses, there is no transfection, or at least no more transfection than that obtained by the injection of naked DNA in the absence of any electric pulse (14, 16). DNA must be present in the tissue when the electric pulses are delivered. It was also shown that long pulses (at least 5-ms duration) are much more efficient than the short 100-μs pulses (14). As a matter of fact, in 1998, the key year in the development of electrogenetherapy, all four published papers on the subject displayed efficient electrotransfer using long pulses: 5 ms in tumors (17), 50 ms in liver (18), and 50 and 20 ms in skeletal muscle (14, 19).

Previous knowledge on the electrotransfer of small molecules anticipated this result. Indeed, for "small" molecules (small in comparison with normal plasmid size), the number of molecules internalized is roughly inversely proportional to the size of the molecules. For different purposes, we transferred different molecules into the same cell type using exactly the same in vitro pulsing conditions. Lucifer yellow (450 Da) transfer resulted in an internal associated concentration equivalent to 100% (compared with the external concentration at the time of the pulse delivery) (6), bleomycin (1,500 Da) transfer resulted in an internal concentration of 33% (20), 21-mer oligonucleotide (about 6,000 Da) transfer resulted in ~10% internal concentration (21), and soluble protein antiricin A chain antibody (about 150,000 Da) transfer resulted in merely 1% internal concentration (22). Thus, the uptake of a 5-kb plasmid (about 3,000,000 Da) by the same mechanism, diffusion across the electropermeabilized membrane, is very improbable.

As mentioned earlier, DNA uptake is easily detected when using long pulses. One explanation for the need of long duration pulses is the Schwan equation (a derivation of the Laplace equation, published by Schwan in 1957 (2)), which shows that electric pulse field strength determines the area of the cell surface that will be affected by the changes in the cell membrane. However, this equation is a static solution and does not take into account the dynamics of the cell membrane structural changes. Duration of the pulses is also an important parameter. In particular, as shown by Teissie and colleagues (23, 24), total pulse duration determines the intensity of the changes within the area defined by the Schwan equation (total pulse duration being the product of the number of pulses and the individual duration of the pulses). Thus, one cannot exclude that the facilitation of large molecule uptake by long pulses results from a very intense membrane change, which is caused by the long pulses and results in a large increase in the permeability coefficients to externally added molecules.

Also, because DNA is highly charged and can electrophoretically move in the presence of an electric field, we suspected the involvement of electrophoretic forces in DNA electrotransfer (14, 16, 25). The importance of the electrophoretic

forces in the electric-pulse-mediated gene transfer has been demonstrated by our team, using combinations of short intense pulses (termed HV pulses for high-voltage pulses) and long low pulses (termed LV pulses for low-voltage pulses) (16, 26). Indeed, we showed that, under appropriate pulse conditions, the HV pulses permeabilize the muscle cells for 5 min and that the LV pulses (which cannot permeabilize the cells), are responsible for the plasmid transfer to muscle fibers. Moreover, we found that DNA need not be present at the time of cell electropermeabilization (16). The DNA can be injected within a given time after the permeabilizing pulses (up to 50 min under our experimental conditions) and it has to be injected always before the electrophoretic LV pulses, which efficiently push the DNA toward the membrane that is still altered from the delivery of the electroporating pulse. What is the structure of the membrane during this period of time (between 5 and 50 min after the delivery of the electroporating pulse)? Obviously, this is not known; however, observations are more in favor of the theory of cell electropermeabilization (in which gradual changes in membrane electroporation and in membrane resealing might occur) than the theory of electroporation (in which passage of large molecules cannot be dissociated from the passage of the small molecules by diffusion through the pores (holes)). According to the cell electropermeabilization theory, during the lipid bilayer structure reconstitution, the diffusion coefficient for the small molecules progressively decreases, limiting the diffusion of these molecules to amounts that will be below the thresholds allowing the detection of these molecules inside the cells. However, if passage across the membrane is forced because external forces are applied to the molecules that have to cross the membrane, these molecules will still be able to enter the cell, which should be the case for DNA. The electric fields provoke the electrophoretic acceleration of the DNA (a highly charged molecule with one net charge per 300 Da); therefore, the DNA is projected toward the cell membrane, facilitating the interaction between the lipid bilayer and the DNA and its ulterior uptake.

The importance of the electrophoretic effects are also sustained by the vectoriality of DNA electrotransfer (that is, the effects are associated with the direction of the field) (27, 28). However, there is also a vectoriality of the cell permeabilization—all the effects of the electric pulses on living cells are vectorial, as already shown in 1984 by Teissie (29), as well as later on (30)—in the case of DNA electrotransfer.

5. In Vivo Electroporation/Electropermeabilization and DNA Electrotransfer

First of all, it is necessary to recall that DNA electrotransfer is not a systemic method of gene transfer, but a strictly local one. There are two reasons for this restriction (in some cases) or advantage (in other situations).

Indeed electric pulses must be delivered, locally, only to small parts of the body, because:

1. The amount of energy stored by the pulse generators and the voltages delivered by these devices,
2. The relationship between the voltage applied (electric field strength) and the amperage (current) delivered (knowing that the higher the distance between the electrodes, the higher the voltage delivered to keep roughly constant the ratio of the voltage applied to the distance between the electrodes, often used as a way to give a value of the electric field),
3. The potential burns at the contact points between the electrodes and the tissue (if very high-voltage pulses are delivered) and
4. The safety of the procedures and the tolerance of the experimental subject, or of the patients.

The second reason is less stringent: It is important to inject the DNA locally because, on account of the large size of the DNA molecule, its concentration is always very low and thus very sensitive to dilution in systemic (e.g. intravenous) administration.

Moreover, it is important to realize that the transfer of DNA is limited to the tissue volumes that are locally exposed to electric fields of sufficient field strength. However, field strength must not be excessive, because excessive electropermeabilization, as already discussed, causes irreversible changes in the cell membrane and, consequently, the absence of the membrane resealing, death of the target cells, and a decrease in the efficacy of DNA electrotransfer.

Bi- and tri-dimensional models of the in vivo field distribution were developed already in 1999 (15), 2000 (31), and thereafter (32). These analytical or numerical models allow us to understand, define, and foresee the volumes of tissue that will actually be electropermeabilized in vivo, as a function of electrode geometry and voltage applied to the electrodes. Thus, they allow us to anticipate the volumes of tissue that can actually be electrotransferred, as well as those that may be damaged by the exposure to excessive electric fields.

The following methods have been developed to qualitatively or quantitatively determine the level of tissue permeabilization:

1. Chemical means using the toxicity of bleomycin (since at low concentrations only the electropermeabilized cells will be killed) (33) or other biological effects of bleomycin (taking advantage of the pseudoapoptotic properties of this molecule) (31, 34) that may provide even topological information (31, 34),
2. Radioactive means using compounds such as ^{57}Co-bleomycin (33, 35), ^{51}Cr-EDTA (ethylenediamine tetracetic acid) (14), and ^{99}Tc-DTPA (diethylene triamine pentacetate) (36),
3. Fluorescent means using compounds such as propidium iodide (23) or lucifer yellow (6), and
4. Physical means using measurement of impedance or conductance (or both).

The numerical models have been validated by the experimental use of one or more of these tests (for example by Miklavcic et al. (31)). An interesting result that accompanied the work by Miklavcic and colleagues on the in vivo electropermeabilization of the liver was the finding that the diameter of the needles had a

major impact on field distribution in tissues: the thinner the needles, the less homogeneous the electric field distribution (31).

It is also important to note that the exposure to appropriate electric pulses carries consequences at the following three levels:

1. Molecules, provoking the electrophoretic displacement of the charged molecules,
2. Cells, provoking their permeabilization, and
3. Tissues, provoking a vascular lock in the exposed tissue.

The effects of the electric pulses at the level of the tissues are often ignored. In fact, electroporating pulses provoke a vascular lock (that is a transient hypoperfusion) just in the parts of the tissue exposed to the electric pulses, sometimes also with effects on the blood circulation distal to the electropulsed tissue. Vascular lock is a physiological reaction that is histamine-dependent (37). In the skeletal muscle, it lasts for 1–2 min, but the duration can be more pronounced when high-voltage pulses (leading to irreversible electroporation) are used (37). In the liver, the vascular lock lasts longer, but still within the range of a few minutes (38). Conversely, in tumors, the vascular lock lasts for hours (39, 40). In these tissues, the modifications in blood flow could be particularly advantageous for DNA electrotransfer, as this would decrease the washout of the injected DNA (37).

Another very interesting consequence of vascular lock is that there is no bleeding when invasive electrodes (for example, arrays of two or more needles) are removed from the tissue after the pulse delivery. This feature is very important because it allows electrotransfer of the DNA to subcutaneous targets and, even percutaneously, to deeper tissues.

As tissues are characterized by their conductivity and general geometry, it is possible to know the electrical parameters necessary to get electropermeabilization of their cells. Of course, convenient electrical parameters can also be reached "empirically." As a matter of fact, gene transfer by electric means has been tested and successfully achieved in a large number of tissues in many animal species, including the usual laboratory species (mice, rats, rabbits), cattle, pets, and exotic animal species, as extensively reviewed, for example, in 2004 by André and Mir (41), and in 2005 by Mir et al. (42). Many of these trials have been performed even without prior detailed knowledge of the electrical parameters required to achieve reversible electropermeabilization in the tissues of these species. This of course means that a certain level of cell electropermeabilization has been empirically achieved in these tissues. However, it is important that more emphasis be given to the determination of the actual electropermeabilization levels and volumes and to the use of the models of field distribution to perform better electrotranfers and to avoid misinterpretations. For example, when it was proposed that long-term expression after electrotransfer in the skeletal muscle is dependent on the transfection of the satellite cells (43), this observation was partly true and partly wrong. Indeed, the use of electric pulses that are too intense causes irreversible electroporation and, thus, kills all the mature fibers (because of their large diameter

(refer to the Schwan equation) while the satellite cells (which have a much smaller diameter) are reversibly permeabilized, transfected, and survive. Under these conditions, only the signal brought by the satellite cells will be observed a few days after the DNA electrotransfer, but there are many other published works, performed under conditions that do not hamper the survival of mature fibers, in which long-term gene expression has been observed in the absence of muscle regeneration (and thus in the absence of massive muscle damage).

Another interesting example is provided by conclusions arrived at after DNA electrotransfer using a couple of thin needles as electrodes (44): The authors of this work suggested that there is no influence of the voltage applied under these conditions. This idea is apparently true if only overall gene expression is considered for voltages between 25 and 75 V under their experimental settings. However, because needles generate a very inhomogeneous electric field distribution, particularly if very thin needles are used (31), models allow predictions that, at the lowest of the efficient voltages applied, the electrotransferred volumes will be located close to the needles and almost no tissue will be damaged by the electrotransfer. On the contrary, at the highest of the efficient voltages applied (giving the same overall transfection level), the electrotransferred volumes will be located far from the needles, because all the vicinity of the needles will be irreversibly electroporated and killed, resulting in severe damage to the tissue (necrosis around the places of needle insertion). Thus, both the volumes transfected and the tissue necrosis will be very different even though the global level of gene expression will be similar. Only the association of validated models, appropriate electrode geometry, and appropriate applied voltages may ensure a maximal volume of transfection with minimal damages of the tissues.

6. Is Electroporation/Electropermeabilization Feasible on Humans?

The electrochemotherapy example demonstrated that permeabilizing electric pulses can be safely and repeatedly delivered to humans. Electrochemotherapy is the electroporation/permeabilization-assisted delivery of drugs to the cells of solid tumors. In 1986, using trains of eight short (100 µs) square wave electric pulses, we found experimental conditions under which we could expose cells in culture and have only about 10% loss of cell viability with 98% permeabilization of the surviving cells (6). A new pharmacology using transiently (electro)permeabilized cells was then possible (45, 46). The main finding in 1987 was that the in vitro toxicity of bleomycin was incredibly increased, up to hundreds or thousands of times, by the transient and reversible electropermeabilization of the cells (45). The first preclinical trials on transplanted and spontaneous tumors in mice showed that this effect is present in vivo as well, because the same electric pulses were able to increase the antitumor efficacy of

bleomycin by at least a thousand-fold (47, 48). Ulterior preclinical trials in many places showed that more than 40 different histological types of tumors could be successfully treated by this procedure, associating a physical perturbation of the cell membrane permeability and nonpermeant drugs such as bleomycin or low-permeant drugs such as cisplatin. The bases in vivo are the same as those determined in vitro, as efficacy is actually associated with the complete electropermeabilization of the tumors (49–51). It is noteworthy that, provided the drugs are injected before the pulses delivery, electrochemotherapy requires only the permeabilization of the cells by the electric pulses. Because bleomycin and cisplatin are very small molecules, they are able to diffuse from the extracellular liquid to the inside of the cell as soon as the cells are made permeable by the electric pulses, without any other electrically dependent manipulation. Thus, trains of eight short pulses of 100 µs have been found to be sufficient for the treatment of any kind of solid tumor in animals and patients, as the initial clinical trials immediately showed that the efficacy found during the preclinical trials could be reproduced in cancer patients (52–55).

Electrochemotherapy—on the basis of its principles, which includes cell electropermeabilization, which may be achieved in all the tissue types, and the use of drugs acting directly on the DNA, whatever the pattern of expression of the cells—is a safe and efficient treatment for cutaneous and subcutaneous metastases of any origin. The efficacy of electrochemotherapy has been proven in all the clinical trials published: on average, 75–80% of the treated nodules went into complete regression (total disappearance) in a few weeks, and about 10% more went into partial regression, an excellent result since most of the nodules treated in the initial trials were located in already irradiated areas, unresponsive to usual chemotherapy, and often unresectable by conventional surgery (56, 57). Recently the Standard Operating Procedures (SOP) were prepared within the EU funded ESOPE (European Standard Operating Procedures for Electrochemotherapy and Electrogenetherapy) project (58). These SOP include the recommendations for anesthesia to be delivered to the patients. Indeed, we evaluated the level of pain provoked by the short and intense permeabilizing pulses as well as the usual anesthetic procedures. EMLA cream was found insufficient to attenuate the disagreeable painful sensations linked to the electric pulses delivery, while local anesthesia using lidocaïne injections or general sedation was able to almost completely attenuate these sensations. The same recommendations apply for electrogenetherapy (G. Sersa and J. Gehl, personal communications).

7. Mechanisms and Safe Treatment Options for the Future Use in Humans

As discussed earlier, the electric pulses permeabilize the cells and electrophoretically drive the DNA towards the permeabilized membrane of the target cells. In an attempt to define the exact roles of the electric pulses in gene therapy by

electroporation, we used combinations of short intense electric pulses and long LV pulses. The short pulses were termed HV pulses because of the high voltage applied. Their duration was fixed to 100 μs since we could demonstrate that in vivo, like in vitro, trains of eight pulses of this duration are sufficient to provoke the reversible electroporation of the cells in the tissues (at appropriate electric field strengths). Typically, these are the pulses used to treat tumor nodules in cancer patients by electrochemotherapy. To determine this permeabilization level, several methods can be used, as already described. Whenever it is possible, we recommend the Chromium EDTA method designed by Dr. Julie Gehl, as it is one of the simplest, most reliable, and most powerful methods (15, 59).

As also reported earlier, the combination of HV and LV pulses seems to be the safer and more efficient method (L.M. Mir et al., in preparation), even though there is not yet much experience using these pulses. Nevertheless, several devices can already deliver these trains of pulses, even though until now they have not been the most used. The PulseAgile® generators from Cyto Pulse Sciences Inc. (Glen Burnie, MD, USA) provide high versatility for choosing parameters for whatever sequence of pulses. The Cythorlab™ from Aditus (Lund, Sweden) is a device that stops the train of pulses delivered when electropermeabilization is reached; it does so by means of a feedback control based on impedance, which may be of some interest in vitro (except that in vitro variability is reduced, and once conditions are found, they can be easily reproduced, without the need for a continuous individual monitoring of each sample) but of no interest in vivo. Indeed, in vivo, the pulses provoke muscle contractions, which change the relative positions of the electrodes with respect to the treated tissues. Then the measures of impedance after each pulse will reflect more the changes in the electrode position than the changes in the electrical properties of the tissue. Only real time measurements of the changes in tissue electrical properties during the pulses may be of interest to control the pulses.

The Cliniporator™ from IGEA (Carpi, Italy) is the first CE-labeled pulse generator fully designed for clinical purposes. It provides the user with a complete immediate feedback that allows the user to verify whether the treatment has been successfully delivered. The Cliniporator is already used to treat patients' tumor nodules by electrochemotherapy in several European countries. It has been validated by the European consortium ESOPE, which prepared the SOPs for electrochemotherapy. Thus, safe and efficient procedures, as well as appropriate safety equipment, are available to allow the translation of the impressive and interesting results of the large number of preclinical trials (41, 42) into the clinics in the near future (60).

As a matter of fact, apart from the four clinical trials announced by R. Heller and M. Fons at the 2006 annual meeting of the American Society for Gene Therapy, the ESOPE consortium (including L.M. Mir, G. Sersa, and J. Gehl among many other investigators) is also testing the transfer of reporter genes in humans (personal communication). Unfortunately, all these clinical trials use different types of pulses (short, long, combinations, etc.) and future harmonization of the procedures will be necessary.

8. Conclusions

This chapter has presented the author's personal view and comprehension of current knowledge of DNA electrotransfer, particularly the manipulation of tissues by electrical pulses. It is not a review on the electric-pulse-mediated DNA transfer or other methods of nonviral gene transfer. In conclusion, it would be good to draw on the lessons of the past and the present for the future development of this technology not only in the laboratories but also, and mainly, in the clinics.

From the past, there is no doubt that cell electroporation is the method of choice to transduce bacterial cells, for which it is still largely used. It could have also been the method of choice for eukaryotic cells. Very simple devices using exponentially decaying pulses were proposed to the users, even though a large loss of viability is associated with these types of pulses. Unfortunately, these very simple devices became largely popular, which disseminated the idea that electroporation of eukaryotic cells is directly associated with a large decrease in the viability of the treated cells. The market and profits were probably very interesting to the companies distributing such devices, but they did not actually contribute to a good hearing of the method apart from bacterial transduction.

Now, there are net improvements in eukaryotic cell electric-pulse-mediated transfection both in vitro and in vivo. There is still a real interest from companies that try to make in vitro electrotransfection popular and to allow it to compete with the chemical methods that have been developed in parallel. There are companies developing sophisticated equipment in which the user may control all of the experimental parameters. Unfortunately, there are also other companies that transform DNA electrotransfer into an obscure and almost cabbalistic technology because they do not want to make public the electrical parameters used by their devices or the composition of the mediums proposed to improve the efficacy of DNA electrotransfer. Let us hope and let us believe that, for the in vivo settings, this beautiful technique will not be inhibited by this type of restriction. Electric pulses, as recalled in the introduction, carry many societal fears. To be accepted by society for the clinical use of this new method, it will be necessary not only to communicate its principles but also to make public all preclinical and clinical data and the totality of the procedural details.

The possibility that electric-pulse-mediated gene transfer becomes a routinely used nonviral gene method is more than real, once safety and efficiency are demonstrated by appropriate clinical trials. Safety is probably the main concern, and, as presented in this chapter, pulse delivery in vivo can be particularly well controlled if operators know what to do. The development of this technology at the clinical levels will need the guarantee that safety is respected. There is indeed no doubt that efficacy will be improved by a better knowledge of the expression of the genes conveyed by plasmids. Improvements in gene expression (and thus in efficacy) will come by the amelioration of the promoter, insulators, sequences of attachment to the matrix, inclusion of eukaryotic replication origins, use of minicircles, etc. Under these perspectives, nonviral transfer and, in particular the

electric-pulse-mediated gene transfer, may be one of the techniques for clinical gene transfer in the future, because of its efficacy and safety.

Acknowledgments L.M. Mir thanks all his colleagues for stimulating discussions and Dr. Ruggero Cadossi for his comments on the manuscript.

References

1. Neumann, E., Schaefer-Ridder, M., Wang, Y., and Hofschneider, P.H. (1982) Gene transfer into mouse lyoma cells by electroporation in high electric fields. *EMBO J.* **1**, 841–845.
2. Schwan, H.P. (1957) Electrical properties of tissue and cell suspensions. *Adv. Biol. Med. Phys.* **5**, 147–209.
3. Kotnik, T. and Miklavcic, D. (2000) Analytical description of transmembrane voltage induced by electric fields on spheroidal cells. *Biophys. J.* **79**, 670–679.
4. Gimsa, J. and Wachner, D. (2001) Analytical description of the transmembrane voltage induced on arbitrarily oriented ellipsoidal and cylindrical cells. *Biophys. J.* **81**, 1888–1896.
5. Teissie, J., Knutson, V.P., Tsong, T.Y., and Lane, M.D. (1982) Electric pulse-induced fusion of 3T3 cells in monolayer culture. *Science.* **216**, 537–538.
6. Mir, L.M., Banoun, H., and Paoletti, C. (1988) Introduction of definite amounts of nonpermeant molecules into living cells after electropermeabilization: direct access to the cytosol. *Exp. Cell. Res.* **175**, 15–25.
7. Chang, D.C. and Reese, T.S. (1990) Changes in membrane structure induced by electroporation as revealed by rapid-freezing electron microscopy. *Biophys. J.* **58**, 1–12.
8. Davalos, R.V., Mir, I.L., and Rubinsky, B. (2005) Tissue ablation with irreversible electroporation. *Ann. Biomed. Eng.* **33**, 223–231.
9. Miller, L., Leor, J., and Rubinsky, B. (2005) Cancer cells ablation with irreversible electroporation. *Technol. Cancer Res. Treat.* **4**, 699–705.
10. Teissie, J., Golzio, M., and Rols, M.P. (2005) Mechanisms of cell membrane electropermeabilization: a minireview of our present (lack of?) knowledge. *Biochim. Biophys. Acta.* **1724**, 270–280.
11. Lopez, A., Rols, M.P., and Teissie, J. (1988) 31P NMR analysis of membrane phospholipid organization in viable, reversibly electropermeabilized Chinese hamster ovary cells. *Biochemistry.* **27**, 1222–1228.
12. Tieleman, D.P., Leontiadou, H., Mark, A.E., and Marrink, S.J. (2003) Simulation of pore formation in lipid bilayers by mechanical stress and electric fields. *J Am. Chem. Soc.* **125**, 6382–6383.
13. Tarek, M. (2005) Membrane electroporation: a molecular dynamics simulation. *Biophys. J.* **88**, 4045–4053.
14. Mir, L.M., Bureau, M.F., Gehl, J., et al. (1999) High-efficiency gene transfer into skeletal muscle mediated by electric pulses. *Proc. Natl. Acad. Sci. USA.* **96**, 4262–4267.
15. Gehl, J., Sorensen, T.H., Nielsen, K., et al. (1999) In vivo electroporation of skeletal muscle: threshold, efficacy and relation to electric field distribution. *Biochim. Biophys. Acta.* **1428**, 233–240.
16. Satkauskas, S., Bureau, M.F., Puc, M., et al. (2002) Mechanisms of in vivo DNA electrotransfer: respective contributions of cell electropermeabilization and DNA electrophoresis. *Mol. Ther.* **5**, 133–140.
17. Rols, M.P., Delteil, C., Golzio, M., Dumond, P., Cros, S., and Teissie, J. (1998) In vivo electrically mediated protein and gene transfer in murine melanoma. *Nat. Biotechnol.* **16**, 168–171.
18. Suzuki, T., Shin, B.C., Fujikura, K., Matsuzaki, T., and Takata, K. (1998) Direct gene transfer into rat liver cells by in vivo electroporation. *FEBS Lett.* **425**, 436–440.

19. Aihara, H. and Miyazaki, J. (1998) Gene transfer into muscle by electroporation in vivo. *Nat. Biotechnol.* **16**, 867–870.

20. Poddevin, B., Orlowski, S., Belehradek, J., Jr., and Mir, L.M. (1991) Very high cytotoxicity of bleomycin introduced into the cytosol of cells in culture. *Biochem. Pharmacol.* **42 (Suppl.)**, S67–S75.

21. Bazile, D., Mir, L.M., and Paoletti, C. (1989) Voltage-dependent introduction of a d[alpha]octothymidylate into electropermeabilized cells. *Biochem. Biophys. Res. Commun.* **159**, 633–639.

22. Casabianca-Pignède, M.-R., Mir, L.M., Le Pecq, J.-B., and Jacquemin-Sablon, A. (1991) Stability of antiricin antibodies introduced into DC-3F Chinese hamster cells by electropermeabilization. *J Cell. Pharmacol.* **2**, 54–60.

23. Rols, M.P. and Teissie, J. (1998) Electropermeabilization of mammalian cells to macromolecules: control by pulse duration. *Biophys. J.* **75**, 1415–1423.

24. Teissie, J. and Ramos, C. (1998) Correlation between electric field pulse induced long-lived permeabilization and fusogenicity in cell membranes. *Biophys. J.* **74**, 1889–1898.

25. Bureau, M.F., Gehl, J., Deleuze, V., Mir, L.M., and Scherman, D. (2000) Importance of association between permeabilization and electrophoretic forces for intramuscular DNA electrotransfer. *Biochim. Biophys. Acta.* **1474**, 353–359.

26. Satkauskas, S., Andrè, F., Bureau, M.F., et al. (2005) Electrophoretic component of electric pulses determines the efficacy of in vivo DNA electrotransfer. *Hum. Gene. Ther.* **16**, 1194–1210.

27. Faurie, C., Phez, E., Golzio, M., et al. (2004) Effect of electric field vectoriality on electrically mediated gene delivery in mammalian cells. *Biochim. Biophys. Acta.* **1665**, 92–100.

28. Phez, E., Faurie, C., Golzio, M., Teissie, J., and Rols, M.P. (2005) New insights in the visualization of membrane permeabilization and DNA/membrane interaction of cells submitted to electric pulses. *Biochim. Biophys. Acta.* **1724**, 248–254.

29. Teissie, J. and Blangero, C. (1984) Direct experimental evidence of the vectorial character of the interaction between electric pulses and cells in cell electrofusion. *Biochim. Biophys. Acta.* **775**, 446–448.

30. Teissie, J. and Rols, M.P. (1993) An experimental evaluation of the critical potential difference inducing cell membrane electropermeabilization. *Biophys. J.* **65**, 409–413.

31. Miklavcic, D., Semrov, D., Mekid, H., and Mir, L.M. (2000) A validated model of in vivo electric field distribution in tissues for electrochemotherapy and for DNA electrotransfer for gene therapy. *Biochim. Biophys. Acta.* **1523**, 73–83.

32. Sel, D., Mazeres, S., Teissie, J., and Miklavcic, D. (2003) Finite-element modeling of needle electrodes in tissue from the perspective of frequent model computation. *IEEE Trans. Biomed. Eng.* **50**, 1221–1232.

33. Belehradek, J., Jr., Orlowski, S., Ramirez, L.H., Pron, G., Poddevin, B., and Mir, L.M. (1994) Electropermeabilization of cells in tissues assessed by the qualitative and quantitative electroloading of bleomycin. *Biochim. Biophys. Acta.* **1190**, 155–163.

34. Tounekti, O., Pron, G., Belehradek, J., Jr., and Mir, L.M. (1993) Bleomycin, an apoptosis-mimetic drug that induces two types of cell death depending on the number of molecules internalized. *Cancer. Res.* **53**, 5462–5469.

35. Poddevin, B., Belehradek, J., Jr., and Mir, L.M. (1990) Stable [57Co]-bleomycin complex with a very high specific radioactivity for use at very low concentrations. *Biochem. Biophys. Res. Commun.* **173**, 259–264.

36. Engstrom, P.E., Persson, B.R., and Salford, L.G. (1999) Studies of in vivo electropermeabilization by gamma camera measurements of (99m)Tc-DTPA. *Biochim. Biophys. Acta.* **1473**, 321–328.

37. Gehl, J., Skovsgaard, T., and Mir, L.M. (2002) Vascular reactions to in vivo electroporation: characterization and consequences for drug and gene delivery. *Biochim. Biophys. Acta.* **1569**, 51–58.

38. Ramirez, L.H., Orlowski, S., An, D., et al. (1998) Electrochemotherapy on liver tumours in rabbits. *Br. J. Cancer.* **77**, 2104–2111.

39. Sersa, G., Cemazar, M., Parkins, C.S., and Chaplin, D.J. (1999) Tumour blood flow changes induced by application of electric pulses. *Eur. J. Cancer.* **35**, 672–677.
40. Sersa, G., Cemazar, M., Miklavcic, D., and Chaplin, D.J. (1999) Tumor blood flow modifying effect of electrochemotherapy with bleomycin. *Anticancer. Res.* **19**, 4017–4022.
41. Andrè, F. and Mir, L.M. (2004) DNA electrotransfer: its principles and an updated review of its therapeutic applications. *Gene. Ther.* **11 (Suppl. 1)**, S33–S42.
42. Mir, L.M., Moller, P.H., Andrè, F., and Gehl, J. (2005) Electric pulse-mediated gene delivery to various animal tissues. *Adv. Genet.* **54**, 83–114.
43. Peng, B., Zhao, Y., Lu, H., Pang, W., and Xu, Y. (2005) In vivo plasmid DNA electroporation resulted in transfection of satellite cells and lasting transgene expression in regenerated muscle fibers. *Biochem. Biophys. Res. Commun.* **338**, 1490–1498.
44. Liu, F. and Huang, L. (2002) A syringe electrode device for simultaneous injection of DNA and electrotransfer. *Mol. Ther.* **5**, 323–328.
45. Orlowski, S., Belehradek, J., Jr., Paoletti, C., and Mir, L.M. (1988) Transient electropermeabilization of cells in culture. Increase of the cytotoxicity of anticancer drugs. *Biochem. Pharmacol.* **37**, 4727–4733.
46. Orlowski, S. and Mir, L.M. (1993) Cell electropermeabilization: a new tool for biochemical and pharmacological studies. *Biochim. Biophys. Acta.* **1154**, 51–63.
47. Mir, L.M., Orlowski, S., Belehradek, J., Jr., and Paoletti, C. (1991) Electrochemotherapy potentiation of antitumour effect of bleomycin by local electric pulses. *Eur. J. Cancer.* **27**, 68–72.
48. Belehradek, J., Jr., Orlowski, S., Poddevin, B., Paoletti, C., and Mir, L.M. (1991) Electrochemotherapy of spontaneous mammary tumours in mice. *Eur. J. Cancer.* **27**, 73–76.
49. Miklavcic, D., Beravs, K., Semrov, D., Cemazar, M., Demsar, F., and Sersa, G. (1998) The importance of electric field distribution for effective in vivo electroporation of tissues. *Biophys. J.* **74**, 2152–2158.
50. Gothelf, A., Mir, L.M., and Gehl, J. (2003) Electrochemotherapy: results of cancer treatment using enhanced delivery of bleomycin by electroporation. *Cancer. Treat. Rev.* **29**, 371–387.
51. Mir, L.M. (2001) Therapeutic perspectives of in vivo cell electropermeabilization. *Bioelectrochemistry.* **53**, 1–10.
52. Mir, L.M., Belehradek, M., Domenge, C., et al. (1991) [Electrochemotherapy, a new antitumor treatment: first clinical trial]. *CR Acad. Sci. III.* **313**, 613–618.
53. Belehradek, M., Domenge, C., Luboinski, B., Orlowski, S., Belehradek, J., Jr., and Mir, L.M. (1993) Electrochemotherapy, a new antitumor treatment. First clinical phase I-II trial. *Cancer.* **72**, 3694–3700.
54. Domenge, C., Orlowski, S., Luboinski, B., et al. (1996) Antitumor electrochemotherapy: new advances in the clinical protocol. *Cancer.* **77**, 956–963.
55. Heller, R., Jaroszeski, M.J., Glass, L.F., et al. (1996) Phase I/II trial for the treatment of cutaneous and subcutaneous tumors using electrochemotherapy. *Cancer.* **77**, 964–971.
56. Marty, M., Sersa, G., Garbay, J.R., et al. (2006) Electrochemotherapy - an easy, highly effective and safe treatment of cutaneous and subcutaneous metastases: results of the ESOPE (European Standard Operating Procedures of Electrochemotherapy) study. *Eur J Cancer Supplements.* **4**, 3–13.
57. Sersa, G. (2006) The State-of-the-art of electrochemotherapy before the ESOPE study; advantages and clinical uses. *Eur J Cancer Supplements.* **4**, 52–59.
58. Mir, L.M., Gehl, J., Sersa, G., et al. (2006) Standard Operating Procedures of the Electrochemotherapy: Instructions for the use of bleomycin or cisplatin administered either systemically or locally and electric pulses delivered by the Cliniporator™ by means of invasive or non-invasive electrodes. *Eur J Cancer Supplements.* **4**, 14–25.
59. Gehl, J. and Mir, L.M. (1999) Determination of optimal parameters for in vivo gene transfer by electroporation, using a rapid in vivo test for cell permeabilization. *Biochem. Biophys. Res. Commun.* **261**, 377–380.
60. Heller, L.C. and Heller, R. (2006) In vivo electroporation for gene therapy. *Hum Gene Ther.* **17**, 890–897.

Chapter 2
Mechanism by Which Electroporation Mediates DNA Migration and Entry into Cells and Targeted Tissues

Marie-Pierre Rols

Abstract Cell membranes can be transiently permeabilized under application of electric pulses that allow hydrophilic therapeutic molecules, such as anticancer drugs and DNA, to enter into cells and tissues. This process, called *electropermeabilization* or *electroporation,* has been rapidly developed over the last decade to deliver genes to tissues and organs, but there is a general agreement that very little is known about what is really occurring during membrane electropermeabilization. It is well accepted that the entry of small molecules, such as anticancer drugs, occurs through simple diffusion while the entry of macromolecules, such as DNA, occurs through a multistep mechanism involving the electrophoretically driven association of the DNA molecule with the destabilized membrane and then its passage across the membrane. Therefore, successful DNA electrotransfer into cells depends not only on cell permeabilization but also on the way plasmid DNA interacts with the plasma membrane and, once into the cell, migrates toward the nuclei.

Keywords: gene transfer, gene expression, membrane, electric field, electroporation, electropermeabilization

1. Introduction

The administration of naked nucleic acids into cells and tissues can be considered the simplest and safest method of gene delivery (1); however, one drawback of this method for gene therapy is the low efficiency of gene expression. Therefore, different strategies have been developed over the years in this field, based on the use of viruses as biological vectors and on the development of chemical or physical methods. The common goal of these methods is to transfer DNA into cells via cell membrane modification. Virus and chemical vectors fuse with the lipid membrane and are endocytosed. Physical vectors transiently destabilize the membrane, creating leaky structures or pores. One of the biggest successes of gene therapy came with the use of viruses for the treatment of severe combined immunodeficiency disease, but the clinical trial had to be suspended because the virus caused leukemia,

in turn due to the insertion of the therapeutic gene near an oncogene (2–4). Currently, there is still a need and a challenge to develop an efficient and secure method of gene transfer.

The use of electric pulses as a safe tool to deliver therapeutic molecules to tissues and organs has been rapidly developed over the last decade. The method refers to the transient increase in the permeability of cell membranes when submitted to electric field pulses. This process is commonly known as electropermeabilization or electroporation (5–7). Hydrophilic molecules that are otherwise nonpermeant, such as the highly toxic drug bleomycin, can gain direct access to the cytosol of cells. A cancer treatment modality, electrochemotherapy, has emerged (8–10), and it has been successfully used in clinical trials for cancer treatment (11–13). Besides drugs, electropermeabilization can be used to deliver a wide range of potentially therapeutic agents, including proteins, oligonucleotides, RNA, and DNA (14). At present, it represents one of the most widespread techniques used in molecular genetics. In vivo electrotransfer is of special interest, as it is the most efficient nonviral strategy of gene delivery and also because of its low cost, ease of realization, and safety. The most widely targeted tissue is skeletal muscle. The strategy is not only promising for the treatment of muscle disorders, but also for the systemic secretion of therapeutic proteins. Vaccination and oncology gene therapy are other fields of application (15). Electrogenetherapy is relevant in a variety of research and clinical settings, including cancer therapy, modulation of pathogenic immune responses, and delivery of therapeutic proteins and drugs (16). This method, together with the capacity to deliver large DNA constructs, greatly expands the research and clinical applications of in vivo DNA electrotransfer (17–20).

However, the mechanisms underlying cell membrane permeabilization and associated gene transfer are not completely understood. Understanding these concepts is of importance not only for the in vitro use of the method in terms of efficiency but also for the in vivo use of the technique in terms of safety. The successful electrotransfer of plasmid DNA into cells depends on the way the cell membrane has been permeabilized, reversibly or not, and on the way DNA interacts with the membrane and is transported from the plasma membrane toward the nuclear envelope. The focus of this chapter is to describe the different aspects of what is known of the mechanism of membrane permeabilization and associated gene transfer and, by doing so, what are the actual limits of the DNA delivery into cells.

2. Basics of Cell Membrane Electropermeabilization

2.1. Membrane Electropermeabilization or Membrane Electroporation?

As its name suggests, the basic theory of *electroporation* is that the electric field causes the formation of pores or holes in the membrane (5, 21). Theoretical models have been proposed to explain the mechanism of this reversible membrane

electropermeabilization and its potential to allow the access of nonpermeant molecules inside the cells. The formation of pores is energetically favorable because they allow the reduction of the electrostatic energy of the system. It is hindered by the line tension needed to create their circumference and it is thus only at a critical value of the field strength that their formation is possible. Roughly speaking, the poration process is an activated one, similar to nucleation, where the surface tension favors pore formation while the line tension tends to suppress it. The effect of the electric field is to effectively increase the surface tension and, thus, lower the activation barrier (6).

It is widely accepted, however, that the standard theory of electroporation has problems (22). For instance, these pores have never been observed, but large pores, arising from primary pores, have indeed been detected in cells pulsed under hypoosmotic conditions (23). An alternative theory, which is less favored because of conflicts with experimental results, is based on an electro-mechanical instability where the electric field compresses the membrane. At present, computer simulations at the microscopic length scale are becoming possible; however, the complexity and, more importantly, the size of the system that can be simulated are still rather limited (24). Molecular dynamics have suggested that electropores could be generated, but they have been obtained under field conditions larger than those experimentally required to induce reversible membrane permeabilization (25). The geometry of the membrane and its mechanical properties such as surface to volume ratio, rigidity, and composition, for example the role played by microdomains, must be taken into account, as well as the surface tension and line tension associated with the pores. Indeed, only a few experimental data concerning the molecular changes involved in membrane electropermeabilization have been reported.^{31}P NMR studies performed on both model membranes and mammalian cells suggested a reorganization of the polar head group region of the phospholipids, leading to a weakening of the hydration layer (26, 27). Transbilayer reorientation of phospholipid probes has been reported in the human erythrocyte membrane, suggesting an increase in the flip-flop of phospholipids as a direct consequence of electropermeabilization (28).

Altogether, and even if the term electroporation is commonly used among biologists, the term electropermeabilization should be preferred so as to prevent any molecular description of the phenomenon.

2.2. Membrane Electropermeabilization Is Controlled by the Transmembrane Voltage

It has been known for more than 30 years that, as far as membrane permeabilization is concerned, the key effect of electric field on cells is a position-dependent change in the resting transmembrane potential difference $\Delta\Psi_0$ of their plasma membrane. The electrically induced potential difference $\Delta\Psi_E$, which is the difference between

the potential inside the cell Ψ_{in} and the potential outside the cell Ψ_{out}, at a point M on the cell surface, is given by the equation

$$\Delta\Psi_{E}(t) = \Psi_{in} - \Psi_{out} = -f\, g(\lambda)\, rE\cos\theta(M)\left[1 - e^{-t/\tau}\right] \qquad (1)$$

where t is the time, f is related to the shape of the cell, g depends on the conductivities λ of the membrane, cytoplasm, and extracellular medium, r is the radius of the cell, E is the field strength, $\theta(M)$ is the angle between the normal to the membrane at the position M and the direction of the field, and τ is the membrane charging time (29). The field-induced potential difference is added to the resting potential (30, 31):

$$\Delta\Psi = \Delta\Psi_{o} + \Delta\Psi_{E} \qquad (2)$$

Being dependent on the angular parameter θ, the field effect is position dependent on the cell surface; therefore, the side of the cell facing the anode is hyperpolarized while the side of the cell facing the cathode is depolarized. This theoretical prediction has been experimentally verified by using a voltage-sensitive fluorescent dye (32). The transmembrane potential of a cell exposed to an electric field is, therefore, a critical parameter for successful cell permeabilization, and depends on the cell size, shape, and orientation (33, 34).

2.3. Membrane Electropermeabilization Is Controlled by Electric Field Parameters

Indeed, permeabilization occurs only on the part of the membrane where the potential difference has been brought to its critical value (31, 35). This value has been evaluated to be of the order of 200–300 mV whatever the cell type (36, 37). Permeabilization is, therefore, controlled by the field strength. This means that a field intensity, E, larger than a critical value, E_{p}, must be applied. E_{p} is dependent on the size of the target cells. It ranges from values close to 100 V/cm in large cells such as myotubes to 1–2 kV/cm in bacteria (36). Large cells are, therefore, more sensitive to lower field strengths than are small ones. Electric field values, therefore, have to be adapted to each cell line in order not to affect their viability. The field strength triggers permeabilization when E is larger than E_{p}, and it controls the area of the cell surface which is affected (38). From Eq. 1, it is clear that for field intensities close to E_{p}, permeabilization is only present for θ values close to 0 or π. Under that condition, only the localized parts of the membrane surface facing the electrodes are affected. However, within these permeabilized cell caps, the extent of permeabilization is not a function of the field strength (37, 39, 40); instead it is controlled by the pulse number and duration (40). So, membrane permeabilization occurs only at electric field values E higher than the threshold value E_{p}, whatever the pulse number and the pulse duration. Increasing E above E_{p} leads to an increase in the area where

permeabilization takes place, and, in that particular area, the extent of permeabilization is determined by the pulse number and duration.

This electro-induced permeabilization of the cell membrane can be quantified in terms of the flow F_S of molecules S diffusing through the plasma membrane. Fick's law and experimental data obtained in the case of the release of ATP from Chinese hamster ovary (CHO) cells allowed the establishment of the following equation

$$F_s(t) = P_s x(N,T)A/2(1 - E_p/E)\Delta S e^{-k(N,T)t} \tag{3}$$

where P_S is the permeation coefficient of the molecule S across the membrane, x—a function which depends on the pulse number N and the pulse duration T—represents the probability of permeabilization ($0 < x < 1$), A is the cell surface, E is the applied electric field intensity, E_p is the threshold for permeabilization, ΔS is the concentration difference of S between cell and external medium, k is the time constant of the resealing process, and t is the time after the pulse (39). Such a concept leads to the notion of *membrane domains* involved in electropermeabilization: macrodomains where permeabilization can take place, in which the area is determined by the pulse intensity according to $A/2$ ($1 - E_p/E$), and, within that macrodomain, *microdomains* where permeabilization actually can take place, whose number and size depend on the pulse number and the pulse duration according to the x function. The molecular characteristics of these domains, in terms of lipid composition, organization, asymmetry, and dynamics, remain an open question.

2.4. Membrane Electropermeabilization Is a Fast and Localized Process

The use of videomicroscopy allows visualization of the permeabilization phenomenon at the single cell level. Propidium iodide can be used as a probe for small molecules. Its uptake in the cytoplasm is a fast process that can be detected during the seconds following electric pulses application. In less than a minute, it appears at the nuclei level. Moreover, exchange across the pulsed cell membrane is not homogeneous on the whole cell membrane. It occurs at the sides of the cells facing the electrodes in an asymmetrical way (Fig. 2.1). It is more pronounced at the anode facing side of the cells than at the cathode one, i.e., in the hyperpolarized area than in the depolarized one (41), in agreement with the earlier mentionned theoretical considerations and Eqs. 1 and 2.

Electropermeabilization can therefore be described as a three-step process with respect to the electric field (EF).

1. Before EF: membrane acts as a barrier that prevents the free exchange of hydrophilic molecules between cell cytoplasm and external medium.
2. During EF: when reaching a threshold value, the transmembrane potential increase induces the formation of local transient permeable structures that allow the exchange of molecules.

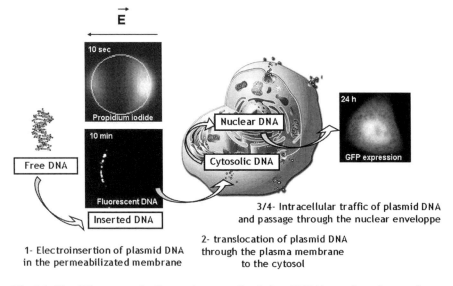

Fig. 2.1 The different steps leading to electro-mediated plasmid DNA transfer and expression

3. After EF: resealing occurs. Membrane permeability to small molecules is present, with a lifetime ranging from seconds to minutes depending on EF conditions and the temperature (39, 42). After resealing, the uptaken solutes are sequestered inside the treated cell.

In larger molecules (MW above 4 kDa), direct transfer to the cytoplasm is observed only if macromolecules, such as proteins or DNA, are present during the electric pulse, but proteins added after the permeabilizing electric field can indeed enter the cell via endocytosis-like processes (43, 44). Gene expression is observed only when DNA is present during application of the pulses; therefore, the mechanism of macromolecule uptake is different from the one observed for small molecules.

3. Basics of Plasmid DNA Electrotranfer and Expression

3.1. Theoretical Considerations

Although the first pioneering report on plasmid DNA electrotransfer in cells was published more than 20 years ago by Eberhard Neumann (21), the mechanisms that mediate DNA electrotransfer remain to be elucidated, and different scenarios can be proposed. The simplest one is that the membrane is permeabilized, and then the DNA is pushed through the putative electropores by the

electrophoretic effect, which may be necessary to overcome the problem of passing a charge into the relatively low dielectric medium of the membrane. Another possibility is that the electric field leads to an aggregation of ion pumps which open and permit the passage of the DNA. Finally, another scenario is that the DNA forms a charged vesicle which is internalized by the cell, as in endocytosis.

In the context of studies on model membranes, DNA interactions with lipid bilayers have indeed been studied. DNA injection by a micropipette, to a part of a giant unilamellar vesicle, resulted in membrane topology transformations, which can be monitored using phase contrast microscopy (45, 46). DNA-induced endocytosis was observed in the absence of any electric field. A possible mechanism for DNA/lipid membrane interaction is DNA encapsulation within an inverted micelle included in the lipid membrane. High molecular mass DNA was efficiently taken up by large unilamellar vesicles exposed to a short pulse of electric field (0.1–1 ms) with an intensity as high as 12.5 kV/cm, which indicated that DNA was taken up as a result of the electrostimulated formation of endosome-like vesicles rather than via field-induced membrane pores (47). Other data report that electrotransfer of DNA through the lipid bilayer could be mediated by transient complexes between DNA and the lipids in the pore edges of elongated, electropercolated hydrophilic pore zones (48). Moreover, the association of DNA with a lipid bilayer greatly facilitates the transport of small ions. This association suggests a locally conductive DNA/lipid interaction zone where parts of the DNA strand may be transiently inserted in the bilayer, leaving other parts of the DNA probably protruding out from the outer surface of the bilayer. DNA is not only transiently inserted in, but also actually electrophoretically pulled through, the permeabilized zones onto the other membrane side, finally leaving the bilayer structure basically intact (49).

3.2. DNA Electrotransfer Depends not only on Membrane Permeabilization but also on DNA Electrophoresis

Electrotransfection, i.e., plasmid DNA electrotransfer into cells and then its expression, is detected at electric field values leading to plasma membrane permeabilization. Milliseconds-long pulses are generally required to obtain efficient gene expression while keeping intact the cell viability by limiting the electric field intensities required when short pulses are used (42, 50, 51). Nevertheless, electrotransformation can be obtained with short strong pulses (21). Transfection threshold values are the same as those for cell permeabilization (52).

The transfection efficiency TE obeys the following equation

$$TE = K \ N \ T^{2.3} \ (1 - E_p / E) \ f(ADN) \tag{4}$$

where K is a constant. The dependence on the plasmid concentration $f(ADN)$ is rather complex, and high levels of plasmids are toxic (53). The effect of pulse

duration appears to be crucial, and so pulse duration appears to be a key parameter for efficient gene expression in cells and tissues (21, 54).

Electrically induced DNA uptake by cells is a vectorial process with the same direction as DNA electrophoresis in an external electric field (55, 56). The transfection efficiency is significantly higher in cell monolayers facing the cathode when compared with those exposed to field pulses of the reverse direction. This higher efficiency is due to the contribution of the electrophoresis to the translocation of the polyanionic plasmid DNA across the electropermeabilized cell membrane (57). The polarity of the electric field has, therefore, a direct effect on transfection. Adherent cells facing the cathode exhibit much higher transfection yield as well as gene expression than do cells facing the anode (57). This dependance of the transfection efficiency on the direction of the field might be due to the involvement of the electrophoretic force in the translocation of the negatively charged DNA molecule (55). While cell permeabilization is only slightly affected by reversing the polarity of the electric pulses or by changing the orientation of pulses, transfection level increases are observed. These last effects are due to an increase in the cell membrane area where DNA interacts (58).

The use of a two-pulse technique allowed the separation of the two effects provided by an electric field: membrane electropermeabilization and DNA electrophoresis. The first pulse (high voltage, microsecond duration) creates permeabilization efficiently. The second pulse of much lower amplitude, but substantially longer (millisecond duration), does not cause permeabilization and transfection by itself but instead enhances TE by about one order of magnitude. In vivo, the effect of electrophoresis is not completely understood. A recent publication indicates that electrophoresis cannot play an important role in gene electrotransfer (59), but many studies show the benefit of the combination of short high-voltage and long low-voltage pulses, allowing to evidence the necessity of association of cell electropermeabilization and convenient electrophoretic transport of DNA toward or across the permeabilized membrane within the tissue (60–62). DNA electrotransfer has indeed been achieved in tibialis cranialis muscles of C57BL/6 mice by using such long but low-intensity pulses (63, 64). Other data from experiments performed in the skin also demonstrate that the combination of such electric pulses is an efficient protocol to enhance DNA expression (65).

3.3. DNA Electrotransfer Is a Multistep and Localized Process

Fluorescent plasmids allowed monitoring of the interaction of nucleic acids with electric field treated cells. Negatively charged DNA molecules migrate when submitted to an electric field. In a low electric field regimen, the DNA simply flows around the membrane toward the anode; however, beyond a critical field value $(E > E_{p})$ the DNA interacts with the membrane but only at the pole opposite the

cathode (as opposed to the case for small molecules where both poles become permeable to these molecules) (41). Clusters or aggregates of DNA are seen to form, but once the field is cut, the growth of these clusters is stopped. The DNA/ membrane interaction is not homogeneously distributed in the permeabilized areas facing the cathode but is present in membrane microdomains ranging from 0.1 to 0.5 μm (52). These observations are consistent with a process in which plasmids interact with electropermeabilized parts of the cell surface because of their interfacial electrophoretic accumulation (66, 67). Owing to the good correlation between visualization of DNA accumulation at the membrane and gene expression, these results are consistent with a multistep process of DNA electrotransfer (Fig. 2.1):

(1) During EF, the plasma membrane is permeabilized on the sides of the cell facing the electrodes, and the negatively charged DNA migrates electrophoretically toward the plasma membrane on the cathode side where it is accumulated and interacts. Plasmid DNA is "trapped" in vesicles strongly associated with the plasma membrane.

(2) After EF, a translocation of the plasmid to the cytoplasm must take place, leading to gene expression. This second step, including plasmid DNA diffusion in the cytosol and its passage through the nuclear pores, remains rather misunderstood.

New directions of research are now needed to characterize membrane domains involved in electrotransfer of molecules. DNA transfer occurs through microdomains present in the electropermeabilized cell membrane. Their size is in the same range as the so-called rafts domains. Lipid rafts are plasma membrane microdomains enriched in sphingolipids and cholesterol. These domains have been suggested to serve as platforms for various cellular events, such as signaling and membrane trafficking (68, 69). One can wonder whether they are involved in DNA electrotransfer (70).

3.4. Gene Expression from DNA Electrotransfer Is Hindered by Internal Membrane

If the first steps of gene electrotransfection, i.e., migration of the plasmid DNA toward the electropermeabilizated plasma membrane and its interaction with it, come to be described and, therefore, represented as guidelines to improve gene electrotransfer, the successful expression of the gene from the plasmid DNA depends on its subsequent migration into the cell. Therefore, diffusion properties of plasmid DNA, metabolic instability of plasmid DNA in cells and tissues, as well as plasmid DNA nuclear translocation represent cell limiting factors that have to be considered (71).

The cytoplasm is composed of a network of microfilament and microtubule systems and a variety of subcellular organelles bathing in the cytosol. The mesh-like structure of the cytoskeleton, the presence of organelles, and the high protein concentration impose an intensive molecular crowding of the cytoplasm, which

limits the diffusion of large-sized macromolecules. The translational mobility of macromolecule smaller than 500–750 kDa is only three- to fourfold slower than in water, but it is markedly impeded for larger molecules (72). Mobility of plasmid DNA is negligible in the cytoplasm of microinjected myotubes (73), but successful in vivo DNA expression can be obtained by electric fields. These discordant results might be explained by the disassembly of the cytoskeleton network that may occur during electropermeabilization (74) and reinforce the idea that the cytoplasm constitutes a diffusional barrier to gene transfer.

Stability of plasmid DNA can be quantitatively assessed by microinjection. Such experiments revealed that 50% of the DNA is eliminated in 12 h from HeLa and COS cells and in 4 h from myotubes (75). Cytosolic elimination of plasmid DNA cannot be attributed only to cell division since degradation is observed in cell cycle arrested cells. In tissues, radiolabeled plasmid progressively leaves muscles and is degraded as soon as 5 min after plasmid injection, with or without electrotransfer. While a major part of plasmid DNA is rapidly cleared and degraded, the electro-transferable pool of plasmid DNA represents a very small part of the amount injected and belongs to another compartment where it is protected from endogenous DNases (76).

Finally, besides the cytoskeleton, the nuclear envelope represents the last obstacle to the expression of the plasmid DNA. The inefficient nuclear uptake of plasmid DNA from the cytoplasm was recognized more than 20 years ago. While molecules smaller than 40 kDa can diffuse through the nuclear pore complexes, larger molecules must contain a specific targeting signal, the nuclear localization sequence, to traverse the nuclear envelope. The significant size of plasmid DNA (2–10 MDa) makes it unlikely that the nuclear entry occurs by passive diffusion. Dividing cells are highly transfectable when compared with quiescent ones, suggesting that DNA enters the nucleus upon the disassembly of the nuclear envelope during mitosis. Cell synchronization also affects gene delivery by electric field (77–79), reinforcing the statement that the melting of the nuclear membrane facilitates direct access of plasmid DNA to the nucleus.

Therefore, clear limits of efficient gene expression by electric pulses are due to cytoplasm crowding and transfer through the nuclear envelope. The new challenges are to overcome these limiting steps. The dense latticework of the cytoskeleton impedes free diffusion of DNA; however, since transfections do work, there must be mechanisms by which DNA circumvents cytoplasmic obstacles. One possibility is that plasmids become cargo on cytoskeletal motors, much like viruses do, and move to the nucleus in a directed fashion. Electrotransfered plasmid DNA, containing specific sequences, like most viruses, could then use the microtubule network and its associated motor proteins to traffic through the cytoplasm to the nucleus (80).

Another alternative could come from nanosecond pulsed electric fields. New findings indicate that very short (10–300 ns) but high pulses (up to 300 kV/cm) extend classical electropermeabilization to include events that primarily affect intracellular structures and functions. As the pulse duration is decreased below the plasma membrane charging time constant (see Eq. 1), plasma membrane effects decrease and intracellular effects predominate (81, 82). When used in conjugation

with classical electropermeabilization, nanopulses can increase gene expression. The idea is to perform classical membrane permeabilization allowing plasmid DNA electrotransfer, and then 30 min later to specifically permeabilize the nuclear envelope by using nanopulse EF. In this way, it will become possible to not only electropermeabilize cells but also electromanipulate them.

4. Conclusion

Clear differences of processes by which molecules of different sizes translocate across the electropermeabilized membrane have been observed. While small soluble molecules could rather freely cross the permeabilized membrane for a time much longer than the duration of the electric pulse application, DNA transfer involves complex steps, including interaction with the membrane and migration into the cytosol. If the effects of the electric field parameters are about to be elucidated (pulse strength higher than a threshold value, long pulse duration for efficient gene expression), the associated destabilization of the membrane which is a stress for the cells and may affect the cell viability is yet to be clearly described. Moreover, it becomes evident that extracellular as well as intracellular barriers compromise the transfection efficiency. Studies are necessary to understand the cascade of events triggered by electropermeabilization at the tissue level where new constraints coming from tissue organization are present, such as the inhomogeneity of the electric field and the intercellular distribution of DNA (83).

The challenge for electro-mediated gene therapy is to pinpoint the rate limiting steps in this complex process and implement strategies to overcome the barriers encountered by therapeutic plasmid DNA.

Acknowledgments Many thanks are due to M. Golzio, C. Faurie, E. Phez, J. Teissié, and also to the financial sponsors—the CNRS, the Association Française sur les Myopathies, and the Fondation pour la Recherche Médicale.

References

1. Wolff, J.A. and Budker, V. (2005) The mechanism of naked DNA uptake and expression. *Adv. Genet.* **54**, 3–20.
2. Hacein-Bey-Abina, S., Von Kalle, C., Schmidt, M., et al. (2003) LMO2-associated clonal T cell proliferation in two patients after gene therapy for SCID-X1. *Science.* **302**, 415–419.
3. Hacein-Bey-Abina, S., Le Deist, F., Carlier, F., et al. (2002) Sustained correction of X-linked severe combined immunodeficiency by ex vivo gene therapy. *N. Engl. J. Med.* **346**, 1185–1193.
4. Bester, A.C., Schwartz, M., Schmidt, M., Garrigue, A., et al. (2006) Fragile sites are preferential targets for integrations of MLV vectors in gene therapy. *Gene. Ther.* **13**, 1057–1059.
5. Neumann, E., Sowers, A.E., and Jordan, C.A. (1989) *Electroporation and electrofusion in cell biology.* Plenum, New York.

6. Weaver, J.C. (1995) Electroporation theory. Concepts and mechanisms. *Methods Mol. Biol.* **55**, 3–28.

7. Teissie, J., Eynard, N., Vernhes, M.C., et al. (2002) Recent biotechnological developments of electropulsation. A prospective review. *Bioelectrochemistry.* **55**, 107–112.

8. Mir, L.M., Belehradek, M., Domenge, C., et al. (1991) [Electrochemotherapy, a new antitumor treatment: first clinical trial]. *CR Acad. Sci. III.* **313**, 613–618.

9. Mir, L.M., Orlowski, S., Belehradek, J., Jr., and Paoletti, C. (1991) Electrochemotherapy potentiation of antitumour effect of bleomycin by local electric pulses. *Eur. J. Cancer.* **27**, 68–72.

10. Belehradek, M., Domenge, C., Luboinski, B., Orlowski, S., Belehradek, J., Jr., and Mir, L.M. (1993) Electrochemotherapy, a new antitumor treatment. First clinical phase I-II trial. *Cancer.* **72**, 3694–3700.

11. Gehl, J. (2003) Electroporation: theory and methods, perspectives for drug delivery, gene therapy and research. *Acta. Physiol. Scand.* **177**, 437–447.

12. Gothelf, A., Mir, L.M., and Gehl, J. (2003) Electrochemotherapy: results of cancer treatment using enhanced delivery of bleomycin by electroporation. *Cancer Treat. Rev.* **29**, 371–387.

13. Mir, L.M., Glass, L.F., Sersa, G., Teissie, J., et al. (1998) Effective treatment of cutaneous and subcutaneous malignant tumours by electrochemotherapy. *Br. J. Cancer.* **77**, 2336–2342.

14. Golzio, M., Rols, M.P., and Teissie, J. (2004) In vitro and in vivo electric field-mediated permeabilization, gene transfer, and expression. *Methods.* **33**, 126–135.

15. Scherman, D., Bigey, P., and Bureau, M.F. (2002) Applications of plasmid electrotransfer. *Technol. Cancer Res. Treat.* **1**, 351–354.

16. Bloquel, C., Fabre, E., Bureau, M.F., and Scherman, D. (2004) Plasmid DNA electrotransfer for intracellular and secreted proteins expression: new methodological developments and applications. *J. Gene Med.* **6 (Suppl 1)**, S11–S23.

17. Trezise, A.E., Buchwald, M., and Higgins, C.F. (1993) Testis-specific, alternative splicing of rodent CFTR mRNA. *Hum. Mol. Genet.* **2**, 801–802.

18. Miklavcic, D., Semrov, D., Mekid, H., and Mir, L.M. (2000) A validated model of in vivo electric field distribution in tissues for electrochemotherapy and for DNA electrotransfer for gene therapy. *Biochim. Biophys. Acta.* **1523**, 73–83.

19. Gehl, J., Sorensen, T.H., Nielsen, K., et al. (1999) In vivo electroporation of skeletal muscle: threshold, efficacy and relation to electric field distribution. *Biochim. Biophys. Acta.* **1428**, 233–240.

20. Gilbert, R.A., Jaroszeski, M.J., and Heller, R. (1997) Novel electrode designs for electrochemotherapy. *Biochim. Biophys. Acta.* **1334**, 9–14.

21. Neumann, E., Schaefer-Ridder, M., Wang, Y. and Hofschneider, P.H. (1982) Gene transfer into mouse lyoma cells by electroporation in high electric fields. *EMBO J.* **1**, 841–845.

22. Teissie, J., Golzio, M., and Rols, M.P. (2005) Mechanisms of cell membrane electropermeabilization: a minireview of our present (lack of?) knowledge. *Biochim. Biophys. Acta.* **1724**, 270–280.

23. Chang, D.C. and Reese, T.S. (1990) Changes in membrane structure induced by electroporation as revealed by rapid-freezing electron microscopy. *Biophys. J.* **58**, 1–12.

24. Tarek, M. (2005) Membrane electroporation: a molecular dynamics simulation. *Biophys. J.* **88**, 4045–4053.

25. Tieleman, D.P. (2004) The molecular basis of electroporation. *BMC Biochem.* **5**, 10.

26. Stulen, G. (1981) Electric field effects on lipid membrane structure. *Biochim. Biophys. Acta.* **640**, 621–627.

27. Lopez, A., Rols, M.P., and Teissie, J. (1988) 31P NMR analysis of membrane phospholipid organization in viable, reversibly electropermeabilized Chinese hamster ovary cells. *Biochemistry.* **27**, 1222–1228.

28. Haest, C.W., Kamp, D., and Deuticke, B. (1997) Transbilayer reorientation of phospholipid probes in the human erythrocyte membrane. Lessons from studies on electroporated and resealed cells. *Biochim. Biophys. Acta.* **1325**, 17–33.

29. Bernhardt, J. and Pauly, H. (1973) On the generation of potential differences across the membranes of ellipsoidal cells in an alternating electrical field. *Biophysik.* **10**, 89–98.

30. Mehrle, W., Hampp, R., and Zimmermann, U. (1989) Electric pulse induced membrane permeabilization. Spatial orientation and kinetics of solute efflux in freely suspended and dielectrophoretically aligned plant mesophyll protoplasts. *Biochim. Biophys. Acta.* **978**, 267–275.

31. Kotnik, T. and Miklavcic, D. (2000) Analytical description of transmembrane voltage induced by electric fields on spheroidal cells. *Biophys. J.* **79**, 670–679.

32. Hibino, M., Itoh, H. and Kinosita, K., Jr. (1993) Time courses of cell electroporation as revealed by submicrosecond imaging of transmembrane potential. *Biophys. J.* **64**, 1789–1800.

33. Valic, B., Golzio, M., Pavlin, M., et al. (2003) Effect of electric field induced transmembrane potential on spheroidal cells: theory and experiment. *Eur. Biophys. J.* **32**, 519–528.

34. Pucihar, G., Kotnik, T., Valic, B., and Miklavcic, D. (2006) Numerical determination of transmembrane voltage induced on irregularly shaped cells. *Ann. Biomed. Eng.* **34**, 642–652.

35. Hibino, M., Shigemori, M., Itoh, H., Nagayama, K., and Kinosita, K., Jr. (1991) Membrane conductance of an electroporated cell analyzed by submicrosecond imaging of transmembrane potential. *Biophys. J.* **59**, 209–220.

36. Teissie, J. and Rols, M.P. (1993) An experimental evaluation of the critical potential difference inducing cell membrane electropermeabilization. *Biophys. J.* **65**, 409–413.

37. Gabriel, B. and Teissie, J. (1997) Direct observation in the millisecond time range of fluorescent molecule asymmetrical interaction with the electropermeabilized cell membrane. *Biophys. J.* **73**, 2630–2637.

38. Schwister, K. and Deuticke, B. (1985) Formation and properties of aqueous leaks induced in human erythrocytes by electrical breakdown. *Biochim. Biophys. Acta.* **816**, 332–348.

39. Rols, M.P. and Teissie, J. (1990) Electropermeabilization of mammalian cells. Quantitative analysis of the phenomenon. *Biophys. J.* **58**, 1089–1098.

40. Gabriel, B. and Teissie, J. (1999) Time courses of mammalian cell electropermeabilization observed by millisecond imaging of membrane property changes during the pulse. *Biophys. J.* **76**, 2158–2165.

41. Golzio, M., Teissie, J., and Rols, M.P. (2002) Direct visualization at the single-cell level of electrically mediated gene delivery. *Proc. Natl. Acad. Sci. USA.* **99**, 1292–1297.

42. Rols, M.P. and Teissie, J. (1998) Electropermeabilization of mammalian cells to macromolecules: control by pulse duration. *Biophys. J.* **75**, 1415–1423.

43. Glogauer, M., Lee, W., and McCulloch, C.A. (1993) Induced endocytosis in human fibroblasts by electrical fields. *Exp. Cell. Res.* **208**, 232–240.

44. Rols, M.P., Femenia, P. and Teissie, J. (1995) Long-lived macropinocytosis takes place in electropermeabilized mammalian cells. *Biochem. Biophys. Res. Commun.* **208**, 26–35.

45. Angelova, M.I. and Tsoneva, I. (1999) Interactions of DNA with giant liposomes. *Chem. Phys. Lipids.* **101**, 123–137.

46. Angelova, M.I., Hristova, N., and Tsoneva, I. (1999) DNA-induced endocytosis upon local microinjection to giant unilamellar cationic vesicles. *Eur. Biophys. J.* **28**, 142–150.

47. Chernomordik, L.V., Sokolov, A.V., and Budker, V.G. (1990) Electrostimulated uptake of DNA by liposomes. *Biochim. Biophys. Acta.* **1024**, 179–183.

48. Hristova, N.I., Tsoneva, I., and Neumann, E. (1997) Sphingosine-mediated electroporative DNA transfer through lipid bilayers. *FEBS Lett.* **415**, 81–86.

49. Spassova, M., Tsoneva, I., Petrov, A.G., Petkova, J.I., and Neumann, E. (1994) Dip patch clamp currents suggest electrodiffusive transport of the polyelectrolyte DNA through lipid bilayers. *Biophys. Chem.* **52**, 267–274.

50. Kubiniec, R.T., Liang, H., and Hui, S.W. (1990) Effects of pulse length and pulse strength on transfection by electroporation. *Biotechniques.* **8**, 16–20.

51. Liang, H., Purucker, W.J., Stenger, D.A., Kubiniec, R.T. and Hui, S.W. (1988) Uptake of fluorescence-labeled dextrans by 10T 1/2 fibroblasts following permeation by rectangular and exponential-decay electric field pulses. *Biotechniques.* **6**, 550–552, 554, 556–558.

52. Wolf, H., Rols, M.P., Boldt, E., Neumann, E., and Teissie, J. (1994) Control by pulse parameters of electric field-mediated gene transfer in mammalian cells. *Biophys. J.* **66**, 524–531.

53. Rols, M.P., Coulet, D., and Teissie, J. (1992) Highly efficient transfection of mammalian cells by electric field pulses. Application to large volumes of cell culture by using a flow system. *Eur. J. Biochem.* **206**, 115–121.

54. Heller, R., Jaroszeski, M., Atkin, A., et al. (1996) In vivo gene electroinjection and expression in rat liver. *FEBS Lett.* **389**, 225–228.

55. Klenchin, V.A., Sukharev, S.I., Serov, S.M., Chernomordik, L.V., and Chizmadzhev Yu, A. (1991) Electrically induced DNA uptake by cells is a fast process involving DNA electrophoresis. *Biophys. J.* **60**, 804–811.

56. Sukharev, S.I., Klenchin, V.A., Serov, S.M., Chernomordik, L.V., and Chizmadzhev Yu, A. (1992) Electroporation and electrophoretic DNA transfer into cells. The effect of DNA interaction with electropores. *Biophys. J.* **63**, 1320–1327.

57. Muller, K.J., Horbaschek, M., Lucas, K., Zimmermann, U., and Sukhorukov, V.L. (2003) Electrotransfection of anchorage-dependent mammalian cells. *Exp. Cell Res.* **288**, 344–353.

58. Faurie, C., Phez, E., Golzio, M., et al. (2004) Effect of electric field vectoriality on electrically mediated gene delivery in mammalian cells. *Biochim. Biophys. Acta.* **1665**, 92–100.

59. Liu, F., Heston, S., Shollenberger, L.M., et al. (2006) Mechanism of in vivo DNA transport into cells by electroporation: electrophoresis across the plasma membrane may not be involved. *J. Gene Med.* **8**, 353–361.

60. Satkauskas, S., Bureau, M.F., Puc, M., et al. (2002) Mechanisms of in vivo DNA electrotransfer: respective contributions of cell electropermeabilization and DNA electrophoresis. *Mol. Ther.* **5**, 133–140.

61. Satkauskas, S., Andre, F., Bureau, M.F., Scherman, D., Miklavcic, D., and Mir, L.M. (2005) Electrophoretic component of electric pulses determines the efficacy of in vivo DNA electrotransfer. *Hum. Gene Ther.* **16**, 1194–1201.

62. Bureau, M.F., Gehl, J., Deleuze, V., Mir, L.M., and Scherman, D. (2000) Importance of association between permeabilization and electrophoretic forces for intramuscular DNA electrotransfer. *Biochim. Biophys. Acta.* **1474**, 353–359.

63. Mir, L.M., Bureau, M.F., Gehl, J., et al. (1999) High-efficiency gene transfer into skeletal muscle mediated by electric pulses. *Proc. Natl. Acad. Sci. USA.* **96**, 4262–4267.

64. Aihara, H. and Miyazaki, J. (1998) Gene transfer into muscle by electroporation in vivo. *Nat. Biotechnol.* **16**, 867–870.

65. Pavselj, N. and Preat, V. (2005) DNA electrotransfer into the skin using a combination of one high- and one low-voltage pulse. *J. Control. Release.* **106**, 407–415.

66. Phez, E., Faurie, C., Golzio, M., Teissie, J., and Rols, M.P. (2005) New insights in the visualization of membrane permeabilization and DNA/membrane interaction of cells submitted to electric pulses. *Biochim. Biophys. Acta.* **1724**, 248–254.

67. Rols, M.P. (2006) Electropermeabilization, a physical method for the delivery of therapeutic molecules into cells. *Biochim. Biophys. Acta.* **1758**, 423–428.

68. Brown, D.A. and London, E. (2000) Structure and function of sphingolipid- and cholesterol-rich membrane rafts. *J. Biol. Chem.* **275**, 17221–17224.

69. Brown, D.A. and London, E. (1998) Functions of lipid rafts in biological membranes. *Annu. Rev. Cell Dev. Biol.* **14**, 111–136.

70. Phez, E., Cezanne, L., Charpentier, A., et al. (2005) Can lipid domains control membrane permeabilization and DNA uptake in cells submitted to electric field? *Eur. Biophys. J.* **34**, 553.

71. Lechardeur, D. and Lukacs, G.L. (2002) Intracellular barriers to non-viral gene transfer. *Curr. Gene Ther.* **2**, 183–194.

72. Seksek, O., Biwersi, J., and Verkman, A.S. (1997) Translational diffusion of macromolecule-sized solutes in cytoplasm and nucleus. *J. Cell Biol.* **138**, 131–142.

73. Dowty, M.E., Williams, P., Zhang, G., Hagstrom, J.E., and Wolff, J.A. (1995) Plasmid DNA entry into postmitotic nuclei of primary rat myotubes. *Proc. Natl. Acad. Sci. USA.* **92**, 4572–4576.

74. Rols, M.P. and Teissie, J. (1992) Experimental evidence for the involvement of the cytoskeleton in mammalian cell electropermeabilization. *Biochim. Biophys. Acta.* **1111**, 45–50.

75. Lechardeur, D., Sohn, K.J., Haardt, M., et al. (1999) Metabolic instability of plasmid DNA in the cytosol: a potential barrier to gene transfer. *Gene Ther.* **6**, 482–497.

76. Bureau, M.F., Naimi, S., Torero Ibad, R., et al. (2004) Intramuscular plasmid DNA electrotransfer: biodistribution and degradation. *Biochim. Biophys. Acta.* **1676**, 138–148.

77. Takahashi, M., Furukawa, T., Nikkuni, K., et al. (1991) Efficient introduction of a gene into hematopoietic cells in S-phase by electroporation. *Exp. Hematol.* **19**, 343–346.

78. Schwachtgen, J.L., Ferreira, V., Meyer, D., and Kerbiriou-Nabias, D. (1994) Optimization of the transfection of human endothelial cells by electroporation. *Biotechniques.* **17**, 882–887.

79. Golzio, M., Teissie, J., and Rols, M.P. (2002) Cell synchronization effect on mammalian cell permeabilization and gene delivery by electric field. *Biochim. Biophys. Acta.* **1563**, 23–28.

80. Vaughan, E.E. and Dean, D.A. (2006) Intracellular trafficking of plasmids during transfection is mediated by microtubules. *Mol. Ther.* **13**, 422–428.

81. Schoenbach, K.H., Beebe, S.J., and Buescher, E.S. (2001) Intracellular effect of ultrashort electrical pulses. *Bioelectromagnetics.* **22**, 440–448.

82. Beebe, S.J., White, J., Blackmore, P.F., Deng, Y., Somers, K., and Schoenbach, K.H. (2003) Diverse effects of nanosecond pulsed electric fields on cells and tissues. *DNA Cell Biol.* **22**, 785–796.

83. Pavselj, N., Bregar, Z., Cukjati, D., Batiuskaite, D., Mir, L.M., and Miklavcic, D. (2005) The course of tissue permeabilization studied on a mathematical model of a subcutaneous tumor in small animals. *IEEE Trans. Biomed. Eng.* **52**, 1373–1381.

Chapter 3
Applicator and Electrode Design for In Vivo DNA Delivery by Electroporation

Dietmar Rabussay

Abstract As in vivo electroporation advances from the preclinical phase to clinical studies and eventually to routine medical practice, the design of electroporation devices becomes increasingly important. Achieving safety and efficacy levels that meet regulatory requirements, as well as user and patient friendliness, are major design considerations. In addition, the devices will have to be economical to manufacture. This chapter will focus on the design of applicators and electrodes, the pieces of hardware in direct contact with the user and the patient, and thus key elements responsible for the safety and efficacy of the procedure. The two major foreseeable applications of the technology in the DNA field are for gene therapy and DNA vaccination. Design requirements differ considerably for these applications and for the diseases to be treated or prevented. In addition to the trend of device differentiation, there is also a trend to build devices capable of performing both the step of delivering the DNA to the target tissue and the subsequent step of electroporation. This chapter presents the electrical and biological principles underlying applicator and electrode design, gives an overview of existing devices, and discusses their advantages and disadvantages. The chapter also outlines major design considerations, including regulatory pathways, and points out potential future developments.

Keywords: electroporation electrodes, electroporation applicators, in vivo DNA delivery, in vivo electroporation, clinical electroporation, electroporation therapy, DNA vaccines, gene therapy

1. Introduction

Interest in electroporation for in vivo transfection continues to increase, driven by medical applications such as gene therapy and DNA vaccines. As other delivery methods have fallen short of meeting the requirements for safe, effective, and economical in vivo delivery, the proven and potential advantages of electroporation have increasingly become more attractive. In vivo electroporation was first

S. Li (ed.), *Electroporation Protocols: Preclinical and Clinical Gene Medicine.*
From *Methods in Molecular Biology, Vol. 423.*
© Humana Press 2008

introduced in 1987 for intratumoral delivery of an anticancer drug (1) but has been used for DNA delivery only since 1996 (2–4). Recently, the first clinical trials applying in vivo electroporation have been initiated.

Properly designed applicators and electrodes are crucial for the safety, efficacy, and patient acceptance of electroporation therapy. Colloquially, the terms "applicator" and "electrode" are often used interchangeably, but it is useful to distinguish between the two. An applicator basically consists of a handle and power cord, with the electrodes permanently or exchangeably inserted into the handle; however, in some designs the handle also serves as a DNA injection device or is combined with an injection device.

The two essential requirements for successful DNA transfer by electroporation include the presence of the DNA at the treatment site in sufficient concentration and the application of an appropriate electrical field pulse covering the treatment site. The DNA is delivered to the treatment site by mechanical means, most frequently by injection. The electrical pulse originates from a generator which provides a voltage

Table 3.1 Functions of the generator and applicator

Generator[a]
Individual pulse parameter
Voltage
Pulse length
Wave form
Polarity
Multiple-pulse (single-train) parameter
Number of pulses
Frequency (Hz_T)
Electrode switching patterns (X_T)
Multiple-train parameter
Pulses per train
Frequency (Hz_{MT})
Electrode switching patterns (X_{MT})
Applicator
Electrodes[b]
Parameter
Field strength (V/cm): a function of electrode distance
Field homogeneity: a function of electrode shape
Field orientation: a function of electrode shape and relative position
Field depth (for meander-type electrodes): a function of electrode distance and shape
Current, current density and resistance: a function of electrode surface area
Handle[c]
Facilitates safe and ergonomical placement of electrodes at desired anatomical site
Helps ensure proper contact with target tissue

[a]Controls pulse and train parameters
[b]Determine electric field and current parameters
[c]It may be part of an integrated DNA injection-electroporation device

output to the electrodes. The resultant voltage (potential difference) between the electrodes generates an electrical field encompassing the volume between the electrodes and extending for a short distance beyond. The optimal pulse and applicator parameters are essentially the same for different plasmid DNAs but may vary depending on the target tissue.

Table 3.1 summarizes the basic functions of the generator, applicator, and electrodes. The generator determines the two main parameters of the pulse, its amplitude (output voltage) and pulse length. In addition, the generator controls the pulse shape (e.g., square wave or exponential decay wave), pulse polarity, and pulse pattern.

The electrodes make contact with the target tissue and transform the voltage from the generator into an electric field. The geometry (shape and distance) of the electrodes determines the strength, shape, and homogeneity of the electrical field and, thus, the efficacy as well as the side effects of the voltage pulse.

This chapter will almost exclusively concentrate on the delivery of plasmid DNA and will only mention experiences from using electroporation for the delivery of other drugs where such experiences are instructive for improving DNA delivery. The focus will be on the safety, efficacy, and side effects of in vivo electroporation as a function of the various applicator and electrode designs. Pulse parameters, aspects of user- and patient-friendliness, and economical and regulatory considerations will also be discussed.

2. Delivery of DNA into Tissue

As a first step in efficiently delivering plasmid DNA into cells of living organisms, the DNA is normally delivered into the interstitial space of the target tissue. One exception is the delivery of DNA by gene gun, which delivers at least some DNA-coated particles directly into cells. Delivery to the interstitial space can be accomplished by the following methods summarized in Table 3.2: (1) Needle and syringe method is most commonly used to inject DNA solution either directly into the target tissue (muscle, tumor, skin, and others), or indirectly through intravenous or intra-arterial injection. (2) Needle-free injection uses a high-speed fluid jet which either delivers DNA into the skin or penetrates all the way into the underlying muscle, depending on the amount of pressure employed. (3) The so-called hydrodynamic method, a variation of regular intravenous injection, entails the rapid injection of a relatively large volume of DNA solution. This method has been used for delivery into muscle and certain internal organs. (4) Catheters of various designs have been used to infuse DNA into blood vessels and their walls or, indirectly, into the tissues served by those vessels. (5) Beds of microneedles, iontophoresis, electroporation of the stratum corneum, and ultrasound have been used to deliver DNA intradermally.

Skeletal muscle has been the most frequently used target tissue, followed by tumor tissue, skin, and a variety of other tissues. DNA delivery into muscle, tumor tissue, and skin has predominantly been performed by needle and syringe, although needle-free injection

Table 3.2 Methods and devices for DNA delivery to the interstitial space

Method or device	DNA delivery		Location of EP site[a] relative to injection site	Reference
	Route	Target tissue		
Needle and syringe	i.m.	Skeletal muscle	a.t.c.o. injection site	5
	i.t.	Tumor	a.t.c.o. injection site or entire tumor	6,7
	i.d.	Epidermis and/or dermis	a.t.c.o. injection site	8
	i.a.	Various	Distant target tissue	2
Fluid jet	i.d.	Dermis	a.t.c.o. injection site	9
	i.m.	Skeletal muscle	a.t.c.o. injection site	9
Hydrodynamic delivery	i.v.	Muscle, liver, other	Distant target tissue	10
Electroporation catheter	i.a.	Vessel wall	Distant from catheter insertion site	11, 12
Microneedles, iontophoresis, electroporation of stratum corneum, ultrasound	i.d.	Epidermis and/or dermis	At or near delivery site	13–15

[a]*EP site* electroporation site, *i.m.* intramuscular, *i.t.* intratumoral, *i.d.* intradermal, *i.v.* intravenous, *i.a.* intraarterial, *a.t.c.o.* around the center of

is more effective (9). It is worth noting that tumors differ from other tissues in that they display relatively high interstitial pressure, which tends to make intravenous or intra-arterial delivery less effective than intratumoral injection. To our knowledge, only needle and syringe but not needle-free injection has been used on tumors.

Delivery of DNA into and through the skin by methods other than injection is made difficult by the stratum corneum, the outermost layer of the skin consisting mostly of dead keratinocytes. Relying on natural pathways through the stratum corneum, such as sweat glands and hair follicles, iontophoresis has been moderately effective in transporting charged molecules, including DNA, into skin. Subjecting the stratum corneum to electroporation pulses creates additional pathways (aqueous pores) and increases transport across the stratum corneum. Ultrasound enhances penetration of DNA into skin as well. Conceptually the most elegant method for intradermal delivery may be the use of beds of DNA-coated microneedles, which deliver DNA as they penetrate through the stratum corneum into the epidermis and upper dermis. A drawback of this method is the relatively small amount of DNA or RNA that can be delivered.

2.1. Nonintegrated Injection and Electroporation Systems

Among the methods and devices listed in Table 3.2, only some have undergone development for eventual use in humans. Except for the systems using coated microneedles and fluid jets, essentially all systems listed in Table 3.3 use needle and

Table 3.3 Nonintegrated and integrated injection and electroporation systems

Nonintegrated systems	Integrated systems
MedPulser DDS (N&S) (16)*	Elgen (N&S) (18)*
MedPulser DETS (N&S) (16)*	TriGrid (N&S) (19)
CliniPorator (N&S) (17)	EKD (N&S) (20)
	EasyVax (DNA-coated microneedles) (13)
	Jet DDS (fluid jet) (21)

Numbers in parentheses refer to references; devices labeled with an asterisk* are presently used in human DNA therapy studies

N&S needle and syringe, *DDS* DNA delivery system, *DETS* DNA electroporation therapy system

syringe injection in one form or another. Some of these systems use common hand-injection of DNA with needle and syringe; after injection the needle is withdrawn, needle electrode arrays surrounding the injection site are inserted, and electroporation pulses are delivered. We have called these systems nonintegrated because they use separate injection and electroporation devices. Two MedPulser systems (Genetronics, Inc., a subsidiary of Inovio Biomedical Corporation), which are presently in DNA therapy phase I clinical trials, and the CliniPorator system (IGEA srl) fit this category (Table 3.3, column 1). The MedPulser DD system was specifically designed for intramuscular DNA delivery and is being evaluated in a phase I anticancer DNA vaccine study. Both the MedPulser DET (DNA electroporation therapy) system and the CliniPorator were designed primarily for delivering anticancer drugs into tumors but are also useful as DNA delivery platforms. The DET system is presently used for intratumoral DNA delivery in two human immunotherapy trials. Advantages of nonintegrated systems include simple design and handling, as well as flexibility and precision in terms of DNA injection and electroporation. Especially the latter two features are important for treating tumors of various shapes and consistencies.

2.2. Integrated Injection and Electroporation Systems

This group of systems combines injection and electroporation devices into one unit (Table 3.3, column 2); however, DNA injection and electroporation are still performed in two subsequent steps. The Elgen system (Inovio AS) uses two syringes with needles mounted in parallel on a device which advances the needles into the target muscle at a controlled speed and injects DNA through the needles as they penetrate into the muscle. Once insertion and injection have stopped, the needles serve as electrodes. The TriGrid system (Ichor MS) also uses a regular needle and syringe which is inserted into a hand-held device together with a disposable electrode array surrounding the injection needle. The EKD system (ADViSYS, Inc.) features a central injection port surrounded by five needle electrodes. CytoPulse Sciences, Inc., has developed the EasyVax system using a bed of DNA-coated microneedles which are inserted into the skin to a minimal

depth and also serve as electrodes. Finally, Inovio Biomedical Corporation has conceptualized a device combining fluid jet injection and electroporation, with the fluid jet doubling as an electrode, or two fluid jets functioning as two electrodes. Integrated systems (with the exception of the Elgen) are less threatening to the patient because the needles are less visible. More important, these systems have the advantage over nonintegrated systems that the injection site and the electrodes are automatically aligned, thus eliminating the possibility of misalignment. However, sterile assembly and disassembly of integrated systems makes them more cumbersome than nonintegrated ones. Certain aspects of the systems listed in Table 3.3 will be discussed in more detail in sect. 5.2.

3. Electrode and Applicator Design Process

An essential first step in the design process is to clearly define the intended use of the device or system. It is also helpful to explicitly state which uses are not intended. A thorough understanding of the intended use will help to avoid later disappointments and difficulties ranging from the initial design stage all the way to the marketplace.

The second step is to carefully evaluate and define the requirements the future device or system will need to meet. This process should also include documentation of requirements that have been considered but were eliminated for specific reasons. Requirement definition should include, in that order, the following:

1. Regulatory requirements and the intended regulatory path; or a reasoned analysis as to why the intended use of the product is exempt from regulations
2. Biomedical performance and test specifications
3. Engineering design and test specifications

Major design considerations concerning these points will be discussed in sect. 4. After incorporating those considerations in the design and development plan, the actions listed below will lead to an investigational device or system and eventually to a commercial product:

1. Choose one or more designs to be made into prototypes.
2. Validate that the design as implemented in the prototypes actually meets the proposed requirements.
3. Build devices according to the planned manufacturing process.
4. Verify that the product meets physical and functional specifications and can be manufactured consistently.

In many cases, the design process will be iterative and changes in specifications will become necessary or desirable. However, a clear definition of the product's purpose and specifications at the outset should make the design and development process expedient and cost effective.

4. Major Design Considerations

The major design considerations should be driven by ethical and regulatory imperatives, user and patient friendliness, and requirements for commercial success. Although the Food and Drug Administration (FDA) is primarily interested in ensuring the safety and efficacy of both investigational and commercial products, patient- and user-friendliness, as well as commercial viability, are additional vital aspects that must be addressed during the design phase of the product.

4.1. Regulatory Pathways

Depending on the intended use of the device, the design and development process may be regulated by one or several agencies, or not at all. Devices intended for use with nonfood animals, e.g., research and pet animals, are generally exempt from regulation. Electroporation devices to be used for DNA delivery to food animals may be regulated by the United States Department of Agriculture (USDA) and/or the FDA. For devices delivering DNA to humans, several regulatory pathways are possible. If one seeks approval for the device only, a 510K or premarket approval (PMA) application may be chosen. However, clearance obtained via this pathway will not automatically allow the device to be used with DNA species that have not obtained clearance in conjunction with their delivery by that particular device. In regulatory terms, the FDA treats the use of an electroporation device for delivery of DNA to humans as a drug-device combination product. That means, gene therapy or DNA vaccine products delivered via electroporation will be evaluated (as a minimum) by the Center for Device and Radiation Health (CDRH) and the Center for Biologics Evaluations and Research (CBER) of the FDA. The combination products must also be cleared by the Recombinant DNA Advisory Committee (RAC), usually before the FDA will start processing an application. Thus, to pursue investigational or commercial clearance for a DNA drug to be delivered by electroporation, it may be preferable to choose the Initial New Drug (IND) application route and have the device manufacturer provide the necessary device information package. Alternatively, the required device documentation may be satisfied by cross-referencing an existing FDA document, such as a Device Master File, or a relevant IND application containing the device information.

4.2. Safety Considerations

Both the safety of the applicator and its electrodes, as well as the effects of electroporation itself on the target tissue, need to be considered.

4.2.1. Applicators and Electrodes

Risks associated with the applicator/electrodes include risks related to sharp and electrical injuries, microbial contamination (infection), and toxicity from incompatible materials. To avoid accidental sharp injuries of both the operator and the patient, appropriate shielding elements should be designed to protect against sharp electrode needle tips.

The risk of electrical shock is mitigated by the fact that applicators and their electrodes are not, and should not be, grounded. Therefore, any shock will essentially be limited to the tissue between the positive and negative contact points, e.g., the tissue between two electrodes, and will not travel through the body to the ground. Safety features should be built into the power supply to prevent accidental discharge. Since some electroporation pulses deliver high currents (up to several amperes), both the electrodes and the connections to the pulse generator must be able to conduct such currents without excessive heating, sparking, or frying. In this context, it is also important that the electrodes are designed to make good and even contact with the target tissue to avoid local tissue damage either from heating at high-resistance electrode/tissue interphases or from high current densities in lower-resistance spots. Although this is generally not a problem with invasive electrodes (except in heterogeneous tissue), it can be problematic with surface electrodes (see sect. 5.1).

Electrodes, and, for certain applications, the entire applicator must be sterile. Designs that do not allow sterilization of the applicator or require manipulations that easily lead to contamination of the sterile electrodes should be improved. Keeping all parts that come in contact with the patient sterile during applicator assembly seems to be difficult with some of the integrated designs listed in Table 3.3.

Finally, applicator and electrode materials, both plastics and metals, should be biocompatible. For example, the charge transferred during pulses is sufficient to dissolve quantities of toxic nickel, chromium, and iron from invasive stainless steel electrodes that may cause at least local toxicity. Some people are allergic to nickel and other heavy metals. The MedPulser DDS (DNA delivery system) therefore uses gold-coated electrodes.

4.2.2. Effects of Electroporation on Normal Tissue and Physiology

Extensive evidence has accumulated over the past 15 years that electroporation effects on normal tissues do not raise safety concerns. From numerous studies in small and large animals and in humans, it is evident that cell killing in vivo, at electroporation parameters highly efficacious for drug and DNA delivery, is substantially lower than cell killing in vitro. It appears that cells in normal tissue are less vulnerable and/or are better able to recover from electroporation than are cells suspended in buffer or media. Histological changes are visible after electroporation, but they are generally mild to moderate and transient. For example, histological changes in muscle and skin after electroporation within a

relatively broad range of conditions (multiple pulses of 100 V for 60 ms, or 1,100 V for 0.1 ms) include moderate inflammation, minor necrosis, and minor fibrosis, which, except for traces of fibrosis, disappear between 20 and 40 days after electroporation. Even muscle and skin tissue in healing incision wounds are only slightly affected by multiple 510-V, 100-µs pulses. Systematic and complete studies as to which electroporation parameters, or combinations of parameters, contribute to these histological changes are not available, but enough is known to minimize these changes when it is desirable to do so. The severity of changes increases with higher voltage and the number and duration of pulses (22, 23).

In summary, given a certain electrical pulse and certain electrical properties of the target tissue, electrode design parameters that affect safety in terms of electrical effects on the treated tissue include the following: (1) electrode distance, determining nominal and actual field strength; (2) electrode shape, determining the three-dimensional field shape, i.e., the spatial field strength distribution; and (3) electrode surface area, determining resistance, current, current density, and charge density (see also Table 3.1). Safety will be compromised if, for a given set of pulse parameters, the electrode distance is too short (leading to high field strength, sparking), the electrode shape includes sharp edges, protrusions, or warps (resulting in spots of high field strength, sparking, or spots of high current density), or the electrode surface is too large or too small (causing excessive current, current density, or charge density). The same factors that compromise safety will generally also lessen therapeutic efficacy.

Physiological effects of electroporation are essentially limited to muscle contractions and a certain degree of pain (see sect. 4.4.2). Skeletal muscle contractions are very brief, while contractions of the smooth muscle cell layer of larger blood vessels, which occur during a fraction of electroporations, can last from seconds to 1 min. Interestingly, electroporation with electroporation catheters under conditions efficacious for the delivery of DNA or drugs into vessel walls does not affect normal heart function or blood chemistry (11).

4.3. Efficacy

In the context of DNA therapies, the efficacy of electroporation is measured by the extent it improves a particular therapy over DNA administration without electroporation. Thus, optimal electroporation conditions will differ for different types of treatment. For many gene therapy applications, high levels of gene expression over a long period of time in the absence of an immune response against the transgene product will be most efficacious. On the other hand, when employing electroporation for the delivery of DNA vaccines, the strongest possible immune response is desired. In this case, amplitude and duration of gene expression may be less important than the adjuvant effect of electroporation.

For both gene therapy and DNA vaccination, it is important that as much of the injected DNA as possible is located within the effective field produced by the electrodes. Our knowledge about how and how quickly DNA spreads and diffuses in the different target tissues and the effects different injection techniques have on DNA distribution is still insufficient. The timing of electroporation relative to injection also influences transfection efficiency, because of time-dependent diffusion and degradation of the DNA. Moreover, it is important to treat an appropriate volume of target tissue to produce the desired amount of transgene product. The size and distance of the electrodes are limited by constraints in current and voltage that can be applied safely. Therefore, if larger volumes need to be treated than can be electroporated by a given electrode configuration, repeated electroporations at multiple treatment sites may be necessary.

For effective gene therapies, electrodes generating a highly homogeneous field of relatively low strength will be advantageous. Such a field minimizes histological changes ("cell and tissue damage") and thus promotes high and long-lasting gene expression while dampening the risk of a strong immune response. Fields of higher strength and lesser homogeneity are fine for DNA vaccinations, as long as the cell and tissue damage is transient. These fields will cause some inflammation, result in shorter gene expression, and provoke a stronger immune response.

4.4. User- and Patient-Friendliness

4.4.1. User-Friendliness

The goal is to design applicators that are easy to use, minimize the chance for user mistakes, and are error-tolerant.

These criteria ask for a design that fulfills functional requirements in the simplest way possible. Simplicity generally enhances safety. The applicators in Table 3.3 are adequate for gene therapy and therapeutic vaccine applications but, with the possible exception of the EasyVax system, are clearly too cumbersome for mass vaccinations. Even for gene therapy and therapeutic vaccinations, the integrated devices seem unnecessarily complex. Instead of assembling and disassembling the mix of sterile and nonsterile, disposable and reusable components, it appears simpler to use nonintegrated devices which are easier and faster to use, less expensive, and less prone to breaching sterility. The advantages of the integrated devices (sect. 2.2) may not be sufficiently compelling to outweigh their operational disadvantages. That being said, the future device that will meet the requirements for mass vaccinations will probably be a vastly simplified integrated device that will require a minimum of handling.

Table 3.4 Types of electroporation electrodes

Electrode type	Target tissue	Notes
Surface electrodes a. Plate electrodes b. Meander and ring electrodes c. Microneedle arrays	Skin and organ surfaces	"a" and "b" can be used for electroporating the stratum corneum, and cells in the epidermis and dermis, preferentially in connection with intradermal DNA injection. "c" mechanically penetrates the stratum corneum and electroporates skin cells. Effective electric field depth of surface electrodes is only several millimeters

Needle electrodes
a. Needle pair
b. Needle arrays
  – Square, rectangular, or rhomboid arrays (4 needles)
  – Rows and grids of needles
  – Circular arrays (5, 6, or more needles)
	Muscle, tumor, skin, and other tissues	Needle arrays can cover relatively large volumes and areas of target tissue to several centimeters in depth. If mounted on appropriate handles, catheters, or endoscopic devices, internal organs can be reached

Catheter electrodes	Blood vessels and hollow organs	A catheter with a porous balloon and coil electrodes on the distal end of the catheter shaft is advanced in the hollow organ to the treatment site where the DNA and electroporation pulses are delivered (Fig. 3.7)
Other electrode configurations	Skin, muscle, blood vessels, eye, other tissue	See text
Induction "electrode"	Any tissue within the reach of sufficiently strong magnetic fields	The electric field is generated by induction from a coil with a fast-varying electrical current. Very high current and frequency requirements have prevented practical use

4.4.2. Patient-Friendliness

Here the goal is to minimize pain and muscle twitching, two inherent side effects of the procedure. Different types of electrodes (Table 3.4) generate different levels of pain. When applied properly, surface electrodes are the least painful and generate only mild pain sensations (14). Catheter electrodes are predicted to be also relatively benign, although these electrodes have not been tested in humans. Invasive electrodes such as needle electrodes cause higher but tolerable pain. The remainder of this section will focus on invasive electrodes.

Only during the last 1–2 years has it become clear that pain is not an obstacle for the broad use of electroporation, even in the absence of anesthesia. This conclusion was reached from studies with healthy volunteers and the experiences in four ongoing clinical studies wherein all patients treated with the MedPulser systems and the Elgen system, respectively (approximately 100 patients by July 2007), have returned for repeated sessions of DNA immunotherapy and DNA vaccinations.

One source of pain is the insertion of injection and electrode needles, the numbers and dimensions of which vary for the different electroporation systems (sect. 5.2). Of course, the pain increases with the number, thickness, and length of the needles; therefore, one design goal is to minimize needle number and dimensions while maintaining safety and efficacy.

The other source of pain is the electric pulse. The strength of the sensation depends on a multitude of factors and, as any pain perception, can vary considerably from person to person. As far as electrode design is concerned, three parameters determine the pain generated. (1) Electrode surface area (in the case of needle electrodes, needle diameter and insertion depth) determines the contact area with the tissue and, at a given voltage, the amount of current flowing. Pain increases with greater contact area and current, although not linearly. (2) Electrode shape determines current density. The pain is greater at the spots of higher current density. From a pain perspective, flat opposing plate electrodes are ideal. (3) Electrode distance affects the tissue volume and, thus, the number of sensory elements exposed to current.

An example of how pain sensation can be reduced is the development of the MedPulser DDS. This system is based on the MedPulser EPT (electroporation therapy) system, the parent of both MedPulser systems now used for DNA delivery (Table 3.3). The EPT system (designed for the delivery of bleomycin into tumors) uses a six-needle array electrode of 3-cm-long 22-GA needles (needle distance = 0.86 cm) and six bipolar, rotating pulses of 1,130 V, 100 μs, at 4 Hz. After redesigning the system for DNA delivery with the goal of applying it without anesthesia, the DDS now uses a four-needle array electrode of 1.5-cm-long 26-GA needles (needle distance = 0.43 cm), and 2 monopolar pulses of 106 V, 60 ms, at 4 Hz. This redesign was performed in several steps, ensuring at every step that electroporation efficiency did not suffer. First, the voltage and pulse length were optimized for DNA delivery and determined to be 212 V and 60 ms, respectively. Second, the number and polarity of pulses was reduced to two monopolar pulses, eliminating the pain of 4 pulses. Third, the needle diameter and length were reduced (less insertion pain, less current, and less treated tissue volume). Fourth, the number of needles was reduced to four (less insertion pain, smaller tissue volume affected). Finally, the needle distance was shortened by one half, which further reduced the tissue volume affected. Halving the distance also allowed the voltage to be cut in half while keeping the field strength (246 V/cm) the same. Reduction of the voltage entailed a proportional reduction of current and current density and, thus, a further reduction in pain (16, 24).

5. Electrodes and Applicators

The purpose of the electrodes is to make contact with the target tissue, conduct the voltage pulse to the tissue, and transform it into electrical fields localized between and around the electrodes. The role of applicators is to allow safe handling of the electrodes and to facilitate placement of the electrodes at the chosen target site. An overview of different types of electrodes used for in vivo electroporation is given in Table 3.4.

5.1. Surface Electrodes

Besides plate and meander electrodes, other types of surface electrodes have been designed, which will not be discussed here. Microneedle electrodes have been included in this section because their penetration depth is minimal and their performance characteristics resemble more closely those of surface electrodes rather than needle electrodes.

5.1.1. Plate Electrodes

Conductive parallel plates are best suited to generate close to homogeneous fields, provided the resistance of the target tissue between the plates is rather uniform. A practical design uses plates mounted on a Vernier caliper, which allows the electrode distance to be read after the target tissue has been squeezed between the plates. This, in turn, allows one to calculate the voltage to be applied in order to achieve the desired field strength (nominal field strength = V/cm). Smaller versions of plate electrodes ("paddle electrodes") are also commercially available. Plate electrodes have been used to electroporate skin or tissue under the skin, including subcutaneous tumor nodules. In mice, entire limbs, or portions thereof, may be squeezed between plates for electroporation of both skin and muscle tissue. However, the thickness of tissue that can be electroporated by plate electrodes is limited to about 0.5 cm. The voltage required for generating effective fields for electroporating thicker tissue folds becomes prohibitive since the resistance of the stratum corneum, even after electroporation, is relatively high. If the voltage was increased to the required levels, the resulting high current densities in the newly formed pathways would result in unacceptable skin damage (25). Therefore, plate electrodes are of limited use for electroporation in larger animals with thick skin and for cells deeper below the surface (26). Successful initial treatments of certain cutaneous and subcutaneous malignancies were performed with plate electrodes, but superficial skin burnings were observed (27, 28). Sersa et al. (29) improved efficacy by rotating the position of the electrodes and, thus, the field by 90° between pulses. A four-plate electrode design by Heller et al. is simpler to use and increases electroporation efficiency when compared with caliper electrodes (30).

Transcutaneous electroporation effects and side effects vary with the condition of the skin which differs depending on species, anatomical location, age, and environmental conditions. Dry skin reduces surface currents and enhances electroporation efficiency.

5.1.2. Meander and Ring Electrodes

Similar to plate electrodes, the purpose of meander and ring electrodes can be twofold: to create pathways across the stratum corneum and to electroporate cells within or below the skin. DNA delivery to skin cells can be accomplished by injecting DNA intradermally, followed by electroporation of the skin. An alternative though less effective way is to use electroporation for both creating new pathways across the stratum corneum and delivery into cells. In this case the limiting step is the diffusion or transport of DNA through the new pathways.

A meander electrode consists of opposing finger electrodes of alternating polarity on a flexible base (Fig. 3.1) (31). The electrodes are placed on the skin, uniform contact with the skin being important. When pulses of appropriate voltage (50–100 V) are applied, electrical fields are generated between the electrodes. A strong field enhancement takes place in the highly resistive but thin stratum corneum, which leads to the formation of pores across the stratum corneum into the highly conductive epidermis. This process is analogous to the formation of pores across the high-resistance cell membrane into the highly conductive cytosol. Electrostatic and equipotential field lines generated by three adjacent finger electrodes in contact with skin, before poration of the stratum corneum, are shown in Fig. 3.2. After poration, the depth range of the electric field depends on the number and size of newly generated pathways (pores), the spacing of the electrodes, and the voltage applied. Wider spacing and higher voltage will increase depth range; however, achievable depth range is usually limited to a few millimeters because of skin damage that would be caused by higher voltage and, thus, higher current density. Also, the advantage of shallow fields and low pain associated with

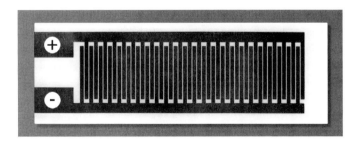

Fig. 3.1 Meander electrodes. Positive and negative finger electrodes are alternately arranged on a flexible support

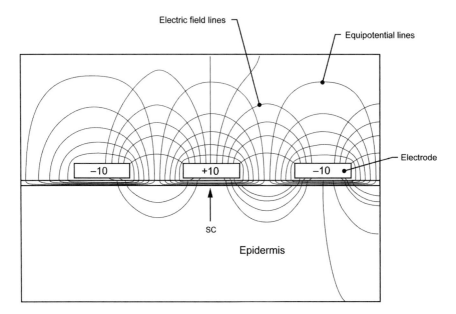

Fig. 3.2 Equipotential and electrostatic field lines of meander electrodes placed on top of the stratum corneum. Calculations were performed for the following specific resistivities: medium surrounding the meander electrodes, $1,000\,\Omega$ cm; stratum corneum, $6 \times 10^8\,\Omega$ cm; viable epidermis, $10^5\,\Omega$ cm. Potential difference between electrodes: ± 10 V. Distance between electrode and stratum corneum: 10 μm. (From (31), with permission by the publisher)

surface electrodes will disappear with higher voltage because more sensory and motor nerves will be activated by deeper fields.

Ring electrode arrays consist of many small concentric pairs of conductive rings on a soft, flexible, nonconductive base which conforms to the surface to be electroporated (32). Electrical fields are generated between the inner and outer ring electrodes of each circular element. Ring electrodes maximize the edge effect (high field strength at the edge of electrodes) and, thus, achieve higher electroporation efficiency at equal applied voltage when compared with meander electrodes; however, their manufacture is more difficult.

5.1.3. Microneedle Arrays

An interesting device is the EasyVax applicator, which contains many DNA-coated microneedles designed to pierce the stratum corneum to deliver DNA vaccines intradermally (Table 3.3). Provided enough DNA can be delivered, this method will circumvent the need for intradermal injection by other means. Since the needles are spaced closely, the applied voltage can be low, thus minimizing pain. It will be interesting to see how the device fares in clinical trials.

5.2. Needle Electrodes

The limitation of surface electrodes to shallow fields can be overcome by using invasive electrodes, most frequently in the form of needles. However, catheter electrodes (sect. 5.3) and some electrodes discussed in sect. 5.4 can also be considered invasive.

5.2.1. Needle Pair

The first in vivo electroporation ever for the delivery of bleomycin into tumor cells was performed in 1987 with a needle pair electrode (1) as was one of the earliest in vivo DNA deliveries (2). Since then, needle pairs have been used for DNA transfer into many kinds of tissues, particularly muscle and tumor tissue in small animals. The field generated by needle pairs is highly divergent, showing the characteristic hour-glass shape (Fig. 3.3). As mentioned earlier, the actual field strength deviates considerably from the nominal one (applied V/electrode distance). Figure 3.4 demonstrates the actual vs. nominal field strength in the center location between the needles for different types and dimensions of needle arrays. The six-needle array shows the least dip in the center field strength and was also most effective in bleomycin delivery to tumor cells (33).

The Elgen system (Table 3.3) uses a needle pair that is used for DNA injection into muscle prior to pulse delivery. Its relatively high efficiency is attributed to the location of most of the injected DNA in the needle tracks, directly adjacent to the needle surfaces where the field strength is the highest.

5.2.2. Four-Needle Arrays: Square, Rectangular, and Rhomboid Configurations

The field generated by two parallel needle pairs of same polarity arranged in a square yields good coverage of the area within the needles and some area outside the square. Switching the field orientation by 90° after the first pulse(s) provides even more homogeneous electroporation coverage (31).

The MedPulser DDS (16) uses a rectangular rather than a square four-needle array (Fig. 3.5) to increase the homogeneity and strength of the resultant field relative to the field generated by a square array. Note the higher field strength in the center of the four-needle array when compared with that in the center of the needle pair (see legends of Figs. 3.3 and 3.5).

The TriGrid system (19) employs four needles placed at the corners of a rhombus to better accommodate its triangular pulsing pattern (two needles of same polarity vs. one needle of opposite polarity).

5.2.3. Circular Needle Arrays

Six-needle Array: In this array, needles are arranged at the corners of a regular hexagon. Consecutive pulses are applied using two parallel needle pairs at a time

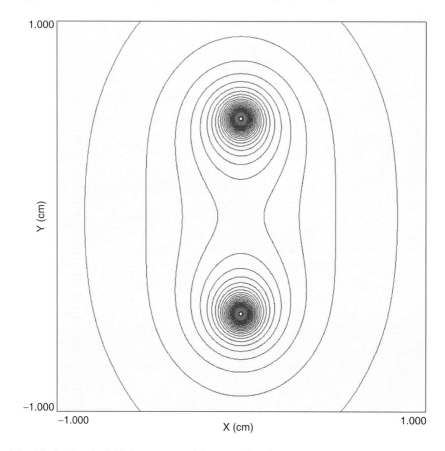

Fig. 3.3 Isoelectric field lines generated by a needle pair (cross section perpendicular to the z-axis at midplane). Needle diameter = 0.2 mm; needle distance = 5.0 mm; resistance of medium between needles = 100 Ω; applied voltage = 100 V. Contours of electrical field strength (V/cm) are plotted at intervals of 10 V/cm. Field strength at the center between needles is 42.6 V/cm. For higher or lower applied voltages, the field strength changes proportionately, but the field pattern remains the same. (Courtesy of Field Precision LLC, www.fieldp.com)

and switching to a new double-pair positioned 60° clockwise from the previous double pair (31). Thus, after pulsing three double pairs, the hexagonal area has been covered completely and effectively (Figs. 3.4 and 3.6).

The MedPulser DET system (Table 3.3) uses its needle array and pulsing scheme for both bleomycin and DNA delivery, although this system is not ideal for DNA delivery. However, this system (now in a phase III clinical trial) has the advantage of having been used to treat over 500 cancer patients with local electroporation-enhanced chemotherapy and, thus, was considered to be most acceptable by the FDA for DNA studies as well. Because of the extensive use in tumor therapy, six-needle arrays have been incorporated into a variety of handles to facilitate the treatment of tumors in various anatomical locations (31). One of the electrode arrays of the CliniPorator is also a 6-needle array (17).

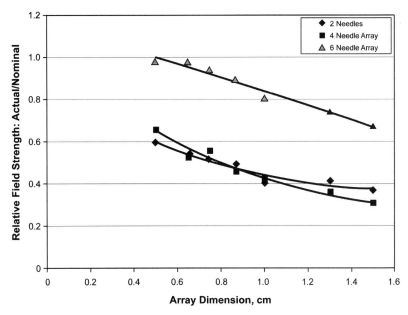

Fig. 3.4 Ratio of actual vs. nominal field strength for comparable arrays of two needles, four needles (square), and six needles (hexagon). (From (31), with permission by the publisher)

Five-Needle Arrays: The needle array of the EKD system consists of a central needle for DNA injection and five needle electrodes surrounding the center needle (20). The five needle electrodes are fired two needles at a time in a rotating pattern.

A different five-needle array is remarkable for the pulses it has been employed to deliver, although it has not been used for DNA delivery. The array consists of a center needle of one polarity surrounded by four needles of opposite polarity forming a square around the center needle. It has been used to deliver high voltage, high field strength, ultrashort pulses (3 kV, 7.5 kV/cm, 300 ns) to tumor tissue. To avoid the problem of sparking and flashover that can be triggered by such pulses, the needles protrude from a Teflon base and the skin is coated with vegetable oil prior to insertion of the needles into the target tissue (34).

5.2.4. Rows and Grids of Needles

Multiple square or rectangular needle arrays can be arranged to form two parallel rows of needles to cover larger treatment areas and/or to more densely cover an area with electrodes. Needle rows have been used for electroporating muscle, skin, and cutaneous malignancies (35). Multiple squares have also been arranged two-dimensionally to create electrode grids of any desired size and shape (31).

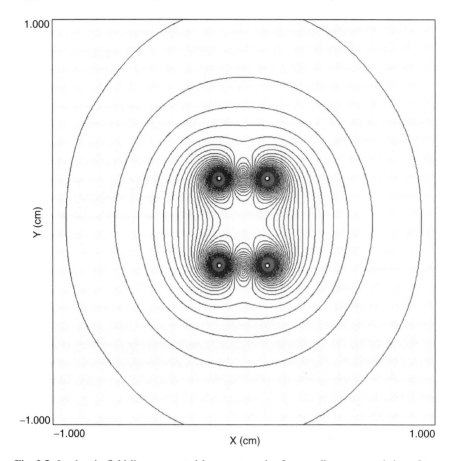

Fig. 3.5 Isoelectric field lines generated by a rectangular four-needle array consisting of two parallel needle pairs of same polarity (cross section perpendicular to *z*-axis at midplane). Needle diameter = 0.2 mm; needle distance within a pair = 4.3 mm; distance between pairs = 2.5 mm; resistance of medium between pairs = 100 Ω; applied voltage = 100 V. Contours of electrical field strength (V/cm) are plotted at intervals of 10 V/cm. Field strength at the center of the array is 155.9 V/cm. For higher or lower applied voltages the field strength changes proportionately while the field pattern remains the same. (Courtesy of Field Precision LLC, www.fieldp.com)

5.3. Catheter Electrodes

The use of electroporation catheters is an intriguing concept for the delivery of DNA and drugs into blood vessel walls to treat vascular diseases. These catheters have undergone an evolution which for now has culminated in the device shown in Fig. 3.7. The proximal end of the catheter has ports for a guide wire, electrode wires, and DNA infusion. The distal end, having a microporous balloon with one spiral electrode inside and two spiral electrodes adjacent to the outer ends of the balloon, is advanced via the guide wire through the vessel to the site of treatment. Infusion of DNA solution results in inflation of the balloon and DNA seepage into

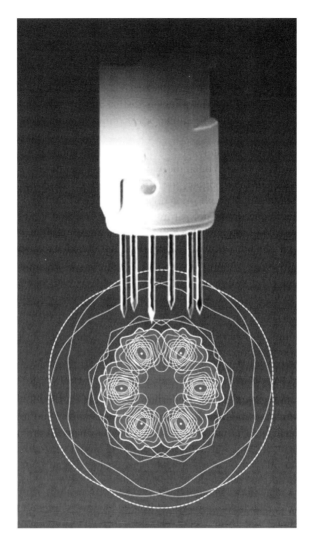

Fig. 3.6 Six-needle array with a cross section of the electrical fields produced at midplane of the array. Lines represent iso-field lines generated by a rotating six-pulse cycle (see text). Every point within the strong white line surrounding the six needles is subjected to a field strength of at least 600 V/cm. Distance between active electrodes: 0.86 cm; applied voltage: 1,130 V

the balloon-vessel interphase. Pulses are delivered which create fields that electroporate cells in the vessel wall, allowing DNA to be taken up. The transgene product may evoke local, regional, or systemic effects. Delivery of heparin into lesions produced by angioplasty has completely prevented restenosis in an animal model (11). Many candidate genes for fighting vascular diseases have been identified and are waiting for an effective and safe delivery method. In the author's opinion, catheter electroporation has much to offer for vascular DNA and drug delivery but has not received adequate attention.

(a)

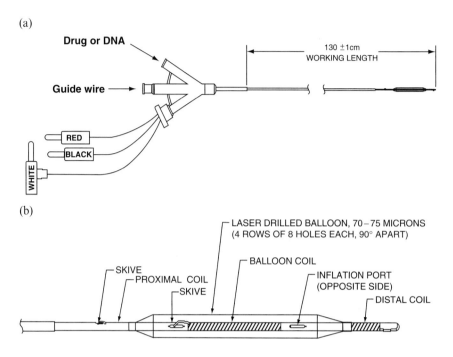

(b)

Fig. 3.7 Electroporation catheter. (**a**) Catheter with ports at the proximal end for drug or DNA infusion, guide wire, and wires leading to electrodes. The porous balloon with electrodes is at the distal end. (**b**) Detailed drawing of porous balloon with coiled-wire electrodes around the catheter shaft inside and outside the balloon

5.4. Other Electrode Configurations

5.4.1. Eye Electrodes

A specialized electrode has been developed to treat the cause of glaucoma. This disease occurs when the filtering channel of the eye becomes blocked by benign tissue growth. Conventional delivery of cytostatic agents has at best been temporarily successful. Electroporative delivery of cytostatica in an animal model has shown promising longer-term efficacy (36). Transfection of the ciliary muscle of the eye has also been demonstrated (37).

5.4.2. Electrode for External Drug Delivery to Blood Vessels

Certain vascular surgical procedures, e.g., insertion of vessel grafts, could benefit from in vivo drug or gene delivery to blood vessel walls from the outside rather than from the inside via catheters. A device which bathes the dissected blood vessel in DNA solution and allows electroporation of the vessel wall has been successfully tested in animals but has not yet been tested in humans (38).

5.4.3. Fluid Jet Electrode

Several devices are on the market for needle-free intradermal and intramuscular injections. These devices shoot a high-pressure, thin fluid jet containing vaccine or therapeutic agent into the tissue. Provided the conductivity of the fluid jets is made sufficiently high, these fluid jets can serve as electrodes. Since this concept has not been investigated experimentally, it is not clear whether essentially simultaneous DNA injection and electroporation will yield effective transfection (21).

5.4.4. Whole Organ Electroporation

For the electroporative treatment of certain genetic diseases, e.g., cystic fibrosis, whole organ electroporation is either a prerequisite or highly desirable (for a discussion of approaches under consideration for airway gene therapy, see (39)). Electroporation of whole organs has been demonstrated in small animals (40), but scale-up for the treatment of humans has not been achieved under well-controlled conditions. However, whole organ electroporation is feasible as judged from electroporation of the heart which occurs during defibrillation (41). In addition to in vivo electroporation, whole organ electroporation is attractive as an *ex vivo* technology for transfecting organs with immunosuppressive genes, such as IL-10, prior to transplantation, with the aim to minimize transplant rejection.

5.4.5. Induction "Electrodes"

In principle, electroporation can be achieved by inducing the necessary electrical fields in the tissue to be treated by applying external magnetic fields (42). This electrodeless approach is particularly attractive for treating areas that cannot be easily reached with conventional electrodes. Implementation of this treatment modality has been hampered by the difficulty to generate sufficiently strong localized magnetic fields in a manner compatible with routine treatment requirements.

6. Conclusions

It took about 8 years from the first in vivo DNA deliveries reported in 1996 to the start of the first clinical trial at the end of 2004. During that time period, early research prototypes have evolved into the first generation of applicators, electrodes, and pulse generators suitable for clinical use. Three different models are now in clinical trials for DNA delivery (Table 3.3). Additional trials with these and other devices are poised to follow soon. The potential advantages of electroporation for in vivo DNA delivery over other approaches have gained in attractiveness as the

feasibility of other methods has become less certain and two of the main reservations against electroporation have been eliminated: it is now accepted that the level of pain associated with electroporation is well tolerated by patients and that electroporation does not cause significantly higher integration frequencies of plasmid DNA into genomic DNA than do other methods. The next important milestone will be the demonstration of safety and efficacy in clinical trials.

A trend towards different design requirements for different in vivo electroporation applications is clearly evident. Three major application groups that differ considerably in applicator and electrode requirements, as well as pulse conditions, include applications for gene therapy, therapeutic vaccinations, and mass-use prophylactic vaccinations. The major driving force comes from human applications of the technology, but there is also a veterinary market emerging for both farm animals and pets. For the second-generation designs to become more effective and more patient- and user-friendly, many opportunities for improvement exist. Better coordination of applicator and pulse design can further reduce pain. Exploration and development of induced electrical fields may lead to third-generation "electrodeless" devices. A better understanding and control of DNA delivery to tissue prior to electroporation will allow the design of more effective electrodes. The overwhelming focus, so far, on skeletal muscle and tumors as target tissues can be widened to include other promising tissues such as skin for vaccination, liver for gene therapy, and blood vessels for treatment of vascular diseases. Finally, major improvements or findings in classical electroporation, or new offshoots such as nanopulse technology, may open new and surprising therapeutic possibilities.

References

1. Okino, M. and Mohri, H. (1987) Effects of a high-voltage electrical impulse and an anticancer drug on in vivo growing tumors. *Jpn. J. Cancer Res.* **78**, 1319–1321.
2. Nishi, T., Yoshizato, K., Yamashiro, S., et al. (1996) High-efficiency in vivo gene transfer using intraarterial plasmid DNA injection following in vivo electroporation. *Cancer Res.* **56**, 1050–1055.
3. Zhang, L., Li, L., Hofmann, G.A., and Hoffman, R.M. (1996) Depth-targeted efficient gene delivery and expression in the skin by pulsed electric fields: an approach to gene therapy of skin aging and other diseases. *Biochem. Biophys. Res. Commun.* **220**, 633–636.
4. Heller, R., Jaroszeski, M., Atkin, A., et al. (1996) In vivo gene electroinjection and expression in rat liver. *FEBS Lett.* **389**, 225–228.
5. Li, S. (2004) Electroporation gene therapy: new developments in vivo and in vitro. *Curr. Gene Ther.* **4**, 309–316.
6. Liu, J., Xia, X., Torrero, M., Barrett, R., Shillitoe, E.J., and Li, S. (2006) The mechanism of exogenous B7.1-enhanced IL-12-mediated complete regression of tumors by a single electroporation delivery. *Int. J. Cancer.* **119**, 2113–2118.
7. Jaroszeski, M.J., Heller, L.C., Gilbert, R., and Heller, R. (2004) Electrically mediated plasmid DNA delivery to solid tumors in vivo. *Methods Mol. Biol.* **245**, 237–244.
8. Zhang, L., Widera, G., and Rabussay, D. (2004) Enhancement of the effectiveness of electroporation-augmented cutaneous DNA vaccination by a particulate adjuvant. *Bioelectrochemistry.* **63**, 369–373.

9. Babiuk, S., Baca-Estrada, M.E., Foldvari, M., et al. (2003) Needle-free topical electroporation improves gene expression from plasmids administered in porcine skin. *Mol. Ther.* **8**, 992–998.

10. Wolff, J.A. and Budker, V. (2005) The mechanism of naked DNA uptake and expression. *Adv. Genet.* **54**, 3–20.

11. Dev, N.B., Hofmann, G.A., Dev, S.B., and Rabussay, D.P. (2000) Intravascular electroporation markedly attenuates neointima formation after balloon injury of the carotid artery in the rat. *J. Intervent. Cardiol.* **13**, 331–338.

12. Seidler, R.W., Allgauer, S., Ailinger, S., et al. (2005) In vivo human MCP-1 transfection in porcine arteries by intravascular electroporation. *Pharm. Res.* **22**, 1685–1691.

13. http://www.cytopulse.scom/.

14. Wallace, M.S., Ridgeway, B., Jun, E., Schulteis, G., Rabussay, D., and Zhang, L. (2001) Topical delivery of lidocaine in healthy volunteers by electroporation, electroincorporation, or iontophoresis: an evaluation of skin anesthesia. *Reg. Anesth. Pain Med.* **26**, 229–238.

15. McCreery, T.P., Sweitzer, R.H., Unger, E.C., and Sullivan, S. (2004) DNA delivery in vivo by ultrasound. *Methods Mol. Biol.* **245**, 293–298.

16. Rabussay, D., Dev, N., Fewell, J., Smith, L.C., Widera, G., and Zhang, L. (2003) Enhancement of therapeutic drug and DNA delivery into cells by electroporation. *J. Phys. D: Appl. Phys.* **36**, 348–363.

17. http://www.igea.it/.

18. Tjelle, T.E., Salte, R., Mathiesen, I., and Kjeken, R. (2006) A novel electroporation device for gene delivery in large animals and humans. *Vaccine.* **24**, 4667–4670.

19. http://www.ichorms.com/.

20. http://www.advisys.net/.

21. Hofmann, G.A., Rabussay, D., and Zhang, L. (2003) Method of electroporation-enhanced delivery of active agents. *U.S. Patent* 6,520,950.

22. Rabussay, D.P., Nanda, G.S., and Goldfarb, P.M. (2002) Enhancing the effectiveness of drug-based cancer therapy by electroporation (electropermeabilization). *Technol. Cancer Res. Treat.* **1**, 71–82.

23. Babiuk, S., Baca-Estrada, M.E., Foldvari, M., et al. (2004) Increased gene expression and inflammatory cell infiltration caused by electroporation are both important for improving the efficacy of DNA vaccines. *J. Biotechnol.* **110**, 1–10.

24. Rabussay, D., Widera, G., Zhang, L., et al. (2004) Toward the development of electroporation for delivery of DNA vaccines to humans. *Mol. Ther.* **9**, S209.

25. Prausnitz, M.R. (1996) The effects of electric current applied to skin: a review for transdermal drug delivery. *Adv. Drug. Deliv. Rev.* **18**, 395–425.

26. Domenge, C., Orlowski, S., Luboinski, B., et al. (1996) Antitumor electrochemotherapy: new advances in the clinical protocol. *Cancer.* **77**, 956–963.

27. Mir, L.M., Belehradek, M., Domenge, C., et al. (1991) [Electrochemotherapy, a new antitumor treatment: first clinical trial.] *C.R. Acad. Sci. III.* **313**, 613–618.

28. Heller, R., Jaroszeski, M.J., Glass, L.F., et al. (1996) Phase I/II trial for the treatment of cutaneous and subcutaneous tumors using electrochemotherapy. *Cancer.* **77**, 964–971.

29. Sersa, G., Cemazar, M., Semrov, D., and Miklavcic, D. (1996) Changing electrode orientation improves the efficacy of electrochemotherapy of solid tumors in mice. *Bioelectrochem. Bioenerg.* **39**, 61–66.

30. Heller, L.C., Jaroszeski, M.J., Coppola, D., McCray, A.N., Hickey, J., and Heller, R. (2007) Optimization of cutaneous electrically mediated plasmid DNA delivery using novel electrode. *Gene Ther.* **14**, 275–280.

31. Hofmann, G.A. (2000) Instrumentation and electrodes for in vivo electroporation. In: Jaroszeski, M.J., Heller, R. and Gilbert, R. (eds.). *Electrically mediated delivery of molecules to cells*. Methods in molecular medicine, vol. 3. Humana, Totowa, NJ, pp. 37–61.

32. Hofmann, G.A., Rabussay, D., and Tonnessen, A. (2001) Apparatus and method for the delivery of drugs and genes into tissue. *U.S. Patent* 6,192, 270.

33. Gilbert, R.A., Jaroszeski, M.J. and Heller, R. (1997) Novel electrode designs for electrochemotherapy. *Biochim. Biophys. Acta.* **1334**, 9–14.
34. Nuccitelli, R., Pliquett, U., Chen, X., et al. (2006) Nanosecond pulsed electric fields cause melanomas to self-destruct. *Biochem. Biophys. Res. Commun.* **343**, 351–360.
35. Gehl, J. and Geertsen, P.F. (2000) Efficient palliation of haemorrhaging malignant melanoma skin metastases by electrochemotherapy. *Melanoma Res.* **10**, 585–589.
36. Oshima, Y., Sakamoto, T., Nakamura, T., et al. (1999) The comparative benefits of glaucoma filtering surgery with an electric-pulse targeted drug delivery system demonstrated in an animal model. *Ophthalmology.* **106**, 1140–1146.
37. Bloquel, C., Bejjani, R., Bigey, P., et al. (2006) Plasmid electrotransfer of eye ciliary muscle: principles and therapeutic efficacy using hTNF-alpha soluble receptor in uveitis. *FASEB J.* **20**, 389–391.
38. Martin, J.B., Young, J.L., Benoit, J.N., and Dean, D.A. (2000) Gene transfer to intact mesenteric arteries by electroporation. *J. Vasc. Res.* **37**, 372–380.
39. Davies, J.C. and Alton, E.W.F.W. (2005) Airway gene therapy. In: Hall, J. (ed.). *Non-viral vectors for gene therapy* (2nd edn.)*: Part 2*. Advances in genetics, vol. 54. Elsevier, pp. 291–314.
40. Machado-Aranda, D., Adir, Y., Young, J.L., et al. (2005) Gene transfer of the Na+, K+-ATPase beta1 subunit using electroporation increases lung liquid clearance. *Am. J. Respir. Crit. Care Med.* **171**, 204–211.
41. Nikolski, V.P. and Efimov, I.R. (2005) Electroporation of the heart. *Europace.* **7**, S146–S154.
42. Hofmann, G.A. (2000) Electrporetic gene and drug therapy. *U.S. Patent* 6,132,419.

Chapter 4
Electrode Assemblies Used for Electroporation of Cultured Cells

Leda Raptis and Kevin L. Firth

Abstract Electroporation was initially developed for the introduction of DNA into cells which grow in suspension and was performed in a cuvette with two flat electrodes on opposite sides. Different configurations were subsequently developed for the electroporation of adherent cells in situ, while the cells were growing on nonconductive surfaces or a gold-coated, conductive support. We developed an assembly where the cells grow and are electroporated on optically transparent, electrically conductive indium-tin oxide (ITO). This material promotes excellent cell adhesion and growth, is inert and durable, and does not display spontaneous fluorescence, making the examination of the electroporated cells by fluorescence microscopy possible. The molecules to be electroporated are added to the cells and introduced through an electrical pulse delivered by an electrode placed on top of the cells. We describe several electrode and slide configurations which allow the electroporation of large numbers of cells for large-scale biochemical experiments or for the detection of changes in cell morphology and biochemical properties in situ, with control, nonelectroporated cells growing on the same type of ITO-coated surface, side by side with the electroporated ones. In a modified version, this technique can be adapted for the study of intercellular, junctional communication; the pulse is applied in the presence of a fluorescent dye, such as lucifer yellow, causing its penetration into the cells growing on the conductive half of the slide, and the migration of the dye to the nonelectroporated cells growing on the nonconductive area is microscopically observed under fluorescence illumination. An assembly is also described for the electroporation of sensitive cells without the use of an upper electrode.

Keywords: lucifer yellow, electroporation, in situ

1. Introduction

Electroporation was initially developed for the introduction of DNA into cells which grow in suspension (1, 2). Cells were placed in a cuvette which was a modified version of a plastic spectrophotometer cuvette with a flat aluminum

electrode on two opposing sides (Fig. 4.1). The gap varied from 0.1 to 1 cm to accommodate different numbers of cells. This configuration was also used for the electroporation of adherent cells, i.e., the cells were placed in suspension following trypsinization or EGTA treatment and vigorous pipetting.

2. Electroporation Assemblies for Adherent Cells

It has long been established that the detachment of adherent cells from their substratum can cause significant metabolic alterations, which can make the cells very sensitive to further treatments (3). Later, the value of electroporation for the introduction of other molecules such as peptides, proteins, or drugs was recognized, most importantly for the study of signal transduction (4–6). In this case, the efficient incorporation of the material without cellular damage is an especially crucial requirement since no disturbance to cellular physiology can be tolerated. In addition, cell detachment for electroporation in suspension is not appropriate when working with polarized cells or organized cellular structures, such as epithelia, whose architecture would be disrupted by the detachment process.

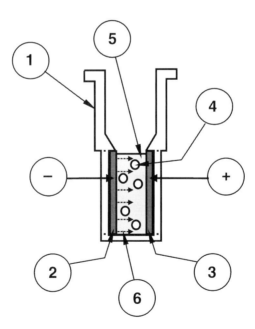

Fig. 4.1 Electroporation in suspension. In the original design, cells [4] were electroporated while suspended in a cuvette [1]. Two opposing sides of the cuvette carry planar electrodes ([2] and [3]), made of aluminum or stainless steel. The gap between the electrodes varied from 0.1 to 1 cm. The electroporation solution [5] carried the molecules to be introduced. [6] denotes electric field lines

A number of approaches have been undertaken to circumvent these problems. Initially, adherent cells were grown on Cytodex™ microcarrier beads which can be suspended in a conventional electroporation cuvette (7). This method is rather inefficient and does not lend itself to detailed microscopic examination of electroporated cells. Later, electroporation was conducted while the cells were growing on a flat, nonconductive support. A simple approach initially undertaken consisted of two platinum wire electrodes positioned at the sides and in contact with a monolayer of cells growing on a (nonconductive) glass cover slip (Fig. 4.2A) (8, 9). Subsequently, the positive and negative electrodes were placed above the cell monolayer (10). In this case, to achieve a sufficiently uniform electrical field intensity over the whole electroporated area, the electrodes had to be at a significant distance from the cells, which substantially increases the volume of solution required (Fig. 4.2B). A different approach was developed by Sedivy et al. (11), wherein the cells were grown on a porous membrane made of polyester or polyethylene tetra phthalate, which was placed in a conventional cuvette or sandwiched between two electrodes placed above and below the membrane in a dedicated setup (Fig. 4.2C).

A different assembly was described by Jen et al. (12). Cells were grown on a pair of interdigitated electrodes of titanium and gold (Fig. 4.3A), and the DNA added to the cells in electroporation medium. The DNA was at first electrophoretically concentrated in the vicinity of the cells by a low-voltage field generated using an electrode placed on top of the cells and one side of the interdigitated electrodes (Fig. 4.3B). Then the current was switched to the two interdigitated electrodes to perform the electroporation with the field directed sideways, as shown in Fig. 4.3C. In all of these cases, the uniformity of poration and volume of solution required were limiting factors for most applications.

3. Electroporation on ITO

Glass coated with thin films which are transparent and conductive to electricity, such as indium-tin oxide (ITO), has had several industrial applications for a number of years, dating back to the Second World War. Certain airplanes at that time had windows made of such material, as do today's coast-guard ships, so that the window pane can be heated to avoid ice build-up. More recently, they are widely used for liquid crystal diode displays. We investigated the possibility of using ITO-coated glass for in situ electroporation of adherent cells (4) because it is inert, conductive, transparent, and commercially available at surface resistivities of up to 2Ω/sq. In our experience, ITO promotes excellent cell adhesion and growth, and its use in electroporation offers the possibility of ready examination of the cells due to their extended morphology. In addition, unlike gold-coated glass, ITO is very durable and not disturbed by cell growth. An added advantage is that, unlike a number of plastics, ITO does not display spontaneous fluorescence, at least at the wavelengths

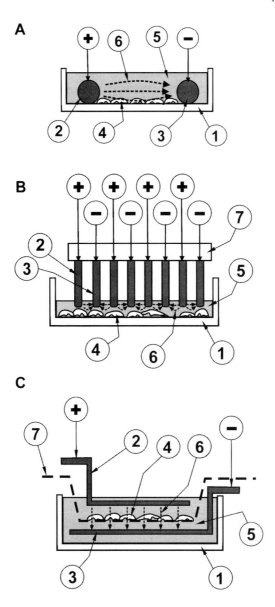

Fig. 4.2 Electroporation of adherent cells on a nonconductive substrate. Cells [4] were grown on a glass cover slip or petri dish [1], and the pulse delivered through two parallel platinum wire electrodes ([2] and [3]), in contact with the cell layer (**A**) (8, 9) or at a small distance from it (**B**) (10). The electroporation solution [5] carried the molecules to be introduced. [6], electric field lines. [7], holder carrying the electrodes. (**C**) Electroporation of adherent cells on a porous membrane. Cells [4] were grown on a porous membrane [7] held in a plastic petri dish [1], and the pulse delivered with two electrodes ([2] and [3]) placed above and below the cell layer (11)

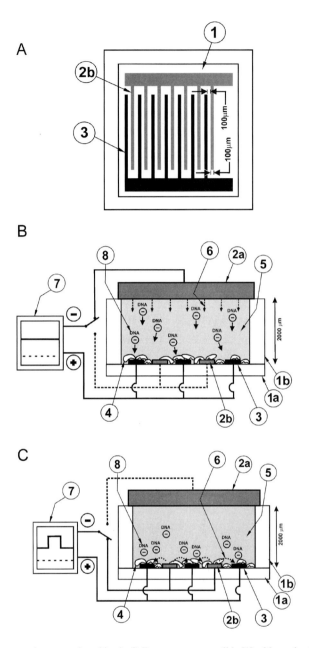

Fig. 4.3 Electroporation on a microchip. **A**. Cells are grown on a slide [1] with a pair of interdigitated electrodes [3] and [2b], and the DNA added to the cells in electroporation medium (12). **B**. The cells are grown on a slide as in (A). An electrode [2a] is placed on top of the cells [4] and kept in place with a spacer [1b]. The DNA is at first electrophoretically concentrated in the vicinity of the cells by a low-voltage direct current field generated using the top electrode and one side of the interdigitated electrodes [3], as shown by the *arrows*. [7], direct current pulse generator. **C**. Once the DNA is concentrated in the vicinity of the cells [4], an electric pulse is delivered by switching the current to the interdigitated electrodes [2b] and [3] to perform the electroporation with the field directed sideways, as shown by the *arrows*. [7], capacitor-discharge pulse generator

of a number of commonly used dyes, such as lucifer yellow, fluorescein, or rodamine, making the examination of the electroporated cells by fluorescence microscopy possible. Another advantage is that it is possible to attach DNA to this surface for direct introduction into the cell, in a microarray format (13). A wide variety of nonpermeant molecules, such as peptides (14–16), oligonucleotides (15, 17), radioactive nucleotides (18, 19), proteins (20), DNA (4), or prodrugs (21), have been successfully introduced using this approach. These can be introduced alone or in combination, at the same or different times, and in growth-arrested cells or cells at different stages of their division cycle. After the introduction of the material, the cells can be either extracted or biochemically analyzed, or their cellular morphology and biochemical properties can be examined in situ. In a modified version, this assembly can be used for the study of intercellular, junctional communication. In this review, we describe a number of electrode and slide combinations that can be used to introduce a variety of nonpermeant molecules to a very high proportion of cultured, adherent cells instantly and with minimal disturbance to their phenotype (6, 22).

3.1. The Instrument

The elements of a system for electroporating cultured adherent cells in situ includes the ITO-coated glass on which the cells grow, a container for holding the electroporation medium in place, and an electric pulse generator (Fig. 4.4). Concerning the

Fig. 4.4 Electroporation electrode and slide assembly for *in situ* electroporation on indium-tin oxide. **A**. Side view: The slide [1] with the ITO coating [1a], electrodes [2 and 3], and their holder [6] and cells [4] are shown. For clarity purposes, the slope of the underside of the negative electrode is exaggerated. Note that the thickness of the ITO coating is exaggerated to better demonstrate the current paths. Four slide configurations are described (B–E). *Dotted lines* point to the positions of the negative [8] and positive [7] electrodes during the pulse. Light shading denotes the ITO coating. Darker shading [5] denotes the Teflon frame. **B**. Fully conductive slide assembly for use in large-scale biochemical experiments. In the setup shown, cell growth area can be up to 7 × 15 mm², but larger slides and electrodes offer larger areas, up to 32 × 10 mm². [*a*], area of cell growth and electroporation. (From (5), reproduced with permission.) **C**. Partly conductive slide assembly, with electroporated [*a*] and non-electroporated [*c*] cells growing on the same type of ITO-coated surface. [*b*], area where the conductive coating has been stripped, exposing the non-conductive glass underneath. Stripping extends to area [*d*]. Cells growing in areas *b* and *c* are not electroporated. [7] and [8], positions of the positive and negative electrodes, respectively, during pulse application. (From (5), reproduced with permission.) **D**. Partly conductive slide assembly for use in the examination of gap junctional, intercellular communication. [*a*], area of electroporated cells. [*b*], cells growing on non-conductive glass. *Arrow* points to the transition line between conductive and non-conductive areas. (From (28, 31), reproduced with permission.) **E**. Partly conductive slide assembly for use in the examination of gap junctional, intercellular communication, especially suitable for cells which are difficult to grow. [*a*], area of electroporated cells. [*b*], cells growing on non-conductive glass. Arrow points to the transition line between conductive and non-conductive areas. (From (35), reproduced with permission)

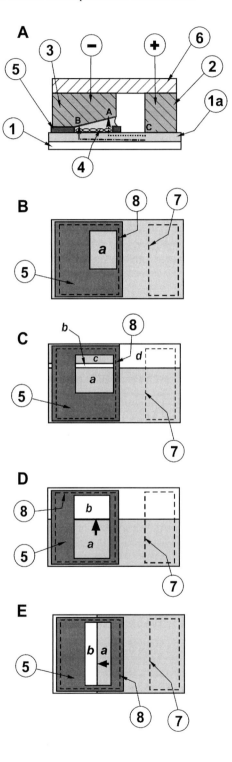

pulses commonly used, their intensity, duration, frequency, number, and polarity have been shown to be very important parameters. A variety of pulse types have been successfully employed, such as capacitor-discharge, square-wave, and oscillating radiofrequency fields for different applications (23). Capacitor-discharge type of electrical fields is the simplest to produce and have been used widely. The pulses required for electroporation of adherent cells growing on ITO are considerably lower in intensity and duration than for cells in suspension, possibly because larger amounts of current flow through an extended cell.

Figure 4.4 outlines the slide and electrode assembly. Cells are grown on a glass slide [1] coated with ITO [1a] in a window cut in an insulating Teflon frame [5]. The pulse is delivered by a stainless steel negative electrode [3] positioned on the cells, resting on a Teflon frame and a positive contact bar [2] placed on the conductive coating as shown (Fig. 4.4A). Although many electrode arrangements can result in some of the cells receiving the optimal electroporation conditions, to obtain a uniform electrical field intensity over the entire cell growth area requires the following geometry. The thin, conductive ITO coating exhibits a significant amount of electrical resistance, and so to achieve uniformity of the field intensity and poration over the entire area below the negative electrode, the undersurface of the negative electrode must rise in the direction of the positive contact bar at an angle proportional to the resistance of the coating (Fig. 4.4A). In this manner, the thickness of the fluid layer is greatest in the region closest to the positive contact bar and narrowest at a point most distant from the positive contact bar, so that the combined resistance of the ITO surface and fluid is the same over the entire cell growth area (Fig. 4.4A, path A to C vs. path B to C). Extensive previous experience indicated that uniform electroporation using slides with a surface resistivity of $2\,\Omega/sq$ requires the negative electrode to have an angle of ~1.5° with the slide surface, which for a Teflon frame of thickness 0.279 mm and a cell growth area of $32 \times 10\,mm^2$, translates into (140 μL of solution being held in place by surface tension. ITO-coated glass of lower conductivity, such as $20\,\Omega/sq$, is less expensive; however, because of the higher surface resistivity, the angle of the negative electrode has to be ~4.4°, which translates into a volume of ~280 μL for the same cell growth area, with a proportional increase in the cost of the material to be introduced. Gold-coated glass, which is highly conductive, can also be used, but this material is not as durable and is considerably more expensive (24).

To introduce a variety of peptides, oligonucleotides, or drugs into large numbers of cells for large-scale biochemical experiments, such as examination by Western blotting (5), thin layer chromatography (25), or other biochemical techniques, fully conductive slides can be employed (Fig. 4.4B). Cells are plated on the conductive surface [*a*], and prior to pulse application, the growth medium is removed, the solution with the material to be introduced is added to the cells, the negative electrode is set in place, the electrical pulse is applied, and the cells are microscopically observed and biochemically analyzed. This basic setup, using fully conductive slides, has been used successfully by a number of laboratories (15–17, 21, 26). For the detailed examination of cell morphology or gene expression in situ, different variations of this approach have been undertaken, as described below.

3.2. Partly Conductive Slides for the Study of Morphological Effects or Biochemical Changes In Situ

To precisely assess small background changes in morphology or gene expression, the presence of nonelectroporated cells side by side with electroporated ones can offer a valuable control. This control can be achieved by growing the cells on a conductive slide from parts of which the coating has been removed, thus exposing the bare, nonconductive glass underneath (27–29). The ITO coating is in general resistant to acids, but if the oxides are reduced to their corresponding metals, the coating is easily removed. In practice, stripping is achieved by etching with hydrochloric acid in the presence of metallic Zn powder. However, a tinge of the ITO-coating (Fig. 4.4C–E, area *a* vs. *b*), combined with the more effective immunostaining of cells growing on ITO (possibly due to a chemical attraction of different immunocytochemistry reagents to the coating), can create problems in the interpretation of results regarding gene expression levels. In addition, a number of cell types may grow slightly better on the conductive, ITO-coated glass than on the nonconductive area, possibly because the ITO-coated surface is less smooth than glass, thus providing a better anchorage for the growth of adherent cells (30). As a result, cell density may be slightly higher on the conductive than on the etched side, which could have important implications if cell growth effects are being studied. It follows that, to assess the effect of the introduced material, it is important to compare the staining and morphology of electroporated cells with nonelectroporated ones, while both are growing on the same type of surface. This was achieved by plating the cells on a slide where the conductive coating was removed in the pattern shown in Fig. 4.4C (27). A thin line of plain glass [*b*] separates the electroporated and control areas while etching extends to area [*d*], so that there is no electrical contact between the positive contact bar and area [*c*]. Application of the pulse results in electroporation of the cells growing in area [*a*] exclusively, while cells growing in area [*b*] or [*c*] do not receive any current. In this configuration, electroporated cells [*a*] are being compared with nonelectroporated ones [*c*], while both are growing on ITO-coated glass. Since the coating is ~1,600 Å thick for glass of surface resistivity 2 Ω/sq, this transition line does not alter the growth of cells across it and is clearly visible microscopically, even under a cell monolayer (5). Coatings with higher resistivity are thinner, so that this line is not as clearly delineated but is still visible.

3.3. Electroporation on a Partly Conductive Slide for the Assessment of Gap Junctional, Intercellular Communication

Gap junctions are membrane channels serving as conduits between the interiors of cells and are frequent targets of a variety of signals stemming from growth factors or oncogenes. Gap junctional, intercellular communication (GJIC) is commonly examined by the microinjection of a fluorescent dye, such as lucifer yellow, into the cells,

followed by observation of its migration to the neighboring cells under fluorescence illumination. We developed a modification of in situ electroporation where junctional permeability can be precisely examined, by using a setup in which cells are grown on a glass slide half of which is coated with ITO (Fig. 4.4D, E). An electric pulse is applied in the presence of the fluorescent dye causing its penetration into the cells growing on the conductive part of the slide, and the migration of the dye to the nonelectroporated cells growing on the nonconductive area is microscopically observed under fluorescence illumination. This technique has been applied upon a large variety of adherent cell types, including primary human lung carcinoma cells (28, 31–33).

In the initial design (28), cells were grown in a window of $7 \times 4\,mm^2$ with a conductive area of $4 \times 4\,mm^2$, so that the transition line between conductive and nonconductive sides runs through the middle of the slide and is 4 mm in length (Fig. 4.4D). This configuration is satisfactory for GJIC examination of most established cell lines. However, a substantial number of cells are required for the examination of their junctional communication to be possible, because the cells need to be situated precisely at the transition line, which is a relatively small proportion of the total area. Many cell types, e.g., primary human tumor cells, are difficult to grow and they senesce within a few days after they are placed in culture (33). Senescence has been shown to dramatically reduce GJIC (34, 35); hence, to obtain a true measurement of their junctional communication in vivo, the chances of the cells attaching at this line must be maximized, so that GJIC can be examined within a few days after surgery before cells approach their limits of life in culture.

A slide configuration that successfully addresses this problem is shown in Fig. 4.3E. The conductive coating is removed from the distal part of the slide, and the frame is precisely placed as shown. In this setup, since the difference in distance between the proximal and distal ends of the conductive surface relative to the positive electrode is only ~2 mm, it is very easy to achieve uniform electrical field intensity over the entire area and along the transition line. The width of the conductive strip should at least be equal to the length of two cells, while the width of the non conductive part must be greater than the greatest possible distance traveled by the dye. For rat F111 cells which have extensive GJIC (28), a width of 1.5–2 mm for area [b] is sufficient to examine the gap junctional communication.

3.4. Examination of Gap Junctional, Intercellular Communication by In Situ Electroporation on Two Coplanar Electrodes

The above technique has been employed extensively for GJIC examination. However, a number of cell types do not adhere sufficiently well and may detach because of the turbulence created as the top electrode is removed after electroporation (36), which makes GJIC examination problematic. To overcome these problems, we have designed an assembly which permits efficient electroporation without the use of a top electrode (Fig. 4.5, ref. 37). Cells are grown on two

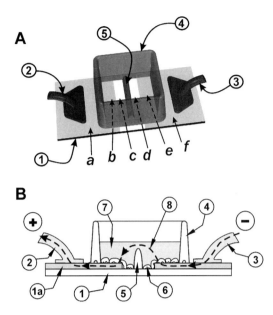

Fig. 4.5 Electroporation in situ in the absence of an upper electrode. **A**. Top view. Cells are grown on two coplanar electrodes of indium-tin oxide [*b*] and [*e*], which are supported by the same glass slide substrate [1]. Fluid is contained in a region above the cells by a plastic enclosing wall [4]. A "dam" of 2 mm [5] is attached to the inner sides of the enclosing wall and to the bare glass in between the two electrodes ([*c*], [*d*]). The two electrodes are connected to the pulse source [2] and [3], outside the cell growth area. **B**. Side view. The coating [1a] with the cells [6] and the direction of the current are shown (Raptis et al., manuscript in preparation). Note that the thickness of the coating is exaggerated to demonstrate the direction of current flow. [7], fluid containing the molecules to be introduced. [8], electric field lines

coplanar electrodes of ITO [*a, b* and *e, f*], supported by the same glass slide substrate [1]. The two electrodes are formed by removing the ITO from a straight line of 0.3 mm in width across the middle of the region on which the cells will be plated [*c, d*]. Fluid is contained in a region above the cells by a plastic enclosing wall [4]. A "dam" of 2 mm [5] is attached to the inner sides of the enclosing wall and the stripped line. The two electrodes are connected to the pulse source [2 and 3], outside the cell growth area. In this configuration, the two halves of the slide are electrically isolated until the electroporation medium is added over the dam [5], thus establishing an electrical contact between the fluid pools that fill the wells on either side of the dam, where the cells grow. Extensive experimentation showed that with a dam [5] of a height greater than 0.5 mm the electrical field experienced by the cells on the ITO near the transition line is essentially perpendicular to the ITO surface (Fig. 4.5B). As a result, the gradient of fluorescence observed in cells growing on the nonconductive sides of the transition lines [*c, d*] can only be due to the movement of lucifer yellow through gap junctions from the electroporated cells on the ITO-coated side of the line (Fig. 4.5).

3.5. Upscaling

The configuration described in Fig. 4.4B has been extensively used for the introduction of a large variety of nonpermeant molecules and is adequate for cell growth areas of up to $32 \times 10\,mm^2$ (5, 14, 15, 17, 21). However, a large number of important signal transducers are present in very small amounts in the cell, and hence a large number of cells may be required to obtain a detectable signal. Moreover, previous experience demonstrated that in most instances electroporation must be conducted under a high concentration of material (5, 14). As a result, for a number of applications, a 32×10-mm^2 cell growth area may not be sufficient. In addition, the cost of the peptide may be prohibitively high, even if multiple slides are used per experiment, in which case some of the solution could be reused. A simple scale-up of this electrode arrangement introduces problems associated with the resistance of the conductive coating, since the amount of electric current required for effective electroporation increases with the area electroporated (25). This is more acute when the slide is coated with materials to improve cell adhesion, such as CelTak (Collaborative Research), polylysine, or collagen, in which case the voltage required is slightly higher. When attempting to electroporate cell growth areas larger than $32 \times 10\,mm^2$, we experienced arcing at the point where the positive contact bar contacts the ITO coating, with burning of the coating and destruction of the slide, resulting in a dramatic reduction in the reproducibility of electroporation conditions. This problem can not be completely alleviated by using multiple pulses of lower voltage instead of one stronger pulse. In addition, the volume of solution required for an area of $50 \times 30\,mm^2$, equivalent to two 3-cm petri dishes, is increased to ~1.13 mL, with a concomitant increase in the cost of the material. Moreover, owing to the larger gap at the edge nearest the positive contact bar, this volume of solution cannot be held in place by surface tension. In addition positioning this large electrode on the cells and removing air bubbles (which prevent current flow) are exceedingly difficult. Therefore, a crucial modification was introduced in the design of electrodes and slides wherein large numbers of cells can be effectively electroporated using a minimal volume of material.

Previous results indicated that only a small proportion of the material added to the cells is taken up with the electroporation process. Therefore, we designed an assembly with a narrow, moveable electrode that electroporates a "strip" of cells at a time (Fig. 4.6); in this configuration, only the cells immediately below the negative electrode are electroporated by a given pulse of electricity. After electroporation of the first strip of cells [2a], the electrode is translocated laterally [2b], dragging the solution under it by surface tension, so that a new strip of cells is electroporated using mostly the same solution (Fig. 4.6A, B). The electric circuit formed during pulse delivery starts at the negative electrode [2a or 2b] and passes through the electroporation fluid [5] and the cells [4], along the conductive slide surface [1a], to the two positive contact bars [3a, 3b], one on each side of the slide. The two positive contact bars form parallel circuit paths, both carrying current from the conductive surface. Since only a part of the area is electroporated at a time, the

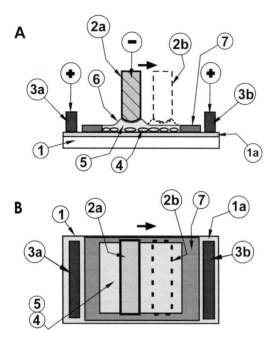

Fig. 4.6 Upscaling, double positive contact bar design for *in situ* electroporation on indium-tin oxide. **A**. Side view. The cells to be electroporated [4] are grown on a glass slide [1] coated with ITO [1a]. The negative electrode [2a, 2b] is a narrow steel bar mounted across the width of the slide, with the ends resting on the Teflon frame, present on all four sides [7]. The underside of the negative electrode is inclined in both directions toward positive contact bars [3a, 3b], such as to optimize the uniformity of electrical field. The solution containing the material to be introduced is added and the electric pulses applied, which pass through a circuit formed by the negative electrode [2a or 2b], the electroporation solution [5] and the cells [4], the conductive surface of the slide [1a], and the two positive contact bars [3a, 3b]. [6], meniscus of the electroporation solution. The electrode is translocated laterally and the procedure repeated, until the whole cell growth area is electroporated. Note that only the area of cells immediately below the electrode is electroporated at any one time. **B**. Top view: The outline of the conductive slide with a Teflon frame [7] in place to define the area of cell growth and electroporation [8] is indicated. The slide is placed in a petri dish (not shown) to maintain sterility. (From (37), reproduced with permission)

electric current intensity can be dramatically reduced. To achieve uniform electrical field intensity, the underside of the negative electrode is given a shape with an incline towards each of the positive contact bars. In this case, the combined resistance of the conductive coating plus the electroporation medium below the negative electrode yields an area with sufficiently uniform electric field intensity. The optimal contour was determined experimentally, using lucifer yellow as an indicator of cell permeation. The results showed that a 35-mm radius can produce even electroporation over the entire surface under the electrode. Using this assembly, an area of $32 \times 10\,mm^2$ can be electroporated with a 2.5-mm-wide electrode, using as little as $30\,\mu L$ of solution, rather than $140\,\mu L$ that would be required using the original design (Fig. 4.4B) (38, 39). An added advantage of this setup is that, since the

electric field strength is constant for a given voltage at any position of the electrode across the cell growth area, up to four strips of cells can be electroporated at different voltages on the same slide, in preliminary trials invariably performed to determine the optimal voltage by electroporation of lucifer yellow (38, 39). In this case, it is also possible to detect slight changes in the morphology of cells electroporated at different voltages and compare with control nonelectroporated cells, side by side on the same slide.

In conclusion, electroporation can be used for the delivery of a large variety of nucleic acids, proteins, peptides, or other nonpermeant molecules. The use of the appropriate electrode configuration offers a powerful technique for the study of signal transduction, gap junctional communication, drug development, and a large variety of other applications.

Acknowledgments The financial assistance of the Canadian Institutes of Health Research, the Canadian Breast Cancer Research Alliance, the Natural Sciences and Engineering Research Council of Canada (NSERC), the Cancer Research Society Inc., and the Department of Defense Breast Cancer Research Program (BCRP-CDMRP) is gratefully acknowledged. We are grateful to Heather Brownell, Evangelia Tomai, Adina Vultur, Rozanne Arulanandam, and Aikaterini Anagnostopoulou for many helpful discussions.

References

1. Potter, H., Weir, L. and Leder, P. (1984) Enhancer-dependent expression of human kappa immunoglobulin genes introduced into mouse pre-B lymphocytes by electroporation. *Proc. Nat. Acad. Sci. U.S.A.* **81**, 7161–7165.
2. Neumann, E., Schaefer-Ridder, M., Wang, Y., and Hofschneider, P. H. (1982) Gene transfer into mouse lyoma cells by electroporation in high electric fields. *EMBO J.* **7**, 841–845.
3. Matsumura, T., Konishi, R., and Nagai, Y. (1982) Culture substrate dependence of mouse fibroblasts survival at 4°C. *In Vitro.* **18**, 510–514.
4. Raptis, L. and Firth, K.L. (1990) Electroporation of adherent cells *in situ*. *DNA Cell Biol.* **9**, 615–621.
5. Raptis, L. (2000) Specific inhibition of growth factor-stimulated ERK1/2 activation in intact cells by electroporation of a Grb2-SH2 binding peptide. *Cell Growth Differ.* **11**, 293–303.
6. Raptis, L., Vultur, A., Brownell, H.L., and Firth, K.L. (2006) Dissecting pathways: *in situ* electroporation for the study of signal transduction and gap junctional communication. In: Celis, J.E. (ed.). *Cell biology: a laboratory handbook*. Academic, San Diego, CA, pp. 341–354.
7. Potter, H. and Cooke, S.W.F. (1992) Gene transfer into adherent cells growing on microbeads. In: Chang, D.C., Chassy, B.M., Saunders, J.A. and Sowers, A.E. (eds.). *Guide to electroporation and electrofusion*. Academic, San Diego, CA, pp. 201–208.
8. Kwee, S., Nielsen, H.V., and Celis, J.E. (1990) Electropermeabilization of human cultured cells grown in monolayers. Incorporation of monoclonal antibodies. *Bioelectrochem. Bioenerg.* **23**, 65–80.
9. Zheng, Q. and Chang, D.C. (1991) High-efficiency gene transfection by *in situ* electroporation of cultured cells. *Biochim. Biophys. Acta.* **1088**, 104–110.
10. Boitano, S., Dirksen, E.R., and Sanderson, M.J. (1992) Intercellular propagation of calcium waves mediated by inositol trisphosphate. *Science.* **258**, 292–295.
11. Yang, T.A., Heiser, W.C., and Sedivy, J.M. (1995) Efficient *in situ* electroporation of mammalian cells grown on microporous membranes. *Nucl. Acids. Res.* **23**, 2803–2810.

12. Jen, C.P., Wu, W.M., Li, M., and Lin, Y.C. (2004) Site-specific enhancement of gene transfection utilising an attracting electric field for DNA plasmids on the electroporation chip. *J. microelectromech. sys.* **13**, 947–955.
13. Yamauchi, F., Kato, K., and Iwata, H. (2005) Layer-by-layer assembly of poly(ethyleneimine) and plasmid DNA onto transparent indium-tin oxide electrodes for temporally and spatially specific gene transfer. *Langmuir.* **21**, 8360–8367.
14. Giorgetti-Peraldi, S., Ottinger, E., Wolf, G., Ye, B., Burke, T.R., Jr., and Shoelson, S.E. (1997) Cellular effects of phosphotyrosine-binding domain inhibitors on insulin receptor signalling and trafficking. *Mol. Cell. Biol.* **17**, 1180–1188.
15. Boccaccio, C., Ando, M., Tamagnone, L., et al. (1998) Induction of epithelial tubules by growth factor HGF depends on the STAT pathway. *Nature.* **391**, 285–288.
16. Bardelli, A., Longati, P., Gramaglia, D., et al. (1998) Uncoupling signal transducers from oncogenic MET mutants abrogates cell transformation and inhibits invasive growth. *Proc. Nat. Acad. Sci. U.S.A.* **95**, 14379–14383.
17. Gambarotta, G., Boccaccio, C., Giordano, C., Ando, M., Stella, M.C., and Comglio, M.C. (1996) Ets up-regulates met transcription. *Oncogene.* **13**, 1911–1917.
18. Boussiotis, V.A., Freeman, G.J., Berezovskaya, A., Barber, D.L., and Nadler, L.M. (1997) Maintenance of human T cell anergy: blocking of IL-2 gene transcription by activated Rap1. *Science.* **278**, 124–128.
19. Raptis, L., Vultur, A., Tomai, E., Brownell, H.L., and Firth, K.L. (2006) *In situ* electroporation of radioactive nucleotides: assessment of Ras activity and ^{32}P-labelling of cellular proteins. In: Celis, J.E. (ed.). *Cell biology: a laboratory handbook.* Academic, San Diego, CA, pp. 329–339.
20. Nakashima, N., Ross, D.W., Xiao, S., et al. (1999) The functional role of crk II in actin cytoskeleton organization and mitogenesis. *J. Biol. Chem.* **274**, 3001–3008.
21. Marais, R., Spooner, R.A., Stribbling, S.M., Light, Y., Martin, J., and Springer, C.J. (1997) A cell surface tethered enzyme improves efficiency in gene-directed enzyme prodrug therapy. *Nat. Biotechnol.* **15**, 1373–1377.
22. Brownell, H.L., Lydon, N., Schaefer, E., Roberts, T.M., and Raptis, L. (1998) Inhibition of epidermal growth factor-mediated ERK1/2 activation by *in situ* electroporation of nonpermeant [(alkylamino)methyl]acrylophenone derivatives. *DNA Cell Biol.* **17**, 265–274.
23. Chang, D.C. (1989) Cell poration and cell fusion using an oscillating electric field. *Biophys. J.* **56**, 641–652.
24. Wegener, J., Keese, C.R., and Giaever, I. (2002) Recovery of adherent cells after *in situ* electroporation monitored electrically. *Biotechniques.* **33**, 348–352.
25. Brownell, H.L., Firth, K.L., Kawauchi, K., Delovitch, T.L., and Raptis, L. (1997) A novel technique for the study of Ras activation: electroporation of [α^{32}P]GTP. *DNA Cell Biol.* **16**, 103–110.
26. Boussiotis, V.A., Freeman, G.J., Berezovskaya, A., Barber, D.L., and Nadler, L.M. (1997) Maintenance of human T cell anergy: blocking of IL-2 gene transcription by activated Rap1. *Science.* **278**, 124–128.
27. Firth, K.L., Brownell, H.L., and Raptis, L. (1997) Improved procedure for electroporation of peptides into adherent cells *in situ. Biotechniques.* **23**, 644–645.
28. Raptis, L., Brownell, H.L., Firth, K.L., and MacKenzie, L.W. (1994) A novel technique for the study of intercellular, junctional communication: electroporation of adherent cells on a partly conductive slide. *DNA Cell Biol.* **13**, 963–975.
29. Raptis, L., Liu, S.K.W., Firth, K.L., Stiles, C.D., and Alberta, J.A. (1995) Electroporation of peptides into adherent cells *in situ. Biotechniques.* **18**, 104–114.
30. Folkman, J. and Moscona, A. (1978) Role of cell shape in growth control. *Nature.* **273**, 345–349.
31. Tomai, E., Brownell, H.L., Tufescu, T., et al. (1998) A functional assay for intercellular, junctional communication in cultured human lung carcinoma cells. *Lab. Invest.* **78**, 639–640.

32. Brownell, H.L., Narsimhan, R.P., Corbley, M.J., Mann, V.M., Whitfield, J.J., and Raptis, L. (1996) Ras is involved in gap junction closure in mouse fibroblasts or preadipocytes but not in differentiated adipocytes. *DNA & Cell Biol.* **15**, 443–451.

33. Tomai, E., Brownell, H.L., Tufescu, T., Reid, K., and Raptis, L. (1999) Gap junctional communication in lung carcinoma cells. *Lung Cancer.* **23**, 223–231.

34. Xie, H.Q., Huang, R., and Hu, V.W. (1992) Intercellular communication through gap junctions is reduced in senescent cells. *Biophys. J.* **62**, 45–47.

35. Raptis, L., Tomai, E., and Firth, K.L. (2000) Improved procedure for examination of gap junctional, intercellular communication by *in situ* electroporation on a partly conductive slide. *Biotechniques* **29**, 222–226.

36. Anagnostopoulou, A., Vultur, A., Arulanandam, R., et al. (2006) Differential effects of Stat3 inhibition in sparse *vs* confluent normal and breast cancer cells. *Cancer Lett.* **242**, 120–132.

37. Anagnostopoulou, A., Cao, J., Vultur, A., Firth, K., and Raptis L. (2007) Examination of gap junctional, intercellular communication by in situ electroporation on two co-planar indium-tin oxide electrodes. *Molecular Oncology.* **1**, 226–231.

38. Raptis, L., Balabo, V., Hsu, T., et al. (2003) *In situ* electroporation of large numbers of cells using minimal volumes of material. *Anal. Biochem.* **317**, 124–128.

39. Tomai, E., Vultur, A., Balboa, V., et al. (2003) *In situ* electroporation of radioactive compounds into adherent cells. *DNA Cell Biol.* **22**, 339–346.

Chapter 5
Formulations for DNA Delivery via Electroporation In Vivo

Khursheed Anwer

Abstract The importance of DNA formulation in safe and efficient electrogene transfer is increasingly recognized as electroporation technology enters into clinical development. A phenomenal increase in naked DNA delivery by electroporation offers new opportunities for nonviral gene therapies previously considered difficult because of insufficient delivery. However, significant tissue damage related to harsh electroporation conditions raises serious safety concerns with the use of electroporation in healthy tissues, which limits its current applications to only nonhealthy tissues such as tumors. DNA formulations designed to minimize tissue damage or enhance expression at weaker electric pulses have been examined to address these concerns. These include formulations fortified with the addition of transfection reagent(s), membrane-permeating agents, tissue matrix modifiers, targeted ligands, or agents modifying electrical conductivity or membrane stability to enhance delivery efficiency or reduce tissue damage. These advancements in DNA formulation could prove to be useful in improving the safety of electroporation protocols for human applications.

Keywords: electroporation, gene delivery, polymer, liposomes, DNA formulation

1. Introduction

Several DNA formulations for in vivo gene electroporation have been described in the literature. DNA formulation in physiological saline has been the most extensively studied formulation for electroporation. DNA electroporation in saline has been shown to enhance transfection efficiency in several tissues, producing both local and systemic levels of therapeutic proteins. The enhancement of gene electroporation is associated with significant tissue damage directly related to electroporation intensity. Milder electroporation conditions, although less toxic, are transfectionally inefficient. Several formulation strategies have been examined to reduce electroporation toxicity without affecting transfection activity. These approaches include DNA formulation in polymers, liposomes, high salt concentration,

S. Li (ed.), *Electroporation Protocols: Preclinical and Clinical Gene Medicine.*
From *Methods in Molecular Biology, Vol. 423.*
© Humana Press 2008

tissue matrix modifiers, cell permeability enhancers, membrane stabilizers, and cell-targeted systems. This chapter describes the in vivo properties of various DNA formulations used in electroporation.

2. Standard Saline Formulation for DNA Electroporation

DNA in saline (naked DNA) is the most commonly used formulation for in vivo gene electroporation (1). In skeletal muscle, electroporation enhancement of *luciferase* gene transfer was 10,000-fold over nonelectroporated control (2, 3). The enhancement of luciferase activity was observed in both small and large animal species. Histochemical analysis of *β-galactosidase* plasmid electroporated muscle showed a larger transfection area per muscle and a higher plasmid copy number per muscle cell when compared with nonelectroporated muscle (4). Muscle electroporation with FGF_1 plasmid also showed significantly larger transfection area in electroporated muscle as compared to nonelectroporated muscle (2). An average of 710 ± 122 muscle fibers were FGF_1-positive in electroporated muscle and 4 ± 2 muscle fibers were FGF_1-positive in nonelectroporated muscle. In monkeys, three out of four muscle fibers were FGF_1 positive in electroporated group and none were FGF_1-positive in the nonelectroporated group.

Skeletal muscle has an excellent capacity to express transgene products and secrete them into systemic circulation following gene transfer (5); however, full benefit of this property of muscle has not been fully achieved because of suboptimal delivery (nonviral vectors) (6) or host immune response (viral vectors) (7). A substantial improvement in muscle delivery with the use of electroporation has renewed interest in muscle tissue for systemic protein therapy. Several therapeutic proteins have been expressed from skeletal muscle and secreted into systemic circulation at substantial concentrations with the use of electroporation. Electroporation of mouse muscle with secretory *alkaline phosphatase* (*SEAP*) plasmid produced systemic levels of SEAP that were up to 120-fold higher than those achieved with *SEAP* plasmid alone (3, 8). Intramuscular injection of *erythropoietin* plasmid in mouse leg produced systemic levels of erythropoietin that were 100-fold higher than those from *erythropoietin* plasmid alone (9, 10). Electroporation of *interleukin* (*IL*)-5 plasmid DNA into mouse tibialis muscle produced 20 ng IL-5/mL while the nonelectroporated delivery produced only 0.2 ng IL-5/mL in the blood (4). Electroporation of mouse muscle with *IL-12* plasmid produced 1,500 pg of IL-12 per injected muscle and 170 pg IL-12/mL in the blood (3). In comparison, only background IL-12 levels were detectable in nonelectroporated group. The huge improvement in muscle delivery (up to 10,000-fold over naked DNA) compared with other nonviral gene delivery systems (at best 10-fold over naked DNA) opens up new product opportunities for muscle-based gene therapy.

Gene delivery into solid tumors after direct injection of formulated or naked DNA preparations is generally low because of a multitude of delivery barrier characteristics of tumor complexity. Tumor electroporation significantly enhances DNA

delivery into solid tumors. Electroporation of *luciferase* DNA into mouse and human tumors produced 10- to 1,200-fold increases in luciferase expression when compared with tumors injected with *luciferase* DNA alone (11). In another tumor study, electroporation gave high levels of luciferase expression while naked DNA injection alone failed to produce detectable luciferase activity (12). The magnitude of transfection enhancement in solid tumors is influenced by the electroporation protocol used. Tumor electroporation by six-needle electrodes (100-μs pulses, 1,500 V/cm) produced a 21-fold enhancement over control while tumor electroporation by caliper electrodes (5,000-μs pulses, 800 V/cm) produced a 42-fold enhancement (13). The transfection efficiency of DNA electroporation was compared with that of nonelectroporation methods including, liposome-DNA complexes and integrin-liposome-DNA complexes in different tumors (14). The electroporation delivery was found to be superior to all other test methods. The maximal enhancement in transfection efficiency by electroporation was up to 30-fold over naked DNA, 5- to 10-fold over liposome-DNA complexes, and over 100-fold over integrin-liposome-DNA complexes. Electroporation produced detectable gene expression in every tumor type while nonelectroporated methods were effective only in some tumors. Fluorescent microscopy of GFP-plasmid-transfected tumors shows larger and brighter transfection areas in electroporated tissue than in nonelectroporated tissue (15). In another CT26 study, tumor electroporation with *IL-12* plasmid produced 100 pg of IL-12 per tumor whereas DNA injection alone failed to yield a measurable expression (16). Tumor electroporation with IL-12 and *IL-18* plasmid DNA produced significantly higher tumor inhibition responses than did plasmid administration without electroporation (12). In B16 melanoma, tumor interferon-γ levels in *IL-12*-plasmid-electroporated tumors were 5- to 10-fold higher than in nonelectroporated tumors (17), which suggests that IL-12 receptor signaling pathways are unaltered by electroporation. Tumor electroporation did not increase cytokine concentration in blood circulation, suggesting that the effect of electroporation was highly localized in the tumor. The efficiency of IL-12 delivery by electroporation was quite comparable to adenoviral vector (18); however, the serum IL-2 levels were 50 times lower than in the adenovirus-treated animals, demonstrating highly localized expression in electroporated tissue. Transfection of human dendritic cells by electroporation was comparable to expression obtained with adenoviral or retroviral vectors (19). In another study, the delivery and anticancer efficacy of MBD2 antisense DNA in electroporated tumors were comparable to the adenovirus-treated groups (20).

The benefit of in vivo electroporation was also observed in skin tissue. Electroporation enhancement of luciferase transfection, when compared with that in nonelectroporated group, was up to 16-fold in mouse skin and up to 83-fold in pig skin (21). The enhancement by electroporation was independent of DNA dose and the pulsing protocol. The intradermal delivery of *IL-12* plasmid in mouse skin led to systemic levels of IL-12 (22). The IL-12 levels were 10-fold higher in electroporated animals when compared with those in nonelectroporated animals (22). The erythropoietin levels in blood after plasmid delivery into rat skin were 2-fold higher in electroporated groups when compared with nonelectroporated groups (23).

Compared with that in the solid tissues, gene electroporation in vascular tissues has not been investigated extensively, presumably because of the practical challenges associated with the procedure. Nishi et al., however, have described a novel method of vascular gene delivery by combining in vivo electroporation and intraarterial injection (24). In this method, *β-galactosidase* (*β-gal*) plasmid was injected into an internal carotid artery supplying blood to tumor implants in brain, and then the tumors were directly electroporated with a caliper electrode. The electroporated tumor showed significant β-gal expression while nonelectroporated tumors did not yield transgene expression.

These studies demonstrate that electroporation of solid tissues following injection of DNA in saline results in significant accumulation of therapeutic proteins in the injected tissues and, in some instances, in the systemic circulation depending upon the type of transfected tissue. The dramatic enhancement in naked DNA transfection by electroporation is also associated with considerable tissue damage that undermines the safety of this electroporation protocol. Histochemical analysis of the electroporated tissues shows massive inflammation and significant rise in Creatine Phosphokinase (CPK) levels in blood circulation, indicating muscle damage (25). These undesirable effects of electroporation must be reduced before advancing this procedure into the clinic. Several formulation approaches have been examined to enhance electroporation efficiency and lower toxicity. These formulation approaches are described in the following sections.

3. Formulation Strategies to Improve DNA Electroporation

3.1. Polymers

DNA formulation with certain types of polymers has been found to enhance electroporation efficiency and, in some cases, reduce treatment-related toxicity. Anionic polymers, including poly-L-glutamate, polyacrylic acid, poly-L-aspartate, dextran sulfate, and pectin, have been examined for their ability to enhance electroporation-mediated gene transfer in skeletal muscle (26–30). DNA formulation with poly-L-glutamate increases the electrogene transfer by 4- to 12-fold depending upon the electroporation conditions, animal species, and target tissues. The enhancement of DNA electroporation by poly-L-glutamate is not significantly influenced by the polymer size or concentration; however, a solution of 15–50-kDa poly-L-glutamate (6 mg/mL) has been reported to give the most consistent results (26). In higher species (dogs and pigs), a lower concentration of poly-L-glutamate (0.01 mg/mL) was most effective and least toxic (28). Electroporation of skeletal muscle with *factor IX* or *erythropoietin* plasmid formulated with poly-L-glutamate (6 mg/mL) produced a significant increase in the circulating levels of factor IX and erythropoietin in mice and dogs (26, 27). Plasmid/poly-L-glutamate formulations encoding for the extracellular and transmembrane domains of the protein product of *Her-2/neu* oncogene at 10-week intervals produced complete clearance of neoplastic progression

and protection from mammary carcinoma (29). Combination of this *Her-2-neu/*poly-L-glutamate vaccine with recombinant IL-12 protein therapy further augmented the therapeutic response in cancer-bearing animals (30). The mechanism of poly-L-glutamate enhancement of electroporation has not been fully understood. Poly-L-glutamate decelerates DNA clearance and enhances DNA stability and DNA retention in mouse muscle (26). Facilitation of DNA uptake into the cell and escape from the lysosomal compartment have also been postulated in the mechanism of poly-L-glutamate action (31). Poly-L-glutamate appeared to be nonimmunogenic and nontoxic for in vivo applications in animals and humans (32).

Formulation of *β-gal* plasmid with SP107 polymer significantly increased electroporation efficiency but did not reduce treatment-related inflammatory activity (33). The enhancement of transfection efficiency was higher at low-intensity electric pulses than at high-intensity ones. Formulation of DNA with another nonionic polymer, poloxamer 188, did not influence electroporation efficiency, but significantly reduced treatment-related increase in creatine phosphokinase activity (25), suggesting that the polymer protects tissue from electroporation damage. Poloxamer 188 has been shown to reduce muscle trauma and promote membrane sealing following electric shock (34). Poly(vinylpyrrolidone) (PVP) is another noncondensing polymer that is used in DNA formulation for electroporation. PVP had previously been shown to promote DNA delivery into skeletal muscle (35). The effect of PVP formulation on electrogene transfer of melanoma antigen *gp100* was examined in a mouse melanoma model (36). About 40% of tumor-bearing mice immunized intramuscularly with gp100 plasmid/PVP rejected tumor challenge when compared with only 9% of mice immunized with gp100 plasmid/saline. However, PVP formulation did not enhance electroporation delivery of *factor IX* plasmid in mouse skeletal muscle (27).

3.2. Liposomes

DNA complexes of cationic liposomes were electroporated into several histologically distinct mouse subcutaneous tumors, and the efficiency of gene transfer was compared with that of naked DNA electroporation (14). Liposomal formulations were transfectionally superior to naked DNA in B16 melanoma, P22 carcinoma, and SaF sarcoma but not in T24 human bladder carcinoma (14) or MC2 mammary carcinoma (37). This variation in tumor response could be due to differences in the state of tumor necrosis, tumor conductivity, or matrix complexity between the different tumors.

Addition of cationic liposomes to transfection medium has been shown to enhance the electroporation of transposon/transposase system in *Spirulina platensis*, a commercially important species of microalgae generally resistant to conventional transformation methods (38). While electroporation alone failed to efficiently transform *S. platensis* with Tn5 transposon/transposase system, addition of 1,2-dioleoyl-3-(trimethylammonium) propane (DOTAP) into electroporation medium improved the transformation efficiency by 100-fold without affecting the biological

integrity of the transformed algae. These studies demonstrate that the efficiency of electrogene transfer can be improved by addition of cationic liposomes in the electroporation medium. The precise mechanism of liposome action on electrogene transfer has not been investigated. A higher interaction of positively charged lipid-DNA complexes with negatively charged cell surfaces could be one of the underlying mechanisms in the lipid enhancement of the electroporation.

Addition of anionic liposomes into the electroporation medium has been found to enhance the delivery of macromolecules into cells. For example, dextran uptake during electroporation was enhanced by 80-fold with the addition of phosphatidylglycerol and phosphatidylcholine into the transfection medium (39, 40). The magnitude of liposome enhancement was dependent on the degree of lipid saturation but independent of polar head group. The effect of anionic liposomes on DNA electroporation has not been investigated.

3.3. Targeted Systems

DNA delivery by electroporation is not target-specific. Several attempts have been made to improve tissue-specific targeting of electroporated DNA with the use of cell-specific ligands (14, 41). Antibodies and other molecular entities that recognize specific cell surface receptors have been conjugated to delivery vehicles to achieve high cell specificity during electroporation. For example, large unilamellar vesicles encapsulating *XGPRT* gene were coated with protein A to target HAL-2-antibody-labeled lymphoid cells prior to electroporation (41). Anti-HLA-antibody-labeled lymphoid cells were electroporated with protein-A-coated liposomes/*XGPRT* DNA complexes to approximately 35% transfection efficiency. In comparison, the same cells electroporated with nontargeted liposomes were transfected to only 5% transfection efficiency. The technical feasibility of in vivo DNA targeting by electroporation has not been fully established. For example, electroporation of integrin-conjugated liposome-DNA complexes yields much lower transfection efficiency than do the nontargeted systems (14). This failure of tumor targeting in vivo could be attributed to poor stability of the targeted complexes in extracellular milieu, altered integrin receptor affinity for integrin ligand, or suboptimal transfection conditions. Hence, the use of targeted ligands is an attractive approach to improve target specificity of electroporation, but its in vivo application has not been fully established.

3.4. Hyaluronidase Treatment

DNA dispersion in muscle is highly restricted because of the rigid collagen- and hyaluronan-rich matrix surrounding muscle fibers. Pretreatment of tissue with hyaluronidase has been shown to improve gene delivery into liver (42) and skeletal

muscles (43). Hyaluronidase treatment prior to electroporation in skeletal muscle produced a substantial increase both in levels and extent of gene transfer in skeletal muscle (44). Hyaluronidase treatment enhanced transfection efficiency at low electric pulses without significantly damaging the muscle structure or function (44). This tissue-protective effect of hyaluronidase has been observed in ischemic myocardium (45) and tissue edema (46). These results demonstrate that hyaluronidase treatment is a useful approach to improve electrogene transfer in higher species where rigid interstitium is a major limitation to plasmid delivery.

3.5. High Salt Formulations

Since the negative charge of DNA could be one of the important factors affecting transfection efficiency, the presence of salt in the DNA vehicle would alter ionic atmosphere, ionic strength, and conductivity of the medium. Therefore, it is reasonable to assume that changing the salt concentration in plasmid solution would influence electroporation efficiency. Lee et al. have examined the effect of various salts and salt concentrations on the efficiency of electrogene transfer in mouse skeletal muscle (47). Plasmid DNA encoding *luciferase* or *β-galactosidase* gene was formulated in water, saline, or phosphate-buffered saline prior to delivery. In the absence of electric pulse, transfection was highest in phosphate-buffered saline, followed by saline and then water. However, in the presence of electric pulse (125 V/cm, two sets of 4 pulses, 50 ms, 1 Hz), this relationship reversed, with the highest expression observed with water, followed by saline and phosphate-buffered saline. The mechanism of this reversal in gene expression pattern is not fully understood. However, histological examination of the electroporated tissue showed higher degree of tissue damage in water injection group. Varying the salt concentration (0–250 mM NaCl) of plasmid formulation had a pronounced effect on the efficiency of electrogene transfer in skeletal muscle. The transfection activity increased with increasing salt concentration, reaching maximal activity at 75 mM NaCl, and then decreasing as the salt concentration was further increased. Histological examinations show higher tissue damage at salt concentrations below 75 mM. The comparison of electrical conductivity and resistance after intramuscular injection showed higher resistance and lower conductivity with water injection than with normal saline injection. Hence, the efficiency of electrogene transfer increases as the ionic strength decreases. Lowering the ionic strength below a threshold (75 mM) lowers tissue conductivity, enlarges "electroleaks" (48), and exerts tissue damage. In the absence of electroporation, high salt concentration enhances transfection efficiency (49).

Increasing the calcium concentration from 2 to 10 mM in the electroporation medium significantly decreased the efficiency of *lac Z* gene transfer in skin and muscle (50). In comparison, electroporation of FITC-labeled oligonucleotide into mouse ovary was significantly enhanced at higher calcium concentrations (51). These opposite effects of calcium could be due to difference in calcium concentrations, nucleic acid size, or tissue.

3.6. Gold Nanoparticles

Application of electromigration field (3 V for 30 s) has been shown to enhance the uptake of DNA-modified gold nanoparticles during cell electroporation (52). Gold nanoparticles devoid of DNA coating were not taken up by cells during electroporation. Transmission electron microscopy of electroporated cells revealed a higher cell surface density of DNA-modified gold particles in cells subjected to electromigration field than in cells that were not, suggesting that the DNA binding to cell surface is crucial for optimal transfection efficiency from electroporation. Formulations that can enhance DNA binding to cell surface in vivo may also enhance electroporation efficiency at weak electric pulses.

3.7. Nucleofection

Transfection of primary cultures of mammalian cells with conventional gene delivery methods, including standard electroporation, is poor. Recently, a gene transfer protocol that combines electroporation with special transfection solutions has been described, which transfects primary cells and other hard-to-transfect cells with high efficiency (53–55). It is believed that this electroporation formulation promotes direct translocation of DNA into the nucleus, producing up to 50% transfection efficiency in otherwise poorly transfectable cells. The chemical composition of this formulation and its in vivo application has not been described.

3.8. Other Chemical Formulants

Satyabhama et al. have examined the effect of DNA formulation with several chemical agents on the efficiency of electrogene transfer in human erythroleukemia cells. These chemical agents included membrane perturbation reagents, anesthetics, lipids, polymers, lysosomal inhibitors, macromolecules, nucleic acid precursors, ethanol, polybrene, and various combinations of these agents (56). Addition of ethanol, polybrene, and PEG 6000 resulted in a 3- to 5-fold enhancement in transfection activity, and various combinations of these chemical agents gave up to 15-fold enhancement in electrogene transfer. The mechanisms of these enhancements are attributed to the ability of these agents to influence DNA trafficking.

4. Summary

Electroporation is a widely recognized method of gene delivery into mammalian tissues. It is a highly efficient method, with delivery efficiency surpassing that of many nonviral vectors. The preclinical development of electroporation in vivo is

Table 5.1 Formulation strategies for improvement of DNA electroporation

Formulant	Gene	Species	Tissue	Effects on electroporation	Reference
Polyglutamate	*β-gal, Luc, FIX, EPO*	Mouse, rat, dog, pig	Skeletal muscle	Increased gene expression	26–30
Poloxamers	*β-gal*	Rat	Skeletal muscle	Increased gene expression and less tissue damage	25, 33
Poly(vinylpyrrolidone)	*FIX*	Mouse	Skeletal muscle	No effect on gene expression	27, 36
	gp 100	Mouse	Solid tumor	Increased gene expression	
Liposomes	*GFP, CAT*	Mouse	Solid tumor	Variable results (tumor-dependent)	14, 37
Targeted ligands (antibody,integrins)	*XGPRT*	Mouse	Lymphoid cells, tumor	Increased gene expression in lymphoid cells, no effect on tumor	14, 41
Hyaluronic acid	*β-gal*	Mouse	Skeletal muscle, liver	Increased gene expression	44
Salt concentration					
NaCl	*Luc, β-gal*	Mouse	Muscle	Optimal expression at 75 m M Na$^+$	47
CaCl$_2$	*β-gal, Oligos*	Mouse	Skin, muscle, ovary	Variable results with Ca^{2+}	50, 51
Nucleofection	*GFP, CAT*	Human, mouse	Leukemia, endothelial and smooth muscle cells	Increased expression	53–55
Chemical stimulants	*CAT*	Mouse, human	Lymphoid and myeloid cells	Moderate enhancement	56
Gold nanoparticles	*Oligos*	Mouse	Osteoblast cells	Increased expression	52

focused on tissues that are easily accessible to electroporation and can withstand electric pulsation. The standard DNA formulation for electroporation is DNA in physiological saline. Under optimal conditions, DNA electroporation in saline yields a 10- to 10,000-fold enhancement in gene delivery efficiency over nonelectroporated controls. This enormous increase in transfection activity, however, accompanies significant tissue damage and local inflammation, which might not be a bad thing to have if the target is cancer. However, for applications in which expression from normal tissues is desired, tissue damage and inflammatory response are not conducive to therapeutic objectives and, therefore, must be minimized. Several formulation strategies have been designed to enhance electroporation efficiency and minimize toxicity (Table 5.1). Encouraging results have been obtained with some approaches, which must be further developed into clinically viable formulations for noncancer applications.

References

1. Li, S. (2004) Electroporation gene therapy: new developments in vivo and in vitro. *Curr. Gene Ther.* **4**, 309–316.
2. Mir, L.M., Bureau, M.F., Gehl, J., Rangara, R., Rouy, D., Caillaud, J., et al. (1999) High-efficiency gene transfer into skeletal muscle mediated by electric pulses. *Proc. Natl. Sci.* **96**, 4262–4267.
3. Li, S., Zhang, X., Xia, X., Zhou, L., Breau, R., Suen, J., et al. (2001) Intramuscular electroporation delivery of IFN-α gene therapy for inhibition of tumor growth located at a distant site. *Gene Ther.* **8**, 400–407.
4. Aihara, H. and Myazaki, J. (1998) Gene transfer into muscle by electroporation in vivo. *Nat. Biotech.* **16**, 867–870.
5. Smith, L.C. and Nordstrom J.L. (2000) Advances in plasmid gene delivery and expression in skeletal muscle. *Current Opin. Mol. Ther.* **2**, 150–154.
6. Wells, D.J. (2006) Viral and non-viral methods of gene transfer into skeletal muscle. Curr. *Opin. Drug Discov. Dev.* **9**, 163–168.
7. Smith, T.A., Mehaffey, M.G., Kayda, D.B., et al. (1993) Adenovirus mediated expression of therapeutic plasma levels of human factor IX in mice. *Nat. Genet.* **5**, 397–402.
8. Rizzuto, G., Cappelletti, M., Mennuni, C.A., et al. (2000) Gene electrotransfer results in a high level transduction of rat skeletal muscle and corrects anemia of renal failure. *Hum. Gene Ther.* **11**, 1891–1900.
9. Kreiss, P., Bettan, M., Crouzet, J., and Scherman, D. (1999) Erythropoietin secretion and physiological effect in mouse after intramuscular plasmid DNA electrotransfer. *J. Gene Med.* **1**, 245–250.
10. Rizzuto, G., Cappelletti, M., Malone, D., et al. (1999) Efficient and regulated erythropoietin production by naked DNA injection and muscle electroporation. *Proc. Natl. Sci.* **96**, 6417–6422.
11. Bettan, M., Ivanov, A., Mir, L.M., Boissiere, F., Delaere, P., and Scherman, D. (2000) Efficient DNA electrotransfer into tumors. *Biochemistry.* **52**, 83–90.
12. Kashida, T., Asada, H., Satoh, E., et al. (2001) In vivo electroporation-mediated transfer of interleukin 12 and interleukin-18 genes includes significant antitumor effects against melanoma in mice. *Gene Ther.* **8**, 1234–1240.
13. Heller, L. and Coppola, D. (2002) Electrically mediated delivery of vector plasmid DNA elicits an antitumor effect. *Gene Ther.* **9**, 1321–1325.

14. Cemazar, M., Sersa, G., Wilson, J., Tozer, G.M., Hart, S.I., and Grosel, A. (2002) Effective gene transfer to solid tumors using different nonviral gene delivery techniques: Electroporation, leptosomes, and integrin-targeted vectors. *Cancer Gene Ther.* **9**, 399–406.
15. Goto, T., Nishi, T., Tamura, T., et al. (2000) Highly efficient electro-gene therapy of solid tumor by using an expression plasmid for the herpes simplex virus thymidine kinase gene. *Proc. Nat. Acad. Sci.* **97**, 354–359.
16. Tamura, T., Nishi, T., Goto, T., et al. (2001) Intratumoral delivery of interleukin 12 expression plasmids with in vivo electroporation is effective for colon and renal cancer. *Hum. Gene Ther.* **12**, 1265–1276.
17. Lucas, M.L., Heller, L., Coppola, D., and Heller, R. (2002) IL-12 plasmid delivery by in vivo electroporation for the successful treatment of established subcutaneous B16.F10 melanoma. *Mol. Ther.* **5**, 668–675.
18. Lohr, F., Lo, D.Y., Zaharoff, D.A., et al. (2001) Effective tumor therapy with plasmid-encoded cytokines combined with in vivo electroporation. *Cancer Res.* **61**, 3281–3284.
19. Lundqvist, A., Noffz, J.G., Pavlenko, M., et al. (2002) Nonviral and viral gene transfer into different subsets of human dendritic cells yield comparable efficiency of transfection. *J. Immunother.* **25**, 3445–3454.
20. Slack, A., Bovenzi, V., Bigey, P., et al. (2002) Antisense MBD2 gene therapy inhibits tumorigenesis. *J. Gene Med.* **4**, 381–389.
21. Drabick, J.I., Malone, J.G., Somiari, S., King, A., and Malone, R. (2001) Cutaneous transfection and immune responses to intradermal vaccination are significantly enhanced by in vivo electropermeabilization. *Mol. Ther.* **3**, 249–255.
22. Heller, R., Schultz, J., Lucas, M.L., et al. (2001) Intradermal delivery of interleukin-12 plasmid DNA by in vivo electroporation. *DNA Cell Biol.* **20**, 21–26.
23. Maruyama, H., Ataka, K., Higuchi, N., Sakamoto, F., Gejyo, F., and Miyazaki, J. (2001) Skin-targeted gene transfer using in vivo electroporation. *Gene Ther.* **8**, 1808–1812.
24. Nishi, T., Yoshizato, K., Yamashiro, S., et al. (1996) High-efficiency in vivo gene transfer using intraarterial plasmid DNA injection following in vivo electroporation. *Cancer Res.* **56**, 1050–1055.
25. Hartikka, J., Sukhu, L., Buchner, C., et al. (2001) Electroporation-facilitated delivery of plasmid DNA in skeletal muscle: plasmid dependence of muscle damage and effect of poloxamer 188. *Mol. Ther.* **4**, 407–414.
26. Nicol, F., Wong, M., MacLaughlin, F.C., Wilson, E., Nordstrom, J.L., and Smith, L.C. (2002) Poly-L-glutamate, an anionic polymer, enhances transgene expression for plasmids delivered by intramuscular injection with in vivo electroporation. *Gene Ther.* **9**, 1351–1358.
27. Fewell, J.G., MacLaughlin, F.C., Mehta, V., et al. (2001) Gene therapy for the treatment of hemophilia B using PINC-formulated plasmid delivered to muscle with electroporation. *Mol. Ther.* **3**, 574–583.
28. Draghia-Akli, R., Khan, A.S., Cummings, K.K., Parghi, D., Carpenter, R.H., and Brown, P.A. (2002) Electrical enhancement of formulated plasmid delivery in animals. *Technol. Cancer Res. Treat.* **1**, 365–372.
29. Quaglino, E., Iezzi, M., Mastini, C., Amici, A., Pericle, F., and Carlo, E.D. (2004) Electroporated DNA vaccine clears away multifocal mammary carcinoma in Her-2/neu transgenic mice. *Cancer Res.* **64**, 2858–2864.
30. Spadaro, M., Ambrosino, E., Iezzi, M., Carlo, E.D., Sacchetti, P., and Curcio, C. (2005) Cure of mammary carcinomas in Her-2 transgenic mice through sequential stimulation of innate (neoadjuvant interleukin-12) and adaptive (DNA vaccine electroporation) immunity. *Clin. Cancer Res.* **11**, 1941–1952.
31. Fuji, T., Suzuki, T., Fujii, M., Hachimori, A., Kondo, Y., and Ohki, K. (1986) Inhibition of microtubule assembly by poly(L-glutamic acid) and the site of its action. *Biochem. Cell Biol.* **64**, 615–621.
32. Maurer, P.H. (1965) Antigenicity of polypeptides (poly-α-amino acids). *J. Immunol.* **95**, 1095–1099.

33. Riera, M., Chillon, M., Aran, J.M., et al. (2003) Intramuscular SP1017-formulated DNA electrotransfer enhances transgene expression and distributes hHGF to different rat tissues. *J. Gene Med.* **6**, 111–118.

34. Block, T.A., Aarsvold, J.N., Mathews, K.L., et al. (1996) The 1995 Lindberg Award. Nonthermally mediated muscle injury and necrosis in electrical trauma. *J. Burn care Rehabil.* **16**, 581–588.

35. Mumper, R.J., Wang, J., Klakamp, S.L., Nitta, H., Anwer, K., and Tagliaferri, F. (1998) Protective, interactive, noncondensing (PINC) polymers for enhanced plasmid distribution and expression in rat skeletal muscle. *J. Control. Release.* **52**, 191–198.

36. Mendiratta, S.K., Thai, G., Eslahi, T., et al. (2001) Therapeutic immunity induced by polyimmunization with melanoma antigens gp100 and TRP-2. *Cancer Res.* **61**, 859–863.

37. Wells, J.M., Li, L.H., Sen, A., Jahreis, G.P., and Hui, S.W. (2000) Electroporation-enhanced gene delivery in mammary tumors. *Gene Ther.* **7**, 541–547.

38. Kawata, Y., Yano, S., Kojima, H., and Toyomizu, M. (2004) Transformation of *Spirulina platensis* strain c1 (*Arthrospira* sp. PCC9438) with Tn5 transposase-transposon DNA-cation liposome complex. *Mar. Biotechnol.* **6**, 355–363.

39. Sen, A., Zhao, Y., Zhang, L., and Hui, S.W. (2002) Enhanced transdermal transport by electroporation using anionic lipids. *J. Control. Release.* **82**, 399–405.

40. Sen, A., Zhao, Y., Zhang, L., and Hui, S.W. (2002) Saturated anionic phospholipids enhance transdermal transport by electroporation. *Biophys. J.* **83**, 2064–2073.

41. Machy, P., Lewis, F., McMillan, L., and Jonak, Z.L. (1988) Gene transfer from targeted leptosomes to specific lymphoid cells by electroporation. *Proc. Natl. Acad. Sci.* **85**, 8027–8031.

42. Dubensky, T.W., Campbell, B.A., and Villarreal, L.P. (1984) Direct transfection of viral and plasmid DNA into the liver or spleen of mice. *Proc. Nat. Acad. Sci.* **81**, 7529–7533.

43. Favre, D., Cherel, Y., Provost, N., et al. (2000) Hyaluronidase enhances recombinant adeno-associated virus (rAAV)-mediated gene transfer in the rat skeletal muscle. *Gene Ther.* **7**, 1417–1420.

44. McMahon, J.M., Signori, E., Wells, K.E., Fazio, E.M., and Wells, D.J. (2001) Optimization of electrotransfer of plasmid into skeletal muscle by pretreatment with hyaluronidase—increased expression with reduced muscle damage. *Gene Ther.* **8**, 1264–1270.

45. Evora, P.R. (2000) Exogenous hyaluronidase induces release of nitric oxide from the coronary endothelium. *J. Thorac. Cardiovasc. Surg.* **120**, 707–711.

46. Johnson, C., Hallgrem, R., and Tufveson, G. (2000) Hyaluronidase can be used to reduce interstitial edema in the presence of heparin. *J. Cardiovasc. Pharmacol. Ther.* **5**, 229–236.

47. Lee, M.J., Cho, S.S., Jang, H.S., et al. (2002) Optimal salt concentration of vehicle for plasmid DNA enhances gene transfer mediated by electroporation. *Exp. Mol. Med.* **34**, 265–272.

48. Sukhorukov, V.L., Mussauer, H., and Zimmermann, U. (1998) The effects of electrical deformation forces on the elctropermeabilization of erythrocyte membrane in low- and high conductivity media. *J. Membr. Biol.* **163**, 235–245.

49. Chesnoy, S. and Huang, L. (2002) Enhanced cutaneous gene delivery following intradermal injection of naked DNA in a high ionic strength solution. *Mol. Ther.* **5**, 57–62.

50. Zhao, Y.G., Lu, H.L., Peng, J.L., and Xu, Y.H. (2006) Inhibitory effect of calcium on in vivo gene transfer by electroporation. *Acta. Pharmacol. Sin.* **27**, 307–310.

51. Suzuki, T., Tsunekawa, J., Murai, A., and Muramatsu, T. (2003) Effect of $CaCl_2$ concentration on the rate of foreign gene transfer and expression by in vivo electroporation in the mouse ovary. *Int. J. Mol. Med.* **12**, 265–368.

52. Jen, C.P., Chen, Y.H., Fan, C.H., Yeh, C.S., Lin, Y.C., and Shieh, D.B. (2004) A nonviral transformation approach in vitro: the design of a gold nanoparticles vector joint with microelectromechanical systems. *Langmuir.* **20**, 1369–1374.

53. Schakowski, F., Buttgereit, P., Mazur, M., Marten, M., Schottker, B., and Gorschluter, M. (2004) Novel non-viral method for transfection of primary leukemia cells and cell lines. *Genet. Vaccines Ther.* **2**, 2.

54. Gresch, O., Engel, F.B., Nesic, D., Tran, T.T., England, H.M., and Hickman, E.S. (2004) New non-viral method for gene transfer into primary cells. *Methods*. **33**, 151–163.
55. Iverson, N., Birkenes, B., Torsdalen, K., and Djurovic, S. (2005) Electroporation by nucleofector is the best nonviral transfection technique in human endothelial and smooth muscle cells. *Genet Vaccines Ther*. **3**, 2.
56. Satyabhama, S. and Epstein, A.L. (1988) Short-term efficient expression of transfected DNA in human hematopoietic cells by electroporation: definition of parameters and use of chemical stimulators. *DNA*. **7**, 203–209.

Chapter 6
Overview of Drug Delivery and Alternative Methods to Electroporation

Sek-Wen Hui

Abstract This chapter provides an overview of the application of electroporation to areas other than gene delivery. These areas include the delivery of drugs and vaccines to tissues and tumors as well as into and through the skin. Achievements and limitations of electroporation in these areas are presented. Alternative physical methods for gene and drug delivery besides electroporation are described. The advantages and drawbacks of electroporation, compared with these methods, are also discussed.

Keywords: electrochemotherapy, transdermal, anesthetics, peptide hormone, vaccine, lipoplex and polyplex, magnetofection, gene gun, ultrasound, laser

1. Introduction

The major application of electroporation has been the delivery of genetic materials. Many aspects of this application will be covered by various chapters in this book; however, the initial conceptual goals of discovery and development of electroporation were not limited to gene delivery. The physical nature of this methodology opens a wide area of applications including drug delivery, protein insertion, fluorescent probe incorporation, initiation of cell fusion, and transdermal delivery just to mention a few. It took many years and numerous publications before the technology was accepted by the biomedical community as a standard methodology. Meanwhile, many competing and complementary physical methods for gene and drug delivery have been developed.

This review chapter begins with an overview of applications of electroporation in areas other than gene delivery. The aim is to offer the reader a broader view of how electroporation can be and has been used. The second part of this chapter provides a perspective of the standing of electroporation as one of the physical delivery methods. Also, the uniqueness as well as the drawbacks of electroporation in these applications are described.

2. What Can Electroporation Deliver Other Than DNA?

Electroporation can be and has been used to deliver a variety of molecules other than DNA for the purpose of anesthesia, cosmetics, chemotherapy, and vaccination. This topic is too vast to be covered in detail in only half a chapter. Fortunately, there are many excellent review articles that provide further details.

2.1. Drug Delivery to Tissues and Tumors

Permeation sites or pores in cell membranes have to be created by electroporation to introduce drugs into cells. Briefly, these pores are created by local, transient electric breakdown of cell membranes. The threshold potential for transient electric breakdown of cell membranes is about 0.5 V. The interior and exterior of cells are normally much more conductive than the cell membranes. When an external electric field is imposed on a cell, the charges (of ions for instance) tend to polarize upon the cell membrane, thus building up a potential gradient, as depicted in Fig. 6.1. For a cell with a 10 µm diameter, the field strength needed to reach and exceed a potential of 0.5 V at each end is about 1,000 V/cm. The time for charge build-up is dictated by the conductivities of the exterior and the content of the cell. The time is normally in the order of a fraction of a millisecond; therefore, the electroporation pulse protocols are usually set at above 1 kV/cm with 1 ms or less in duration. Longer pulse length is sometime used to convey charged molecules by electrophoresis through the newly formed electropores. This condition applies also for contiguous cells and tissues. For electroporation of cell organelles such as nuclei and mitochondria, the smaller size of the target requires pulses of higher field strength and shorter time, usually exceeding 10 kV/cm and in the nanoseconds ranges (1).

By far, the most useful application of electroporation in drug delivery is delivering nonpermeant cancer chemotherapy drugs. Most cancer chemotherapy

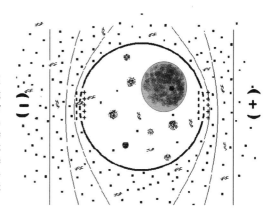

Fig. 6.1 Schematic drawing of a cell in an imposed electric field, showing charge build-up (polarization) near the poles facing the electrodes. Equipotential lines bend around the conductive cell and stack up tightly at the poles, indicating transmembrane potential increases. Electric breakdown permeabilizes the membrane at the poles

drugs are transported through the plasma membrane through a channel or transporter mechanism. Bleomycin, a 1.5-kDa glycopeptide that causes DNA strand breaking, is practically nonpermeant through the plasma membrane. Therefore, its cytotoxic effect on cancer cells is minimal without the help of artificial permeabilization of the plasma membrane. Facilitating drug transport by electroporation was first reported by Okino and Mohri (2). Subsequently, the bleomycin electrochemother-apy method was developed and established by Mir and Orlowski (3) and later expanded by Heller et al. (4) and Sersa and colleagues (5). In bleomycin chemotherapy, treatment was more than 1,000 times more effective with electroporation than without electroporation (6). Other drugs, including cisplatin and actinomycin, have been tried with less impressive improvement, typically only a few folds over treatment without electroporation. Other permeant drugs such as daunorubicin, doxorubincin, 5-fluorouracil, and paclitaxel have no electroporation advantage (6, 7).

Bleomycin electrochemotherapy has been successfully applied to treat melanomas, head and neck squamous cell carcinomas, Kaposi's sarcomas, as well as lung, breast, kidney, and bladder cancers (4, 6). Bleomycin can be introduced intravenously (i.v.) or intratumorally (i.t.), at a dose of about 10–27 U/m^2 (i.v.) or 1 U/cm^3 (i.t.) prior to electroporation. The electroporation protocol is usually 6–8 pulses of 0.1 ms at 1–1.3 kV/cm applied through parallel plate or needle array elec-trodes. A number of electrode designs have been described (8–10). Success rate has been impressive, averaging from 20–30% complete response to 50–60% objective response (8). In certain stage II and III clinical trials, 100% complete recovery has been reported (6). Bleomycin has additional advantages. Its cytotoxicity is higher in cancer tissues than in normal tissues, including arteries and nerves (8). Bleomycin electrochemotherapy induces temporary vasoconstriction, which helps to retain the drug in the tumor tissue (6). Bleomycin electrochemotherapy is, therefore, a prom-ising strategy for one time, local cancer treatment.

2.2. Drug Delivery to and through the Skin

An important application of electroporation to drug delivery is transdermal deliv-ery. Transdermal drug delivery has several potential advantages over other parenteral delivery methods. Apart from the convenience and noninvasiveness, the skin also provides a "reservoir" that sustains delivery over a period of days (11). Furthermore, it offers multiple sites to avoid local irritation and toxicity, yet it can also offer the option to concentrate drugs at local areas to avoid undesirable sys-temic effects. However, at present, the clinical use of transdermal delivery is limited by the fact that very few drugs can be delivered transdermally at a viable rate. This difficulty is because the skin forms an efficient barrier for most molecules, and few noninvasive methods are known to significantly enhance the penetration of this bar-rier. Electroporation is one of the approaches to improve the transdermal delivery by transiently permeabilizing the skin to facilitate drug transport.

2.2.1. Mechanism and Strategy of Transdermal Transport by Electroporation

Mammalian skin has two layers, epidermis and dermis. The epidermis is a stratified squamous keratinizing epithelium. The uppermost stratum of the epidermis is the *stratum corneum* (SC), which consists of about 20 layers of flattened, enucleate, keratin-filled corneocytes surrounded by lamellae of an average of eight lipid bilayers. The SC forms the major barrier to most water-soluble and many hydrophobic drugs and contributes the major portion of the electric resistance of the skin. The mechanism of electroporation depicts the transient electric breakdown site to be the most resistive component of the electric circuit. In transdermal electroporation, the predominant voltage drop of an applied electric pulse to the skin develops across the SC. This voltage distribution favors local electric breakdown (electroporation) of the SC, which is precisely the barrier targeted to be permeabilized to facilitate drug transport.

The application of electroporation to transdermal delivery is a relatively recent development. If the voltage of the applied pulses exceeds a voltage threshold at 75–100 V (equivalent to the breakdown threshold of 8–10 lipid bilayers in the SC), microchannels or "local transport regions" are created through the breakdown sites of the SC (12) (Fig. 6.2). Transdermal transport of calcein was enhanced by electroporation by 2 orders of magnitude, compared with that by diffusion and iontophoresis (13). Subsequently, many small-molecule drugs have been successfully delivered through the skin by electroporation. Transport efficiency for small (mol. wt., ≤1,000) charged molecules (such as protoporphyrin IX), using the same polarity pulses, was more than an order of magnitude higher than that for uncharged molecules (such as protoporphyrin IX methyl ester) or charged molecules with opposite polarity pulses. The results indicated that, besides passive diffusion through electropores,

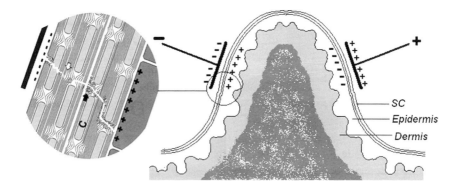

Fig. 6.2 Schematic drawing showing skin and subcutaneous tissue being clamped between a pair of plate electrodes. During the pulse application, charges build up across the *stratum corneum* (SC). When the potential due to charge build-up across the SC exceeds its breakdown potential, the barrier function of the SC is compromised, and molecules may pass through the SC at these sites. Electroporation occurs at both the entrance and exit sites. The pulse current crosses the skin twice. The *inset* shows an enlarged view of the cross-sectional area marked by a *circle*, with a breakdown path through the SC (black and white *arrows*). C denotes corneocytes, *parallel lines* represent lipid lamellae in the SC. Features are not drawn to scale

electrophoretic force of the pulses also contributes to the electroporation-enhanced transport of these charged molecules (14). Transport occurred during the 10–30 min after pulse application was found to be equal to or higher than that which occurred during pulses, indicating that passive diffusion through electropores was also an important transport mechanism (15). Auxiliary current was applied during the recovery period of the skin to further promote drug transport by electrophoretic flow (16, 17). Yet the transport of uncharged or weakly charged molecules was also enhanced by a current of opposite polarity, indicating the action of electroosmotic force (18). This is especially important for delivering larger and uncharged molecules.

2.2.2. Delivery of Low Molecular Weight Compounds, Including Anesthetics, Analgesics, and Antibiotics

A number of drug molecules, which normally are poorly transported through the skin barrier, have been delivered transcutaneously with the aid of electroporation and, in some cases, their transport has been enhanced by additional iontophoresis. The chemicals reportedly transported through the skin by electroporation include negatively charged (19) as well as positively charged fluorescent dyes (14, 20). The Belgian group (21) measured the transport of a large variety of drugs, including metoprolol and pentanyl, by electroporation as a function of applied pulse voltage and duration. The transport efficiency by iontophoresis and electroporation were compared. Subsequently, they reported the transdermal deliveries of flurbiprofen, fentanyl, dihydrotestersteron, alniditan, domperidone, modified oligonucleotides, and other drugs (21–23). Transport of small neutral molecules such as mannitol (21, 24) was also reported. When applied in vivo, the pharmacokinetics and pharmacodynamics of fentanyl were much faster and more effective when delivered by transdermal electroporation than by iontophoresis (25). Vitamin C was delivered transdermally by electroporation for cosmetic purposes (26). An excellent review and summary of transdermal delivery of drugs is given by Vanbever and Preat and, more recently, by Denet et al. (21, 23).

The pro-drug δ-aminolevulinic acid (ALA, a photosensitizer precursor) was delivered by electroporation and iontophoresis into porcine and murine skin. In the skin, ALA was converted metabolically to the photodynamic therapeutic drug protoporphyrin IX (16). Other photosensitizer drugs have also been transported by electroporation (27). Methotrexate (MTX) is a folic acid antagonist with antineoplastic activity and is also used in the treatment of psoriasis by oral and parental but not transdermal administration, because MTX does not readily penetrate the SC. Application of electroporation with anionic lipid enhancers and concurrent iontophoresis at 40°C hyperthermia resulted in an 11-fold enhancement from passive diffusion transport through porcine skin (28). The serum concentration in mice was enhanced by 4.5-fold (29). Another drug for treating psoriasis, cyclosporine A, was delivered into the skin by electroporation, with an 8.5-fold enhancement (30); however, not all drug deliveries can be enhanced by transdermal electroporation (31).

Electroporation-enhanced transdermal transport can also be reversely utilized to sample tissue drug concentration beneath the skin. The time course of free salicylate concentration in the dermal extracellular fluid (ECF), determined using the

electroporation extraction method, was in good agreement with that predicted using a two-compartment pharmacokinetic model after bolus injection (32). The results suggest that electroporation extraction technique may serve as a tool for noninvasive pharmacokinetic studies and monitoring therapeutic drugs and metabolites in the dermal ECF. The advantages of this method, compared with microdialysis and reverse iontophoresis techniques, are the short sampling duration and greater efficiency of drug extraction. This method can also be used to monitor glucose concentration and to deliver insulin across the skin (33).

2.2.3. Delivery of Higher Molecular Weight Compounds, Peptides, and Peptide Antigens as Vaccines

Because the passages created by electroporation of the skin are believed to be structurally restricted (12, 34), there is an upper molecular weight limit of the drugs to be transported in this manner. The molecular weight cutoff has been determined, using FITC-dextrans, to be about 10 kDa (15, 35). Various chemical enhancers have been used to improve iontophoresis, and keratinlytic compounds such as heparin and thiosulfates have been used to enhance transdermal electroporation transport of large molecules (36–38). The presence of an ionic surfactant such as sodium dodecyl sulfate (SDS) reduces the electroporation threshold and significantly improves the transdermal transport of molecules by electroporation (39). Saturated anionic lipids tend to be preferentially retained in the epidermis during electroporation and result in disrupting the lamellar structure of the SC lipids, leading to prolonged lifetime of electropores. As a result, the transport of both charged and neutral macromolecules was enhanced (14, 40).

Bommannan et al. (17) reported the enhanced transport of human luteinizing hormone releasing hormone through heat-stripped human epidermis by electroporation. Larger molecules, including heparin, polylysine (37, 38), antisense polynucleotides (41), lactalbumin, and IgG (36), have been delivered by transdermal electroporation with proper enhancers. The transport of calcium-regulating hormones (42, 43) was found to be increased by applying electroporation and iontophoresis. Anionic lipid formulation has shown significant synergistic effect with electroporation on delivering insulin in vitro (44) and in vivo (18) and has the potential to lower the voltage threshold to an imperceptible level.

Peptides and minigene vaccines are of particular interest since several epitopes of tumor-associated antigens have been employed as cancer vaccines for therapy and prevention. Although small molecular size antigens may be delivered into and through the skin by diffusion or by iontophoresis methods, antigens of higher molecular weight (>1 kDa), such as peptides, DNA, complex carbohydrates, as well as vaccine adjuvants needed to stimulate a strong immune response, are delivered primarily by needle injection at present, unless the SC is removed by tape-striping (45), or by passive delivery through patches of hydrated skin (46) over a period of hours. Needle-free nonadjuvant skin immunization by electroporation has been reported (47). The advantage of using electroporation over passive diffusion routes is timesaving.

In this electroporation study, delivering the antigenic peptide MYR to mice by electroporation resulted in mucosal immunity and specific lymph node cell proliferation. The responses by intradermal injection and by electroporation delivery without adjuvant were not as high as that by intraperitoneal injection with adjuvants. However, the larger size diphtheria toxoid vaccine was not transported across the skin in amounts needed to evoke high immune response. The antigenic OVA-peptide was delivered transcutaneously to mice using the lipid/surfactant formulation and with postpulse electroosmosis (48). An electroporation-transportable oligonucleotide with CpG motif (mol. wt. ~6kDa) was codelivered as an adjuvant. Antigen-specific CTL response to the vaccine delivered by needle-free electroporation/electroosmosis was equivalent to that delivered by intradermal injection with Freund's Complete Adjuvant, as determined by production of interferon-γ in enzyme-linked immunospot (ELISPOT) assay.

2.2.4. Alternative Transdermal Delivery Methods

Recently, several alternative physical methods have emerged to transiently break the SC barrier. The projectile methods use propelled microparticles or micro fluid jet to penetrate the skin barrier (49). Microneedle arrays (50) are inserted through the skin to create pores. "Microporation" and "RF-microchannels" create arrays of pores in the skin by heat and RF ablation (51). Also, ultrasound has been employed to disrupt the skin barrier (52, 53). All these methods have their own advantages and drawbacks. For instance, the particle gun method requires bonding the substance to the carrier particles and releasing them at the target. Microneedle arrays are fragile and nonreusable in general, although breakable and biodegradable needles have been developed. Like electroporation, all these alternative methods involve some temporary skin damage and discomfort to the patient but still are relatively noninvasive when compared with traditional needle injection. Recently, a painless electroporation electrode pad was developed for transdermal delivery (29). A more general and detailed discussion of alternative physical delivery methods is given in the next section.

3. How Electroporation Measures up to Other Physical Delivery Methods

In order for foreign DNA to enter a cell (in vitro) or cells in a tissue (in vivo), an entrance through the cell plasma membrane must first be made. Various ingenious artificial physical and physicochemical methods have been developed to accomplish this step without seriously damaging the target cells. The mechanism, procedures, and efficiency of electroporation have been reviewed extensively in the previous chapters of this book. I shall introduce other physical and physicochemical methods in this chapter for comparison, to show the standing of electroporation among other physical approaches. The comparison is presented in Table 6.1.

Table 6.1 Comparison of in vivo applications of various physical and physicochemical methods to for gene and drug delivery

Method	In vivo tissue penetration	Nucleus penetration	Cell-type specific	Invasiveness pain	Tissue damage	Special equipment
Carrier particles	Local, systemic	No	Endocytosis-dependent	Injection	Mildly toxic	Particle preparation
Hydro-pressure	Surgically accessible	No	See carrier	See carrier	Temporary	Surgery
Magneto-fection	Local	No	See carrier	See carrier	Not by magnet	Magnet
Magnetic nanotubes spearing	Not yet applied	Yes	No	Un-known	Minor	Magnet, nanotubes preparation
Ballistic Particles	mm	Yes	No	Mild	Minor	Gene gun, particle preparation
Microneedle arrays	Skin only	NA	No	Mild	Minor	Microneedle fabrication
Sonoporation +/- ± microbubbles	mm to cm	No	Depends on membrane mechanics	Mild or none	Some degree	Sonic generator, micro-bubble
Laser perforation	mm	Yes	No	Mild, or none	Minor	Laser and fiberoptics
Electroporation ± +/- enhancers	Determined by electrode configure-tion, ~cm	Nucleo-poration only	Depends on cell electrical properties	Mild to severe	Certain degrees	Pulse generator

NA = not applicable

3.1. Micro- and Nanoparticle Carriers

DNA molecules, with negative electric charges, do not normally adhere to negatively charged cell surfaces. DNA will, however, react electrostatically with positively charged molecules or nanoparticles (carriers) to form condensed complexes that can be made to carry neutral or positive charges. These complexes will then be made to adhere to the targeted cell surface to trigger natural endocytosis processes. The commonly used carriers include cationic lipids, cationic polymers, and, in some instances, cationic peptides. Most of these carriers have been well characterized and are extremely effective in vitro. For in vivo applications, these carriers are not immunogenic but are still subject to the retention of the reticulo-endothelial system. Poly(ethylene glycol) has been grafted to these lipids and polymers to render the complexes "stealth" with a hydration shell. Binding ligands have been conjugated to the complexes for targeting specific cells and tissues. These complexes have been used in vivo with efficiency, in some cases, approaching that of viral carriers in terms of the extent of transfection but not in terms of DNA used.

1. *Cationic lipids.* A large variety of cationic lipids, mostly amino derivatives of diacylglycerol and cholesterol, are now commercially available. These cationic lipids are amphiphilic, usually in the form of liposomes, and can be mixed with DNA in given ratios to form liposome-DNA complexes (lipoplexes) of an optimal size and charge (54). For in vitro transfection, micrometer-sized complexes are more efficient, because most cultured cells can be induced to initiate endocytosis of attached micrometer-sized particles (55). For in vivo applications, smaller complex size is usually more efficient (56). Liposomes containing membrane destabilizing components help to release the plasmid DNA (pDNA) from endosomes (57), and pH and oxidation-susceptible components have also been incorporated to facilitate DNA release in the cytosol (56).
2. *Cationic polymers.* Cationic polymers, such as polyethylenimine (PEI), have been used extensively for forming complexes (polyplexes) with pDNA. Like lipoplexes, the cellular uptake of polyplexes is also mainly by endocytosis. The different forms (linear and branched) of PEI have different DNA-reacting characteristics and carrier efficacies for in vitro and in vivo deliveries (58). Recently, biodegradable polymers have been used for in vivo delivery (59, 60). Thermosensitive polymers were used to load and release DNA at different temperatures to effect in situ and delayed delivery (61). Peptides with DNA binding, receptor binding, and membrane disrupting domains have been employed (62).

3.2. Hydro Pressure

Increasing hydrodynamic pressure of an organ by rapidly injecting a large volume of fluid containing naked DNA, lipoplexes, or polyplexes into the circulation and/or occluding the local blood flow was found to increase the expression level (63). It is believed that the sudden increase of hydro pressure in the locale would mechanically

stretch the endothelial lining, thereby facilitating the infusion of DNA and carriers into surrounding tissues and prolonging their dwelling time.

3.3. Magnetic Enhancers

The low efficiency of particle carriers demands a large quantity to be injected in vivo. A method has been developed to concentrate DNA to targeted cells and tissues by complexing them with magnetic particles and, using a magnetic gradient, to confine them locally (64). Ion oxide or other ferromagnetic and paramagnetic particles are coated with polyelectrolytes such as DEAE Dextran or PEI. These coated particles bind to naked DNA, lipoplexes, and polyplexes in salt solution. A static magnet is then used to confine the complexes around the targeted site to increase the local concentration. Uptake is again by endocytosis. The use of a local magnetic gradient increases the transfection efficiency by 2 orders of magnitude (64). It also enables in vivo magnetic field guided targeting. The technique is called magnetofection. It has been applied to deliver antisense oligonucleotides (ODN) and genes in vitro and in vivo to epithelial and endothelial cells (65, 66).

Recently, a new technique using magnetized carbon nanotubes has been reported (67), in which the tips of carbon nanotubes enclose ferromagnetic nickel particles. DNA molecules were conjugated to the derivatized carbon nanotubes. In a magnetic field gradient, the liquid-suspended, DNA-carrying nanotubes were aligned and accelerated, stabbing through the targeted cell membrane to deliver the conjugated DNA to the cytosol and nuclei of targeted cells. The efficiency was higher with the combination of both static and mobile magnetic field gradients. Since the spearing of nanotubes by-passes the endocytosis mechanism of DNA uptake, the transfection efficiency is much higher (10^7-fold) compared with nanotubes alone without magnetic spearing, with efficiency equivalent to those using viral vectors.

3.4. Ballistic Particles

DNA has been delivered into cells through bombardment with DNA-coated particles (using a gene gun). DNA is conjugated to gold or other metallic and non-metallic microparticles, and propelled under high gas pressure into cell layers or tissues through the plasma membrane and into the cytosol and nuclei of targeted cells (68, 69). The tissue penetration depth depends on the construction of the gun nozzle and the propelling gas pressure but is usually limited to millimeters. This method has been applied to cultured cells on plates. In vivo applications are limited to gene delivery to superficial tissues such as the skin and surface tumors. A high pressure design of the gene gun can deliver microparticles to subcutaneous tissues, muscles, and tumors (70), as well as to surgically accessible organs.

3.5. Ultrasound with Microbubble Enhancer

Applying ultrasonic energy to enhance the cellular uptake of naked DNA or lipoplexes has been in practice for some time (71); however, the low-intensity sonic energy was not sufficient to permeabilize the plasma membrane to effect a high-efficiency gene transfer. As the ultrasonic intensity increases, cavitation of microbubbles of the gaseous inclusion in the insoniated medium generates minute shock waves. The shear force of the shock wave with its high-velocity fluid jet is sufficient to permeabilize the plasma membrane in close proximity to the cavitation event. Microbubbles (such as Optison) used in ultrasound imaging enhancement can be used to increase the frequency of the permeabilization event (sonoporation). By this approach, the transfection efficiency was increased 300-fold over that using naked DNA alone (72). If polyplex reagents are used, there was another order of magnitude increase in efficiency, probably due to the adhesion of DNA with carriers to the plasma membrane during sonoporation. Delivery to muscles and skin using sonoporation has been reported (73, 74).

3.6. Focused Laser Beam Perforation

Forming small holes in the plasma membrane of cultured cells with a focused laser beam to facilitate DNA intake was reported two decades ago, with high cell viability and transfection efficiency (75). This method has been improved recently using femtosecond infrared lasers (76). The use of infrared lasers also increased beam penetration depth in muscle tissues (77). The femtosecond laser beam apparently caused little cellular damage. By this method, the transfection efficiency with injected naked DNA is comparable to that by electroporation.

4. Advantages and Disadvantages of Electroporation

The advantage common to all physical methods, including electroporation (excluding methods using particle carriers), is that the transport mechanism does not depend on the uptake functions of the cell; therefore, it can be applied equally well to all cell types and at all stages of the cell cycle. The process is, by itself, biochemically and biologically nontoxic.

All physical methods, including electroporation, involve transient damage of the cell membrane in order for the gene and drug to be transported into cells. Therefore, collateral damage resulting in some cell death is inevitable. This is sometimes referred as "toxicity" of the physical treatment. All physical methods, including electroporation, require sophisticated instruments that seem less convenient than simple injection of naked DNA or with carrier particles.

Both *ex vivo* and in vivo efficiencies of physical methods, in terms of DNA used per successful transfection, are low in comparison with viral methods. The low efficiency of physical methods can be compensated by using higher doses of DNA, but foreign DNA in plasmids has toxicity limits likely to be associated with the unmethylated CpG motifs. This toxicity has been a particular problem associated with transfecting hematopoietic cells by electroporation, which leads to widespread apoptosis (78). Reducing or eliminating CpG motifs in plasmid DNA could reduce the risk (79).

Most physical methods use plasmid DNA. The integration of genes into the cell genome is poor, and the expression is predominantly transient. Long-term expression of DNA introduced by physical and physicochemical methods alone is rare. This aspect may be advantageous for short-term therapy and reducing the risk of accidental infection. It is a serious problem for long-term gene therapy. For long-term expression, hydro pressure or ultrasound has been used to deliver viral vectors (80, 81). On the other hand, self-replicating plasmids and plasmids designed to integrate into mammalian chromosomes have been made to improve the prospect of long-term expression (82, 83).

The following discussion pertains particularly to electroporation in contrast to other physical methods.

1. The applied electric pulses must exceed the cell electroporation threshold of about 1 kV/cm. For a target tissue with 1 cm between electrodes, the voltage required is in the order of 1,000 V. Even with microsecond-long pulses, the pain level can be considerable. In animal experiments, muscle twitching during pulsing is often observed. This also contributes to the unpleasant feeling during pulse application. In this sense, other physical methods are comparatively less painful. For transdermal delivery, the electroporation field needs to penetrate only the SC, the very top layer of the epidermis. The interelectrode distance can be reduced such that the pulse field does not extend beyond the epidermis to stimulate nerve endings. A painless electrode designed for transdermal delivery has been described (29).

2. Collateral damage by electroporation can be serious, compared with some other physical methods. It is known that transfection efficiency decreases once the optimal field strength is exceeded, because of excessive cell death. When electroporation field is applied through the skin using surface plate electrodes, the major potential drop develops across the skin instead of across the targeted subcutaneous tissues. Skin edema is a common consequence. A detailed study of electroporation damage to tumor and normal tissue revealed that, fortunately, normal tissues are less susceptible under the same pulse conditions (8).

3. The recovery of electroporated cells can take from minutes to hours depending on the pulse condition and cell type. During this time, while cells are open to drug and DNA uptake, they are also susceptible to colloidal-osmotic swelling and toxic substance intake. Measures have to be taken to address the swelling and resealing of target cells and tissues (84, 85).

4. Most electroporation protocols aim to permeate only the plasma membranes. Electroporation of the nucleus requires a further step, using higher threshold

voltage and shorter pulse length (nucleoporation) (86). This technique is more complicated than laser beam perforation, ballistic particle, or magnetic carbon nanotubes spearing in which the nuclei are penetrated with the same ease as plasma membranes.

5. For the electroporation field to penetrate subcutaneous tissues and organs, the tissues to be treated can be clamped between plate electrodes, placed between puncturing electrode needle arrays. A flexible clamping electrode has been designed that can be used in catheters of laparoscopes (87). Closely placed side-by-side electrodes with limited field penetration depth have been fabricated for catheter use in gene delivery (9). In these cases, electroporation can reach tissues not accessible by ballistic projectile, ultrasound, and laser beam methods.

6. Although the principle of electroporation is applicable to all cell types, its efficiency does depend on the electrical properties of the cells. Smaller cells require higher field to permeate. This is an important consideration for *ex vivo* gene delivery especially to hematopoietic cells. Cells with less conductive contents (such as adipocytes) are less susceptible. The thresholds for different cells in a heterogeneous tissue would thus vary.

5. Concluding Remarks

In summary, electroporation is one of the physicochemical methods for gene and drug delivery. It is superior in some aspects, but also has several drawbacks. In some cases, as attested by voluminous literature, it is still the preferred method. New developments are expected in the future to make this method even more versatile.

References

1. Beebe, S.J., White, J., Blackmore, P.F., Deng, Y., Somers, K., and Schoenbach, K.H. (2003) Diverse effects of nanosecond pulsed electric fields on cells and tissues. *DNA Cell Biol.* **22**, 785–796.
2. Okino, M. and Mohri, H. (1987) Effects of a high-voltage electrical impulse and an anticancer drug on in vivo growing tumors. *Jpn. J.Cancer Res.* **78**, 1319–1321.
3. Mir, L.M. and Orlowski, S. (1999) Mechanisms of electrochemotherapy. *Adv. Drug Deliv. Rev.* **35**, 107–118.
4. Heller, R., Gilbert, R., and Jaroszeski, M.J. (1999) Clinical applications of electrochemotherapy. *Adv. Drug Deliv. Rev.* **35**, 119–129.
5. Lebar, A.M., Sersa, G., Kranjc, S., Groselj, A., and Miklavcic, D. (2002) Optimisation of pulse parameters in vitro for in vivo electrochemotherapy. *Anticancer Res.* **22**, 1731–1736.
6. Gothelf, A., Mir, L.M., and Gehl, J. (2003). Electrochemotherapy: results of cancer treatment using enhanced delivery of bleomycin by electroporation. *Cancer Treat. Rev.* **29**, 371–387.
7. Kuriyama, S., Matsumoto, M., Mitoro, A., et al. (2000) Electrochemotherapy for colorectal cancer with commonly used chemotherapeutic agents in a mouse model. *Dig. Dis. Sci.* **45**, 1568–1577.

8. Rabussay, D.P., Nanda, G.S., and Goldfarb, P.M. (2002) Enhancing the effectiveness of drug-based cancer therapy by electroporation (electropermeabilization). *Technol. Cancer Res. Treat.* **1**, 71–82.

9. Dev, N.B., Preminger, T.J., Hofmann, G.A., and Dev, S.B. (1998) Sustained local delivery of heparin to the rabbit arterial wall with an electroporation catheter. *Cathet. Cardiovasc. Diagn.* **45**, 337–345.

10. Miklavcic, D., Beravs, K., Semrov, D., Cemazar, M., Demsar, F., and Sersa, G. (1998) The importance of electric field distribution for effective in vivo electroporation of tissues. *Biophys. J.* **74**, 2152–2158.

11. Cullander, C. and Guy, R.H. (1991) Sites of iontophoretic current flow into the skin: identification and characterization with the vibrating probe electrode. *J. Invest. Dermatol.* **97**, 55–64.

12. Pliquett, U., Gallo, S., Hui, S.W., Gusbeth, C., and Neumann, E. (2005) Local and transient structural changes in stratum corneum at high electric fields: contribution of Joule heating. *Bioelectrochemistry.* **67**, 37–46.

13. Prausnitz, M.R., Bose, V.G., Langer, R. and Weaver, J.C. (1993) Electroporation of mammalian skin: a mechanism to enhance transdermal drug delivery. *Proc. Natl. Acad. Sci. U.S.A.* **90**, 10504–10508.

14. Sen, A., Zhao, Y., Zhang, L., and Hui, S.W. (2002) Enhanced transdermal transport by electroporation using anionic lipids. *J. Control. Release.* **82**, 399–405.

15. Murthy, S.N., Sen, A., Zhao, Y.L., and Hui, S.W. (2003) pH influences the postpulse permeability state of skin after electroporation. *J. Control. Release.* **93**, 49–57.

16. Johnson, P.G., Hui, S.W., and Oseroff, A.R. (2002) Electrically enhanced percutaneous delivery of delta-aminolevulinic acid using electric pulses and a DC potential. *Photochem. Photobiol.* **75**, 534–540.

17. Bommannan, D.B., Tamada, J., Leung, L., and Potts, R.O. (1994) Effect of electroporation on transdermal iontophoretic delivery of luteinizing hormone releasing hormone (LHRH) in vitro. *Pharm. Res.* **11**, 1809–1814.

18. Murthy, S.N., Zhao, Y., Hui, S.W., and Sen, A. (2006) Synergistic effect of anionic lipid enhancer and electroosmosis for transcutaneous delivery of insulin. *Int. J. Pharm.* **326(1/2)**, 1–6.

19. Chen, T., Langer, R., and Weaver, J.C. (1998) Skin electroporation causes molecular transport across the stratum corneum through localized transport regions. *J. Invest. Dermatol. Symp. Proc.* **3**, 159–165.

20. Johnson, P.G., Gallo, S.A., Hui, S.W., and Oseroff, A.R. (1998) A pulsed electric field enhances cutaneous delivery of methylene blue in excised full-thickness porcine skin. *J. Invest. Dermatol.* **111**, 457–463.

21. Denet, A.R., Vanbever, R., and Preat, V. (2004) Skin electroporation for transdermal and topical delivery. *Adv. Drug. Deliv. Rev.* **56**, 659–674.

22. Jadoul, A., Bouwstra, J., and Preat, V.V. (1999) Effects of iontophoresis and electroporation on the stratum corneum. Review of the biophysical studies. *Adv. Drug Deliv. Rev.* **35**, 89–105.

23. Vanbever, R. and Preat, V.V. (1999) In vivo efficacy and safety of skin electroporation. *Adv. Drug Deliv. Rev.* **35**, 77–88.

24. Vanbever, R., Lecouturier, N., and Preat, V. (1994) Transdermal delivery of metoprolol by electroporation. *Pharm. Res.* **11**, 1657–1662.

25. Vanbever, R., Langers, G., Montmayeur, S., and Preat, V. (1998) Transdermal delivery of fentanyl: rapid onset of analgesia using skin electroporation. *J. Control. Release.* **50**, 225–235.

26. Zhang, L., Lerner, S., Rustrum, W.V., and Hofmann, G.A. (1999) Electroporation-mediated topical delivery of vitamin C for cosmetic applications. *Bioelectrochem. Bioenerg.* **48**, 453–461.

27. Tamosiunas, M., Bagdonas, S., Didziapetriene, J., and Rotomskis, R. (2005) Electroporation of transplantable tumour for the enhanced accumulation of photosensitizers. *J. Photochem. Photobiol. B.* **81**, 67–75.

28. Wong, T.W., Zhao, Y.L., Sen, A., and Hui, S.W. (2005) Pilot study of topical delivery of methotrexate by electroporation. *Br. J. Dermatol.* **152**, 524–530.

29. Wong, T.W., Chen, C.H., Huang, C.C., Lin, C.D., and Hui, S.W. (2006) Painless electroporation with a new needle-free microelectrode array to enhance transdermal drug delivery. *J. Control. Release.* **110**, 557–565.

30. Wang, S., Kara, M., and Krishnan, T.R. (1998) Transdermal delivery of cyclosporin-A using electroporation. *J. Control. Release.* **50**, 61–70.

31. Huang, J.F., Sung, K.C., Hu, O.Y., Wang, J.J., Lin, Y.H., and Fang, J.Y. (2005) The effects of electrically assisted methods on transdermal delivery of nalbuphine benzoate and sebacoyl dinalbuphine ester from solutions and hydrogels. *Int. J. Pharm.* **297**, 162–171.

32. Murthy, S.N., Zhao, Y.L., Hui, S.W., and Sen, A. (2005) Electroporation and transcutaneous extraction (ETE) for pharmacokinetic studies of drugs. *J. Control. Release.* **105**, 132–141.

33. Murthy, S.N., Zhao, Y., Marlan, K., Hui, S.W., Kazim, L., and Sen, A. (2006) Lipid and electroosmosis enhanced transdermal delivery of insulin by electroporation. *J. Pharm. Sci.* **95**, 2041–2050.

34. Gallo, S.A., Sen, A., Hensen, M.L., and Hui, S.W. (1999) Time-dependent ultrastructural changes to porcine stratum corneum following an electric pulse. *Biophys. J.* **76**, 2824–2832.

35. Lombry, C., Dujardin, N., and Preat, V. (2000) Transdermal delivery of macromolecules using skin electroporation. *Pharm. Res.* **17**, 32–37.

36. Zewert, T.E., Pliquett, U.F., Vanbever, R., Langer, R., and Weaver, J.C. (1999) Creation of transdermal pathways for macromolecule transport by skin electroporation and a low toxicity, pathway-enlarging molecule. *Bioelectrochem. Bioenerg.* **49**, 11–20.

37. Weaver, J.C., Vanbever, R., Vaughan, T.E., and Prausnitz, M.R. (1997) Heparin alters transdermal transport associated with electroporation. *Biochem. Biophys. Res. Commun.* **234**, 637–640.

38. Vanbever, R., Prausnitz, M.R., and Preat, V. (1997) Macromolecules as novel transdermal transport enhancers for skin electroporation. *Pharm. Res.* **14**, 638–644.

39. Murthy, S.N., Sen, A., and Hui, S.W. (2004) Surfactant-enhanced transdermal delivery by electroporation. *J. Control. Release.* **98**, 307–315.

40. Sen, A., Zhao, Y.L., and Hui, S.W. (2002) Saturated anionic phospholipids enhance transdermal transport by electroporation. *Biophys. J.* **83**, 2064–2073.

41. Zewert, T.E., Pliquett, U.F., Langer, R., and Weaver, J.C. (1995) Transdermal transport of DNA antisense oligonucleotides by electroporation. *Biochem. Biophys. Res. Commun.* **212**, 286–292.

42. Medi, B.M. and Singh, J. (2003) Electronically facilitated transdermal delivery of human parathyroid hormone (1–34). *Int. J. Pharm.* **263**, 25–33.

43. Chang, S.L., Hofmann, G.A., Zhang, L., Deftos, L.J., and Banga, A.K. (2000) The effect of electroporation on iontophoretic transdermal delivery of calcium regulating hormones. *J. Control. Release.* **66**, 127–133.

44. Sen, A., Daly, M.E., and Hui, S.W. (2002) Transdermal insulin delivery using lipid enhanced electroporation. *Biochim. Biophys. Acta.* **1564**, 5–8.

45. Klimuk, S.K., Najar, H.M., Semple, S.C., Aslanian, S., and Dutz, J.P. (2004) Epicutaneous application of CpG oligodeoxynucleotides with peptide or protein antigen promotes the generation of CTL. *J. Invest. Dermatol.* **122**, 1042–1049.

46. Guerena-Burgueno, F., Hall, E.R., Taylor, D.N., et al. (2002) Safety and immunogenicity of a prototype enterotoxigenic *Escherichia coli* vaccine administered transcutaneously. *Infect. Immun.* **70**, 1874–1880.

47. Misra, A., Ganga, S., and Upadhyay, P. (1999) Needle-free, non-adjuvanted skin immunization by electroporation-enhanced transdermal delivery of diphtheria toxoid and a candidate peptide vaccine against hepatitis B virus. *Vaccine.* **18**, 517–523.

48. Zhao, Y.L., Murthy, S.N., Manjili, M.H., Guan, L.J., Sen, A., and Hui, S.W. (2006) Induction of cytotoxic T-lymphocytes by electroporation-enhanced needle-free skin immunization. *Vaccine.* **24**, 1282–1290.

49. Babiuk, S., Baca-Estrada, M.E., Foldvari, M., et al. (2003) Needle-free topical electroporation improves gene expression from plasmids administered in porcine skin. *Mol. Ther.* **8**, 992–998.

50. Mikszta, J.A., Alarcon, J.B., Brittingham, J.M., Sutter, D.E., Pettis, R.J., and Harvey, N.G. (2002) Improved genetic immunization via micromechanical disruption of skin-barrier function and targeted epidermal delivery. *Nat. Med.* **8**, 415–419.

51. Bramson, J., Dayball, K., Evelegh, C., Wan, Y.H., Page, D., and Smith, A. (2003) Enabling topical immunization via microporation: a novel method for pain-free and needle-free delivery of adenovirus-based vaccines. *Gene Ther.* **10**, 251–260.

52. Mitragotri, S. and Kost, J. (2004) Low-frequency sonophoresis: a review. *Adv. Drug Deliv. Rev.* **56**, 589–601.

53. Doukas, A.G. and Kollias, N. (2004) Transdermal drug delivery with a pressure wave. *Adv. Drug Deliv. Rev.* **56**, 559–579.

54. Xu, Y., Hui, S.W., Frederik, P., and Szoka, F.C., Jr. (1999) Physicochemical characterization and purification of cationic lipoplexes. *Biophys. J.* **77**, 341–353.

55. Hui, S.W., Langner, M., Zhao, Y.L., Ross, P., Hurley, E., and Chan, K. (1996) The role of helper lipids in cationic liposome-mediated gene transfer. *Biophys. J.* **71**, 590–599.

56. Dauty, E., Behr, J.P., and Remy, J.S. (2002) Development of plasmid and oligonucleotide nanometric particles. *Gene Ther.* **9**, 743–748.

57. Xu, Y. and Szoka, F.C., Jr. (1996) Mechanism of DNA release from cationic liposome/DNA complexes used in cell transfection. *Biochemistry.* **35**, 5616–5623.

58. Kircheis, R., Wightman, L., and Wagner, E. (2001) Design and gene delivery activity of modified polyethylenimines. *Adv. Drug Deliv. Rev.* **53**, 341–358.

59. Bikram, M., Lee, M., Chang, C.W., Janat-Amsbury, M.M., Kern, S.E., and Kim, S.W. (2005) Long-circulating DNA-complexed biodegradable multiblock copolymers for gene delivery: degradation profiles and evidence of dysopsonization. *J. Control. Release.* **103**, 221–233.

60. Zhao, Z., Wang, J., Mao, H.Q., and Leong, K.W. (2003) Polyphosphoesters in drug and gene delivery. *Adv. Drug Deliv. Rev.* **55**, 483–499.

61. Jeong, B., Kim, S.W., and Bae, Y.H. (2002) Thermosensitive sol-gel reversible hydrogels. *Adv. Drug Deliv. Rev.* **54**, 37–51.

62. Rittner, K., Benavente, A., Bompard-Sorlet, A., et al. (2002) New basic membrane-destabilizing peptides for plasmid-based gene delivery in vitro and in vivo. *Mol. Ther.* **5**, 104–114.

63. Liu, F. and Huang, L. (2001) Improving plasmid DNA-mediated liver gene transfer by prolonging its retention in the hepatic vasculature. *J. Gene Med.* **3**, 569–576.

64. Plank, C., Anton, M., Rudolph, C., Rosenecker, J., and Krotz, F. (2003) Enhancing and targeting nucleic acid delivery by magnetic force. *Expert. Opin. Biol. Ther.* **3**, 745–758.

65. Krotz, F., de Wit, C., Sohn, H.Y., Zahler, S., Gloe, T., Pohl, U., and Plank, C. (2003) Magnetofection–a highly efficient tool for antisense oligonucleotide delivery in vitro and in vivo. *Mol. Ther.* **7**, 700–710.

66. Gersting, S.W., Schillinger, U., Lausier, J., et al. (2004) Gene delivery to respiratory epithelial cells by magnetofection. *J. Gene Med.* **6**, 913–922.

67. Cai, D., Mataraza, J.M., Qin, Z.H., et al. (2005) Highly efficient molecular delivery into mammalian cells using carbon nanotube spearing. *Nat. Methods.* **2**, 449–454.

68. Mahvi, D.M., Sheehy, M.J., and Yang, N.S. (1997) DNA cancer vaccines: a gene gun approach. *Immunol. Cell Biol.* **75**, 456–460.

69. Kitagawa, T., Iwazawa, T., Robbins, P.D., Lotze, M.T., and Tahara, H. (2003) Advantages and limitations of particle-mediated transfection (gene gun) in cancer immuno-gene therapy using IL-10, IL-12 or B7-1 in murine tumor models. *J. Gene Med.* **5**, 958–965.

70. Dileo, J., Miller, T.E., Jr., Chesnoy, S., and Huang, L. (2003) Gene transfer to subdermal tissues via a new gene gun design. *Hum. Gene Ther.* **14**, 79–87.

71. Kim, H.J., Greenleaf, J.F., Kinnick, R.R., Bronk, J.T., and Bolander, M.E. (1996) Ultrasound-mediated transfection of mammalian cells. *Hum. Gene Ther.* **7**, 1339–1346.

72. Lawrie, A., Brisken, A.F., Francis, S.E., Cumberland, D.C., Crossman, D.C., and Newman, C.M. (2000) Microbubble-enhanced ultrasound for vascular gene delivery. *Gene Ther.* **7**, 2023–2027.

73. Liang, H.D., Lu, Q.L., Xue, S.A., et al. (2004) Optimisation of ultrasound-mediated gene transfer (sonoporation) in skeletal muscle cells. *Ultrasound Med. Biol.* **30**, 1523–1529.

74. Yang, L., Shirakata, Y., Tamai, K., et al. (2005) Microbubble-enhanced ultrasound for gene transfer into living skin equivalents. *J. Dermatol. Sci.* **40**, 105–114.
75. Kurata, S., Tsukakoshi, M., Kasuya, T., and Ikawa, Y. (1986) The laser method for efficient introduction of foreign DNA into cultured cells. *Exp. Cell Res.* **162**, 372–378.
76. Tirlapur, U.K. and Konig, K. (2002) Targeted transfection by femtosecond laser. *Nature.* **418**, 290–291.
77. Zeira, E., Manevitch, A., Khatchatouriants, A., et al. (2003) Femtosecond infrared laser—an efficient and safe in vivo gene delivery system for prolonged expression. *Mol. Ther.* **8**, 342–350.
78. Li, L.H., McCarthy, P., and Hui, S.W. (2001) High-efficiency electrotransfection of human primary hematopoietic stem cells. *FASEB J.* **15**, 586–588.
79. Yew, N.S., Zhao, H., Przybylska, M., et al. (2002) CpG-depleted plasmid DNA vectors with enhanced safety and long-term gene expression in vivo. *Mol. Ther.* **5**, 731–738.
80. Chen, S., Shohet, R.V., Bekeredjian, R., Frenkel, P., and Grayburn, P.A. (2003) Optimization of ultrasound parameters for cardiac gene delivery of adenoviral or plasmid deoxyribonucleic acid by ultrasound-targeted microbubble destruction. *J. Am. Coll. Cardiol.* **42**, 301–308.
81. Ding, Z., Fach, C., Sasse, A., Godecke, A., and Schrader, J. (2004) A minimally invasive approach for efficient gene delivery to rodent hearts. *Gene Ther.* **11**, 260–265.
82. Tanaka, S., Iwai, M., Harada, Y., et al. (2000) Targeted killing of carcinoembryonic antigen (CEA)-producing cholangiocarcinoma cells by polyamidoamine dendrimer-mediated transfer of an Epstein-Barr virus (EBV)-based plasmid vector carrying the CEA promoter. *Cancer Gene Ther.* **7**, 1241–1250.
83. Keravala, A., Groth, A.C., Jarrahian, S., et al. (2006) A diversity of serine phage integrases mediate site-specific recombination in mammalian cells. *Mol. Genet. Genomics.* **276**, 135–146.
84. Abidor, I.G., Li, L.H., and Hui, S.W. (1994) Studies of cell pellets: II. Osmotic properties, electroporation and related phenomena: membrane interactions. *Biophys. J.* **67**, 427–435.
85. Lee, R.C., River, L.P., Pan, F.S., Ji, L., and Wollmann, R.L. (1992) Surfactant-induced sealing of electropermeabilized skeletal-muscle membranes in vivo. *Proc. Nat. Acad. Sci. U.S.A.* **89**, 4524–4528.
86. Lenz, P., Bacot, S.M., Frazier-Jessen, M.R., and Feldman, G.M. (2003) Nucleoporation of dendritic cells: efficient gene transfer by electroporation into human monocyte-derived dendritic cells. *FEBS Lett.* **538**, 149–154.
87. Soden, D.M., Larkin, J.O., Collins, C.G., et al. (2006) Successful application of targeted electrochemotherapy using novel flexible electrodes and low dose bleomycin to solid tumours. *Cancer Lett.* **232**, 300–310.

Chapter 7
Nanoelectroporation: A First Look

Raji Sundararajan

Abstract As the medical field moves from treatment of diseases with drugs to treatment with genes, safe and efficient gene delivery systems are needed to make this transition. One such safe, nonviral, and efficient gene delivery system is electroporation (electrogenetherapy). Exciting discoveries by using electroporation could make this technique applicable to drug and vaccine delivery in addition to gene delivery. Typically, milli- and microsecond pulses have been used for electroporation. Recently, the use of nanosecond electric pulses (10–300 ns) at very high magnitudes (10–300 kV/cm) has been studied for direct DNA transfer to the nucleus in vitro. This article reviews the work done using high intensity, nanopulses, termed as nanoelectroporation (nano-EP), in electroporation gene delivery systems.

Keywords: nanosecond pulses, electroporation, membrane breakdown, calcium burst, apoptosis, caspase

1. Introduction

Many potential drugs developed to treat cancer and other diseases have found limited success because of the lack of efficient and safe delivery systems that allow the molecules to cross the cell plasma membranes. Because of the difficulty in passing through both the hydrophilic and hydrophobic portions of the lipid bilayer, the membranes are impermeable to these molecules. Electroporation, or electropermeabilization (EP), is a technique that utilizes precisely controlled electric fields of short duration and high intensities to open up transient pores (aqueous pathways) through semipermeable membranes and tissues, allowing targeted delivery of therapeutic materials, including drugs, antibodies, and genes (DNA) (1–3). EP offers a 100- to 1,000-fold improved therapeutic benefit, compared with using a drug alone, and is gaining acceptance as a viable technique to enhance the efficacy of drug delivery for cancer treatment, gene transfer, and similar applications in biology, biotechnology, and medicine (4–6). Electroporation-mediated

chemotherapy, or electrochemotherapy (ECT), is a viable alternative to conventional cancer treatments, as evidenced by successful phase I and phase II clinical trials for various cancers, such as skin cancer, lymphomas, squamous-cell carcinomas, testicular carcinomas, and malignant pleural effusions (1, 5, 6). There are isolated reports of cases of vocal chord and oral cancer, pancreatic cancer, brain tumors, and hepatic metastases that have been successfully treated using this technique. EP is effective for treating breast cancer cases, when surgery, radiotherapy, or chemotherapy failed (7).

EP is a physical phenomenon resulting from the interaction of plasma membrane with an electric field, and it can be applied to all histological types of solid tumors. EP increases the permeability of the cell membranes by dielectric breakdown, allowing temporary access to the cell interior (reversible electroporation). The cell membrane is resealed once the application of pulses is stopped. It is generally accepted that membrane breakdown occurs if the induced membrane voltage reaches a value of about 1 V at room temperature. EP has gained common acceptance in many areas of biotechnology and medicine because it is more controllable, reproducible, and efficient than alternative chemical and viral techniques.

Typically, 1,300-V/cm, 100-μs pulses are used for skin cancer trials (1, 5). Most studies have used 6 or 8 pulses with an interval of 1 s (1 Hz) (1, 2, 5). For gene therapy, lower intensity and longer pulses, such as 125 V/cm with 25-ms pulses (8) and 200 V/cm with 10-ms pulses (9) have been found to be effective. Lately, ultrashort pulses with nanosecond (ns) durations and extremely high electric fields in a range between 10 and 300 kV/cm have been used for electroporation (10–14). This type of electroporation is referred to as nanoelectroporation (nano-EP). Under these conditions, the plasma membrane acts as a short circuit, allowing the pulse to directly manipulate the internal organelles of the cell (15). Hallmarks of apoptosis, including phosphatidylserine (PS) translocation, and caspase activation, have been observed with high intensity, nanopulse electroporation. With these effects, it is expected that ultrashort pulses of submicrosecond or nanosecond durations could be used for additional applications in biotechnology and medicine, specifically for cancer or gene therapy. The use of high-intensity DC pulses with durations shorter than the time constant of membrane charging (1 μs) can offer valuable insight into the biophysical mechanisms involved in the electropermeabilization of the cells (10–12). However, because of the complexity of the design and the high cost involved in constructing a nanoelectroporation system, there are no commercially available nanopulsers. Technical advancements in the field of pulsed power equipment (16, 17) have enabled a few researchers to conduct nano-EP studies, in which the efficiency of ultrashort electric pulses for the reversible permeabilization of mammalian cells has been explored. To this end, nanopulse generators, delivering unipolar square pulses with electric field strengths up to 300 kV/cm and as small as 10 ns, were built and tested. Various cells, such as human Jurkat T lymphoblasts, HL-60, and mouse myeloma cells were studied using these pulsers.

2. Frequency Response of Cells

Charges accumulate at the plasma membrane of a biological cell when a voltage pulse is applied, and the induced potential across the membrane is increased (15, 18, 19). Depending upon the magnitude, duration, and the frequency of the voltage applied, the cell membranes can temporarily break down, creating pores that can eventually be resealed. The membrane potential V_m is given as the following equation (15):

$$V_m = 1.5 \times E \times R \times \cos(\delta)/[1 + R \times G_m (\rho_i + 0.5 \times \rho_a) \times (1/(1 + j\omega T))] \qquad (1)$$

where ρ_i and ρ_a are resistivities inside and outside the cell, respectively; R is the cell radius; δ is the angle between the electric field E and the radius vector; ω is the radian frequency, which is equal to $2\pi f$, where f is the frequency; T is the time constant; and G_m is the membrane conductance. Assuming that the cytoplasm and the external medium are purely resistive, the influence of their dielectric constants can be neglected. If the conductance is also neglected, the membrane potential is given as the following equation:

$$V_m = 1.5 \times E \times R \times \cos(\delta)/(1 + j\omega T); \text{ magnitude,}$$
$$V_m = 1.5 \times E \times R \times \cos(\delta)/\sqrt{(1 + \omega T)^2} \qquad (2)$$

Equation 2 can be considered as the general expression for the induced plasma membrane potential at all frequencies. When $\omega T \ll 1$ or $f = 0$, as in the DC case, Eq. 2 becomes the following well-known frequency-independent expression:

$$V_m = 1.5 \times E \times R \times \cos(\delta) \qquad (3)$$

At very low frequencies, the internal field strength is zero if the membrane conductance is neglected, and the cell interior is shielded by the capacitive membrane. Thus, the membrane receives the total potential applied to the cell, thereby enhancing the external field strength to the membrane field strength level by several folds. On the other hand, at very high frequencies, the membrane capacitance is short-circuited and the total potential applied to the cell by the external field is available to the cytoplasm (15) and the internal organelles. Thus, the nano-EP effect can "reach inside" the cells, disrupting internal membranes and causing calcium release, apoptosis, and other events normally associated with internal cellbiological signaling (11). A simulation study of the cell behavior at various frequencies correlates very well with this concept (20). At low frequencies, the outer plasma membrane with large capacitance is affected, and, at high frequencies, the outer membrane acts like a short and the applied voltage appears mainly across the interior of the cell. Thus, at low frequencies, the potential across the interior is very small. As the frequency increases, the input voltage is applied across the internal organelles of the cell. Figure 7.1 illustrates this concept. This theory also correlates very well with the recent nanosecond pulse results of Beebe et al. (11). Using 10–300-ns pulses and tens of kV/cm fields, they observed that, as the

(a)

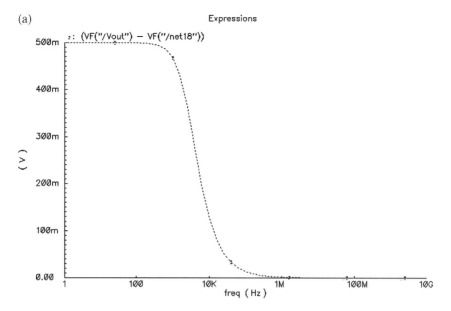

(b)

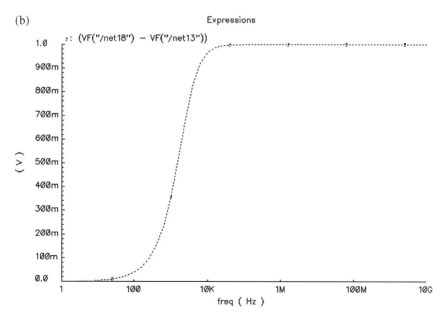

Fig. 7.1 Frequency response across the outer cell membrane (**a**) and across the interior of the cell (**b**). At low frequencies, the outer plasma membrane has more induced voltage. At high-frequency electric field, the outer membrane is shorted and the input voltage is applied across the inner membrane. At low frequencies there is no voltage across the nucleus. As the frequency is increased to megahertz range or nanosecond pulse range, we can gain control over the nucleus (20)

pulse durations decrease, the effect on the external plasma membrane decreases and the effects on intracellular signal transduction mechanisms increase. Thus, the interaction of electric fields and biological cells also depends upon the frequency of the voltage applied. This is due to the frequency dependency (dielectric dispersion) of the relative permittivity (dielectric constant) and conductivity (or resistivity) of the tissues on the total frequency range, from a few hertz to several gigahertz (21).

3. Nanoelectroporation

High-frequency, nanosecond pulses of high intensity can reach inside cells, disrupting internal membranes and causing calcium release, apoptosis, and other events normally associated with internal cellbiological signaling (10, 11). When applied in physiological media, they can produce significant voltages across intracellular membranes, such as those of the nucleus, mitochondria, storage vacuoles, Golgi compartments, and the endoplasmic reticulum, without irreversibly porating the plasma membrane. The primary effect of nanopulses is the perturbation of internal organelles, with little or no effect on the plasma membrane (Fig. 7.2) (22). The induction of apoptosis, including PS externalization, calcium bursts, caspase activation, etc., can be considered as secondary effects. Thus, nanosecond pulses can physically manipulate intracellular structures in a precise manner with a wide range of amplitudes, durations, and patterns, extending the reach of the electric field past the cytoplasmic membrane to the internal membranes of the cell. Even when the plasma membrane is reversibly permeabilized, the extent of permeation

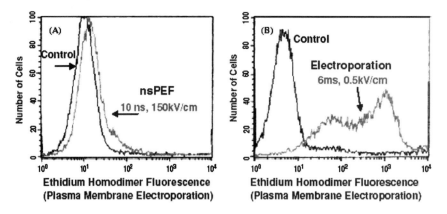

Fig. 7.2 Comparison of nano-EP and classical plasma membrane electroporation in human Jurkat cells (22). Cells were exposed to a single 10-ns pulse at 150kV/cm and a single 6-ms pulse at 500V/cm in the presence of ethidium homodimer-1 and were analyzed by flow cytometry. As expected, the 10-ns pulse did not have any significant effect on the plasma membrane, while the 6-ms pulse at an energy density of only 50% of the 10-ns pulse resulted in considerable increase in ethidium homodimer-1 fluorescence, indicating significant disruption of the plasma membrane

and the resultant physiological and morphological changes are different from those seen with conventional electroporation (11, 22).

Although nano-EP has its own specific characteristics, like classical electroporation, its effects also depend on the magnitude, duration, and number of pulses, type of cell, the conductivity of the medium, etc., as observed from the noticeable differences in the responses of the Jurkat cells and the mouse myeloma cells (13).

4. Nano-EP Effects

4.1. Apoptosis

Apoptosis, also known as programmed cell death or the selective suicide of cells, can be induced in a number of ways, such as nutrient exhaustion, growth factor deprivation, toxin accumulation (23), cross-linking of the CD95 molecule, ionizing radiation, glucocorticoids, cytotoxic T cell activity, long duration current, serum removal, and a variety of other means (24–26). Apoptosis is characterized by specific morphological and biochemical alterations of the cell, including membrane blebbing, chromatin condensation, DNA fragmentation, and the degradation of intracellular proteins such as poly(ADP-ribose) polymerase (PARP), lamin, and others (24–27). Apoptosis is also induced by electric pulses of both classical electroporation and nano-EP (10, 13, 14, 22, 24). Induction of apoptosis was reported by Zimmermann's group, where Jurkat cells suspended in R-10 culture medium were subjected to a single, exponentially decaying electric pulse of 40-μs duration with intensities ranging from 3.6 to 8.1 kV/cm. Treatment of cells with pulses above 4.5 kV/cm readily induced DNA fragmentation (24). The degree of fragmentation increased with an increase in field intensity. Similar nanopulse-induced DNA fragmentation was also observed in HL-60 cells in a voltage-dependent fashion (10, 22).

Apoptosis in human Jurkat and HL-60 cells were observed by Schoenbach's group with the application of pulses of 10–300 ns, and up to 300 kV/cm (11, 22, 28). Flow cytometry was used to characterize apoptosis induction using annexin-V binding and caspase activation. Table 7.1 shows details of apoptosis observed in a study where Jurkat cells were exposed to three 60-ns, 60-kV/cm pulses with 1-s intervals (11). Flow cytometry with ethidium homodimer and annexin-V-FITC was used for the analysis. The markers were added 5, 10, and 30 min after pulsing. Before pulsing, 89% of the cells were normal and nonapoptotic, as evidenced by their intact plasma membrane (no ethidium fluorescence), and there was no phosphatidylserine on the outer cell membrane, as evidenced by the absence of annexin-V-FITC fluorescence. Five minutes after pulsing, 80% of the cells were apoptotic as evidenced by intact cell membrane (no ethidium fluorescence) and translocation of phosphatidylserine to the outer cell membrane as evidenced by annexin-V-FITC fluorescence. The apoptosis decreased at longer durations of 10 and 30 min. However, the decrease in apoptosis was accompanied by an increase in necrosis at these durations. Similar results were also obtained

Table 7.1 Nanopulse-induced apoptosis in Jurkat cells—analysis by flow cytometry (11)

Item	% Live cells	% Apoptotic cells	% Necrotic
Control	84.5 ± 1.9	12.3 ± 2.5	1.8 ± 0.5
Pulsed after 5 min	4.2 ± 0.7	82.2 ± 1.3	12.8 ± 1.4
Pulsed after 30 min	2.1 ± 0.9	70.0 ± 2.8	26.8 ± 3.7

Cells were exposed to three 60-ns, 60-kV/cm pulses (1–2 s intervals)

for the other nanopulse conditions: 300 ns at 26 kV/cm and 10 ns at 150 kV/cm strengths. The intensity of the apoptotic markers was more at 300-ns pulses and less at 10-ns pulses, compared with that at 60-ns pulses. These results illustrate that nano-EP affects the cell interior and activates signaling pathways for apoptosis induction (28).

Apoptosis was also observed by Vernier et al. in their studies using Jurkat cells with 10-ns and 20–40-kV/cm pulses (13, 14). In their studies (13), they observed different sensitivities to the two cell types they used, the human Jurkat cells and the mouse myeloma cells, affirming the fact that electroporation (conventional or nano) parameters vary between cell types and that the parameters need to be optimized for each cell type.

4.2. Phospholipid Phosphatidylserine Externalization

A common feature of the apoptotic process is the induced translocation of the phospholipid phosphatidylserine (PS) to the outer leaflet of the plasma membrane (25). PS-dependent signaling is coupled to the final common pathway of apoptosis, i.e., the caspase-driven dismantling of the cell, thus allowing for effective phagocytosis and clearance of all corpses (29). In cellular plasma membranes, PS is normally located on the internal leaflet of the lipid bilayer. The internal PS is externalized when the cells undergo apoptosis (30). Thus, apoptosis is associated with the externalization of PS in the plasma membrane and subsequent recognition of PS by specific macrophage vectors (29).

Induction of PS externalization by nano-EP was reported by Vernier et al. (13, 31). More than 30% of cells showed evidence of PS externalization when assayed immediately after exposure to fifty 7-ns, 25-kV/cm pulses at 20 Hz. The externalization was studied using antiannexin-V-FITC and flow cytometry. Detection of external annexin V indicates the translocation of PS to the external leaflet of the membrane lipid bilayer due to the high-intensity nanopulses. Fewer than 5% of control cells (no pulse exposure) showed annexin-V-FITC binding. Flow cytometry of the cells subjected to nanosecond pulses indicated that exposure of mammalian cells to nanosecond, high-intensity electric pulses produced a rapid, dose-dependent translocation of PS to the external face of the cytoplasmic membrane (13). Recent experiments with HCT116 colon carcinoma cells have

shown that the translocation of PS can, when induced by nanosecond pulses, be reversible (32).

4.3. Caspases

Caspases are a group of cysteine proteases that play a crucial role in apoptotic pathways induced by a variety of stimuli. These enzymes act as important messengers and lead to the disassembly of the cell. Caspases are activated either by a ligand binding to a death receptor, which leads to rapid induction of initiator caspases, or by mild cytotoxic stimuli, which stimulate the release of cytochrome c and apoptosis-inducing factor from mitochondria in a protracted manner. Activated caspases cleave a variety of intracellular proteins, including major structural elements, a number of protein kinases, and the DNA repair machinery, thereby disrupting cell survival pathways. Deregulation of apoptotic pathways and caspase activity contributes to a large number of pathological conditions, including neurodegenerative disorders, autoimmune diseases, and cancer. Hence, caspases have become the primary targets for therapeutic interventions in these diseases.

Nanopulsed electric fields can induce caspase activation along with other intracellular effects, as reported by Beebe et al. and Vernier et al. (11, 13, 33). Human Jurkat cells were pulsed with 5-ns pulses at 1-s intervals with various intensities and durations (Fig. 7.3), and were analyzed by flow cytometry after a 20-min incubation with the cell permeable, irreversible, fluorescent inhibitor of active caspases, VAD-fmk-FITC. Figure 7.3 shows the presence of active caspases in intact cells (33). The longer the pulses, the greater are the observed caspase activities (13, 33). Caspase-active cells were reduced in size and had intact membranes.

4.4. Calcium Bursts

The initiation of calcium bursts due to nano-EP has also been reported by various researchers (14, 28). Human Jurkat T lymphoblast cells loaded with a calcium-sensitive fluorochrome calcium green and suspended in RPMI growth medium containing propidium iodide ($5 \mu g/mL$) showed no morphological signs of electroporation when exposed to ten 30-ns, 25-kV/cm pulses (14). However, a marked intensification of calcium green fluorescence was seen within seconds of pulse exposure, indicating an increase in the intracellular calcium ion concentration ($[Ca^{2+}]i$). The calcium green fluorescence intensification was uniform across the cell. No single or multiple release points were noticed. The calcium green fluorescence increased everywhere inside the cell

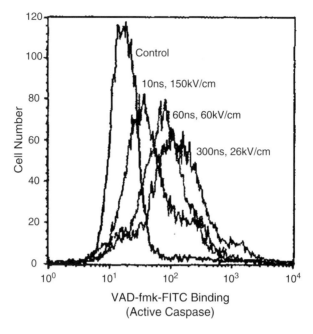

Fig. 7.3 Caspase activation in Jurkat cells due to nano-EP (33). The VAD-fmk-FITC binding indicates active caspase in intact cells

at once, within the limits of the resolution of these observations (~100 ms and 0.4 μm). It is presumed that this phenomenon must have originated from the cell and was not due to the propagation of a calcium wave from a disturbance in the plasma membrane (14). A detailed study of the source of the calcium burst speculates that the nanopulses at high intensity could have stimulated the release of calcium from ER compartments.

Experiments with Jurkat cells loaded with calcium green marker showed that calcium release is coupled to PS externalization (28). The pulses applied to the cells were 30 ns long at an amplitude of 25 kV/cm. Although PS externalization is an apoptosis marker, it does not necessarily lead to cell death. Additional results were obtained by Beebe et al. (22). In this study, nanopulsed HL-60 cells released significant amounts of intracellular calcium, with 8- to 10-fold increase in magnitude after exposure to a single 60-ns, 15-kV/cm pulse. This release was identified using uridine triphosphate (UTP). The threshold for the release of Ca^{2+} was field-intensity-dependent. There was a 4-fold increase with 4-kV/cm nanopulse and a 15-fold increase with a 15-kV/cm pulse, compared with base level in the presence of extracellular Ca^{2+}. Again, it was speculated that nano-EP induced calcium mobilization from intracellular reserves, followed by possibly capacitive-induced increases in calcium from the extracellular environment through the plasma membrane.

4.5. Cytochrome C Release

To verify whether the nanopulse-induced apoptosis is based on a mitochondrial-dependent mechanism, the presence of cytochrome c in the cytoplasm was studied by Beebe et al. (11). Here, Jurkat cells were exposed to 3-ns pulses with 1-s intervals. The various pulses applied were as follows: 10 ns, 150 kV/cm; 60 ns, 60 kV/cm; and 300 ns, 26 kV/cm. The energy density was approximately the same (1.7 J/cc/pulse) in all these cases. Cell extracts were separated into cytosol and mitochondrial fractions and analyzed for cytochrome c by immunoblot analysis. The results indicated the presence of cytochrome c in the nanopulsed cells (Fig. 7.4) (11). Cytochrome c release from the mitochondria into the cytoplasm suggests that the apoptosis induced by the nano-EP is indeed dependent on mitochondria. However, it is not clear whether mitochondria are primary or secondary targets. More studies in this line would help clarify this mechanism.

4.6. Sodium Entry into the Cell

Since Ca^{2+} bursts have been observed with nano-EP, it is of practical interest to verify whether a similar phenomenon also applies to sodium, i.e., diffusion of sodium through the nanopores. For this purpose, sodium green-loaded cells were exposed to both micro-second and nanosecond pulses of 5 kV/cm for 30 µs and 25 kV/cm for 30 ns, respectively (14). Electroporation due to conventional microsecond pulses caused a measurable sodium influx, while no noticeable sodium entry was observed with nano-EP (14).

5. Nanopulsers

High-voltage, nanosecond pulse generators are an integral part of the nano-EP research. Because of very short time durations integrated with the high voltages required for these applications, it is very challenging to design and build a success-

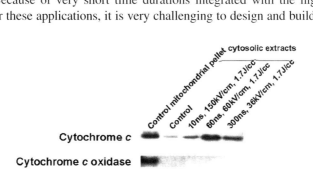

Fig. 7.4 Immunoblot analysis of Jurkat cells for cytochrome c release into the cytoplasm due to nano-EP. Cells were exposed to three 10-ns, 150-kV/cm; 60-ns, 60-kV/cm; and 300-ns, 36-kV/cm pulses at 1-s intervals (11)

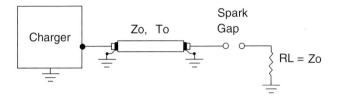

Fig. 7.5 Pulse-forming network (PFN) concept used for nanopulser. Shown here is a single-ended transmission PFN with characteristic impedance Z_0 and transmission delay T_0 (34)

fully functioning nanosecond pulser with minimum oscillations and reflections in the waveform. Advancement in pulsed power technology (16, 17) has enabled the development of a number of nanopulsers using the Blumlein pulse-forming network (PFN) concept (Fig. 7.5) (34). Various first- and second-generation nanosecond pulse generators for biological applications were designed and developed primarily by three research groups (11, 12, 28, 31, 34–37). The various nano-EP pulsers developed include the following models: 10–300 ns, up to 300-kV/cm, Blumlein pulser; 4-kV, 5-ns, 10-Ω load, inductive adder charged cable pulser; 10-kV, 5-ns, 20-Ω load, water Blumlein minipulser; 4-kV, 5-ns, 10-Ω load, ceramic Blumlein minipulser; 400-V, 5-ns, MOSFET-based micropulser for microscope slide load; 10-kV, 10-ns, 1-kHz, nanosecond flashlamp for real-time microscopy of cell electroperturbation; 15-kV, 11-ns pulser; and 15-kV, and 45-, 65-, and 95-ns pulser. The electric field strength is calculated as $E = V/d$, where V is the voltage in kilovolts and d is the distance in centimeters between the electrodes. Another simple, but efficient MOSFET-based nanosecond pulse generator was designed and developed at Arizona State University (38). A simulation study of the influence of various parameters, such as the effect of load on nanosecond rise and fall times, etc., was also carried out (39).

5.1. 60- and 300-ns Pulsers

Two nanopulsers, 60 ns, 60 kV/cm and 10–300 ns, up to 300 kV/cm, were developed using a Blumlein pulse-forming network (PFN) (40). The pulse width of these generators is determined by the length (L) of their transmission lines. A pressurized spark gap was used for fast switching in these pulsers. The 60-kV/cm Blumlein PFN consisted of two 50-Ω coaxial cables to give a total impedance of 100 Ω. The energy stored in the cables is transferred into the matched load in the form of a high-voltage rectangular pulse, whose duration is determined by the cable length and the propagation speed of the electromagnetic waves in the cable dielectric. The load (cuvette with biological cells) must match the source impedance for obtaining the desired waveform. The spark gap that serves as a fast electrical switch has a closing time of less than 10 ns. This design reduces jitter in the wave to less than 5%. With a maximum voltage generation of 2 kV across 1/3-mm electrode gap, this pulser could generate a maximum electric field of (2 kV/(1/3 mm)=) 60 kV/cm. The hold-off voltage of the cable connectors will limit this maximum value to some extent. This

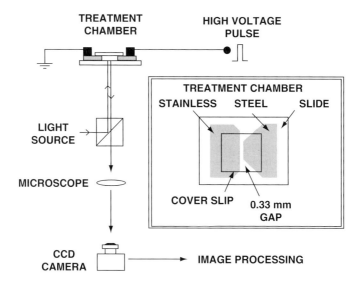

Fig. 7.6 Experimental setup of the microscope-based nanopulser system (35, 40). Cells are placed between the two stainless steel electrodes and are covered by a cover slip

nanopulser was connected to the cell chamber containing the cuvette by using a microscope slide arrangement to obtain real-time imaging. Figure 7.6 illustrates the experimental setup (35, 40). An Olympus IX70 inverted microscope was used for this purpose. The chamber comprises a 51×76-mm^2 glass microscope slide with two 0.1-mm-thick stainless steel electrodes attached to the slide surface with silicone adhesive. The two electrodes were placed 0.33 mm apart to form a 5-mm-long channel. Cell suspension (40–70 μL) was placed in this channel. A standard 0.17-mm-thick glass slip was used to cover the electrode gap. The channel was aligned with the microscope light path, and solid copper contacts attached to the high-voltage power supply were laid on the stainless steel electrodes.

5.2. MOSFET-Based 1-kV Blumlein Pulser

Figure 7.7 shows the circuit setup and schematic of another nanosecond pulser, developed by the group mentioned earlier (28). This model also utilizes a Blumlein configuration using high-frequency cables, just as in the earlier model and several of their other pulse generators. A high-speed, 1-kV MOSFET was used as a fast switch in this pulser. Using a very narrow electrode gap of 100 μm, this pulser can generate up to 100-kV/cm electric pulses. This very narrow electrode gap was used to study single cells under a microscope (Fig. 7.7a). The waveform is shown in Fig. 7.7b. Its rise time is about 3 ns and the length (duration) of the pulse is less than 10 ns. This group built a number of other pulsers with various pulse widths, types of switches, and PFNs, using a similar configuration (28, 41).

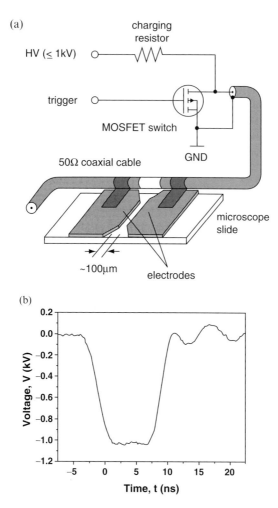

Fig. 7.7 (**a**) MOSFET + Blumlein-based, 1-kV, nanosecond pulser (28). With an electrode gap of 100 μm, it is possible to develop electric fields of up to 1 kV/100 μm, which equals 100 kV/cm in this "Microreactor." Cells in suspension are placed between the electrodes. MOSFET is used as a fast switch using Blumlein circuit as shown. (**b**) Nanosecond voltage pulse generated using the earlier-mentioned nanopulser

5.3. MOSFET-Based 400-V, 30-ns Nanopulser (Micropulser)

A number of nanopulsers were also developed by a second research group (31, 34, 37). Some of these were based on Blumlein configuration, similar to that of the first research group. A MOSFET-based, inductive pulser with a balanced, coaxial-cable pulse-forming network (PFN) and spark gap switch was designed and developed to provide trapezoidal electrical voltage pulses to cell suspensions in growth medium

contained in commercial rectangular cuvettes with 1-mm electrode separation (31). The same system, with a reconfigured PFN, was developed to deliver pulses to a custom exposure chamber built for checking bacterial spores in paper envelopes. In this case, the pulse developed was 150 ns long, with fields up to 50 kV/cm and a 7-ns rise time at a frequency of 2 kHz. The electrodes were a copper plate positioned over a copper ground plane. The ground plane was covered by a Delrin sheet to reduce corona and prevent arcing at high pulse frequencies.

For microscopic observations, cells were placed in a rectangular channel of 100 μm width, 25 μm depth, and 15 mm length. This channel was formed by two

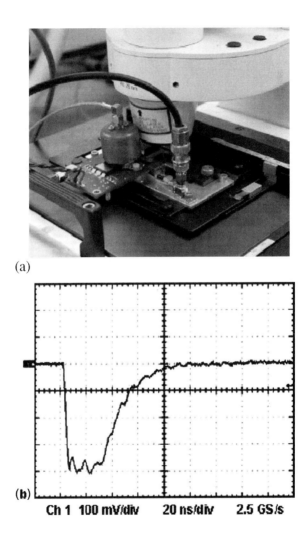

(a)

(b)

Ch 1   100 mV/div   20 ns/div   2.5 GS/s

Fig. 7.8 (a) Fast RF MOSFET-based micropulser and microscope-slide-based pulse chamber mounted on the stage of a fluorescence microscope. (b) Micropulser voltage waveform: 370 V, 30 ns (31)

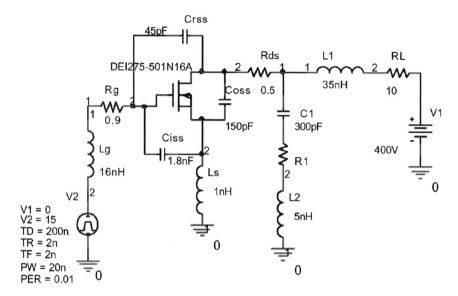

Fig. 7.9 Charging circuit for the MOSFET-based micropulser delivering nanopulses (37)

precision-cut strips of platinum foil attached to a glass microscope slide with paraffin. The platinum strips serve both as electrodes and as the walls of the exposure chamber. The fast radio frequency (RF) MOSFET micropulser was mounted on the microscope stage for delivery of pulses directly to the microchamber electrodes (Fig. 7.8a) (31). Figure 7.8b shows the waveform generated. The micropulser RF schematic is shown in Fig. 7.9 (37). Details of other pulsers developed by this group can be found in (34, 42).

5.4. 10–100-ns Pulse Generators

Two different square wave nanopulsers based on PFN were developed, again using PFN (12). The first pulser was built with a 1-m-long strip line as the PFN to generate pulses of a constant length of 11 ± 0.1 ns. Longer pulses were developed using a PFN consisting of several coaxial cables in parallel. The cables varied in length, and hence, the pulse lengths also varied. Thus, using 3-, 5-, and 9-m cables for the PFN, it was possible to generate 45-, 65-, and 95-ns pulses, respectively. The pulser used a 30-kV DC power supply for charging; therefore, the load voltage across the matching load will be 50% of the input voltage, 15 kV in this case. A spark-gap operating in the self-breaking mode was used as the fast switch to discharge the stored energy. The matched load was a high-voltage resistor connected in parallel with a 1-mm electrode gap cuvette that holds 100 μL of cell suspension.

The maximum electric field that could be generated was 150 kV/cm. The pulse voltage was monitored using a 500-MHz Tektronix digital storage oscilloscope (12).

5.5. *MOSFET-Based 400-V, 75-ns Nanopulser*

A MOSFET-based nanopulser was designed to drive a 50-Ω resistive load with 0–400-V, 75-ns pulses (38). The circuit includes three stages: (1) an integrated variable frequency and variable duration pulse width signal source, (2) an inverter/driver stage, and (3) a power buffer consisting of a passively loaded single power MOSFET. The circuit construction is physically optimized to reduce distortion and retain signal integrity while minimizing reflected power from the circuit and load. The driver used is a MC33151 (On Semiconductor) integrated 15-ns rise time, dual, inverting Schmitt trigger input with totem pole outputs. The buffer is an IXYS DE275-501N16A RF MOSFET with 500 V, 16 A with a 6-ns rise time. The MOSFET is biased into cutoff, and the circuit is designed to maximize rise time and operated common source for optimal efficiency. The voltage sources are linear designs with the signal source and driver supplies actively regulated. Since nanosecond pulsers are not commercially available, this circuit was designed and built using components that are mass manufactured for standard consumer applications in a very simple, low cost circuit that will function well as a research-quality high-voltage nanopulser for electroporation applications. The circuit was implemented using exclusively off-the-shelf devices and was constructed using commercially available RF groundplane microstrip prototype components. Figure 7.10a shows the construction, and Fig. 7.10b illustrates the waveform obtained for an input voltage of 400 V. Using 0.1-cm cuvettes, it is possible to generate fields of up to 4 kV/cm magnitude with this pulser. A 1,000-V version of this nanopulser is under development.

6. Discussion and Conclusion

The prediction that electric fields of appropriate magnitudes at high frequencies should be capable of coupling electrical energy into the interior, and therefore, into the internal organelles of biological cells was made by Schwan in 1985 (43). However, it is only in the past few years with the advancements in the pulsed power technology that sufficiently short and intense pulses of electric fields have been applied to cells to evoke the interesting and potentially very useful effects that are possible by selective manipulation of such intracellular compartments. The latest technological advances have enabled today's researchers to design and develop nanosecond pulse generators that can operate at high voltages. Nanosecond (10–300 ns), high-intensity (10–300 kV/cm) electric pulses have been shown to induce increases in the intracellular concentration of cytosolic calcium and the translocation of phosphatidylserine to the outer layer of the plasma membrane in

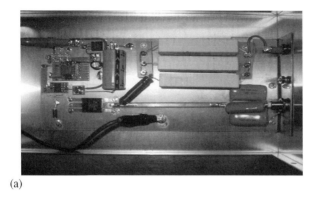

(a)

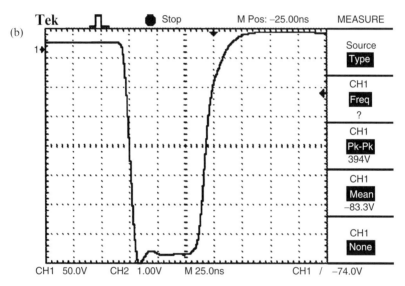

Fig. 7.10 (**a**) 400-V, 75-ns nanopulser prototype with ground plane and solder mount on surfboard PCB-transmission line construction (**b**). Output voltage pulse waveform with 50-Ω load-Input voltage applied was 408 V and output voltage obtained was 394 V. Horizontal scale is 25 ns/div and the vertical scale is 50 V/div. The pulse width is 70 ns. The fall time is about 10 ns. The waveform was recorded using a Tektronix TDS 210, 60 MHz, 1 GS/s Digital Storage Scope (38)

human Jurkat T lymphoblasts, as observed by various researchers. PS externaliza-tion in living cells exposed to nanosecond, several kilovolts per centimeter electric pulses indicates the development of nanometer-diameter aqueous pores within nanoseconds after application of high-intensity nanopulses, and electrophoretic transport of the anionic PS headgroup along the newly constructed hydrophilic pore surface commences even while pore formation is still in progress (44). PS exposure plays an important role in normal tissues, marking cells for phagocytosis and

removal. Once the basis of this mechanism is understood, activating PS using nanosecond pulses would offer intriguing possibilities for fundamental biophysical investigations and for treatment of diseases. In addition, nano-EP also illustrated that low-conductivity media (whose conductivity is lower than that of the cytosol) can greatly improve the electropermeabilization via the electrical stretching force exerted on the cell membrane as in classical electroporation (12).

Nanopulse-induced calcium bursts occurred within milliseconds and PS externalization occurred within minutes of the pulse exposure. Caspase activation and other indicators of apoptosis followed these initial apoptotic signals. Pulse-induced PS translocation was observed even in the presence of caspase inhibitors. Similar observations were also made using ultrasound pulses (25, 26, 45). Using low-intensity ultrasound pulses (25), they observed apoptosis, PS translocation, and other effects similar to nano-EP. The high-intensity, nanopulse electroporation effects differ markedly from those of classical electroporation. The electroperturbations produced by the nanopulses are due to the intracellular disturbances of physiological equilibria, while the integrity of the external plasma membrane is maintained. Thus, nano-EP offers new ways to manipulate the contents of living cells.

The results reviewed here clearly demonstrate that nanosecond pulses of high magnitudes can be used for injecting membrane-impermeable molecules into mammalian cells without significantly affecting the external plasma membrane. Nanosecond, high-intensity pulses (with low energy) extend the reach of an external electric field to the nuclear and mitochondrial membranes, to the nucleoplasm and the mitochondrial matrix, and to the membranes and contents of storage vacuoles and other intracellular compartments. Manipulation of these compartments can trigger apoptosis and other cell-controlling signals. Thus, nano-EP also has the potential for selective electromanipulation of intracellular membrane-bound organelles and for targeting malignant cells in a mixed population of cancerous and healthy cells (12, 13, 28, 35). With appropriate pulse generators, it is possible to perform in situ electropermeabilization of organelles within the cell without deteriorating the plasma membrane. This technique could be used for the manipulation of the mitochondrial genome, which is essential for the analysis of molecular pathology and for treatment of a large group of mitochondrial disorders (12).

Acknowledgments The author is very grateful to Josh Hutcheson, School of Chemical and Biomolecular Engineering of Georgia Institute of Technology, for his excellent review of the manuscript.

References

1. Heller, R., Jaroszeski, M., Grass, L., et al. (1996) Phase I/II trial for the treatment of cutaneous and subcutaneous tumors using electrochemotherapy. *Cancer.* **77**, 964–971.
2. Gehl, J. and Geertsen, P.F. (2000) Efficient palliation of haemorrhaging malignant melanoma skin metastases by electrochemotherapy. *Melanoma Res.* **10**, 1–5.
3. Mir, L.M., Bureau, M.F., Gehl, J., et al. (1999) High efficiency gene transfer into skeletal muscle mediated by electric pulses. *Proc. Natl. Acad. Sci. U.S.A.* **96**, 4262–4267.

Part II
In Vitro Targeted Gene Delivery via Electroporation

25. Ashush, H., Rozenszajn, L.A., Blass, M., et al. (2000) Apoptosis induction of human myeloid leukemic cells by ultrasound exposure. *Cancer Res.* **60**, 1014–1020.

26. Feril, L.B. and Kondo, T. (2004) Biological effects of low intensity ultrasound: the mechanism involved, and its implications on therapy and biosafety of ultrasound. *J. Radiat. Res.* **45**, 479–489.

27. Kroemer, G., Dallaporta, B., and Resche-Rigon, M. (1998) The mitochondrial death/life regulator in apoptosis and necrosis. *Annu. Rev. Physiol.* **60**, 619–642.

28. Schoenbach, K.H., Joshi, R.P., Kolb, J.F., et al. (2004) Ultrashort electrical pulses open a new gateway into biological cells. *Proc. IEEE.* **92**, 1122–1137.

29. Tyurina, Y.Y., Serinkan, F.B., Tyurint, V.A., et al. (2004) Lipid antioxidant, etoposide, inhibits phosphatidylserine externalization and macrophage clearance of apoptotic cells by preventing phosphatidylserine oxidation. *J. Biol. Chem.* **279**, 6056–6064.

30. Naito, M., Nagashima, K., Mashima, T., and Tsuruo, T. (1997) Phosphatidylserine externalization is a downstream event of interleukin-1b-converting enzyme family protease activation during apoptosis. *Blood.* **89**, 2060–2066.

31. Vernier, P.T., Sun, Y., Marcu, L., Carft, C.M., and Gundersen, M.A. (2004) Nanoelectropulse-induced phosphatidylserine translocation. *Biophys. J.* **86**, 4040–4048.

32. Hall, E.H., Schoenbach, K.H., and Beebe, S.J. (2005) Nanosecond pulsed electric fields (nsPEF) induce direct electric field effects and biological effects on human colon carcinoma cells. *DNA Cell Biol.* **24**, 283–291.

33. Beebe, S.J., Fox, P.M., Rec, L.J., Somers, K., Stark, R.H., and Schoenbach, K.H. (2002) Nanosecond pulsed electric field (nsPEF) effects on cells and tissues: apoptosis induction and tumor growth inhibition. *IEEE Trans. Plasma Sci.* **30**, 286–292.

34. Behrend, M., Kuthi, A., Xianyue, G., et al. (2003) Pulse generators for pulsed electric field exposure of biological cells and tissues. *IEEE Trans. Dielectr. Electr. Insul.* **10**, 820–825.

35. Buescher, E.S. and Schoenbach, K.H. (2003) Effects of submicrosecond, high intensity, pulsed electric fields on living cells—intracellular electromanipulation. *IEEE Trans. Dielectr. Electr. Insul.* **10**, 788–795.

36. Vernier, P.T., Thu, M.M.S., Marcu, L., Craft, C.M., and Gundersen, G.A. (2004) Nanosecond electroperturbation—mammalian cell sensitivity and bacterial spore resistance. *IEEE Trans. Plasma Sci.* **32**, 1620–1625.

37. Behrend, M., Kuthi, A., and Vernier, P.T. (2002) Micropulser for real-time microscopy of cell electroperturbation. In: *Proceedings of the 25th international IEEE power modulator symposium*, Hollywood, CA, pp. 358–361.

38. Chaney, A. and Sundararajan, R. (2004) Simple MOSFET-based high voltage nanosecond pulse circuit. *IEEE Trans. Plasma Sci.* **32**, 1919–1924.

39. Sundararajan, R., Shao, J., Soundararajan, E., Gonzales, J., and Chaney, A. (2004) Performance of solid state high voltage pulsers for biological applications-a preliminary study. *IEEE Trans. Plasma Sci.* **32**, 2017–2025.

40. Hair, P.S., Schoenbach, K.H., and Buescher, E.S. (2003) Sub-microsecond, intense pulsed electric field applications to cells show specificity effects. *Bioelectrochemistry.* **61**, 65–72.

41. Deng, J., Schoenbach, K.H., Buescger, E.S., Hair, P.S., Fox, P.M., and Beebe, S.J. (2003) The effects of intense submicrosecond electrical pulses on cells. *Biophys. J.* **84**, 2709–2714.

42. Yinghua, S., Vernier, P.T., Behrend, M., Marcu, L., and Gundersen, M.A. (2005) Electrode microchamber for noninvasive perturbation of mammalian cells with nanosecond pulsed electric fields. *IEEE Trans. Nanobiosci.* **4**(5), 277–283.

43. Schwan, H.P. (1985) Dielectric properties of cells and tissues. In: Chiabrera, A., Nicolini, C., and Schwan, H.P. (ed.). *Interactions between electromagnetic fields and cells.* Plenum, New York, pp. 75–97.

44. Vernier, P.T., Ziegler, M.J., Sun, Y., Chang, W.V., Gundersen, M.A., and Tieleman, D.P. (2006) Nanopore formation and phosphatidylserine externalization in a phospholipids bilayer at high transmembrane potential. *J. Am. Chem. Soc.* **128**, 6288–6289.

45. Feril, L.B., Kondo, T., Takaya, K., and Riesz, P. (2004) Enhanced ultrasound-induced apoptosis and cell lysis by a hypotonic medium. *Int. J. Radiat. Biol.* **80**, 165–175.

4. Dev, S.B., Rabussay D.P., Widera G., and Hofmann, G.A. (2000) Medical applications of electroporation. *IEEE Trans. Plasma Sci.* **28**, 206–223.

5. Heller, R., Gilbert, R., and Jaroszeski, M.J. (2000) Clinical trials for solid tumors using electrochemotherapy. In: Jaroszeski, M., Heller, R., and Gilbert, R. (eds.). *Electrochemotherapy, electrogenetherapy, and transdermal delivery*. Humana, New Jersey, pp. 137–156.

6. Gothelf, A., Mir, L., and Gehl, J. (2003) Electrochemotherapy: results of cancer treatment using enhanced delivery of bleomycin by electroporation. *Cancer Treat. Rev.* **29**, 371–387.

7. Rodriguez-Cuevas, S., Barroso-Bravo, S., Almanza-Estrada, J., Cristobal-Marinez, L., and Gonzalez-Rodriguez, E. (2001) Electrochemotherapy in primary and metastatic skin tumors: phase II trial using intralesional bleomycin. *Arch. Med. Res.* **32**, 273–276.

8. Jaroszeski, M., Gilbert, R., Nicolau, C., and Heller, R. (2000) Delivery of genes in vivo using pulsed electric fields. In: Jaroszeski, M., Heller, R., and Gilbert, R. (eds.). *Electrochemotherapy, electrogenetherapy, and transdermal delivery*. Humana, New Jersey, pp. 173–186.

9. Martin, J.B., Young, J.L., Benoit, J.N., and Dean, D.A. (2000) Gene transfer to intact mesenteric arteries by electroporation. *J. Vasc. Res.* **37**, 372–380.

10. Schoenbach, K.H., Beebe, S.J., and Buescher, E.S. (2001) Intracellular effect of ultrashort electrical pulses. *Bioelectromagnetics.* **22**, 440–448.

11. Beebe S.J., Fox, P.M., Rec, L.J., Willis, E.L., and Schoenbach, K.H. (2003) Nanosecond, high-intensity pulsed electric fields induce apoptosis in human cells. *FASEB J.* **17**, 1493–1495.

12. Muller, K.J., Sukhorukov, V.L., and Zimmermann, U. (2001) Reversible electropermeabilization of mammalian cells by high-intensity, ultra-short, pulses of submicrosecond duration. *J. Membr. Biol.* **184**, 161–170.

13. Vernier, P.T., Li, A., Marcu, L., Craft, C.M., and Gundersen, M.A. (2003) Ultrashort pulsed electric fields induce membrane phospholipids translocation and caspase activation: differential sensitivities of Jurkat T lymphoblasts and rat glioma C6 cells. *IEEE Trans. Dielectr. Electr. Insul.* **10**, 795–809.

14. Vernier, P.T., Sun, Y., Marcu, L., Salemi, S., Craft, C., and Gundersen, M.A. (2003) Calcium bursts induced by nanosecond electric pulses. *Biochem. Biophys. Res. Commun.* **310**, 286–295.

15. Schwan, H.P. (1989) Dielectrophoresis and rotation of cells. In: Neumann, E., Sowers, A.E., and Jordan, C.A. (eds.). *Electroporation and electrofusion in cell biology*. Plenum, New York, pp. 3–21.

16. Kristiansen, M. and Hagler, M.O. (1987) Pulsed power systems. In: Meyers, R.A. (ed.). *Encyclopedia of physical science and technology*. Academic, Orlando, FL, pp. 410–419.

17. Schoenbach, K.H., Peterkin, F.E., Alden, W.A., Beebe S.J. (1997) The effect of pulsed electric fields on biological cells: experiments and applications. *IEEE Trans. Plasma Sci.* **25**, 284–292.

18. Mussauer, H., Sukhorukov, V.L., Haase, A., and Zimmermann, U. (1999) Resistivity of red blood cells against high intensity, short-duration electric field pulses induced by chelating agents. *J. Membr. Biol.* **170**, 121–133.

19. Schoenbach, K.H., Katsuki, S., Stark, R.H., Buescher, E.S., Beebe, S.J. (2002) Bioelectrics—new applications for pulsed power technology. *IEEE Trans. Plasma Sci.* **30**, 293–300.

20. Ellappan, P. and Sundararajan, R. (2005) A simulation study of the electrical model of biological cells. *J. Electrostat.* **63**, 297–307.

21. Schwan, H.P. (1989) Dielectric properties of tissues and biological materials: a critical review. *Crit. Rev. Biomed. Eng.* **17**, 25–104.

22. Beebe, S.J., Blackmore, P.F., White, J., Joshi, R.P., and Schoenbach, K.H. (2004) Nanosecond pulsed electric fields modulate cell function through intracellular signal transduction mechanisms. *Physiol. Meas.* **25**, 1077–1093.

23. Mastrangelo, A.J., Hardwick, J.M., and Betenbaugh, M.J. (2000) Overexpression of bcl-2 family members enhances survival of mammalian cells in response to various culture insults. *Biotechnol. Bioeng.* **67**, 555–564.

24. Hofmann, F., Ohnimus, H., Scheller, C., Strupp, W., Zimmermann, U., and Jassoy, C. (1999) Electric field pulses can induce apoptosis. *J. Membr. Biol.* **169**, 103–109.

Chapter 8
Nucleofection of Human Embryonic Stem Cells

Henrike Siemen, Lars Nolden, Stefanie Terstegge,
Philipp Koch, and Oliver Brüstle

Abstract Human embryonic stem cells (HESCs) are widely used as a model system for human cell type specification. Genetic modification forms a valuable tool for HESC technology, as it provides the basis for lineage selection, i.e., the purification of a specific cell type after differentiation. Electroporation is an efficient way to transfect HESCs. Nucleofection is an electroporation-based transfection technique which utilizes cell-type-specific buffer solutions and specific electric settings. Customization of these two parameters has been proven to result in highly efficient gene transfer even in hard-to-transfect cells. We can show that nucleofection surpasses conventional electroporation in efficiency and decreases the experimental effort for transfection of HESCs.

Keywords: human embryonic stem cells, nucleofection, stable transfection, lineage selection

1. Introduction

1.1. Transfection of Human Embryonic Stem Cells

Human embryonic stem cells (HESCs) are derived from the human blastocyst and show the same distinctive characteristics as murine ES cells: ES cells are able to self-renew indefinitely and can give rise to cells of all three embryonic germ layers (1–3). These two properties make ES cells an outstanding tool for biomedical research. Since the derivation of the first HESC line in 1998 (3), there has been an increasing interest in HESC research (for a review on HESC characterization, see [4]). HESC-derived somatic cells hold great promise for the development of novel strategies in drug discovery and regenerative medicine. Nonetheless, more basic research is required for the full understanding of HESC self-renewal and differentiation. One major focus lies on the controlled differentiation of HESCs into specific somatic cell types and the purification of these somatic cell populations.

S. Li (ed.), *Electroporation Protocols: Preclinical and Clinical Gene Medicine.*
From *Methods in Molecular Biology, Vol. 423.*
© Humana Press 2008

A commonly used method is lineage selection, which requires the genetic modification of HESCs. It involves the introduction of a reporter construct consisting of a selectable marker (e.g., an antibiotic resistance gene or a fluorescent marker) under the control of a cell-type-specific promoter. Lineage selection allows for the generation of highly purified HESC-derived somatic cell populations (5–7). This approach can be further refined by homologous recombination, which permits the inactivation of specific genes and the targeted insertion of transgenes into defined loci (8, 9).

Widely used transfection methods for HESCs are lipofection, lentiviral transduction, and electroporation (10–14). Lipofection is easy and quickly accomplished; however, the transfection efficiency is low (10). Also, there seem to be large variations in the efficiency between different reagents. Furthermore, the efficiency of lipofection depends on cell division, and the generation of stably transfected HESC clones is difficult. Transduction of HESCs with lentiviral vectors is highly efficient but requires specific safety precautions. In addition, lentiviral transduction is not practicable to induce homologous recombination. In 2002, Zwaka et al. were the first to employ electroporation of HESCs for gene targeting by homologous recombination (13). Yet, the need for large cell numbers (3×10^7) and large quantities of DNA ($40\,\mu g$) per reaction have left conventional electroporation a suboptimal procedure.

1.2. Nucleofection—A Novel Transfection Technique

The Nucleofector™ technology is an electroporation method developed by the German company Amaxa (www.amaxa.com). Amaxa's goal was to develop an efficient technique for nonviral transfection of primary cells. This goal was achieved by the combination of cell-type-specific electric parameters and transfection buffers. Many hard-to-transfect primary cells and cell lines were successfully nucleofected, e.g., primary neurons (15), naïve T cells (16), and natural killer cell lines (17). Nucleofection, like conventional electroporation, is cell-division-independent. The DNA is delivered directly into the nucleus (Fig. 8.1). The nuclear transfer of the DNA allows for an important mechanism: the stable integration of the transgene into the genome (Fig. 8.2).

The protocol for the nucleofection of HESCs described here is based on our experience with the HESC line H9.2 (18). Using the optimized conditions reported here, we achieved transfection efficiencies of up to 66% (19, Fig. 8.3). However, HESC lines vary biologically and genetically, and so working with different lines probably requires optimization of the nucleofection conditions. Mainly two parameters should be validated and optimized if necessary: the nucleofection solution and the electric settings. Until now, there is no HESC-specific nucleofection kit available. Some researchers employ the mouse ES cell kit for the nucleofection of HESCs (14), which, in our opinion, is not the best choice.

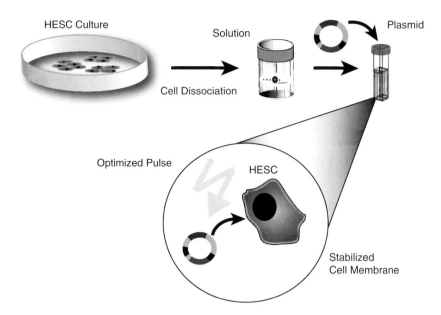

Fig. 8.1 Nucleofection delivers DNA directly into the nucleus of the transfected cell. This is achieved by an optimized pulse length and amplitude for individual cell types. During nucleofection, cell membranes are stabilized by a nucleofection buffer which is customized for the cell type to be transfected

The nucleofection buffers are specifically designed for each cell type, based on individual parameters, including cell size and membrane properties. Mouse and human ES cells differ greatly in size and shape, and choosing Amaxa's cell line kit V results in higher transfection efficiencies than using the mouse ES cell kit (66% vs. 22%). The second variable is the nucleofection program. To determine the optimal program for each HESC line, Amaxa recommends testing the programs A-23, A-27, A-13, A-12, and B-16.

2. Materials

2.1. Cell Culture

1. Mouse embryonic fibroblast (MEF) cell culture medium: Dulbecco's modified Eagle medium (high glucose), 10% (v/v) fetal bovine serum, 1% (v/v) sodium pyruvate, 1mM L-glutamine, and 1% (v/v) nonessential amino acids.
2. Human embryonic stem cell (KO/SR) culture medium: Knockout-Dulbecco's modified Eagle medium (Knockout-DMEM), 20% (v/v) serum replacement, 1%

(v/v) nonessential amino acids, 1 mM L-glutamine, 0.1 mM 2-mercaptoethanol, and fibroblast growth factor 2 (FGF-2; 4 ng/mL).

3. Collagenase IV (1 mg/mL): Collagenase IV is dissolved in Knockout DMEM and sterilized by filtration. It is stored at 4°C for up to 1 week (*see* **Note 1**).

4. Accutase II (*see* **Note 2**) (PAA Laboratories).

5. Matrigel (*see* **Note 3**) (BD Bioscience).

2.2. Nucleofection

1. Nucleofector device (Amaxa).

2. Nucleofection Cell Line Solution Kit V (Amaxa; includes cuvettes, transfer pipettes, transfection buffer, supplement, and a positive control vector pmaxGFP). The kit is stored at room temperature; after addition of supplement to the buffer, the solution should be stored at 4°C and is stable for 1 month.

3. DNA (circular or linear), dissolved in TE buffer (pH 8.5) or de-ionized water (*see* **Note 4**).

3. Methods

3.1. Cell Culture

1. Human ES cells are cultured on mitotically inactivated mouse embryonic fibroblasts (MEF) (*see* **Note 5**). The cells are passaged every 3–4 days (depending on the cell line) using collagenase to maintain small clumps. To avoid feeder contamination in the transfection experiments, HESCs can be cultured for one or more (*see* **Note 6**) passages on Matrigel using MEF-conditioned medium (*see* **Note 7**). Information regarding NIH-approved HESC lines can be obtained at http://stem cells.nih.gov/research/registry/.

3.2. Nucleofection: Preparation and Pulse

1. Nucleofection is performed 2–3 days after the passage of HESCs. The cells are washed twice with Knockout-DMEM and treated with accutase II for about 5 min (on MEFs) or up to 20 min (on Matrigel). HESCs can be dissociated to single cells, but they can also be efficiently transfected as clumps of about 50 cells (*see* **Note 8**). Cells are counted, and the desired number of cells ($1–2 \times 10^6$ per reaction) is pelleted at $120g$ for 3 min. The medium is removed and the cells are resuspended in nucleofection solution V (after addition of the

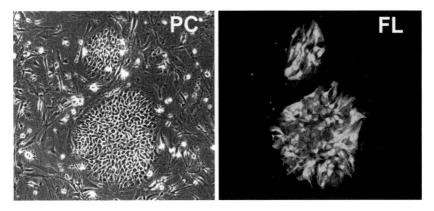

Fig. 8.2 Human embryonic stem cell colonies stably expressing enhanced green fluorescent protein after nucleofection with an EGFP-encoding plasmid. *PC* phase contrast, *FL* epifluorescence illumination for EGFP expression

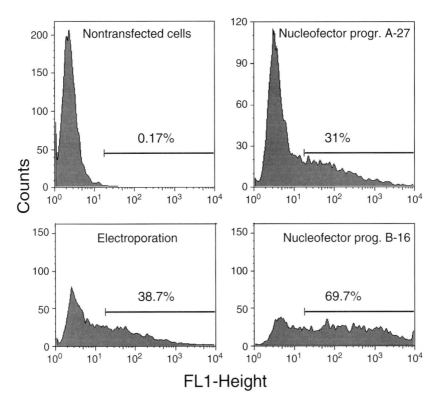

Fig. 8.3 Flow cytometric detection of transgene expression 24 h after nucleofection and conventional electroporation (representative examples). Measurements are based on a threshold value taken from nontransfected cells. (Reproduced from (19) with permission from Mary Ann Liebert Inc.)

supplement) at a concentration of $1–2 \times 10^6$ cells/100 µL. The solution should be at room temperature. Cells should not be kept in the nucleofection solution for more than 15 min.

2. Sterile tubes (one per sample) with at least 500 µL HESC medium are prewarmed to 37°C.

3. The following steps are performed separately for each sample. Plasmid DNA (2–5 µg, volume ≤ 5 µL) is transferred into the nucleofection cuvette. About 100 µL of the cell suspension is added into the cuvette, and the cuvette is inserted into the nucleofector device. Nucleofection is performed using program A-27 or B-16 (*see* **Note 9**). Immediately after the pulse, the cells are transferred to prewarmed KO/SR medium with the plastic pipettes supplied along with the nucleofection kit (*see* **Note 10**). It is recommended to add the prewarmed KO/SR medium to the cell solution in the cuvette and then carefully transfer the solution back into the tube. Tubes are kept at 37°C for about 15 min. These steps are repeated for all samples.

4. The transfected cells are transferred to the prepared 6-well plates (coated with either MEFs or Matrigel).

5. At this point, aliquots can be taken to monitor the cell viability after the pulse. Samples are kept on ice for 1–2 h to allow the membrane to stabilize. Trypan blue is used to determine the number of living cells compared to the number of dead cells.

6. The transgene expression (transient) is determined as early as 3 h after transfection; however, the highest expression is observed between 12–24 h after transfection.

7. Selection for stable transfection is started 48 h after transfection.

4. Notes

1. Collagenase is used to passage HESCs in aggregates. For nucleofection, however, cells may be dissociated to single cells to prevent cell fusion during the application of the electric pulse.

2. Accutase allows the selective removal of HESCs from the feeder layer by careful pipetting. The incubation time with accutase II can vary from 5 to 20 min at 37°C. If the HESCs are grown on MEFs, the incubation time should be as short as possible so as not to detach the MEFs.

3. Matrigel gels quickly at temperatures below and above 4°C. It should be kept on ice all the time while working with it.

4. The purity of the plasmid DNA preparation is crucial for cell survival rate and transfection efficiency. The A260/280 ratio should be at least 1.8 or higher.

5. MEFs should be used not less than 1 day and not more than 3 days after plating.

6. Extended culture on Matrigel has been associated with karyotype instability; therefore, HESCs should be karyotyped regularly when grown on Matrigel for a longer period.

7. To provide HESCs with factors secreted from MEFs, MEF-conditioned medium can be used. MEFs are plated in MEF medium and the medium is changed to KO/SR medium after 24 h. Conditioned KO/SR medium is collected after 24 h, filtered through a 0.2-µm filter, and used immediately or stored at −20°C for later use. FGF-2 is added before use. The FGF-2-supplemented medium is used without further dilution.

8. Cell number is a crucial factor for efficient transfection. If cells are transfected as clumps, an aliquot should be taken, trypsinized, and counted to determine the cell concentration in the sample.

9. It is recommended to include a positive control (plasmid pmaxGFP, supplied with the nucleofection kit). In a previous study (19), program A-27 was found to result in lower transfection efficiency (32%), although the survival rate was higher (81%). In contrast, program B-16 yielded a higher efficiency (66%), but the survival rate was slightly lower (73%).

10. After nucleofection, cells are very sensitive to shear forces. The provided pipette should be used.

References

1. Evans, M.J. and Kaufman, M.H. (1981) Establishment in culture of pluripotential cells from mouse embryos. *Nature*. **292**, 154–156.

2. Martin, G.R. (1981) Isolation of a pluripotent cell line from early mouse embryos cultured in medium conditioned by teratocarcinoma stem cells. *Proc. Natl. Acad. Sci. U.S.A.* **78**, 7634–7638.

3. Thomson, J.A., Itskovitz-Eldor, J., Shapiro, et al. (1998) Embryonic stem cell lines derived from human blastocysts. *Science*. **282**, 1145–1147.

4. Hoffman, L.M. and Carpenter, M.K. (2005) Characterization and culture of human embryonic stem cells. *Nat. Biotechnol.* **23**, 699–708.

5. Schmandt, T., Meents, E., Gossrau, G., Gornik, V., Okabe, S., and Brüstle, O. (2005) High-purity lineage selection of embryonic stem cell-derived neurons. *Stem Cells Dev.* **14**, 55–64.

6. Glaser, T., Perez-Bouza, A., Klein, K., and Brüstle, O. (2005) Generation of purified oligodendrocyte progenitors from embryonic stem cells. *FASEB J.* **19**, 112–114.

7. Li, M., Pevny, L., Lovell-Badge, R., and Smith, A. (1998) Generation of purified neural precursors from embryonic stem cells by lineage selection. *Curr. Biol.* **27**, 971–974.

8. Thomas, K.R. and Capecchi, M.R. (1987) Site-directed mutagenesis by gene targeting in mouse embryo-derived stem cells. *Cell.* **51**, 503–512.

9. Wernig, M., Tucker, K.L., Gornik, V., et al. (2002) Tau EGFP embryonic stem cells: an efficient tool for neuronal lineage selection and transplantation. *J. Neurosci. Res.* **15**, 918–924.

10. Eiges, R., Schuldiner, M., Drukker, M., Yanuka, O., Itskovitz-Eldor, J., and Benvenisty, N. (2001) Establishment of human embryonic stem cell-transfected clones carrying a marker for undifferentiated cells. *Curr. Biol.* **11**, 514–518.

11. Gropp, M., Itsykson, P., Singer, O., et al. (2003) Stable genetic modification of human embryonic stem cells by lentiviral vectors. *Mol. Ther.* **7**, 281–287.

12. Xiong, C., Tang, D.Q., Xie, C.Q., et al. (2005) Genetic engineering of human embryonic stem cells with lentiviral vectors. *Stem Cells Dev.* **14**, 367–377.

13. Zwaka, T.P. and Thomson, J.A. (2003) Homologous recombination in human embryonic stem cells. *Nat. Biotechnol.* **21**, 319–321.

14. Lakshmipathy, U., Pelacho, B., Sudo, K., et al. (2004) Efficient transfection of embryonic and adult stem cells. *Stem Cells.* **22**, 531–543.

15. Dityateva, G., Hammond, M., Thiel, C., et al. (2003) Rapid and efficient electroporation-based gene transfer into primary dissociated neurons. *J. Neurosci. Methods.* **130**, 65–73.

16. Lai, W., Chang, C.H., and Farber, D.L. (2003) Gene transfection and expression in resting and activated murine CD4 T cell subsets. *J. Immunol. Methods.* **282**, 93–102.

17. Maasho, K., Marusina, A., Reynolds, N.M., Coligan, J.E., and Borrego, F. (2004) Efficient gene transfer into the human natural killer cell line, NKL, using the Amaxa nucleofection system. *J. Immunol. Methods*. **284**, 133–140.
18. Amit, M., Carpenter, M.K., Inokuma, M.S., et al. (2000) Clonally derived human embryonic stem cell lines maintain pluripotency and proliferative potential for prolonged periods of culture. *Dev. Biol.* **227**, 271–278.
19. Siemen, H., Nix, M., Endl, E., Koch, P., Itskovitz-Eldor, J., and Brüstle, O. (2005) Nucleofection of human embryonic stem cells. *Stem Cells Dev*. **14**, 378–383.

Chapter 9
Delivery of Whole Tumor Lysate into Dendritic Cells for Cancer Vaccination

Linda N. Liu, Rama Shivakumar, Cornell Allen, and Joseph C. Fratantoni

Abstract Results from multiple human studies have continued to spur the development of dendritic cells (DCs) as therapeutic vaccines for the treatment of cancer, chronic viral infections, and autoimmune diseases. The antigen-specific activity of DCs is dependent on the ability of the DCs to take up and process tumor-associated antigens for presentation to the immune system. Although immature DCs have been shown to naturally take up tumor-associated antigens by phagocytosis, approaches that significantly affect antigen delivery need further evaluation, especially if such methodologies can be demonstrated to result in the elicitation of more robust and comprehensive immune responses. We have developed a rapid, robust, scalable, and regulatory-compliant process for loading DCs with whole tumor lysate. The use of whole tumor lysate facilitates the generation of a more robust immune response targeting multiple unique antigenic determinants in patient's tumors and likely reduces the tumor's potential of immune escape. We demonstrate that DCs electroloaded with tumor lysate elicit significantly stronger antitumor responses both in a tumor challenge model and in a therapeutic vaccination model for preexisting metastasic disease. These effects are observed in a processing scheme that requires 20- to 40-fold lower amounts of tumor lysate when compared with the standard coincubation/coculture methods employed in loading DCs.

Keywords: immunotherapy, tumor lysate loading, dendritic cells

1. Introduction

Electroporation has been established as an efficient method for loading a wide range of cell types (1–7), including DCs (8–10), with various bioactive molecules. Because it is a physical approach causing temporary permeability of the cell membrane, electroporation-mediated loading of macromolecules avoids virus-related complications, and it is especially important when dealing with immunotherapeutics that there will be no cross contamination with viral antigens. The use of

autologous tumor lysate allows vaccinating cancer patients regardless of their HLA haplotypes (10). Many previous studies have shown the feasibility of loading DCs with whole tumor lysate, most commonly by standard coincubation/pulsing, for treating different types of cancers, including solid carcinomas (11–13), myeloma (14), glioma (15), and melanoma (16–18). These studies demonstrate that DCs coincubated with whole tumor lysate can elicit specific antitumor T-cell responses. We hypothesized that electroloading DCs with whole tumor lysate would result in more effective antitumor responses due to active tumor antigen delivery to DCs and perhaps by forcing a class I response.

Syngeneic mouse tumor lysate was obtained by four cycles of quick freeze and thaw of cultured mouse renal carcinoma cells (RENCA) and Lewis lung carcinoma (LLC) cells. To avoid nonspecific, antitumor effects from tissue culture related materials, LLC tumor cells were subcutaneously injected into syngeneic mice. The tumor tissue that grew out was extracted and dissected into small pieces prior to the freeze and thaw procedure. After optimization of the electroloading procedure via fluorescein isothiocyanate (FITC)-conjugated dextran (250 kDa), immature DCs (imDC, 5×10^7 cells/mL) were electroloaded with the whole tumor lysate at 0.5 mg/mL, which was equivalent to the ratio of 10 imDCs to 1 tumor cell. After overnight maturation, RENCA tumor lysate loaded or incubated mDCs were subcutaneously injected into syngeneic Balb/C mice, followed by RENCA tumor challenge 10 days later. Tumor growth was measured in mice (Fig. 9.1).

For the metastases tumor model, C57BL6 mice were first injected with 5×10^5 syngeneic LLC tumor cells via tail vein. The mice later received two shots of LLC lysate electroloaded or incubated mDCs, 1×10^6 cells, at days 3 and 6 after tumor

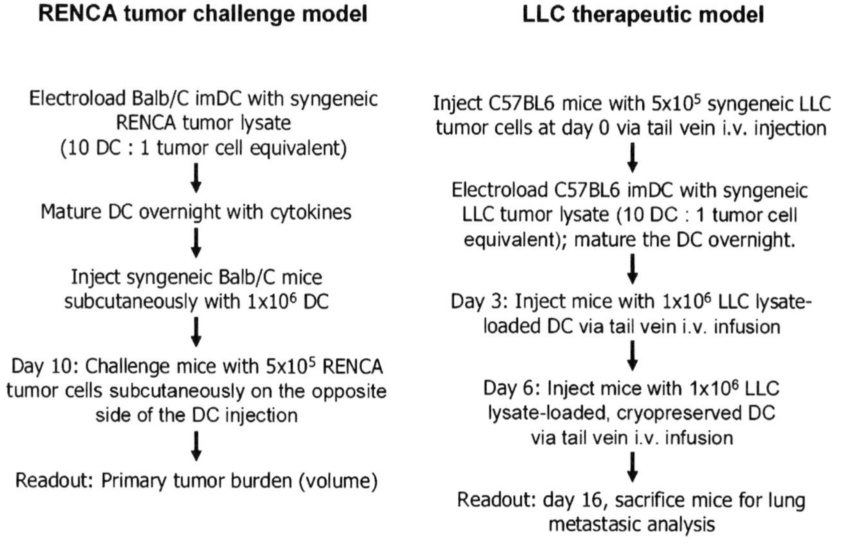

RENCA tumor challenge model

Electroload Balb/C imDC with syngeneic RENCA tumor lysate
(10 DC : 1 tumor cell equivalent)
↓
Mature DC overnight with cytokines
↓
Inject syngeneic Balb/C mice subcutaneously with 1×10^6 DC
↓
Day 10: Challenge mice with 5×10^5 RENCA tumor cells subcutaneously on the opposite side of the DC injection
↓
Readout: Primary tumor burden (volume)

LLC therapeutic model

Inject C57BL6 mice with 5×10^5 syngeneic LLC tumor cells at day 0 via tail vein i.v. injection
↓
Electroload C57BL6 imDC with syngeneic LLC tumor lysate (10 DC : 1 tumor cell equivalent); mature the DC overnight.
↓
Day 3: Inject mice with 1×10^6 LLC lysate-loaded DC via tail vein i.v. infusion
↓
Day 6: Inject mice with 1×10^6 LLC lysate-loaded, cryopreserved DC via tail vein i.v. infusion
↓
Readout: day 16, sacrifice mice for lung metastasic analysis

Fig. 9.1 Schematic flow chart of the in vivo experiments

implantation. Metastasic lesions were measured at day 15 by weighing both lungs from the killed mice (Fig. 9.1). In addition, specific in vitro tumor killing was analyzed using splenocytes isolated from C57BL6 mice that had received LLC lysate electroloaded DCs.

2. Materials

2.1. Cell Culture

1. Renal carcinoma (RENCA) and Lewis lung carcinoma (LLC) were generously provided by EntreMed (Rockville, MD).
2. Cell culture mediam: RPMI 1640 and Dulbecco's modified Eagle's medium (DMEM).
3. Supplemental components for cell culture median: fetal bovine serum (FBS), L-glutamine, nonessential amino acids (NEAA) mixture, sodium pyruvate, and penicillin/streptomycin.
4. MEM vitamin mixture.
5. Cell passage reagent: trypsin-Versene mixture and phosphate-buffered saline (PBS).
6. Cell culture flasks: T25, T75, and T150.

2.2. DC Isolation and Differentiation

1. C57BL6 and Balb/C male mice (6–8 weeks old) were obtained from Jackson Laboratories (Bar Harbor, ME).
2. PBS.
3. Scissors and forceps, kept in sterile beaker with 70% ethanol.
4. 10-mL syringe with 18- GA needle.
5. 100-mm petri dish.
6. Box or equipment for CO_2 asphyxiation.
7. Cell strainer.
8. ACK red blood cell lysis buffer (Quality Biological Inc.).
9. T175 tissue culture flasks.
10. Ultra low attachment (ULA) 6-well plates (Corning).
11. Mouse granulocyte-macrophage colony-stimulating factor (GM-CSF), interleukin (IL)-4, tumor necrosis factor alpha (TNF-α), and interferon gamma (IFN-γ).
12. Prostaglandin E2 (PGE-2) and lipopolysaccharide (LPS).
13. Human serum albumin.
14. AIM-V media.
15. Enzyme-free, PBS-based cell dissociation buffer.

2.3. Electroloading

1. Electroporator: MaxCyte Gene Transfer (GT) system or electroporator from other vendors, for example, BTS EC830 (Inovio) and Gene Pulser II (BioRad).
2. Electroloading buffer (Hyclone).
3. Cell-processing cuvettes (MaxCyte or other vendors).
4. FITC-conjugated dextran (250 kDa, Sigma), suspended in PBS at 10 mg/mL in a 15-mL conical tube wrapped with aluminum foil and stored at 4°C.
5. Water bath set at 37°C.
6. Propidium iodine, 0.5 mg/mL (Roche).
7. FACS staining buffer: PBS containing 2% FBS.
8. FACS Calibur flow cytometer and CellQuest software (BD Bioscience).

2.4. In Vivo Experiment

1. C57BL6 and Balb/C male mice (6–8 weeks old) were obtained from Jackson Laboratories.
2. 1-mL insulin syringes with various gauges of needles for injection.
3. Mechanical caliper for measuring tumor growth.
4. Mouse holder for tail vein injection

2.5. In Vitro Tumor Killing Assay

1. $Na_2[^{51}CrO_4]$.
2. XVIVO-15 media (Cambrex).
3. Mouse recombinant GM-CSF, IL-2, and IL-7.
4. Scintillation counter.
5. 24-well ultra low attachment plates.
6. U-bottomed 96-well plates.
7. Triton X-100.

3. Methods

In this study, syngeneic Balb/C mice were used to demonstrate that RENCA tumor lysate electroloaded mDCs could delay tumor growth. LLC-tumor-bearing syngeneic C56BL6 mice were used to show the therapeutic effects of lysate-electroloaded mDCs on prevention of metastasis. We compared treatment groups using tumor lysate electroloaded, incubated, and control liver lysate electroloaded DCs.

3.1. Preparation of Mouse Bone-Marrow-Derived DCs

1. Six-week-old C57BL6 and Balb/C male mice were obtained from Jackson Laboratories. Maintain the mice under an approved Institutional Animal Care and Use Committee (IACUC) protocol.
2. Asphyxiate the mice in a sealed container by using CO_2 at the age of 8–12 weeks (*See* **Note 1**).
3. After cervical dislocation, use scissors and forceps to isolate both the femur and tibia bone by cleaning away as much of the surrounding muscle and fat tissue as possible and leaving the bones intact (*See* **Note 2**).
4. In a biological safety cabinet, remove one end of the bone by a pair of sterilized scissors, leaving the bone marrow exposed.
5. Flush out the bone marrow from the intact end by a 10-mL syringe with an 18-gauge needle filled with PBS into a 100-mm petri dish.
6. Carefully transfer the collected bone marrow by a pipette and filter it through a cell strainer to remove residual tissue and debris.
7. Harvest the marrow cells by centrifugation for 10 min at $400 \times g$.
8. Aspirate the supernatant and then wash the cells once with PBS, followed by centrifugation for 10 min at $400 \times g$.
9. Afterwards, discard the supernatant, then resuspend the cells with 10 mL ACK red blood cell lysis buffer and incubate the cells at room temperature for 10 min.
10. Wash the cells three times with PBS. After the final PBS wash, resuspend the cell pellet with AIM-V culture media containing 2 mM L-glutamine and 0.2% HSA, and then count the cells with a hemocytometer.
11. Seed the bone marrow cells to T-175 flasks at ~5×10^6 cells/mL, 35 mL/flask in AIM-V culture media containing 2 mM L-glutamine, 0.2% HAS, 25 ng/mL of mGM-CSF, and 12.5 ng/mL of mIL-4.
12. Add fresh mGM-CSF (25 ng/mL) and mIL4 (12.5 ng/mL) to the culture every 2–3 days.
13. After 6–8 days in culture, harvest the imDCs by combining the suspension and adherent cells detached by enzyme-free, PBS-based cell dissociation buffer.
14. Process the imDCs freshly or cryopreserve them in FBS containing 10% dimethyl sulfoxide (DMSO). Use a small aliquot of the cells for surface marker analysis by flow cytometry.

3.2. Preparation of Whole Tumor Lysate

1. Culture murine renal carcinoma cell line (RENCA) in RPMI 1640 supplemented with 10% FBS, 2 mM L-glutamine, 1 mM sodium pyruvate, 1× NEAA, 2× MEM vitamin mix, and 1× penicillin/streptomycin.
2. Culture Lewis lung carcinoma cell line (LLC) in DMEM supplemented with 10% FBS, 2 mM L-glutamine, 1 mM sodium pyruvate, 2× MEM vitamin mix, 1× NEAA, and 1× penicillin/streptomycin.

3. On the day of tumor cell harvesting, after trypsinization, wash the tumor cells extensively (say three times) with PBS.

4. After counting the cell number with a hemocytometer, resuspend the tumor cells in PBS at 1×10^8 cells/mL.

5. To eliminate the potential for tissue-culture contaminants such as FBS and trypsin during lysate preparations, also prepare the whole tumor lysate from tumor tissue freshly removed from syngeneic animals with tumors established by subcutaneous injection of tumor cells. Dissect the tumors to eliminate as much normal tissue as possible, and then finely mince the tumors with razors. Resuspend the minced tissue in PBS. At the same time, also isolate the liver to prepare lysate as control.

6. Quickly freeze the cells by incubating the tube in a dry ice/alcohol bath for 5 min.

7. Quickly thaw the cells by incubating the tube in a 37°C water bath.

8. Quickly freeze and thaw the cells an additional three times.

9. Spin the lysed cells in a benchtop microcentrifuge for 10 min at the highest speed at 4°C, and then transfer the supernatants to sterilized O-ring microcentrifuge tubes.

10. Store the whole tumor lysates at −80°C until use.

11. Quantify the total protein concentration using a commercially available protein assay kit (Pierce), which approximately is 10 mg/mL.

3.3. Optimization of Electroloading imDC

1. Harvest the imDCs by combining floating suspension cells and adherent cells detached by enzyme-free, PBS-based cell dissociation buffer.

2. After centrifugation at $200 \times g$ for 10 min, resuspend the imDCs in 10 mL of electroporation buffer, followed by centrifugation at $200 \times g$ for 10 min.

3. Resuspend the imDC pellet in electroporation buffer at 5×10^7 cells/mL (*See* **Note 3**).

4. Add FITC-conjugated dextran to a final concentration of 0.5 mg/mL.

5. For electroloading, transfer 20–400 µL of imDC and FITC-dextran mixture to a MaxCyte cell-processing chamber and process using the MaxCyte 'imDC' protocol (*See* **Note 4**).

6. After electroporation, detach the chamber and transfer it to a biosafety cabinet. Transfer the electroporated imDCs to a sterile tube, followed by incubation in a 37°C water bath for 20 min.

7. Resuspend the processed imDCs in AIM-V media containing 2 mM l-glutamine, 0.2% HAS, 25 ng/mL mGM-CSF, and 12.5 ng/mL mIL-4 at a final cell concentration of 1×10^6 cells/mL. Plate the cells onto ULA 24-well plates at 1 mL/well.

8. For coincubation control, after finishing the electroloading process, dilute the imDC and FITC-mixture to 1×10^6 cells/mL and plate the same way as the electroporated imDCs.

3.4. Analyzing Electroloading Efficiency

1. Harvest the FITC-dextran modified DC 2–5 h after seeding by pipetting the loosely attached DCs.
2. Wash the plates once with cold PBS and pool them together with the collected DCs.
3. Centrifuge the collected cells at $200 \times g$ for 10 min.
4. After one wash with FACS staining buffer, resuspend the cells in FACS staining buffer and stain the cells with propidium iodine (PI) by adding 1 μg of PI/mL immediately prior to analysis by FACS Calibur flow cytometer and CellQuest software (*See* **Note 5**).
5. An example is shown in Fig. 9.2.

3.5. Preparation of Whole Tumor Lysate Modified mDC

1. Determine the optimal electroporation procedure by the FITC-dextran electroporation to obtain good FITC-dextran uptake and good cell viability.
2. Thaw previously prepared, frozen tumor lysate to room temperature by leaving the lysate-containing microcentrifuge tube in the biosafety cabinet for 20 min.

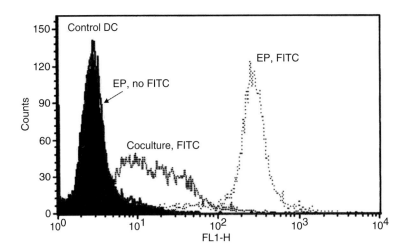

Fig. 9.2 Efficient electroporation-mediated delivery of macromolecule into DC. C57BL6 mouse bone-marrow-derived DCs were either cocultured or electroloaded with FITC-dextran (250 kDa). FACS analysis was performed 3 h after electroporation. Electroporation significantly enhanced FITC-dextran uptake when compared with coculture

3. Harvest the imDCs by combining floating suspension cells and adherent cells that detached briefly by enzyme-free, PBS-based cell dissociation buffer.

4. Centrifuge the imDCs at $200 \times g$ for 10 min, and then resuspend the imDCs in 10 mL of electroporation buffer, followed by centrifugation at $200 \times g$ for 10 min.

5. Resuspend the imDC pellet with electroporation buffer at 5×10^7 cells/mL.

6. Add the thawed whole tumor lysate to the imDC suspension at 0.5 mg/mL, which is equivalent to a ratio of 10 imDCs to 1 tumor cell lysate (*See* **Note 6**).

7. For electroloading, transfer 20–400 µL of imDC and tumor lysate mixture to a MaxCyte cell-processing chamber and process using the MaxCyte 'imDC' protocol.

8. After electroporation, detach the chamber and transfer it to a biosafety cabinet. Transfer the electroporated imDCs to a sterile tube, followed by incubation for 20 min in a 37°C water bath.

9. Later resuspend the processed imDCs in AIM-V media containing 2 mM L-glutamine, 0.2% HAS, 25 ng/mL each of mGM-CSF, mTNF-α, and m-IFNγ, 12.5 ng/mL mIL-4, 1 µg/mL PGE-2, and 10 µg/mL LPS at 5×10^6 cells/mL. Plate the cells onto ultra low attachment (ULA) 6-well plates at 2×10^6 cells/mL and allow the cells to mature overnight in a 37°C, 5% CO_2 incubator.

10. For coincubation control, after all the electroloading processes, dilute the imDC and tumor lysate mixture with the maturation media to 2×10^6 cells/mL and plate the same way as the electroporated imDCs for overnight maturation.

11. For negative control, electroporate the imDCs either without any lysate or with syngeneic liver lysate as described earlier.

12. Harvest the lysate-modified and matured DCs by collecting all the cells in the ULA tissue culture vessels plus a cold PBS wash of the vessels.

13. After centrifugation at $200 \times g$ for 10 min, either directly inject the modified mDCs into animals or cryopreserve them in FBS with 10% DMSO.

3.6. In Vivo Experiments

3.6.1. Whole Tumor Lysate Electroloaded DCs Delay Tumor Growth in a Tumor Challenge Model

1. For the RENCA tumor challenge model, subcutaneously inject Balb/C mouse with 1×10^6 DCs derived from syngeneic mouse bone marrow that have been electroloaded or coincubated with RENCA lysate.

2. As controls, some mice receive no DCs at all or 1×10^6 DCs that have been electroloaded with an equivalent amount of whole liver lysate.

3. Ten days later, harvest RENCA cells from tissue culture by trypsinization.

4. After three washes with PBS, resuspend the RENCA cell pellet with PBS at 2.5×10^7 cells/mL.

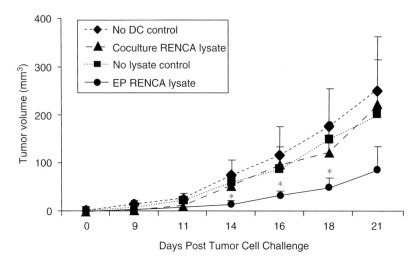

Fig. 9.3 DCs loaded with whole tumor lysate delayed RENCA tumor growth in a tumor challenge model. Balb/C mouse bone-marrow-derived imDCs were either coincubated or electroloaded with syngeneic RENCA tumor lysate at a cell ratio of 10 imDCs to 1 tumor cell equivalent and then matured. Approximately 1×10^6 mock or RENCA cell lysate modified DCs were subcutaneously injected into syngeneic Balb/C mice at the left flank. There were 10 mice in each group. Ten days later, the mice were challenged with 5×10^5 RENCA tumor cells at the side (right) opposite to the DC injection. The size of the tumors was measured starting at 9 days posttumor challenge. Significantly decreased tumor volumes were detected on day 14, 16, and 18 in those mice that received tumor lysate electroloaded DCs (*$p < 0.05$, compared with both no DC and no lysate controls)

5. Carefully inject the mouse with 200 µL (5×10^5 cells) of the RENCA cell suspension subcutaneously at a site different from the that of DC injection.
6. Measure the primary tumor areas twice a week using a mechanical caliper (*See* **Note 7**).
7. Calculate the tumor volumes with the formula $v = \pi ab^2/6$, where a is the longest diameter and b is the diameter perpendicular to 'a', and plot out as in Fig. 9.3.

3.6.2. Whole Tumor Lysate Electroloaded DCs Prevent Metastasis in a Therapeutic Animal Model

1. Harvest the LLC cells from tissue culture by trypsinization.
2. After three washes with PBS, resuspend the LLC cell pellet with PBS at 1×10^6 cells/mL.
3. Inject C57BL6 mouse with 0.5 mL of the live LLC cell suspension (5×10^5 cells) via tail vein by using a 1-mL syringe.
4. Three days later, carefully thaw previously processed, and cryopreserved whole tumor lysate modified mDCs.
5. After a wash with PBS, resuspend the mDCs with PBS at 2×10^6 cells/mL.

6. Divide the LLC-cell-injected mice into four groups (8 mice/group): *a* no treatment, *b* treat with mDC electroloaded with control liver lysate, *c* treat with mDC incubated with LLC whole tumor lysate, *d* treat with mDC electroloaded with LLC whole tumor lysate.

7. Carefully inject 0.5 mL of mDC suspension (1 × 10⁶ DC) via tail vein.

8. Three days later, at day 6, treat each mouse group with a 2nd dose of mDCs.

9. On day 17, kill all mice, including another 5 healthy mice of the same age. Isolate both lungs from each mouse. After briefly washing the lungs with PBS, lay the lungs on kleenex tissue. Weigh the lungs from each mouse.

10. Use the lung weight from the 8 healthy mice that receive no tumor cells as baseline lung weight controls and plot out all the lung weight as shown in Fig. 9.4.

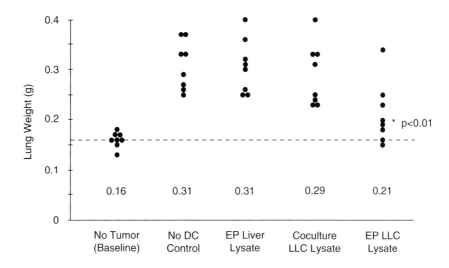

Fig. 9.4 Whole tumor lysate electroloaded DCs prevented Lewis lung metastases in a therapeutic model. At the start of the experiment, 5×10^5 Lewis lung carcinoma (LLC) cells were administered intravenously via tail vein into syngeneic C57BL6 mice. C57BL6 mouse bone-marrow-derived imDCs were either coincubated or electroloaded with LLC whole tumor cell lysate. As a control, imDCs were electroporated with liver lysate. After maturation, the modified mDCs were cryopreserved and stored in liquid nitrogen. Three days after LLC injection, 1×10^6 DCs were thawed and administered by tail vein injection (8 mice/group). There were four groups: *a* as negative control, no DCs were injected; *b* liver lysate electroloaded mDCs; *c* LLC tumor lysate incubated mDCs; *d* LLC tumor lysate electroloaded DCs. After an additional 3 days (day 6), the mice were treated with a second dose of identically processed mDCs (1×10^6) by tail vein injection. On day 15 post-LLC injection, mice were killed and lungs were dissected and weighed. The no tumor control group reflects normal lung weights of mice that were not challenged with any LLC. Administration of DCs that had been electroloaded with LLC lysate caused a significant reduction in LLC lung metastases, as indicated by a significant decrease in lung weights ($^*p < 0.01$)

3.7. In Vitro Tumor Killing Assay

3.7.1. Preparation of Effector Cells

1. Remove spleens from syngeneic mice (*See* **Note 8**).
2. Place the spleens in a 100-mm dish with 30 mL of RPMI 1640 media.
3. Using scissors and a razor blade, slice the spleens into small pieces and later mince 10–20 times using the flat end of the plunger from a 5-mL sterile syringe.
4. Carefully transfer the media, including all the splenocytes, and filter it through a cell strainer to remove residual tissue and debris.
5. Collect the splenocytes by centrifugation at $400 \times g$ for 10 min and later wash the cells twice with PBS.
6. After the final PBS wash, resuspend the splenocytes in XVIVO-15 media and count the cells with a hemocytometer.
7. Then dilute the splenocytes to 2×10^7 cells/mL with XVIVO-15 media supplemented with mGM-CSF (25 ng/mL), mIL2, and mIL7 (10 ng/mL each).
8. Add 500 µL of the splenocyte suspension (1×10^7 cells) to one well of a 24-well, ultra low attachment plate. Divide the wells into four groups and add 0.5 mL of previously processed mDCs (1×10^6 cells/mL), resulting in a cell ratio of 10 responder to 1 stimulator. The four groups are as follows: *a* LLC cell lysate electroloaded mDC, *b* LLC lysate coincubated mDC, *c* mDC electroloaded with liver lysate, and *d* no mDC, just XVIVO media.
9. Restimulate the cells with appropriate mDCs two more times, 1 week apart.
10. One week after the final stimulation, collect the cells from each group via centrifugation at $400 \times g$ for 10 min.
11. After three washes with PBS, resuspend the effector CTL in XVIVO-15.

3.7.2. Preparation of Target Cells

1. Split the LLC cells at 1:5 into T25 flasks in the DMEM complete media 2 days before the killing assay.
2. On the day of the killing experiment, harvest the LLC cells by trypsinization.
3. After a wash with complete DMEM media, count the cells with a hemocytometer.
4. Based on the cell concentration, transfer 1×10^6 LLC cells to a 15-mL conical tube.
5. Centrifuge the cells at $200 \times g$ for 5 min.
6. Carefully aspirate the supernatant and leave ~200 µL of complete media in the tube.
7. Add 20 µL of heat-activated FBS and 100 µL of $Na_2[^{51}CrO_4]$ (1 µCi/µL) to the tube (*See* **Note 9**).
8. After carefully resuspending the cell pellet, place the tube in a 37°C, CO_2 incubator with the cap loosely attached to the tube.
9. Incubate the LLC cells with ^{51}Cr for 1 h.

10. Then, add 5 mL of XVIVO-15 complete media to the ^{51}Cr-labeling tube.
11. Centrifuge the ^{51}Cr-labeled LLC cells at $200 \times g$ for 5 min.
12. Carefully remove and discard the ^{51}Cr-containing supernatant into a radioiso-tope waste container.
13. Wash the ^{51}Cr-labeled LLC cells four times with 5 mL of complete media via centrifugation.
14. After the final wash, resuspend the cells in XVIVO-15 complete media at 1×10^5 cells/mL.

3.7.3. Determining the CTL Activity by Coincubation of Target and Effector Cells

1. Seed 100 µL of the ^{51}Cr-labeled LLC cell suspension into each well on a U-bottomed 96-well plate (1×10^4 cells/well).
2. Add the LLC-specific CTL generated from the in vitro stimulation to the ^{51}Cr-labeled LLC cells at various ratios: 50:1, 10:1, and 1:1. Each ratio includes triplicate wells.
3. As complete cell lysing positive control, add 100 µL of 2% Triton X-100 to the ^{51}Cr-labeled target LLC cell suspension on the U-bottomed 96-well plate (4 wells).
4. As a negative control that measures spontaneous cell lysis (the background), add 100 µL of XVIVO-15 complete media to the ^{51}Cr-labeled target LLC cell suspension on the U-bottomed 96-well plate (4 wells).
5. Spin the 96-well plate briefly at $200 \times g$ for 1 min. For the wells containing LLC-specific CTL, after carefully aspirating the supernatant from the wells, add 200 µL of XVIVO-15 complete media.
6. Place the 96-well plate in a 37°C, CO_2 incubator and incubate the plate for 4 h, followed by brief centrifugation of the plate at $200 \times g$ for 1 min.
7. Then, carefully transfer 100 µL of the supernatant from each well to scintilla-tion tubes containing 2 mL of scintillation fluid.
8. Measure the amount of ^{51}Cr in each tube by a scintillation counter.
9. Average the counts from the triplicate wells.
10. Calculate the LLC tumor specific killing using the following equation: CTL activity = [radiation cpm from experimental wells – cpm from spontaneous ^{51}Cr release from tumor cell alone wells]/[cpm from LLC treated with Triton X-100]. An example is shown in Fig. 9.5.

3.8. Statistical Analysis

A two-tailed Student's t test was performed to compare tumor growth, lung weight, and CTL activity among treatment groups.

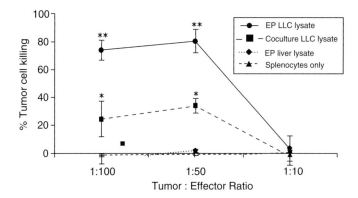

Fig. 9.5 Whole tumor lysate electroloaded mDCs can elicit antitumor immune response. Splenocytes (T cells) were isolated from C57BL6 mice that had previously received two intravenous administrations of 1×10^6 syngeneic, LLC lysate electroloaded mDCs. They were divided into four groups and stimulated with previously processed mDCs at a ratio of 1 DC to 10 T cells: *a* LLC cell lysate electroloaded mDCs, *b* LLC lysate coincubated mDCs, *c* mDC electroloaded with liver lysates, and *d* no mDC at all. The cocultured cells were restimulated once/week for a total of three restimulations with appropriately prepared and cryopreserved mDC. The expanded LLC-tumor-specific CTL were then washed and incubated with ^{51}Cr-labeled intact LLC cells at various ratios of 50:1, 10:1, and 1:1 for a standard in vitro tumor killing assay. The mDCs cocultured with LLC lysate elicited 20–30% specific tumor cell killing (*$p < 0.02$). In contrast, the DCs electroloaded with LLC lysate elicited 70–80% specific tumor cell killing (**$p < 0.005$)

Acknowledgments The authors thank Nicholas Chopas for instrumentation assistance.

4. Notes

1. Older mice (older than 3 months, younger than 1 year) can also be used for bone marrow isolation.
2. Freshly isolated bones need to be kept in PBS to prevent from being air dried. Harvest bone marrow promptly. We usually kill 20 mice at a time. The work is divided between two sites: bone isolation in the animal room and marrow harvesting in the tissue culture room. We process bones from 5 mice each time.
3. imDC concentration of 0.5×10^8 to 1×10^8 cells/mL is used during electroloading.
4. The MaxCyte electroporation instrument system allows for flexible configuration and processing utilizing a sterile closed fluid path (19, 20). The instrumentation and cell processing assemblies are described in MaxCyte's Master File on record with Center for Biologics Evaluation and Research (CBER), the United States Food and Drug Administration (FDA), and Health Canada. The FDA Master File (no. BBMF10702) has been referenced in four instances of review for Investigational New Drug (IND) applications and is currently in use for clinical manufacturing of biological products. MaxCyte's systems are scalable from R&D to clinical/commercial scale and accommodate sample volumes in the 20-μL (1×10^6 cells) to 1,000-mL (1×10^{11} cells) range. The instruments are capable of delivering an optimal amount of energy (40 J/mL) for efficiently electroloading imDCs (with > 90% cells loaded) with multiple molecules without significant

observable loss in cell viability (>90% as assayed by trypan blue exclusion) and with high cell recovery (~80 to 90%), which results in enhanced biological activity for imDCs and in the clinical implementation of such optimized procedures in a current good manufacturing practices (cGMP) environment.

5. FACS analysis is used to identify optimal electroloading conditions to obtain efficient loading while the cells maintain good cell viability measured by FITC-dextran uptake and PI staining exclusion respectively.

6. Various ratios of imDC vs. tumor cells can be tested.

7. One person should be responsible for measuring the tumors from the beginning to the end of the experiment to avoid investigator variation among measurements.

8. We used spleens from the C57BL6 mice that received treatment with LLC tumor lysate electroloaded DCs.

9. All liquid and solid waste starting from this step shall be treated cautiously and discarded in appropriate containers for radioactive materials.

References

1. Neumann, E., Schaefer-Ridder, M., Wand, Y., and Hofschneider, P. (1982) Gene transfer into mouse lyoma cells by electroporation in high electric fields. *EMBO J.* **1**, 841–845.
2. Xie, T.D., Sun, L., and Tsong, T.Y. (1990) Study of mechanisms of electric field-induced DNA transfection. I. DNA entry by surface binding and diffusion through membrane pores. *Biophys. J.* **58**, 13–19.
3. Tsong, T.Y. (1991) Electroporation of cell membranes. *Biophys. J.* **60**, 297–306.
4. Salek, A., Schnettler, R. and Zimmermann, U. (1992) Stably inherited killer activity in industrial yeast strains obtained by electrotransformation. *FEMS Micro Lett.* **75**, 103–109.
5. Li, L.H., Shivakumar, R., Feller, S., et al. (2002) Highly efficient, large volume flow electroporation. *Tech. Cancer Res. Treat.* **1**, 341–349.
6. Weiss, J.M., Shivakumar, R., Feller, S., et al. (2004) Rapid, in vivo, evaluation of anti-angiogenic and anti-neoplastic gene products by non-viral transfection of tumor cells. *Cancer Gene Ther.* **11**, 346–353.
7. Li, S. (2004) Electroporation gene therapy: new developments *in vivo* and *in vitro*. *Curr. Gene Ther.* **4**, 309–316.
8. Breckpot, K., Heirman, C., Neyns, B., and Thielemans, K. (2004) Exploiting dendritic cells for cancer immunotherapy: genetic modification of dendritic cells. *J. Gene Med.* **6**, 1175–1188.
9. Ribas, A. (2005) Genetically modified dendritic cells for cancer immunotherapy. *Curr. Gene Ther.* **5**, 619–628.
10. Weiss, J.M., Allen, C., Shivakumar, R., Feller, S., Li, L.H., and Liu, L.N. (2005) Efficient responses in a murine renal tumor model by electroloading dendritic cells with whole-tumor lysate. *J. Immunother.* **28**, 542–550.
11. Geiger, J.D., Hutchinson, R.J., Hohenkirk, L.F., et al. (2001) Vaccination of pediatric solid tumor patients with tumor lysate-pulsed dendritic cells can expand specific T cells and mediate tumor regression. *Cancer Res.* **61**, 8513–8519.
12. Thurnher, M., Rieser, C., Holtl, L., Papesh, C., Ramoner, R., and Bartsch, G. (1998) Dendritic cell-based immunotherapy of renal cell carcinoma. *Urol. Int.* **61**, 67–71.
13. Schnurr, M., Galambos, P., Scholz, C., et al. (2001) Tumor cell lysate-pulsed human dendritic cells induce a T-cell response against pancreatic carcinoma cells: an in vitro model for the assessment of tumor vaccines. *Cancer Res.* **61**, 6445–6450.
14. Wen, Y.J., Min, R., Tricot, G., Barlogie, B., and Yi, Q. (2002) Tumor lysate-specific cytotoxic T lymphocytes in multiple myeloma: promising effector cells for immunotherapy. *Blood.* **99**, 3280–3285.

15. Parajuli, P. and Sloan, A.E. (2004) Dendritic cell-based immunotherapy of malignant gliomas. *Cancer Invest.* **22**, 479–480.
16. Berard, F., Blanco, P., Davoust, J., et al. (2000) Cross-priming of naive CD8 T cells against melanoma antigens using dendritic cells loaded with killed allogeneic melanoma cells. *J. Exp. Med.* **192**, 1535–1544.
17. Shibagaki, N. and Udey M.C. (2003) Dendritic cells transduced with TAT protein transduction domain-containing tyrosinase-related protein 2 vaccinate against murine melanoma. *Eur. J. Immunol.* **33**, 850–860.
18. Hadzantonis, M. and O'Neill, H. (1999) Review: dendritic cell immunotherapy for melanoma. *Cancer Biother. Radiopharm.* **14**, 11–22.
19. Fratantoni, J.C., Dzekunov, S., Singh, V., and Liu, L.N. (2003) A non-viral gene delivery system designed for clinical use. *Cytotherapy.* **5**, 208–210.
20. Fratantoni, J.C., Dzekunov, S., Wang, S., and Liu, L.N. (2004) A scalable cell-loading system for non-viral gene delivery and other applications. *Bioprocess. J.* **3**, 49–54.

Chapter 10
Delivery of Tumor-Antigen-Encoding mRNA into Dendritic Cells for Vaccination

Annelies Michiels, Sandra Tuyaerts, Aude Bonehill, Carlo Heirman, Jurgen Corthals, and Kris Thielemans

Abstract Antigen-loaded dendritic cells (DCs) have been intensively investigated as potential cellular antitumor vaccines. Several recent reports have indicated that loading DCs with whole tumor derived mRNA or defined tumor-antigen-encoding mRNA represents an effective nonviral strategy to stimulate T cell responses both for in vitro and in vivo models. Here, we describe the electroporation method as a tool for introducing in vitro transcribed capped mRNA into human DCs for tumor vaccination. We use MART-1/Melan-A as a model tumor-associated antigen for the generation of a DC-based vaccine against melanoma cancer. In addition to efficient antigen loading, it is important to obtain a maximal number of potent antigen-presenting cells. Another prerequisite for the development of a DC-based cancer vaccine is to obtain mature DCs. In this chapter, we describe the basic techniques required for the successful genetic modification of DCs by using the mRNA electroporation method.

Keywords: dendritic cells, mRNA, electroporation, cancer immunotherapy, vaccination

1. Introduction

Dendritic cells (DCs) are highly specialized antigen-capturing, antigen-processing, and antigen-presenting cells with the unique capacity to establish and control primary and secondary immune responses. This knowledge has led to the use of DCs as natural adjuvants for vaccinating cancer patients against tumor-specific self-antigens and to break T cell tolerance against these tumor antigens (1, 2). The development of protocols for *ex vivo* generation and modification of DCs provided a rationale for the design and development of DC-based vaccination studies aimed at treating cancer (3). Efficient antigen presentation by DCs is instrumental for establishing an antitumor immune response. Different methods have been used to load DCs with tumor-associated antigens (TAAs) for in vivo vaccination. With regard to clinical applicability, nonviral antigen-loading strategies

have been developed based on transfection of defined or full spectrum tumor mRNA into DCs (4, 5). This approach has the potential for broad clinical applications, as it does not require the knowledge of the MHC haplotype of the patient and the corresponding MHC binding epitopes of the TAA. The ability to modify the TAA with targeting signals such as sig-DCLAMP to obtain better MHC class I and class II presentation may further enhance the tumor-specific immune response (6). Van Tendeloo et al. (7) described the electroporation method as an efficient strategy to load DCs with in vitro transcribed mRNA. In contrast, the modification of human DCs by electroporation with tumor-antigen-encoding DNA remains difficult (7). We have confirmed that DCs can be efficiently electroporated with tumor-antigen-encoding mRNA and that these modified antigen-presenting cells are able to induce an in vitro tumor-antigen-specific T cell response (8–10). Recent studies demonstrated that coelectroporation of DCs with a dsRNA analogue (11) or with selected co-stimulatory molecules (12, 13) improved the stimulation of antigen-specific cytotoxic T lymphocytes and facilitated T helper type 1 polarization of naïve $CD4^+$ T helper cells. Here we describe the electroporation of DCs with capped, in vitro generated MART-1/Melan-A encoding mRNA for vaccination of melanoma patients.

2. Materials

2.1. Cloning

1. pGEM-EGFP plasmid DNA, provided by Dr. E. Gilboa (Duke University Medical Center, Durham, NC).
2. pCMV-sig-LAMP1 plasmid DNA, provided by Dr. P. Hwu (MD Anderson Cancer Center, Houston, TX).
3. pCR2.1 plasmid DNA.
4. MART-1/Melan-A cDNA, provided by Dr. P. Coulie (Ludwig Institute for Cancer Research, Brussels, Belgium).
5. Oligonucleotide primers, synthesized commercially.
6. Restriction enzymes, purchased commercially.
7. Agarose and DNA sequencing equipment.

2.2. RNA Extraction, Purification, and in Vitro Transcription

1. Proteinase K (10 mg/mL).
2. 10% sodium dodecyl sulfate (SDS).
3. Phenol/chloroform, pH 8.0.
4. 3 M sodium acetate, pH 5.2.
5. 100% and 70% ethanol (EtOH).

6. Sterile RNase, DNase, and endotoxin-free water.
7. mMESSAGE mMACHINE®T7 Ultra kit (Ambion, Austin, TX).

2.3. Electroporation

1. RPMI 1640 medium (Cambrex, Verviers, Belgium).
2. 1% heat-inactivated autologous plasma (AP).
3. Recombinant human granulocyte macrophage-colony stimulating factor (GM-CSF; prepared in-house) (*see* **Note 1**).
4. Recombinant human interleukin (IL)-4 (prepared in-house).
5. Serum-free RPMI 1640 medium.
6. Optimix washing solution A (Cell Projects, Kent, UK). Store at room temperature.
7. 2× Optimix solution B (Cell Projects). Store at −20°C.
8. Sterile RNase, DNase, and endotoxin free water.
9. 4-mm electroporation cuvettes (Cell projects).
10. EasyjecT Plus® Multipurpose Electroporation System (Thermo Electron, Kent, UK).
11. Phosphate-buffered saline (PBS; Cambrex).
12. Human serum albumin (HSA; Belgian Red Cross, Belgium).
13. 1-mL syringes (Terumo, Leuven, Belgium).

3. Methods

The methods described here outline (1) the generation of the A27L-mutated MART-1/Melan-A plasmid DNA, (2) the production of in vitro transcribed and capped mRNA, (3) the generation of human DCs, and (4) the electroporation of DCs with MART-1/Melan-A encoding mRNA for tumor vaccination (Fig. 10.1).

3.1. Generation of the A27L-Mutated MART-1/Melan-A Plasmid

For construction of the A27L-mutated MART-1/Melan-A (hereafter referred to as Melan-A) plasmid DNA, we start from the plasmid DNA pGEM-EGFP (*see* **Note 2**), provided by Dr. E. Gilboa. This plasmid DNA contains a 741-bp EGFP encoding fragment, flanked by the 5' and 3' untranslated regions (UTRs) of the *Xenopus laevis* β-globin gene (Fig. 10.2a). The plasmid DNA contains a stretch of 64 adenines downstream of the insert to add a poly(A) tail to the in vitro transcribed RNA. The transcription is controlled by a bacteriophage T7 promoter. At the 3' end of the poly(A) tail, unique *Not*I and *Spe*I restriction enzyme cut sites are present to allow linearization of the

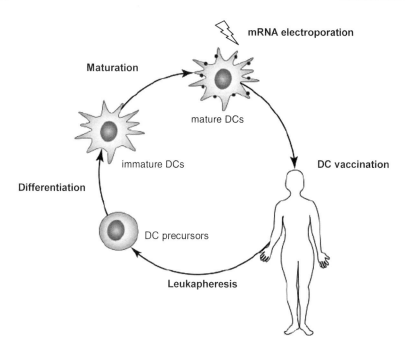

Fig. 10.1 Schematic representation of the production of a dendritic-cell (DC)-based cancer vaccine. DC precursors are collected from the patient by leukapheresis and differentiated into immature DCs, followed by maturation and mRNA electroporation (not necessarily in this order). Finally, these mature, antigen-loaded DCs are injected into the patient

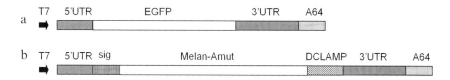

Fig. 10.2 (**a**) Schematic representation of the pGEM-EGFP expression plasmid. The T7 promoter, 5'UTR, 3'UTR, and A64 stretch are shown. (**b**) Schematic representation of the pGEM-sig-Melan-Amut-DCLAMP expression plasmid

plasmid DNA before in vitro transcription. For the cloning of pGEM-sig-DCLAMP, the plasmid DNA pCMV-sig-LAMP1, provided by Dr. P. Hwu is used as a template for polymerase chain reaction (PCR) to amplify the signal sequence (sig) with the primers sig sense 5' CCCCATGG$_{Ncol}$CGGCCCCCGGC 3' and sig antisense 5' GGG GGATCC$_{BamHI}$TCAAAGAGTGCTGA 3', adding a *Nco*I site spanning the start codon at the 5' end and a *Bam*HI site at the 3' end. Complementary DNA (cDNA) extracted from mature human DCs is used as a template for PCR to amplify DC-LAMP. The primers used in this PCR are DC-LAMP sense 5' CACAGGATCC$_{BamHI}$CT

CGTCTGACTACACAATTGTG 3' and DC-LAMP antisense 5' CACAAGATCT$_{Bg}$ $_{lII}$TTAGATTCTCTGGTATCCAGATC 3'. This PCR adds a *Bam*HI site at the 5' end and a stop codon and a *Bgl*II site at the 3' end. Both sig and DC-LAMP are cloned into the pCR2.1 vector, and subsequently, sig is cloned as a *Bam*HI-*Bam*HI fragment into pCR2.1DCLAMP, linearized with *Bam*HI. This cloning step results in pCR2.1sig-DCLAMP with a *Bam*HI site between the sig and DC-LAMP encoding cDNA. The sig-DCLAMP sequence is then excised with *Nco*I and *Eco*RI and cloned into the pGEM-EGFP vector, digested with the same enzymes, resulting in the pGEM-sig-DCLAMP plasmid. The cDNA encoding the human melanoma-associated antigen Melan-A was kindly provided by Dr. P. Coulie. We use PCR to generate *Bam*HI sites at both 5' and 3' boundaries of the gene. The intermediate construct pGEM-sig-DCLAMP is digested with *Bam*HI and the Melan-A gene is inserted. The final plasmid DNA pGEM-sig-Melan-Amut-DCLAMP (Fig. 10.2b) encoding Melan-A with an A27L amino acid mutation (*see* **Note 3**) is created by in situ mutagenesis (GeneEditor™ in vitro Site-Directed Mutagenesis System, Promega, Madison, USA) of cloned wild-type Melan-A using an oligonucleotide containing the desired DNA mutation. The presence of Melan-A with the desired A27L mutation is confirmed by sequencing.

3.2. Production of in Vitro Transcribed Capped mRNA

High-quality, efficiently capped in vitro transcribed mRNA with a poly(A) tail is required for the use in cancer vaccines. To avoid exogenous sources of RNase contamination, always apply RNase-free techniques when working with RNA! Prior to the in vitro transcription, 100 μg of the plasmid DNA pGEM-sig-Melan-Amut-DCLAMP was linearized with 10 μL (100 U) of the restriction enzyme *Spe*I (*see* **Note 4**) in a total volume of 500 μL and purified by using phenol/chloroform extraction and ethanol precipitation as follows:

1. Add 5 μL proteinase K (100×).
2. Add 5 μL 10% SDS.
3. Incubate for 30 min at 50°C.
4. Add an equal volume of phenol/chloroform (pH 8.0) and mix well.
5. Centrifuge for 10 min at $16,000 \times g$ in an Eppendorf microcentrifuge.
6. Transfer the upper layer to a clean, new tube.
7. Add an equal volume of chloroform, vortex, and centrifuge again.
8. Transfer the upper layer to a clean, new tube.
9. Add a 1:10 volume of 3 M sodium acetate (pH 5.2) and 2 volumes of 100% EtOH and mix well.
10. Incubate overnight at −20°C.
11. Centrifuge for 15 min at $16,000 \times g$.
12. Remove supernatant and add 1 mL 70% EtOH.
13. Centrifuge for 5 min at $16,000 \times g$.
14. Resuspend in 100 μL sterile RNase, DNase, and endotoxin-free water.

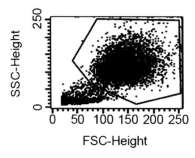 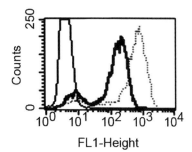

Fig. 10.3 Flow cytometry analysis of EGFP expression in dendritic cells (DCs) after electroporation of EGFP encoding mRNA. EGFP mRNA transcripts containing Anti-Reverse Cap Analog (ARCA) are more highly translated in electroporated DCs. Flow cytometry was performed 24 h after electroporation of DCs to analyze the EGFP expression. DCs were electroporated without mRNA (*thin line*), with EGFP conventional capped mRNA transcripts (*bold line*), or with EGFP ARCA capped mRNA transcripts (*dotted line*)

In vitro transcription is performed with T7 RNA polymerase by using the mMES-SAGE mMACHINE T7 Ultra kit (according to the manufacturer's instructions), which combines an Anti-Reverse Cap Analog (ARCA) with high yield transcription technology to generate RNA transcripts that when translated, result in much higher protein yields than do conventionally capped transcripts (14). Incorporating ARCA into EGFP encoding mRNA transcripts has a strong stimulatory effect on subsequent translation (Fig. 10.3). After in vitro transcription, the remaining plasmid DNA is degraded by DNase treatment to reduce the risk of introducing foreign DNA in the cells. ARCA-mRNA transcripts are routinely checked by agarose gel electrophoresis for correct size and integrity and by measuring the absorbance at 260 nm in a spectrophotometer for quantity and purity (*see* **Note 5**) mRNA is stored at −20°C in small aliquots.

3.3. Generation of Human DCs

Human DCs are specialized, bone-marrow-derived leukocytes and can be generated by different methods (15). Since the DC precursor population in blood is very rare, DC yields are generally too low to obtain sufficient numbers of DCs for vaccination. One way to overcome this problem is to start from proliferating CD34+ progenitor cells. These CD34+ cells can be isolated from blood or bone marrow by positive selection through magnetic separation. When the CD34+ cells are cultured in the presence of GM-CSF in combination with tumor necrosis factor alpha (TNF-α) and/or IL-4 or GM-CSF combined with IL-13, immature DCs develop within 2 weeks. However, the most commonly used cell type for DC generation is the peripheral blood mononuclear cell (PBMC), because this cell type can be collected via buffy coat preparations or leukapheresis. Monocytes

can be easily enriched through adherence to plastic, elutriation, filtration, and positive or negative immunomagnetic selection of CD14$^+$ cells. Classically, immature "myeloid-type" DCs are generated from the enriched monocytes by a 5–7-day culture in the presence of GM-CSF and IL-4 (*see* **Note 6**). To obtain a maximal number of DCs from a single leukapheresis, (*see* **Note 7**) we have designed a concept in which the immature DCs are generated from adherent monocytes in a closed culture system using Cell Factories (3). In order to obtain fully immunostimulatory or effector antigen-presenting cells, it is essential to provide the DCs with the optimal activation signals. DCs can be activated using a wide array of stimuli, such as proinflammatory cytokines and microbial or viral products. These stimuli can be delivered to the DCs by addition to the culture medium or through incorporation into the electroporation process (8, 10, 11).

3.4. Electroporation of MART-1/Melan-A Encoding mRNA into DC for Tumor Vaccination

1. Prepare 6-well plates with 5 mL RPMI medium supplemented with 1% AP, GM-CSFs (1,000 U/mL), and IL-4 (500 U/mL).
2. Adjust the physical parameters of the EasyjecT Plus® apparatus as follows: voltage, 300 V; capacitance, 150 μF, and resistance, 99 Ω, resulting in a pulse time of about 5 ms.
3. Wash 12 × 10^6 DCs (*see* **Note 8**) once with 2 mL serum-free RPMI 1640 medium and once with 2 mL Optimix Washing Solution A.
4. Mix 30 μg mRNA with 100 μL Optimix solution B and adjust to a total volume of 200 μL with sterile RNase-free water.
5. Subsequently, resuspend DCs in RNA solution (*see* **Note 9**) and transfer cell suspension into a 4-mm electroporation cuvette.
6. Insert the cuvette into the electroporation chamber and trigger the pulse.
7. Transfer the DCs immediately after electroporation into the prepared 6-well plate (*see* **Note 10**).
8. Incubate the electroporated DCs at 37°C and 5% CO_2 in a humidified incubator for 1 h to recover.
9. Centrifuge the DCs and resuspend them in 250 μL PBS supplemented with 1% HSA.
10. Transfer the DCs to a 1-mL syringe.

4. Notes

1. All cytokines prepared in-house were animal-protein-free, endotoxin-free, and biologically active, and were titrated against standards obtained from the National Institute for Biological Standards and Control (NIBSC, South Mimms, UK).

2. An in vitro transcription vector can be constructed by subcloning the coding sequence of any gene into the pGEM vector containing the poly(A) tail by standard cloning procedures.

3. The A27L amino acid mutation provides optimal binding affinity with HLA-A*0201 and recognition by T cells (16).

4. Plasmid DNA must be linearized with a restriction enzyme downstream of the insert to be transcribed, since linearized templates give transcripts of defined length and usually higher yields of transcripts. Circular plasmid DNA templates will generate extremely long, heterogeneous RNA transcripts. If there is no restriction site present approximately 50–200 bp downstream of the termination codon of the respective gene, an adequate restriction site must be created by site-directed mutagenesis. Confirm that cleavage is complete.

5. The ratio of the readings at 260 and 280 nm provides an estimate of the purity of RNA with respect to contaminants, such as protein, that absorb the UV light. Pure RNA has an A260/A280 ratio of 1.9–2.1.

6. Leukapheresis is performed with a COBE® Spectra™ (Cobe, Denver, CO) and approximately 8 L of blood is processed. The leukapheresed PBMCs can be washed with a COBE Cell-Processor 2991 to remove contaminating platelets. Samples from the washed cell suspension will be tested for hematocrit, total white blood cells (WBCs), and platelet count.

7. The generation of DCs in the presence of the cytokines interferon (IFN)-β and IL-3 also give rises to phenotypically and functionally immature DCs, but, in our laboratory, this type of DC could not be electroporated with in vitro transcribed mRNA (unpublished data).

8. Manufacture the cellular vaccine in a clean room, following the current guidelines of Good Manufacturing Practice (GMP). During the production of a vaccine, no other biological materials may be processed within the production area.

9. The RNA must be added to the cell suspension shortly before the pulse to prevent degradation due to RNases.

10. During the electroporation pulse, electrolysis of the medium occur. This affects the characteristics of the medium, notably the pH. Therefore, it is absolutely advisable to transfer the electroporated cells to fresh medium within seconds or as soon as possible after the pulse; failure to do so will considerably increase the mortality rate.

References

1. Steinman, R.M. and Dhodapkar, M. (2001) Active immunization against cancer with dendritic cells: the near future. *Int. J. Cancer.* **94**, 459–473.

2. Banchereau, J., Schuler-Thurner, B., Palucka, A.K., and Schuler, G. (2001) Dendritic cells as vectors for therapy. *Cell.* **106**, 271–274.

3. Tuyaerts, S., Noppe, S.M., Corthals, J., et al. (2002) Generation of large numbers of dendritic cells in a closed system using Cell Factories. *J. Immunol. Methods.* **264**, 135–151.

4. Mitchell, D.A. and Nair, S.K. (2000) RNA-transfected dendritic cells in cancer immunotherapy. *J. Clin. Invest.* **106**, 1065–1069.

5. Ponsaerts, P., Van Tendeloo, V.F., and Berneman, Z.N. (2003) Cancer immunotherapy using RNA-loaded dendritic cells. *Clin. Exp. Immunol.* **134**, 378–384.

6. Bonehill, A., Heirman, C., Tuyaerts, S., et al. (2004) Messenger RNA-electroporated dendritic cells presenting MAGE-A3 simultaneously in HLA class I and class II molecules. *J. Immunol.* **172**, 6649–6657.

7. Van Tendeloo, V.F., Ponsaerts, P., Lardon, F., et al. (2001) Highly efficient gene delivery by mRNA electroporation in human hematopoietic cells: superiority to lipofection and passive pulsing of mRNA and to electroporation of plasmid cDNA for tumor antigen loading of dendritic cells. *Blood.* **98**, 49–56.

8. Tuyaerts, S., Michiels, A., Corthals, J., et al. (2003) Induction of influenza matrix protein 1 and MelanA-specific T lymphocytes in vitro using mRNA-electroporated dendritic cells. *Cancer Gene Ther.* **10**, 696–706.

9. Bonehill, A., Heirman, C., Tuyaerts, S., et al. (2003) Efficient presentation of known HLA class II-restricted MAGE-A3 epitopes by dendritic cells electroporated with messenger RNA encoding an invariant chain with genetic exchange of class II-associated invariant chain peptide. *Cancer Res.* **63**, 5587–5594.

10. Michiels, A., Tuyaerts, S., Bonehill, A., et al.(2005) Electroporation of immature and mature dendritic cells: implications for dendritic cell-based vaccines. *Gene Ther.* **12**, 772–782.

11. Michiels, A., Breckpot, K., Corthals, J., et al. (2006) Induction of antigen-specific CD8+ cytotoxic T cells by dendritic cells co-electroporated with a dsRNA analogue and tumor antigen mRNA. *Gene Ther.* **13**, 1027–1036.

12. Grunebach, F., Kayser, K., Weck, M.M., Muller, M.R., Appel, S., and Brossart, P. (2005) Cotransfection of dendritic cells with RNA coding for HER-2/neu and 4-1BBL increases the induction of tumor antigen specific cytotoxic T lymphocytes. *Cancer Gene Ther.* **12**, 749–756.

13. Dannull, J., Nair, S., Su, Z., et al. (2005) Enhancing the immunostimulatory function of dendritic cells by transfection with mRNA encoding OX40 ligand. *Blood.* **105**, 3206–3213.

14. Mockey, M., Gonçalves, C., Dupuy, F.P., Lemoine, F.M., Pichon, C., and Midoux, P. (2006) mRNA transfection of dendritic cells: Synergistic effect of ARCA mRNA capping with Poly(A) chains in cis and in trans for a high protein expression level. *Biochem. Biophys. Res. Commun.* **340**, 1062–1068.

15. Reichardt V.L., Brossart, P., and Kanz, L. (2004) Dendritic cells in vaccination therapies of human malignant disease. *Blood Rev.* **18**, 235–243.

16. Valmori, D., Fonteneau, J.F., Lizana C.M., et al. (1998) Enhanced generation of specific tumor-reactive CTL in vitro by selected Melan-A/MART-1 immunodominant peptide analogues. *J. Immunol.* **160**, 1750–1758.

Chapter 11
Delivery of DNA into Natural Killer Cells for Immunotherapy

Kathrin Schoenberg, Hans-Ingo Trompeter, and Markus Uhrberg

Abstract Natural killer (NK) cells are highly resistant to transfection by conventional methods such as electroporation and lipofection. Recently, we reported the employment of a novel electroporation-based method, called nucleofection, which for the first time enabled efficient nonviral gene transfer into NK cells. In this study, we aimed at developing optimized conditions for the transfection of different NK cell lines as well as primary NK cells. Using EGFP (enhanced green fluorescent protein) or luciferase as reporter genes, suitable buffer conditions as well as instrument settings were defined. The new transfection methodology represents a useful tool for the immunotherapeutic use of NK cells, with the potential to enhance cytotoxicity as well as retarget the specificity of cytotoxic lymphocytes in clinical therapy of cancer and viral infection.

Keywords: nucleofection, NK cells, EGFP, luciferase, immunotherapy

1. Introduction

Natural killer (NK) cells are early-acting lymphocytes that play a crucial role in the control of viral infections as well as tumor development in concert with T- and B cells (1). Moreover, increasing evidence suggests that NK cells are important players in the process of elimination of residual tumor cells and promotion of graft-versus-leukemia (GvL) responses in stem cell transplantation (2). The specificity of NK cells is based on arrays of inhibitory and stimulatory surface receptors, which are able to sense downregulation of major histocompatibility (MHC) class I molecules as well as upregulation of stress-induced ligands, two important signs frequently accompanying pathogenic cellular dysfunctions (3). The two major effector functions exerted by NK cells are, first, cytotoxicity through lysis of susceptible target cells by release of perforin and granzymes from lytic endosomes and, second, production of cytokines such as interferon gamma (IFN-γ) to stimulate natural immune responses as well as adaptive immunity, predominantly of the Th1 type.

Recently, several studies considered the possibility of exploiting the antitumor abilities of allogeneic NK cells for immunotherapeutic use against leukemic cells as well as solid tumors (4). However, more sophisticated therapeutic approaches involving retargeting of NK cells to carcinogenic target cells were hampered by the lack of efficient gene transfer methods and were so far restricted to a few suitable NK cell lines (5). Since NK cells are highly resistant to transfection by conventional electroporation or lipofection protocols, several groups used retroviral approaches to introduce various target genes into NK cell lines (6, 7) or, more recently, primary NK cells (8, 9). Lately, nucleofection was introduced as a nonviral alternative for NK cell transfection (10). For the first time, the method enabled analysis of KIR gene regulation in primary NK cells (11). Nucleofection is a comparatively rapid and simple procedure and provides a viable alternative to retroviral gene transfer. Thus, it will facilitate the clinical use of NK cells, for example, by changing their target specificity through introduction of complementary DNAs (cDNAs) coding for tumor-specific humanized antibodies.

In this chapter, optimized conditions are defined for nucleofection of several NK cell lines as well as for primary NK and T cells.

2. Materials

2.1. Cell Culture

1. NK3.3 cells. The cloned NK cell line was kindly provided by Dr. J. Kornbluth (St. Louis University, MO). NK3.3 cells were maintained in RPMI 1640 supplemented with 25 mM HEPES, 2 mM L-glutamine, 10% fetal calf serum, 10% supernatant of a mixed lymphocyte culture, penicillin (100 U/mL), and streptomycin (0.1 mg/mL).
2. NKL cells. The IL-2-dependent leukemic NK cell line was kindly provided by Dr. P. Parham (Stanford University, CA). NKL cells were cultivated in RPMI 1640 supplemented with 25 mM HEPES, 2 mM L-glutamine, 10% FCS, penicillin (100 U/mL), streptomycin (0.1 mg/mL), and 200 U IL-2/mL.
3. Primary NK cells were cultured in RPMI 1640 supplemented with 25 mM HEPES, 2 mM l-glutamine, 10% FCS, penicillin (100 U/mL), streptomycin (0.1 mg/mL), 5% human AB-serum, and 1000 U IL-2/mL. Primary NK cells as well as NK cell lines were cultured at 37°C and 5% CO_2.

2.2. Isolation and Preparation of NK Cells

1. Ficoll separating solution (Biocoll, Biochrom, Berlin, Germany).
2. Phosphate-buffered saline (PBS) containing 2 mM EDTA (Roth, Karlsruhe, Germany).

3. Ammonium chloride solution (Roth).
4. Cell strainer (BD Biosciences, Heidelberg, Germany).
5. MACS system: MACS human CD56 Multisort Kit, MACS human CD3 microbeads, and MACS LS columns (Miltenyi Biotec, Bergisch Gladbach, Germany).

2.3. Plasmid DNA and Cloning

1. pGL3/GAPDH: The pGL3 basic vector (Promega, Mannheim, Germany) contains the GAPDH promoter inserted in front of the luciferase gene. Additionally, the vector contains an ampicillin resistance gene.
2. pEGFP-N1 (Clontech, Mountain View, CA): This consists of a CMV promoter inserted in front of EGFP (enhanced green fluorescent protein). Additionally, the vector contains a kanamycin resistance gene.
3. LB medium.
4. Plasmid DNA purification kit (Qiagen, Hilden, Germany).
5. Electrocompetent *Escherichia coli* bacteria cells.

2.4. Nucleofection

1. NK cell nucleofector solution (amaxa, Köln, Germany). Additional information regarding the use of modified buffers is available upon request.
2. Nucleofector (amaxa).
3. Cuvettes (0.2 cm).

2.5. Flow Cytometry

1. Monoclonal antibodies for flow cytometric analysis: anti-CD56- or anti-CD16-conjugated with phycoerythrin (PE) and CD3 coupled with PE-Cy5 (all from Beckman Coulter, Krefeld, Germany).
2. Flow cytometry: Cytomics FC 500 Series with CXP Software (all from Beckman Coulter, Krefeld, Germany).

2.6. Luminometry

1. Luciferase reporter gene assay kit (Roche, Grenzach-Wyhlen, Germany).
2. Luminometer MiniLumat9506 (Roche, Grenzach-Wyhlen, Germany).
3. Protein assay kit for normalization of protein content at 750 nm (Berthold, Bad Wildbad, Germany) (BIO-RAD, Munich, Germany).
4. Tubes (3.5 mL).

3. Methods

3.1. Preparation of NK Cells

3.1.1. Isolation of Peripheral Blood Mononuclear Cells

Peripheral blood mononuclear cells (PBMCs) were isolated from fresh human buffy coats, kindly provided by Dr. T. Tonn (Red Cross Blood Donor Service Hessia, Frankfurt, Germany), by sedimentation over Ficoll gradient.

1. Add 15 mL Ficoll to a 50-mL tube and slowly add 30 mL blood (*see* **Note 1**). Centrifuge the sample for 35 min at 440g and room temperature without brake.
2. Collect PBMCs from the interphase, transfer to a new tube, fill up with PBS/EDTA, and centrifuge for 7 min at 250g at 10°C.
3. Discard the supernatant and resuspend the cells with 1 mL PBS/EDTA.
4. To eliminate residual erythrocytes, fill the tube with 10 mL ammonium chloride. Incubate the suspension for 10 min at room temperature and subsequently centrifuge the cells for 7 min at 10°C and 250g.
5. Discard the supernatant and resuspend the cells with PBS/EDTA.

3.1.2. Magnetic Cell Separation

CD56$^+$CD3$^-$ NK cells were enriched from PBMCs by magnetic cell sorting (Miltenyi Biotech).

1. Briefly, incubate PBMCs with anti-CD56 monoclonal antibody (mAb) for 15 min at 4°C and isolate CD56$^+$ lymphocytes by using LS columns (*see* **Note 2**).
2. Deplete CD56$^+$/CD3$^+$ NK-like T cells by using CD3 magnetic microbeads. To reduce washing steps, CD56 magnetic beads are detached at the same time using CD56 release solution.
3. The resulting flow-through contains the enriched fraction of CD56$^+$/CD3$^-$ NK cells, which are stored on ice until use.

3.2. Nucleofection Procedure

1. Prepare 5 × 10^6 NK cells for nucleofection (*see* **Notes 3,4**). Centrifuge the cells for 7 min at 400g and 10°C to wash the cells. Discard the supernatant completely.
2. Prepare 1.5-mL plastic tubes with 3–5 µg plasmid DNA. In each experiment, two negative controls are included: one control without DNA that is nucleofected, one with DNA but without nucleofection.
3. Resuspend cells in 90 µL nucleofection buffer and transfer to DNA-containing tube.

4. Transfer sample to electroporation cuvette and insert into the nucleofector. Use the electrical setting U-01 (*see* **Notes 5,6**). Systematic testing of different nucleofector settings revealed that program U-01 resulted in the highest transfection efficiency, as shown in Fig. 11.1.

5. Following nucleofection, immediately add 500 µL of cell culture medium and transfer the cell suspension to a new 1.5-mL plastic tube. Incubate the cell suspension for 10 min at 37°C.

6. Add the nucleofected cell suspension to 1.5 mL prewarmed NK cell medium in six-well microtiter plates and incubate at 37°C and 5% CO_2 for 4 h for luciferase and 24 h for EGFP (*see* **Note 7**).

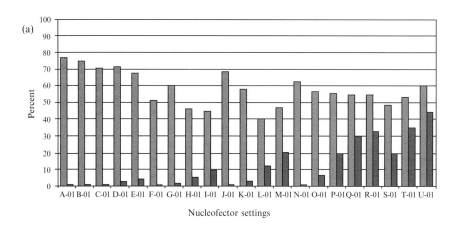

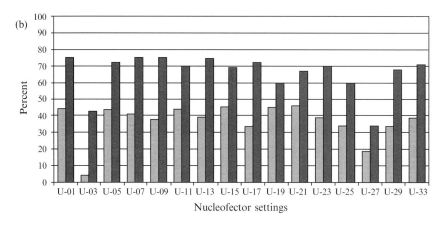

Fig. 11.1 Influence of different instrument settings on nucleofection efficiency. NK3.3 cells were nucleofected with pEGFP-N1, and the percentage of live cells (*light bar*) as well as the transfection efficiency, in percent, of live cells (*dark bar*) were measured by flow cytometry. (**a**) Systematic analysis of the different program settings revealed U as the best program. (**b**) Further analysis of subprograms revealed 01 as the best choice. In summary, the combination of high content of live cells and transfection efficiency was optimal at U-01

3.3. Analysis of Luciferase Activity

1. Resuspend the content of one well, fill in a 2-mL tube, and centrifuge for 7 min, at 300g and 4°C.
2. Discard the supernatant and resuspend the cells in 800 μL PBS/EDTA and centrifuge again using the same parameters.
3. Discard the supernatant and resuspend the cells in 70 μL lysis buffer and incubate the sample for 15 min at 20°C.
4. After incubation, centrifuge the sample for 2 min at 16,500g. Transfer the supernatant in a new tube and keep on ice until measurement.
5. For light emission measurement, mix 20 μL of the sample with 100 μL of assay buffer and measure luminescence.
6. To normalize the samples based on their protein content, light absorption is determined at 750 nm with the DC protein assay (BioRad).

3.4. Analysis of EGFP Activity

1. Nucleofected cells are incubated for 24 h to allow proper translation and protein folding of EGFP molecules.
2. Resuspend the content of one microtiter well and transfer the NK cell suspension in a plastic tube. Centrifuge and wash the cells with PBS/EDTA. Resuspend the cells in 200 μL PBS/EDTA.
3. Measure the transfection rate by flow cytometry (*see* **Note 8**).
4. The transfection efficiencies of different nucleofected primary lymphocyte populations (*see* **Note 9**) are shown in Fig. 11.2.

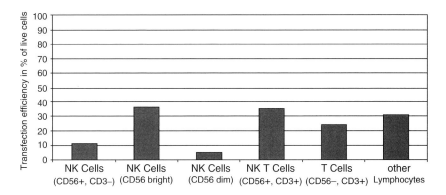

Fig. 11.2 Nucleofection efficiencies of different PBMC-derived lymphocyte populations. PBMCs were nucleofected with pEGFP-N1. After a 1-day cell culture period, cells were stained with anti-CD3-FITC and anti-CD56-PE mAbs and analyzed by flow cytometry. In a representative experiment, nucleofection efficiencies of CD56^{bright} and CD56^{dim} NK cells, NK-like T-cells, T cells, and CD3$^-$CD56$^-$ lymphocytes (non-NK-/non-T-cell fraction) are shown

4. Notes

1. Cell recovery is increased by 1:1 dilution of blood with PBS.
2. Use cooled LS columns and cooled reagents during the whole procedure of magnetic cell separation.
3. Optimal transfection rates are achieved when cells were starved for 2 days (no medium change) before nucleofection.
4. Trypan blue staining determines the dead cell rate in the culture, so that only living cells are used for nucleofection.
5. If the nucleofector shows an error message, check the reaction volume: It must be not more than $100\,\mu L$ and not less than $90\,\mu L$ to get optimal results.
6. Another reason for error messages of the nucleofector are bubbles in the cuvette. Check whether bubbles are inside and, if present, remove them by knocking the cuvette briefly on the bench.
7. Prepare the medium (it has to be prewarmed) in the six-well plates before the nucleofection procedure.
8. Optionally, for characterization of different surface receptors, cells can be counterstained with fluorescent mAbs. If they are coupled to dyes that do not interfere with the EGFP emission spectrum, one or two different surface markers can be measured with a standard flow cytometer. Typically, we use NK-cell- or T-cell-specific mAbs (e.g. anti-CD3, anti-CD56, or anti-CD16), coupled to PE or PE-Cy5.
9. The nucleofection procedure is not only useful for CD56dim NK cells, which are the main NK cell population and have predominantly cytotoxic effector functions, but also works even better for CD56bright NK cells, which are less cytotoxic but highly efficient at producing cytokines. Moreover, NK-like T cells, mainly consisting of effector/memory T cells, are nucleofected with more than 30% efficiency (Fig. 11.2).

References

1. Trinchieri, G. (1989) Biology of natural killer cells. *Adv. Immunol.* **47**, 187–376.
2. Ruggeri, L., Mancusi, A., Perruccio, K., Burchielli, E., Martelli, M.F., and Velardi, A. (2005) Natural killer cell alloreactivity for leukemia therapy. *J. Immunother.* **28**, 175–182.
3. Moretta, L., Bottino, C., Pende, D., Castriconi, R., Mingari, M.C., and Moretta, A. (2006) Surface NK receptors and their ligands on tumor cells. *Semin. Immunol.* **18**, 151–158.
4. Miller, J.S., Soignier, Y., Panoskaltsis-Mortari, A., et al. (2005) Successful adoptive transfer and in vivo expansion of human haploidentical NK cells in cancer patients. *Blood.* **105**, 3051–3057.
5. Uherek, C., Tonn, T., Uherek, B., et al. (2002) Retargeting of natural killer-cell cytolytic activity to ErbB2-expressing cancer cells results in efficient and selective tumor cell destruction. *Blood.* **100**, 1265–1273.
6. Tran, A.C., Zhang, D., Byrn, R., and Roberts, M.R. (1995) Chimeric zeta-receptors direct human natural killer (NK) effector function to permit killing of NK-resistant tumor cells and HIV-infected T lymphocytes. *J. Immunol.* **155**, 1000–1009.
7. Nagashima, S., Mailliard, R., Kashii, Y., et al. (1998) Stable transduction of the interleukin-2 gene into human natural killer cell lines and their phenotypic and functional characterization in vitro and in vivo. *Blood.* **91**, 3850–3861.
8. Becknell, B., Trotta, R., Yu, J., et al. (2005) Efficient infection of human natural killer cells with an EBV/retroviral hybrid vector. *J. Immunol. Methods.* **296**, 115–123.
9. Guven, H., Konstantinidis, K.V., Alici, E., et al. (2005) Efficient gene transfer into primary human natural killer cells by retroviral transduction. *Exp. Hematol.* **33**, 1320–1328.

10. Trompeter, H.I., Weinhold, S., Thiel, C., Wernet, P., and Uhrberg, M. (2003) Rapid and highly efficient gene transfer into natural killer cells by nucleofection. *J. Immunol. Methods.* **274**, 245–256.
11. Trompeter, H.I., Gomez-Lozano, N., Santourlidis, S., et al. (2005) Three structurally and functionally divergent kinds of promoters regulate expression of clonally distributed killer cell Ig-like receptors (KIR), of KIR2DL4, and of KIR3DL3. *J. Immunol.* **174**, 4135–4143.

Chapter 12
Electroporation of Adherent Cells In Situ for the Study of Signal Transduction and Gap Junctional Communication

**Leda Raptis, Adina Vultur, Heather L. Brownell, Evangelia Tomai,
Aikaterini Anagnostopoulou, Rozanne Arulanandam, Jun Cao,
and Kevin L. Firth**

Abstract Cultured adherent cells can be electroporated in situ, as they grow on
a glass slide coated with electrically conductive, optically transparent indium-tin
oxide (ITO). Although the introduction of DNA is a common use, the technique of
electroporation in situ is valuable for studying many aspects of signal transduction.
This is because, under the appropriate conditions, in situ electroporation can be
remarkably nontraumatic, while a large variety of molecules, such as peptides,
oligonucleotides, or drugs, are introduced instantly and into essentially 100% of
the cells, making this technique especially suitable for kinetic studies of effector
activation. Following the introduction of the material, the cells can be either
extracted or biochemically analyzed, or their morphology and gene expression can
be examined by immunocytochemistry. In this chapter, we describe the introduction
of a peptide blocking the Src-homology 2 domain of the adaptor Grb2 to inhibit
the activation of the downstream effector Erk1/2 by EGF. The setup includes
nonelectroporated, control cells growing side by side with the electroporated ones
on the same type of ITO-coated surface. In a modified version, this assembly can
be used very effectively for studying intercellular, junctional communication: cells
are grown on a glass slide half of which is ITO-coated. An electric pulse is applied
in the presence of the fluorescent dye lucifer yellow, causing its penetration into the
cells growing on the conductive part of the slide, and the migration of the dye to
the nonelectroporated cells growing on the nonconductive area is microscopically
observed under fluorescence illumination.

Keywords: electroporation, lucifer yellow, adherent cells, in situ electroporation,
ITO

1. Introduction

Electroporation is a fascinating cell membrane phenomenon with many applications.
In cultured adherent cells, electroporation can be conducted in situ, while cells are
grown on a glass surface coated with electrically conductive, optically transparent

indium-tin oxide (ITO). This coating promotes excellent cell adhesion and growth, allows direct visualization of the electroporated cells, and offers the possibility of detailed microscopic examination because of their extended morphology.

Although the introduction of DNA is a common use, electroporation in situ is valuable for studying many aspects of signal transduction. This is because under the appropriate conditions in situ electroporation can be remarkably nontraumatic. Unlike other techniques of cell permeabilization, it does not affect cell morphology, the length of the G1 phase of serum-stimulated cells (1), the activity of the extracellular signal regulated kinase (Erk1/2 or Erk), or two kinases commonly activated by a number of stress-related stimuli, JNK/SAPK and p38^{hog} (2), presumably because the pores reseal rapidly so that the cellular interior is restored to its original state (3). Besides, unlike calcium-phosphate- or lipofectamine-mediated transfection, which was found to activate the signal transducer and activator of transcription 3 (Stat3), no changes in Stat3 activity were noted by in situ electroporation (4). In addition, unlike other methods of permeabilization, such as streptolysin-O, no activation of nucleotideases was noted by in situ electroporation, which makes this method ideal for the introduction of radioactive nucleotides, such as [α^{32}P]GTP or [γ^{32}P]ATP (5, 6), for measuring the activity of membrane GTPases or for labelling cellular proteins with^{32}P in vivo (7). Finally, under appropriate conditions, the material can be introduced instantly and into essentially 100% of the cells, which makes this technique especially suitable for kinetic studies of effector activation.

The procedures described in this chapter are applicable to a wide variety of nonpermeant molecules, such as peptides (8–10), oligonucleotides (such as antisense DNA, double-stranded DNA decoy oligonucleotides, or small interfering RNA (siRNA) (9, 11, 12)), radioactive nucleotides (5, 6, 13, 14), proteins (15), DNA (1), or drugs (16). These compounds can be introduced alone or in combination, at the same or different times, in growth-arrested cells, cells at different stages of their division cycle, or cells at different stages of differentiation (17). After the introduction of the material, the cells can be extracted and biochemically analyzed, or their morphology and biochemical properties examined in situ. The technique can be applied to a large variety of adherent cell types. Cells which do not adhere well can be grown and electroporated on the same conductive slides coated with CelTak™, fibronectin, or polylysine. In a modified version, this assembly can be used for studying intercellular, junctional communication. In this chapter we describe two applications: the introduction of a peptide blocking the Grb2-SH2 domain to inhibit activation of Erk1/2 by EGF and the examination of gap junctional, intercellular communication by electroporation on a partly conductive slide.

The introduction of peptides to interrupt signalling pathways by using the modification of in situ electroporation described is a powerful approach for the in vivo assessment of the relevance of in vitro interactions. An essentially complete and specific inhibition of EGF-dependent Erk activation can be achieved through electroporation of specific peptides. The stepwise dissection of signalling cascades is essential for the understanding of proliferative pathways, and

the examination of the potential of different peptides to inhibit a specific pathway is the first important step in the development of peptidomimetic drugs for the rational treatment of neoplasia.

2. Materials

1. Cell culture: Cells are grown in Dulbecco's modification of Eagle's medium (DMEM) with 10% calf serum (Life technologies).
2. Calcium-free DMEM (Sigma).
3. Lucifer yellow solution: dilithium salt (Sigma or Molecular Probes). To make a 10-mL solution, dissolve 50 mg Lucifer yellow in 10 mL calcium-free DMEM in a final concentration of 5 mg/mL. Stable at 4°C for several months (*see* **Note 1**).
4. Epidermal growth factor (EGF) solution: EGF (Intergen, Cat. no. 4110–80) is made as a 10,000× stock solution. Dissolve 100 μg of lyophilized EGF in 100 μL sterile water and freeze in 5-μL aliquots. Just prior to the experiment, add 1 μL stock solution to 10 mL calcium-free DMEM (final concentration, 100 ng/mL). The stock solution is stable at −20 or 70°C for up to 2 months.
5. Lysis buffer: 50 mM Hepes, pH 7.4, 150 mM NaCl, 10 mM EDTA, 10 mM $Na_4P_2O_7$, 100 mM NaF, 2 mM sodium orthovanadate, 0.5 mM phenylmethylsulfonyl fluoride (PMSF), aprotinin (10 μg/mL), leupeptin (10 μg/mL), and 1% Triton X-100. (All chemicals can be purchased from Sigma)
6. CelTak™ solution (BD Biosciences, Cat. no. 354240).
7. Rabbit anti-peptide antibodies against phosphorylated Erk1/2 kinase (Biosource International, Cat. No. 44-680). When stored frozen in aliquots, they are stable for more than 5 years. They were used at 1:500 for immunostaining and 1:10,000 for Western blotting. However, both the stability and the final dilutions vary from lot to lot.
8. Cell staining kit, including the goat serum, secondary antibody, and avidin-biotin complex (Vector Labs, Vectastain Kit, Cat. No. PK6101).
9. Peptides: the Grb2-SH2 binding peptide was based on the sequence flanking the Y^{1068} of the EGF receptor (PVPE-Pmp-INQS; mol. wt., 1,123). To enhance stability of the phosphate group, the phosphotyrosine analog phosphono-methylphenylalanine (Pmp), which cannot be cleaved by phosphotyrosine phosphatases yet binds to SH2 domains with high affinity and specificity (18), was incorporated at the position of phosphotyrosine. The Pmp monomer was custom-synthesized by Color your enzyme Inc. (Kingston, ON, Canada). As a control, we used the same peptide containing phenylalanine at the position of Pmp. Peptides were synthesized by the Queen's University Core Facility, using standard Fmoc chemistry. Prepare a solution of 5–10 mg/mL (~5–10 mM) of each peptide in calcium-free DMEM (*see* **Note 2**).
10. System for electroporation in situ: Epizap model EZ-16 (Ask Science Products Inc., Kingston, ON, Canada. www.askscience.com).

11. Inverted, phase contrast and fluorescence microscope, equipped with a filter for lucifer yellow. (e.g. Olympus, model IX70).

3. Methods

The technique of in situ electroporation can be used equally effectively for large-scale biochemical experiments (8–10, 14) or for the detection of biochemical or morphological changes in situ (19). Cells are grown on glass slides coated with conductive and transparent indium-tin oxide. The cell growth area is defined by a

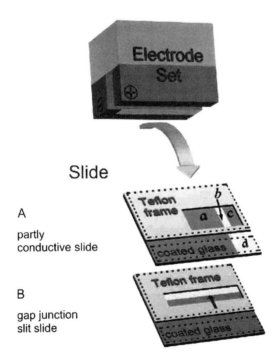

Fig. 12.1 Electroporation electrode and slide assembly. Cells are grown on glass slides coated with conductive and transparent indium-tin oxide (ITO), within a "window" cut into a Teflon frame as shown. The peptide solution is added to the cells and introduced by an electrical pulse delivered through the electrode set which is placed directly on the frame. *Dotted lines* point to the positions of the negative and positive electrodes during the pulse. Two slide configurations are depicted. **A**: partly conductive slide assembly, with electroporated [*a*] and nonelectroporated [*c*] cells growing on the same type of ITO-coated surface. [*b*]: area where the conductive coating has been stripped, exposing the nonconductive glass underneath. Cells growing in areas [*b*] and [*c*] are not electroporated (see Fig. 12.2). Cells in area [*b*] grow on plain glass, while cells in area [*c*] grow on ITO-coated glass. [From (19), reprinted with permission]. **B**: partly conductive slide assembly for use in the examination of gap junctional, intercellular communication. *Arrow* points to the transition line between conductive and nonconductive areas (Fig. 12.3). (From (20), reprinted with permission)

"window" formed with an electrically insulating frame made of Teflon, as illustrated in Fig. 12.1. The slides are placed in 3- or 6-cm petri dishes to maintain sterility. A stainless-steel electrode is placed on top of the cells resting on the frame, and an electrical pulse of the appropriate strength is applied.

3.1. Electroporation of Peptides

To study protein interactions in vivo, protein complexes can be disrupted through the introduction of peptides corresponding to the proteins' point(s) of contact. An example of this approach is described here.

Growth factors such as the epidermal growth factor (EGF) stimulate cell proliferation by binding and activating membrane receptors with cytoplasmic tyrosine kinase domains. This induces receptor autophosphorylation at distinct tyrosine residues, which become docking sites for a number of effector molecules, which are recruited to specific receptors through modules termed Src-homology 2 (SH2) domains [reviewed in (21)]. One such effector is the growth factor binding protein 2 (Grb2), which binds to the receptors for platelet-derived growth factor (PDGF) and EGF, an event activating the Sos/Ras/Raf/Erk pathway, which is central to the mitogenic response stimulated by many growth factors. To determine the functional consequences of disrupting the association of Grb2 with different receptors in vivo, large quantities of a peptide blocking the Grb2-SH2 domain (PVPE-Pmp-INQS) can be delivered into intact, living NIH 3T3 fibroblasts growing on a fully conductive slide, and the extent of signal inhibition can be quantitated by Western blotting using antibodies specific for the phosphorylated, i.e., activated, form of the downstream effector Erk1/2 (19). However, if antibodies suitable for immunocytochemistry are available, then electroporation of the peptide can be performed on cells growing on a partly conductive slide, and the inhibition of Erk activation examined by immunohistochemistry. In this case, in conjunction with measurement of gene product activity, cellular morphology and possible toxic effects can also be examined. This approach can demonstrate the specificity of action of the Grb2-SH2 binding peptide as well as examine the distribution of signal inhibition across the cell layer. A distinct advantage is that it requires a small number of cells, and hence a sub-stantially smaller volume of peptide (~14 μL in the setup shown in Fig. 12.1A),compared with Western blotting (~140 μL for a cell growth area of $32 \times 10 \, mm^2$), which could be a significant consideration, given the cost of the custom-synthesized peptides. To precisely assess small background changes in morphology or gene expression levels, the presence of nonelectroporated cells, side by side with electroporated ones, can offer a valuable control, especially if both are growing on the same type of surface. This is achieved by electroporating the cells on a slide where the conductive coating is stripped in the pattern shown in Fig. 12.1A (22). A thin line of plain glass [b] separates the electroporated and control areas while etching extends to area [d], so that there is no electrical contact between the positive contact bar and area [c]. Application of

the pulse results in electroporation of the cells growing in area [*a*] exclusively, while cells growing in area [*b*] or [*c*] do not receive any pulse. In this configuration, electroporated cells [*a*] are being compared with nonelectroporated ones [*c*], while both are growing on ITO-coated glass. Since the coating is only about 1,600 Å thick, this transition zone does not alter the growth of cells across it and is clearly visible microscopically, even under a cell monolayer (Figs. 12.2 and 12.3).

1. Plate the cells on the slide and place the slide inside a petri dish of appropriate size. Uniform spreading of the cells is very important, since the optimal voltage depends, in part, upon the degree of cell contact with the conductive surface (*see* **Note 3**). Add a sufficient amount of medium (DMEM containing 10% calf serum) to cover the slide (~9 mL for a 6-cm dish). Pipette the cell suspension in the window cut in the Teflon frame (Fig. 12.1A) and place the petri dishes in a tissue-culture incubator till they reach the desired confluence.
2. Prior to the experiment, starve the cells overnight in DMEM without serum (*see* **Note 4**).
3. Aspirate the growth medium and gently wash the cells once with calcium-free DMEM.
4. Add the peptide dissolved in calcium-free DMEM to the cells by using a micropipettor.
5. Place the electrode on top of the cells, resting on the Teflon frame, and clamp it in place. If necessary, the electrode can be sterilized with 80% ethanol before the pulse, and the procedure can be carried out in a laminar-flow hood, using sterile solutions.
6. Apply 3–6 pulses of the appropriate voltage and capacitance (*see* **Note 3**).
7. Remove the electrode set. Since usually only a small fraction of the material enters the cells, the peptide solution may be carefully aspirated and used again.
8. Add serum-free growth medium and incubate the cells for 2–5 min at 37°C to recover.
9. Add EGF to the medium to a final concentration of 100 ng/mL and place in the incubator for 5–10 min. The controls receive the same volume of calcium-free DMEM.
10. Fix the cells with 4% paraformaldehyde and probe with the anti-active Erk antibody according to the manufacturer's instructions.

As shown in Fig. 12.2, electroporation of the Grb2-SH2 blocking peptide totally inhibited EGF-induced Erk activation (panel A, area "*a*"), while the control, phenylalanine-containing peptide had no effect (panel C, area "*a*"). This inhibition was uniform across the cell layer. It is especially noteworthy that the inhibition extends into three to four rows of the adjacent, nonelectroporated cells growing on the nonconductive part of the slide (panel A, squiggly bracket), probably because of movement of the 1,123-Da peptide through gap junctions (23). This finding constitutes compelling evidence that the observed inhibition must be due to the peptide, rather than an artifact of electroporation, since cells in this area did not receive any current. At the same time, as shown by phase contrast microscopy (panel B), there was no alteration in the morphology of the electroporated cells

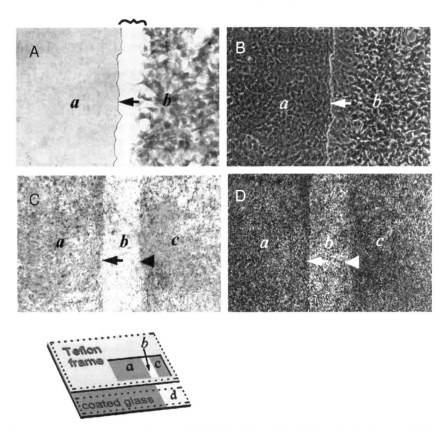

Fig. 12.2 The Grb2-SH2 blocking peptide inhibits EGF-mediated Erk activation in intact, living cells. The Grb2-SH2 blocking peptide (**A** and **B**) or its control, phenylalanine-containing counterpart (**C** and **D**) was introduced by in situ electroporation into NIH 3T3 cells growing on partly conductive slides (*inset*) and growth-arrested in spent medium. Five minutes after pulse application, cells were stimulated with EGF for 5 min, fixed, and probed for activated Erk1/2, and cells from the same field were photographed under bright-field (A and C) or phase contrast (B and D) illumination, respectively. A and B, ×240; C and D, ×40. *Arrow* points to the transition line between the stripped [*b*] and electroporated [*a*] areas, while *arrowhead* points to the line between the control ITO-coated [*c*] and stripped [*b*] areas (*inset*). Cells growing on the left side [*a*] are electroporated, while cells on the stripped zone [*b*] or right side [*c*] of the slide do not receive any pulse. Note that the Grb2-SH2 blocking peptide dramatically reduced the EGF signal (A, *a*), while the degree of Erk activation is the same on both sides of the slide (*a* or *c*) for cells electroporated with the control, phenylalanine-containing peptide (C). In A, the inhibition of the signal extends into about three to four rows of adjacent cells in the nonelectroporated area (squiggly bracket in *b*), probably because of movement of the peptide through gap junctions (23). At the same time, there is no detectable effect upon cell morphology, as shown by phase contrast (B and D). (From (19), reprinted with permission)

under these conditions, suggesting that the observed effect is a result of a specific inhibition, rather than toxic action. As expected, the phenylalanine-containing, control peptide (panels C and D), had no effect upon Erk activation (19).

Examination of the peptide concentration achieved inside the cell indicated that it varied with the size of the peptide and voltage used, but it is 1–5% the concentration applied to the cells (8). We and others have used this technique extensively for the introduction of peptides, peptidomimetics, RNA, drugs, and nucleotides for the study of signal transduction (8–16, 24).

3.2. Electroporation on a Partly Conductive Slide for the Assessment of Gap Junctional, Intercellular Communication

Gap junctions are membrane channels that serve as conduits for the passage of small molecules between the interiors of cells. Gap junctional, intercellular communication is long thought to play an important role in the growth as well as metastatic potential of human cancer cells, while oncogene expression invariably results in a decrease in gap junctional, intercellular communication (GJIC) (25, 26). Junctional permeability is often investigated through microinjection of a fluorescent dye such as lucifer yellow, followed by observation of its migration into neighboring cells, through fluorescence recovery after photobleaching, scrape-loading (27), or other methods. These are invariably time-consuming approaches, requiring expensive equipment, while the mechanical manipulation of the cells may disturb cell-to-cell contact areas, interrupt gap junctions, and cause artifactual uncoupling. These problems can be overcome by an in situ electroporation approach, using a setup in which cells are grown on a glass slide, only parts of which is coated with ITO (Fig. 12.1B). An electric pulse is applied in the presence of lucifer yellow, causing its penetration into the cells growing on the conductive part of the slide, and the migration of the dye to the nonelectroporated cells growing on the nonconductive area is microscopically observed under fluorescence illumination (17, 23, 28). In this way, dye transfer through gap junctions can be precisely quantitated simultaneously and in a large number of cells, without any detectable disturbance to cellular metabolism (1, 3). At the same time, contrary to microinjection or scrape-loading, there is no mechanical manipulation of the cells. This approach, offering the ability to quantitate GJIC, made possible the examination of the link between levels of oncogene expression, transformation, and GJIC, as well as observation of gap junction closure upon adipocytic differentiation (17, 23, 29). This technique was later adapted for GJIC examination of lines established from human lung carcinomas or cells freshly explanted from lung tumor tissues (28, 30).

1. Plate the cells on partly conductive slides in 3-cm petri dishes (Fig. 12.1B) (20).
2. Aspirate the medium. Wash the cells with calcium-free DMEM (*see* **Note 5**).
3. Add the lucifer yellow solution.
4. Place the upper electrode and apply a pulse of appropriate strength, so that the cells growing on the conductive coating at the border with the nonconductive area are electroporated without being damaged. This area receives slightly

larger amounts of current than the rest of the conductive growth surface (*see* **Note 3**).

5. Add calcium-free DMEM containing 10% dialysed serum. Remove the electrode and incubate the cells for 3–5 min in a 37°C, CO_2 incubator (*see* **Note 6**).
6. Wash the unincorporated dye with calcium-free DMEM (*see* **Note 6**).
7. Microscopically examine under fluorescence and phase contrast illumination (Fig. 12.3).
8. Quantitate intercellular communication: Photograph the cells with a 10–20x objective under fluorescence and phase contrast illumination (Fig. 12.3E, F). Identify and mark electroporated cells at the border with the nonconductive area (stars) and fluorescing cells on the nonconductive side (circles), where the dye has transferred through gap junctions. Divide the total number of fluorescing cells on the nonconductive area by the number of electroporated cells along the border with the etched side. The transfer from at least 200 contiguous electroporated border cells is calculated for each experiment (23). A careful kinetic analysis of dye transfer from 10 s to 2 h showed that the observed transfer is essentially complete by 5 min for all lines tested, while fluorescence is eliminated from the cells within ~60 min. After the transfer is complete, cells can be fixed with formaldehyde, in which case fluorescence is retained for several hours.

3.3. Electroporation on ITO in the Absence of a Top Electrode

The earlier-mentioned procedures are adequate for the examination of gap junctional communication of cells which adhere well to solid surfaces such as ITO; however, a large number of cell types tend to detach easily during the placement and removal of the top electrode, especially at high densities. Such cells have to be plated on CelTak™ or other molecules to improve cell adhesion and prevent detachment especially during removal of the electrode. For certain cells, e.g., differentiated adipocytes, even this is inadequate and the removal of the electrode can result in sufficient disturbance as to abolish the differentiation process. To avoid these problems, we developed an assembly where gap junctional communication can be examined without using a top electrode. An added advantage is that in the absence of an upper electrode the recovery of the material after electroporation is very easy and the volume of material required is lesser. This approach can also be employed for the introduction of a variety of molecules, such as peptides or siRNA, followed by detection of changes in gene expression by immunocytochemistry (19). The procedure is the same as in sect. 3.2, the only difference being the geometry of the electrodes; cells are plated on two coplanar electrodes separated by a nonconductive strip, which are supported by the same glass slide, rather than a stainless-steel electrode placed above the cells (Fig. 12.4) (31).

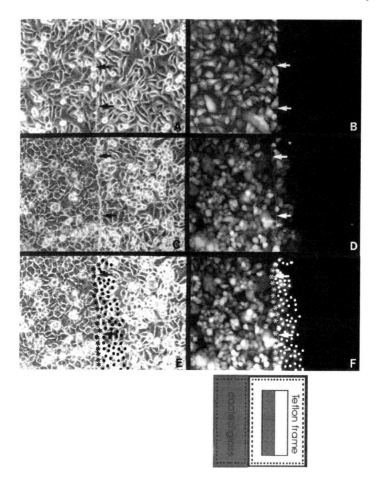

Fig. 12.3 In situ electroporation on a partly conductive slide for the measurement of intercellular, junctional communication. **A** and **B**: Cells established from a human lung carcinoma (line A549) were plated on partly conductive slides (*inset*) and, after reaching confluence, electroporated in the presence of lucifer yellow (5 mg/mL). After washing the unincorporated dye, cells from the same field were photographed under phase contrast (A) or fluorescence (B) illumination. Note the absence of dye transfer through gap junctions. **C** and **D**: Cells established from a human lung carcinoma (line QU-DB) were plated on partly conductive slides, electroporated, washed, and photographed as stated earlier. (C) phase contrast, (D) fluorescence illumination. Note the extensive dye transfer through gap junctions. **E** and **F**: To quantitate intercellular communication, the number of cells in C and D above into which the dye has transferred through gap junctions per electroporated border cell was calculated by dividing the total number of fluorescing cells on the nonconductive side (*white circles*) by the number of cells growing at the border with the conductive coating (*white stars*). To better illustrate the differences, very dense cultures of both types of cells are shown. In all photomicrographs, the left side is conductive. *Arrows* on the nonconductive side point to the interphase between conductive and nonconductive areas. ×200. (From (30), reprinted with permission)

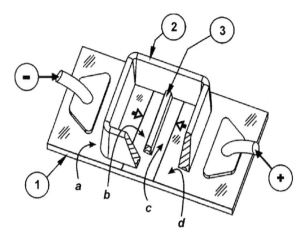

Fig. 12.4 Electroporation in situ in the absence of an upper electrode. Cells are grown on two coplanar electrodes of indium-tin oxide [*a*] and [*d*], which are supported by the same glass slide substrate [*1*]. Fluid is contained in a region above the cells by a plastic enclosing wall [2]. A "dam" of 2 mm [*3*] is attached to the inner sides of the enclosing wall and to the bare glass in between the two electrodes ([*b*], [*c*]). The two electrodes are connected to the pulse source [+] and [−], outside the cell growth area. Two open *arrows* point to the transition lines between conductive and nonconductive areas. The enclosing wall and the dam are shown in a cut-out for clarity (31)

1. Plate the cells in the special chamber (Fig. 12.4).
2. Aspirate the medium. Wash the cells with calcium-free DMEM (*see* **Note 5**).
3. Add the lucifer yellow solution.
4. Apply a pulse of appropriate strength, so that the cells growing on the conductive coating at the border with the nonconductive area are electroporated without being damaged.
5. Add calcium-free DMEM containing 10% dialysed serum and incubate the cells for 3–5 min in a 37°C, CO_2 incubator (*see* **Note 6**).
6. Wash the unincorporated dye with calcium-free DMEM (*see* **Note 6**).
7. Microscopically examine under fluorescence and phase contrast illumination and quantitate as in sect. 3.2.

Acknowledgments The financial assistance of the Canadian Institutes of Health Research (CIHR), the Canadian Breast Cancer Research Alliance, the Natural Sciences and Engineering Research Council of Canada (NSERC), and the Cancer Research Society Inc. is gratefully acknowledged. AV was the recipient of NSERC and Ontario Graduate studentships, a Queen's University Graduate Award (QGA), and a Queen's University travel grant. HB was the recipient of a studentship from the Medical Research Council of Canada and a Microbix Inc. travel award. AA was supported by a QGA and a predoctoral traineeship award from the Department of Defense Breast Cancer Research Program (BCRP-CDMRP, Award #: W81XWH-05-1-0224). RA was the recipient of a CIHR studentship and a QGA. ET was the recipient of an NSERC studentship and an award from the Ontario Government. JC was the recipient of a postdoctoral fellowship from Queen's University. We are grateful to Dr. Erik Schaefer of Biosource Int. for numerous suggestions and valuable discussions.

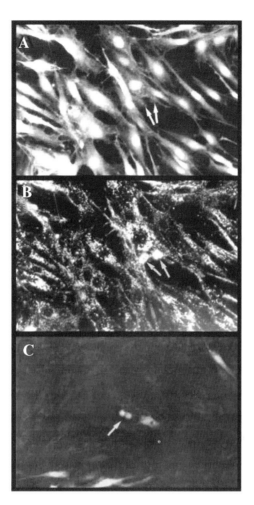

Fig. 12.5 Electroporation of mitotic cells requires a higher voltage when compared with that of cells in other phases of the cycle. **A**: Fluorescence microscopy of rat fibroblasts after introduction of IgG by means of an optimum electrical pulse. Cells were fixed and probed with FITC-conjugated anti-chicken IgG. **B**: Dark-field microscopy of the same field. Note the cell in late anaphase (A and B, *arrow*) which exhibits a very faint fluorescence, although most other cells are fluorescing. **C**: Fluorescence micros-copy of rat fibroblasts treated with an electrical pulse that was stronger than the optimum. The intensely fluorescing mitotic cell (*arrow*) stands out against a background of cells mostly killed by the pulse, apparently because it requires a higher voltage. (From (1), reprinted with permission)

4. Notes

1. The purity of the material to be electroporated is of paramount importance. Substances such as detergents, preservatives, or antibiotics could kill the cells into which they are electroporated, even if they have no deleterious effects if added to the culture medium of nonelectroporated cells. In addition, the material has to be soluble at the high concentrations required. For the examination of gap junctional communication, if the cells are very flat, then higher concentrations

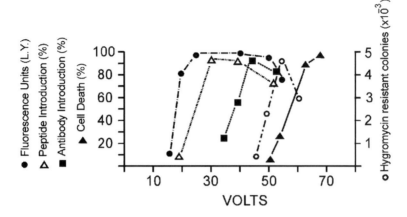

Fig. 12.6 Effect of field strength upon the introduction of different molecules. Six pulses of different voltages were applied to confluent rat F111 fibroblasts growing on a conductive surface of $4 \times 7 \, mm^2$, from a 0.2-µF capacitor, in the presence of lucifer yellow, 5 mg/mL (•); Grb2-SH2 blocking peptide, 5 mg/mL (Δ); chicken IgG, 5 mg/mL (■); or pY3 plasmid DNA, 100 µg/mL, coding for resistance to hygromycin (○). Cells were lysed, and lucifer yellow fluorescence was measured using a model 204A fluorescence spectrophotometer (•), probed with the anti-active Erk antibody (Δ), probed for incorporated IgG with an FITC-coupled anti-IgG antibody (■), or selected for hygromycin resistance (○) (1). Cell killing (▲) was assessed by calculating the plating efficiency of the cells 1 h after the pulse. Note that a wider range of voltages (20–50 V) permits efficient introduction of lucifer yellow with no detectable loss in cell viability, than the introduction of IgG or DNA. Points represent averages of at least three separate experiments. L.Y.: lucifer yellow

of the tracking dye, lucifer yellow, must be used (up to 50 mg/mL) to obtain an accurate assessment. In such cases, purity and solubility are even more important. All solutions must be kept at 37°C, which facilitates pore closure and efficient electroporation.

2. Peptides: The concentration of peptide required varies with the strength of the signal to be inhibited. For example, for the inhibition of the HGF-mediated Stat3 activation in MDCK cells, a concentration of 1 µg/mL of a peptide blocking the SH2 domain of Stat3 (ᴾYVNV) is sufficient (9), while for the inhibition of the EGF-mediated Erk activation in a variety of fibroblasts or epithelial cells, a concentration of 5–10 mg/mL (~5–10 mM) of the Grb2-SH2-blocking peptide is necessary (19). Peptides must be purified by high-pressure liquid chromatography. The pH of the peptide solution must be neutral, as indicated by the color of the DMEM medium where the peptide is dissolved. If it is too acidic, then it must be carefully neutralized with NaOH. In this case, the salt concentration of the no-peptide controls must be adjusted to the same level with NaCl, because a change in conductivity may affect the optimal voltage required (*see* "**Note 3**").

Any peptide soluble in the growth medium can be very effectively electroporated. Good solubility is especially important because the concentration needed for effective signal inhibition can be as high as 10 mg/mL. Nevertheless, at least for certain applications, the inclusion of dimethyl sulfoxide (DMSO) in the electroporation solution at a concentration of up to 5%, which might aid peptide solubility, did not significantly affect the results. However, a number of peptides, e.g., peptides made as fusions with the homeobox-domain or other membrane-translocation sequences, are usually not sufficiently soluble for this application. We have found that, in general, peptides or other substances that can translocate across the membrane cannot be electroporated effectively. This could, in part, be due to their low solubility, but also because they have an affinity for the membrane, so that they cannot penetrate easily during the short time that the pores are open during the electroporation process.

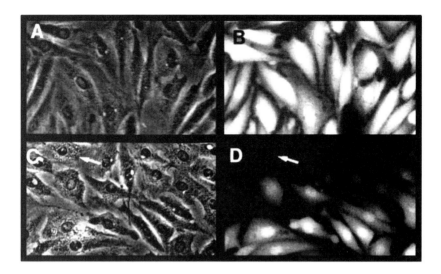

Fig. 12.7. Determination of the optimal voltage. Rat F111 fibroblasts growing on fully conductive slides (4 × 7 mm, *inset*) were electroporated in the presence of Lucifer yellow (5 mg/mL) by using 6 pulses of 30 V (**A** and **B**), or 50 V (**C** and **D**) delivered from a 0.2-µF capacitor. After washing the unincorporated dye, cells were photographed under phase contrast (panels A and C) or fluorescence (panels B and D) illumination. *Arrows* in C and D point to a cell which has been killed by the pulse. Note the dark, pycnotic and prominent nucleus under phase contrast and the flat, nonrefractile appearance. Such cells do not retain any electroporated material, as shown by the absence of fluorescence (D). It is especially striking that cells situated at the corners of the electroporated area received a larger amount of current and have been killed by the pulse. ×240

3. Determination of the optimal voltage and capacitance: electrical field strength and viability are critical parameters for cell permeation. Using a capacitor-discharge pulse generator, it is generally easier to select a discrete capacitance value for a given electroporated area and gap between the conductive coating and the negative electrode and then precisely control the voltage. Both parameters depend upon the size of the electroporated area; larger conductive growth areas necessitate higher voltages and/or higher capacitances for optimal permeation. The optimal pulse strength depends upon the cell type, namely, the degree of cell contact with the conductive surface. Densely growing, transformed cells or cells in a clump require higher voltages for optimum permeation than do sparse, subconfluent cells possibly due to the larger amounts of current passing through an extended cell. Similarly, cells that have been detached from their growth surface by vigorous pipetting prior to electroporation require substantially higher voltages. It is especially striking that cells in mitosis remain intact under conditions in which most cells in other phases of the cycle are permeated (Fig. 12.5) (1). In addition, cells growing and electroporated on CelTak™-coated slides require slightly higher voltages than do cells growing directly on the slide.

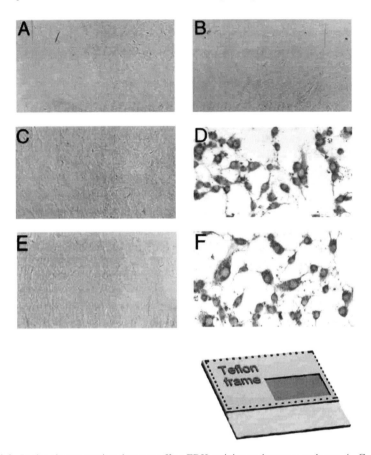

Fig. 12.8 In situ electroporation does not affect ERK activity or the stress pathway. **A**, **C**, and **E**: NIH3T3 cells were plated on fully conductive slides (*inset*, conductive growth area 4 × 7 mm²), growth-arrested in spent medium and electroporated in the absence of peptide (40 V, 0.2 μF, 6 pulses). Ten minutes after the pulse, cells were fixed and stained for activated ERK (A), activated JNK/SAPK (C), or activated p38hog (E), respectively. Electroporated cells were photographed under bright-field illumination. **B**, **D**, and **F**: NIH3T3 cells were plated on conductive slides, treated with UV light for 10 min, fixed and stained for activated ERK (B), activated JNK/SAPK (D), or activated p38hog (F), respectively. ×240. From (3) reprinted with permission

The margins of voltage tolerance depend upon the size and electrical charge of the molecules to be introduced (Fig. 12.6). For the introduction of small, uncharged molecules such as lucifer yellow or peptides, a wider range of field strengths permits effective permeation with minimal damage to the cells, than for the introduction of antibodies or DNA (1, 5). If the material is applied in a medium with a lower salt concentration than DMEM, then the voltages required are lower, possibly due to the simultaneous hypotonic shock to the cells, and to the longer duration of the pulse because of the lower conductivity of the medium. Conversely, electroporation in a hypertonic solution requires higher voltages for optimum permeation. In all cases, cell damage is microscopically manifested by the appearance of dark nuclei under phase contrast illumination, and the absence of fluorescence. This may be more pronounced at the edges of the conductive coating, close to the Teflon frame, because higher amounts of current will flow through these areas (Fig. 12.7C, D; arrow).

Under the appropriate conditions, electroporation was shown not to affect the activity of Erk1/2 or the stress-activated kinases, JNK/SAPK and p38^{hog}. This was shown by probing with antibodies specific for the activated forms of these kinases (Fig. 12.8); no activation of JNK/SAPK or p38^{hog} was found, under conditions of up to 50 V (Fig. 12.8C, E). These kinases were, however, slightly activated at voltages higher than 60 V, when more than 60% of the cells were killed by the pulse (not shown).

4. The margins of voltage tolerance were found to be wider for cells growing in 10% serum than for serum-starved cells. To achieve efficient electroporation, especially for sparsely growing cells, it may be advantageous to grow cells in spent medium, rather than in medium lacking serum entirely. Spent medium is prepared by growing cells to confluence in DMEM with 10% serum, collecting the supernatant 5 days later, and diluting it 1:1 with fresh DMEM.

5. For the determination of GJIC, it is important to wash the dye using a calcium-free solution (growth medium or phosphate-buffered saline). If calcium-containing growth medium is used instead, the values obtained may be reduced, presumably because of the calcium influx, which was shown to interrupt junctional communication.

6. The inclusion of dialysed serum at this point helps pore closure.

References

1. Raptis, L. and Firth, K.L. (1990) Electroporation of adherent cells *in situ*. *DNA Cell Biol.* **9**, 615–621.

2. MacCorkle, R.A. and Tan, T.H. (2005) Mitogen-activated protein kinases in cell-cycle control. *Cell Biochem. Biophys.* **43**, 451–461.

3. Brownell, H.L., Lydon, N., Schaefer, E., Roberts, T.M. and Raptis, L. (1998) Inhibition of epidermal growth factor-mediated ERK1/2 activation by *in situ* electroporation of nonpermeant [(alkylamino)methyl]acrylophenone derivatives. *DNA Cell Biol.* **17**, 265–274.

4. Arulanandam, R., Vultur, A. and Raptis, L. (2005) Transfection techniques affecting Stat3 activity levels. *Anal. Biochem.* **338**, 83–89.

5. Brownell, H.L., Firth, K.L., Kawauchi, K., Delovitch, T.L. and Raptis, L. (1997) A novel technique for the study of Ras activation: electroporation of [α^{32}P]GTP. *DNA Cell Biol.* **16**, 103–110.

6. Tomai, E., Vultur, A., Balboa, V. et al. (2003) *In situ* electroporation of radioactive compounds into adherent cells. *DNA Cell Biol.* **22**, 339–346.

7. Raptis, L., Vultur, A., Brownell, H.L. and Firth, K.L. (2006) Dissecting pathways: *in situ* electroporation for the study of signal transduction and gap junctional communication. In: Celis, J.E., (ed.). *Cell biology: a laboratory handbook*. Academic, NY, pp. 341–354.

8. Giorgetti-Peraldi, S., Ottinger, E., Wolf, G., Ye, B., Burke, T.R., Jr. and Shoelson, S.E. (1997) Cellular effects of phosphotyrosine-binding domain inhibitors on insulin receptor signalling and trafficking. *Mol. Cell. Biol.* **17**, 1180–1188.

9. Boccaccio, C., Ando, M., Tamagnone, L. et al. (1998) Induction of epithelial tubules by growth factor HGF depends on the STAT pathway. *Nature*. **391**, 285–288.

10. Bardelli, A., Longati, P., Gramaglia, D. et al. (1998) Uncoupling signal transducers from oncogenic MET mutants abrogates cell transformation and inhibits invasive growth. *Proc. Nat. Acad. Sci. U.S.A.* **95**, 14379–14383.

11. Gambarotta, G., Boccaccio, C., Giordano, C., Ando, M., Stella, M.C. and Comglio, M.C. (1996) Ets up-regulates met transcription. *Oncogene*. **13**, 1911–1917.

12. Arulanandam, R., Vultur, A. and Raptis, L. (2005) Transfection techniques affecting Stat3 activity levels. *Anal. Biochem.* **338**, 83–89.

13. Boussiotis, V.A., Freeman, G.J., Berezovskaya, A., Barber, D.L. and Nadler, L.M. (1997) Maintenance of human T cell anergy: blocking of IL-2 gene transcription by activated Rap1. *Science*. **278**, 124–128.

14. Raptis, L., Vultur, A., Tomai, E., Brownell, H.L. and Firth, K.L. (2006) *In situ* electroporation of radioactive nucleotides: assessment of Ras activity and ^{32}P-labelling of cellular proteins.

In: Celis, J.E. (ed.). *Cell biology: a laboratory handbook*. Academic, San Diego, CA, pp. 329–339.

15. Nakashima, N., Ross, D.W., Xiao, S. et al. (1999) The functional role of crk II in actin cytoskeleton organization and mitogenesis. *J. Biol. Chem.* **274**, 3001–3008.

16. Marais, R., Spooner, R.A., Stribbling, S.M., Light, Y., Martin, J. and Springer, C.J. (1997) A cell surface tethered enzyme improves efficiency in gene-directed enzyme prodrug therapy. *Nat. Biotechnol.* **15**, 1373–1377.

17. Brownell, H.L., Narsimhan, R.P., Corbley, M.J., Mann, V.M., Whitfield, J.J. and Raptis, L. (1996) Ras is involved in gap junction closure in mouse fibroblasts or preadipocytes but not in differentiated adipocytes. *DNA Cell Biol.* **15**, 443–451.

18. Hodges, R.S. and Smith, J.A. (eds.) (1994) *Peptides: chemistry, structure and biology*. Escom, Leiden.

19. Raptis, L., Brownell, H.L., Vultur, A.M., Ross, G.M., Tremblay, E. and Eliott, B.E. (2000) Specific inhibition of growth factor-stimulated ERK1/2 activation in intact cells by electroporation of a Grb2-SH2 binding peptide. *Cell Growth Differ.* **11**, 293–303.

20. Raptis, L., Tomai, E. and Firth, K.L. (2000) Improved procedure for examination of gap junctional, intercellular communication by *in situ* electroporation on a partly conductive slide. *Biotechniques.* **29**, 222–226.

21. Schlessinger, J. (2000) Cell signaling by receptor tyrosine kinases. *Cell.* **103**, 211–225.

22. Firth, K.L., Brownell, H.L. and Raptis, L. (1997) Improved procedure for electroporation of peptides into adherent cells *in situ*. *Biotechniques.* **23**, 644–645.

23. Raptis, L., Brownell, H.L., Firth, K.L. and MacKenzie, L.W. (1994) A novel technique for the study of intercellular, junctional communication: electroporation of adherent cells on a partly conductive slide. *DNA Cell Biol.* **13**, 963–975.

24. Anagnostopoulou, A., Vultur, A., Arulanandam, R. et al. (2006) Differential effects of Stat3 inhibition in sparse *vs* confluent cells. *Cancer Lett.* **242**, 120–132.

25. Vinken, M., Vanhaecke, T., Papeleu, P., Snykers, S., Henkens, T. and Rogiers, V. (2006) Connexins and their channels in cell growth and cell death. *Cell Signal.* **18**, 592–600.

26. Weinstein, R.S., Merk, F.B. and Alroy, J. (1976) The structure and function of intercellular junctions in cancer. *Adv. Cancer Res.* **23**, 23–89.

27. el-Fouly, M.H., Trosko, J.E. and Chang, C.C. (1987) Scrape-loading and dye transfer: a rapid and simple technique to study gap junctional intercellular communication. *Exp. Cell Res.* **168**, 442–430.

28. Tomai, E., Brownell, H.L., Tufescu, T. et al. (1998) A functional assay for intercellular, junctional communication in cultured human lung carcinoma cells. *Lab. Invest.* **78**, 639–640.

29. Brownell, H.L., Whitfield, J.F. and Raptis, L. (1997) Elimination of intercellular junctional communication requires lower Ras^{leu61} levels than stimulation of anchorage-independent proliferation. *Cancer Detect. Prev.* **21**, 289–294.

30. Tomai, E., Brownell, H.L., Tufescu, T., Reid, K. and Raptis, L. (1999) Gap junctional communication in lung carcinoma cells. *Lung Cancer.* **23**, 223–231.

31. Anagnostopoulou, A., Cao, J., Vultur, A., Firth, K.L. and Raptis, L. (2007) Examination of gap junctional, intercellular communication by *in situ* electroporation on two co-planar indium-tin oxide electrodes. *Molecular Oncology.* **1**, 226–231.

Chapter 13
Delivery of DNA into Adipocytes within Adipose Tissue

James G. Granneman

Abstract Electroporation has been adapted for the transfer of macromolecules into various cells of tissues in vivo. Although mature adipocytes constitute less than 20% of cells residing in adipose tissue, we have found that fat cells are susceptible to selective electrotransfer of plasmid DNA owing to their large size relative to other cells in the tissue. The procedures detailed here permit electrotransfer of plasmid DNA into mature fat cells with greater than 99% selectivity over other cells in the tissue. This "adiporation" technique can be used to image the subcellular targeting of fluorescent bioreporter molecules and to manipulate the activity of specific pathways within adipocytes in situ.

Keywords: adiporation, fluorescence microscopy, obesity, diabetes, in situ imaging

1. Introduction

Adipocytes play a central role in the regulation of systemic energy homeostasis by serving as the principal energy reservoir and as a source of hormones that regulate systemic energy utilization (1, 2). Analysis of fat cell function has relied heavily upon use of primary cell culture and established cell lines, and, although use of cultured cells has provided invaluable information, these in vitro models have several limitations when compared with adipocytes in vivo.

A significant limitation of cultured adipocyte models is inappropriate cell morphology, especially when cells are cultured on two-dimensional plastic surfaces. Subcellular organization of signaling proteins plays a central role in cellular function, yet little is known about the subcellular targeting of proteins and organelles in native adipocytes, and virtually nothing is known about targeting in adipocytes *in adipose tissue*. Therefore, there is a critical need to develop techniques for imaging biological processes in adipocytes in their native environment, namely, adipose tissue. Such a methodology could be to address fundamental questions (e.g., subcellular sites of lipolysis, glucose transport, hormone secretion, etc.) and to validate data obtained in culture models.

S. Li (ed.), *Electroporation Protocols: Preclinical and Clinical Gene Medicine.*
From *Methods in Molecular Biology, Vol. 423.*
© Humana Press 2008

Electroporation has been adapted as a technique for introducing foreign macro-molecules, such as DNA, into cells in vivo, and offers certain advantages over other somatic gene transfer techniques, such as viral vectors, pressure injection, and parti-cle bombardment. We recently found that fat cells are much more susceptible to gene transfer by electroporation, presumably owing to the larger size of these cells relative to stromal cells in the tissue (3). Here, we describe a technique for introducing DNA expression vectors into adipocytes within adipose tissue with greater than 99% selec-tivity. The technique has numerous applications, including visualizing three-dimen-sional cellular morphology in situ, lineage tracing, and functional analysis.

2. Materials

2.1. Plasmids

1. Gene-encoded reporters (e.g., GFP-based fusion proteins), reconstituted at 1 μg/ μL in sterile water (*see* **Note 1**).
2. Qiagen Endo-Free Plasmids kit (Valencia, CA).

2.2. Animals and Surgery

1. C57/Bl6 mice of either sex, 2–4 months old.
2. Avertin (20 mg/mL 2,2,2-tribromoethanol in *tert*-amyl ethanol).
3. 70% ethanol.
4. Surgical instruments (autoclaved)

    – Fine rat-toothed forceps
    – Sterile scalpel with blade (no. 10)
    – 5-O polypropylene or silk surgical suture, needle, and needle holder
    – Hamilton syringe (10 μL) with removable 26-gauge needle.

5. Electroporator (Grass Stimulator, model S88, Natick, MA; or BTX ECM 830 square-wave generator, Holliston, MA).
6. Bipolar electrodes (Tweezertrodes, model 522, BTX, Holliston, MA; Gerald forceps, 1.5 × 8 mm, model ES-300C, Anthony products, Indianapolis, IN).

2.3. Collecting and Processing

1. Paraformaldehyde solution (4%) diluted freshly in phosphate-buffered saline (PBS) from a 40% sealed stock solution (Electron Microscopy Sciences, Hatfield PA).
2. Fine forceps, scissors, and a glass surface (e.g., Petri dish).

2.4. Microscopy

1. Inverted confocal fluorescence microscope with 10×-, 40×-, and 60× objectives.
2. Leiden chamber (Molecular Probes, Eugene, OR).
3. Glass cover slips (22 × 40 mm rectangular).
4. Phosphate-buffered saline.

3. Methods

3.1. Electroporation of Adipose Tissue

1. Anesthetize mouse with an intraperitoneal injection of Avertin (0.5 mL).
2. Remove fur from the interscapular region and disinfect skin with 70% alcohol.
3. Make a 15-mm incision along the midline between and the skin overlying the suprascapular white adipose tissue (SSWAT) and gently dissect so that an approach to this subcutaneous pad could be easily made on either side.
4. Inject 7 μL of plasmid DNA into SSWAT with a Hamilton microsyringe and a 26-GA needle (*see* **Note 2**). Injections are made parallel to the surface of the pad at a depth of 1–2 mm. The needle is first inserted to a length of 5–7 mm, and the injection is made while the needle is slowly withdrawn.
5. Immediately after injection, gather the site of injection with bipolar forceps to a gap width of 1.0 mm.
6. Electroporate with 7 square-wave pulses (50 V for 20 ms), delivered over 3 s. The internal resistance of the Grass stimulator is set at 250 Ω.
7. Close incision with 2–3 sutures.

3.2. Localizing and Dissecting Transfected Cells

1. Dissect electroporated tissue 1–7 days after electrotransfer. Remove only the top 2–4 mm of tissue and place on a glass cover slip. Nick tissue every 5–6 mm to flatten and thin tissue, then fix in 4% paraformaldehyde for 2–4 h. Place fixed tissue in PBS at 4°C until microscopic analysis.
2. Cut tissue into pieces ~6 × 6 × 4 mm³.
3. Place sample on 22 × 40-mm² cover slips, and view specimen with 10× objective and fluorescence filter sets appropriate for fluorescent reporter.
4. Using forceps, tease apart transfected from nontransfected regions. Transfected cells are usually at or near the surface of the tissue, and so be sure to inspect both surfaces before discarding tissue.

3.3. Microscopic Analysis

1. Place dissected SSWAT in a Leiden chamber with 100 μL of PBS. Place a second cover slip on top of the tissue and gently compress the tissue against the lower cover slip.
2. Cells are typically imaged with a 40× dry (0.9 NA) or 60× water-immersion (1.2 NA) objective lens using spinning disc or laser scanning confocal microscopy (*see* **Notes 3–5**). Postcapture processing and image analysis is performed using free (NIH imageJ) or commercial (ImageProPlus, Media Cybernetics) software.

4. Notes

1. Various reporter plasmids can be used depending on the experimental questions being addressed. We have successfully imaged reporters targeted to nuclei (DsRed fused to HNF4), mitochondria (Nrbf1 fused to EYFP) and lipid droplets (perilipin, Abhd5, rab5, and caveolin 2β fused to EYFP or ECFP).
2. Electroporations can be made in each (right or left) SSWAT pad. In this way, control and experimental plasmids can be tested in the same animal. For example, it is possible to electroporate constitutively active or dominant negative constructs of specific signaling pathways (Fig. 13.1). Analysis is usually limited to single cells, typically using an optical read-out.

Mito-EGFP   Mito-EGFP + CA-Gsα

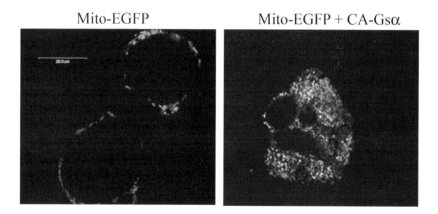

Fig. 13.1 Expression of mitochondrially targeted enhanced green fluorescent protein (EGFP) in control adipocytes and adipocytes expressing constitutively active Gsα. Adipose tissue was electroporated with a plasmid encoding a fusion of Nrbf1 and EGFP to allow visualization of mitochondria, with empty vector (*left panel*), or vector encoding constitutively active mutant (Q227L) of Gsα, the GTP-binding protein that activates adenylyl cyclase. Tissue was harvested after 3 days and imaged by confocal microscopy. Shown are single optical slices made near the equator of the cells. In control cells, mitochondria occupy the sparse cytoplasm between the nonfluorescent lipid core and plasma membrane. Constitutive activation of cAMP-protein kinase A signaling pathway fragments the core lipid droplet and stimulates massive mitochondrial biogenesis

3. Because of the high degree of light scatter, traditional wide-field fluorescence microscopy is not usually acceptable for imaging transfected bioreporters in adipose tissue. In contrast, single or multiphoton laser-scanning or spinning-disc confocal microscopy yields high resolution images.
4. Excessive fixation or prolonged storage can increase tissue autofluorescence.
5. It is sometimes useful to establish the pattern and level of the bioreporter expression fusion relative to the endogenous protein. This can be accomplished using single or double immunohistochemistry of whole mounts (3) or paraffin sections (4).

References

1. Fruebis, J., Tsao, T.S., Javorschi, S., et al. (2001) Proteolytic cleavage product of 30-kDa adipocyte complement-related protein increases fatty acid oxidation in muscle and causes weight loss in mice. *Proc. Natl. Acad. Sci. U.S.A.* **98**, 2005–2010.
2. Zhang, Y., Proenca, R., Maffei, M., Barone, M., Leopold, L., and Friedman, J.M. (1994) Positional cloning of the mouse obese gene and its human homologue. *Nature.* **372**, 425–432.
3. Granneman, J.G., Li, P., Lu, Y., and Tilak, J. (2004) Seeing the trees in the forest: selective electroporation of adipocytes within adipose tissue. *Am. J. Physiol. Endocrinol. Metab.* **287**, 574–582.
4. Moore, H.P., Silver, R.B., Motillo, E., Bernlohr, D.A., and Granneman, J.G. (2005) Perilipin targets a novel pool of lipid droplets for lipolytic attack by hormone-sensitive lipase. *J. Biol. Chem.* **280**, 43109–43120.

Part III
In Vivo Targeted Gene Delivery
via Electroporation

Chapter 14
Delivery of DNA into Muscle for Treating Systemic Diseases: Advantages and Challenges

Capucine Trollet, Daniel Scherman, and Pascal Bigey

Abstract An efficient and safe method to deliver DNA in vivo is a requirement for several purposes, such as the study of gene function and gene therapy applications. Among the different nonviral delivery methods currently under investigation, in vivo DNA electrotransfer has proven to be one of the most efficient and simple methods. This technique is a physical method of gene delivery consisting of a local application of electric pulses after injection of DNA.

This technique can be applied to almost any tissue of a living animal, including tumors, skin, liver, kidney, artery, retina, cornea, or even brain, but the focus of this review will be on electrotransfer of plasmid DNA into skeletal muscle and its possible therapeutic uses for systemic diseases. Skeletal muscle is a good target for electrotransfer of DNA because of the following features: a large volume of easily accessible tissue, an endocrine organ capable of expressing several local and systemic factors, and muscle fibers as postmitotic cells have a long lifespan, which allows long-term gene expression.

In this review, we will describe the main characteristics of DNA electrotransfer, including toxicity and safety issues related to this technique. We will focus on the important possible therapeutic applications of electrotransfer for systemic diseases demonstrated in animal models in the recent years, in the fields of monogenic diseases, tissue-specific diseases, metabolic disorders, immune-system-related diseases, and cancer. Finally, we will discuss the advantages and challenges of this technique.

Keywords: DNA electrotransfer, electroporation, muscle, gene therapy, safety

1. Introduction

Two major groups of vectors are used for gene therapy: viral and nonviral. Viruses are considered to be very efficient vehicles for gene transfer; however, their use is limited by safety concerns, such as immune response, possible mutagenesis, and carcinogenesis, and high production costs. Considering these limitations, the delivery of therapeutic genes to target cells upon direct in vivo administration of non viral

S. Li (ed.), *Electroporation Protocols: Preclinical and Clinical Gene Medicine.*
From *Methods in Molecular Biology, Vol. 423.*
© Humana Press 2008

vectors, i.e., plasmids, is of great value for the development of gene therapy. Unfortunately, the use of plasmids is plagued by poor transfer efficiency, intracellular penetration, nuclear localization, low expression levels, and immunostimulatory properties of plasmid DNA. Among the different nonviral strategies currently under study, in vivo electroporation has proven to be one of the most efficient and simplest methods which could be applied to gene therapy and as a laboratory tool to study gene function. This technique is particularly efficient for gene transfer into the skeletal muscle, allowing its possible use as an endocrine organ for the secretion of therapeutic proteins.

This chapter will describe skeletal muscle DNA electroporation applications in animal models of systemic diseases and its advantages and challenges.

2. Electroporation for DNA Delivery

The nonpermeable lipophilic cell membrane controls the exchange of molecules between the cytoplasm and the external medium. Only a few hydrophobic molecules are able to cross the lipid bi-layer, while others can enter via specific transporter systems; however, most of the hydrophilic molecules are unable to enter the cell. Since the initial reports by Neumann et al. (1), the use of electricity to mediate the delivery of molecules to cells in vitro is now a routine technique. By applying short and intense electric pulses, it is possible to transiently permeabilize the cell membrane, facilitating the entry of any molecule. This technique has subsequently been used on living animals in the early 1990s.

The first relevant in vivo application was demonstrated by cellular uptake into tumors of the antibiotic and chemotherapeutic agent bleomycin (2). A better penetration of bleomycin was obtained by applying electric pulses to tumors, leading to an enhanced cytotoxicity. Since then, this technique of electrochemotherapy (3) has also been applied with another anticancer drug, cisplatin, in clinical trials on human malignant melanoma skin metastases (4) and cutaneous tumor lesions of breast cancer (5) as well as for the treatment of horses (6).

Besides electrochemotherapy, the last few years have seen electric-pulse-mediated plasmid DNA transfer become a rapidly emerging and promising technique, under the name of in vivo DNA electroporation or electrotransfer (7–10). Electrotransfer is based on DNA injection into a targeted tissue, followed by the application onto the tissue of a defined set of electric pulses. This technique can be applied to any tissue, but the skeletal muscle is particularly suitable.

The mechanism by which electrotransfer occurs has not been completely elucidated (see (11) for a review). It is generally believed to be composed mostly of cell permeabilization and DNA uptake through electrophoresis, but the molecular mechanisms underlying in vivo DNA electrotransfer are still controversial. It is probably a multistep process involving to some degree the following processes: DNA injection and distribution, cell permeabilization, DNA transfer facilitated by DNA electrophoresis, and possibly passive diffusion through defects in the membrane due to the

electric pulses. Unravelling the details and contributions of each phase of this complex process will permit the design of more effective electrotransfer strategies.

In order to practically carry out in vivo electrotransfer, a plasmid solution in isotonic saline (NaCl, 150 mM) is injected into the skeletal muscle with a syringe, and electric pulses are then delivered by means of two electrodes placed on either side of the injection site (electrodes can be either needles or plates). As far as the electric parameters are concerned, various efficient protocols have been published by different groups with either low-voltage (100–300 V), long-duration (4–50 ms) pulses (12), or high-voltage (400–1,200 V/cm), short-duration (95–300 μs) pulses (13). As electrical parameters depend greatly on the delivery method, a great diversity of protocols has been used (14).

3. Muscle as a Secretory Organ

Skeletal muscle is the most widely targeted tissue for electroporation for the following reasons:

– It constitutes a large and easily accessible volume of tissue in which DNA electroporation is very efficient.
– Muscle fibers have a long lifespan as they are postmitotic, potentially allowing long-term expression (more than a year) in transfected cells (in the absence of regeneration due to injury or cytotoxic immune response (15)).
– Skeletal muscle is made up of thousands of cylindrical muscle fibers bound together by connective tissue through which run blood vessels and nerves. This muscle has an abundant blood vascular system (16), and, therefore, skeletal muscle is able to produce secreted proteins with functional posttranslational modifications, which can easily reach the blood circulation (for a review, see (17)).
– Cotransfection of multiple unlinked genes can be easily performed by electroporation (18).

Indeed, the persistence of DNA in an episomal state for months and the ability of skeletal muscle to secrete proteins allow multiple therapeutic approaches, such as direct gene transfer for muscle disorders, DNA vaccination, or systemic delivery of therapeutic proteins, to be considered.

3.1. Long-Term Expression

An important issue for gene transfer applications is the level and duration of gene expression. Different kinetics of gene expression have been described after DNA electrotransfer into skeletal muscle. Long-term expression has been observed for a variety of transgenes, such as human secreted alkaline phosphatase (hSEAP), the luciferase reporter gene (15), human factor IX (hFIX) in SCID mice (19), and

murine erythropoietin (mEpo) in immunocompetent mice (20). By studying luciferase expression with a charged-coupled device camera (which allows an in vivo kinetic study without killing the animals), it was observed that gene expression increased with time during the first few days, and then stayed stable for at least 70 days (21). Expression lasted up to a year (19), raising hopes in the gene therapy field.

It is not clearly established why expression in skeletal muscle lasts so long. The general belief is that it is due to the nondividing, multinucleated mature muscle fibers transfected during the electrotransfer process. Recent work by Peng et al. (22) suggests that early expression (3–5 days) is indeed due to transfection of mature myofibers, but most of these cells usually died within 2 weeks because of the damage caused by the electric field, leading to only transient gene expression. However, electrotransfer also led to activation and DNA uptake of satellite cells, which then repair tissue damage by regenerating muscle. The conclusion was that a major contribution to long-term gene expression in skeletal muscle after DNA electrotransfer came from these regenerated myofibers which express the transfected DNA and not from the initial mature transfected cells. This is the only report of such a hypothesis, and the experimental conditions used long and toxic electric pulses (50 ms). Several teams have been working to find milder electrical parameters which minimize muscle damage without a reduction in gene expression, and this work is discussed in the next section.

3.2. Toxicity

The application of brief electric pulses to tissues in order to increase their cellular permeability to DNA might be associated with tissue damage. Although electrical parameters causing minimal muscle damage yet high levels of transfection efficiency have now been well established and applied successfully to therapeutic gene transfer in muscle, cell necrosis and decreased gene expression have been reported after excessive electric field intensities. We describe here the main damage observed in muscle tissue shortly after DNA electrotransfer and the reasons for the same.

Histology studies of muscle after electrotransfer treatment confirm that muscle damage is maximal within the first 7 days. In central areas affected by electroporation, tissue damage is mostly characterized by muscle lesions containing necrotic myofibers and is heavily populated with inflammatory cells. Shortly after treatment, the muscles start to regenerate from satellite cells (22), and by 3 weeks, the muscle has completely recovered (normal structure of muscle fibers, lesions no longer evident).

Several studies have investigated the different factors involved in gene-electro-transfer-associated muscle damage. Cell permeabilization has been very often proposed as one of the main factors of toxicity, as the external media diffuses into the cells, modifying their internal media composition. Gehl and Mir established that the toxicity of electrotransfer was correlated with the degree of cell permeabilization (23). Cell necrosis and decreased gene expression have been reported at

excessive electric field intensities, but these effects can be minimized by determining thresholds for each species. Also, muscle necrosis increases with the cumulative duration of the pulses (24). Recently, an in vivo NMR imaging system has been developed to visualize and evaluate the electropermeabilized zones during electric-field-mediated drug or gene delivery (25) and, thereby, assess more precisely the efficiency and toxicity of different protocols.

The toxicity of DNA electrotransfer could also be directly related to the amount of injected DNA, as was reported in several studies. Briefly, electrotransfer induced plasmid-dependent muscle lesions containing necrotic myofibers, although electrotransferred muscles were indistinguishable from nontreated controls at day 56 (26), and it seems that muscle damage mainly arises from the intracellular presence and expression of plasmid DNA (27). More precisely, necrotic fibers never expressed the transgene (24), and only a few positive muscle cells could be detected inside the damaged area whereas strong expression was observed in adjacent intact cells (27). It has also been suggested that muscle damage is related to an increase in transfection efficiency. Recently, Bertrand et al. (28) used transgenic mice harbouring a transgene under the control of a fiber-specific and nerve-dependent promoter and showed that electroporation induced only a very transient phenotypic and morphological alteration of the muscle fibers. However, the process led to a profound but transient alteration of muscle transcriptional status. In muscle fibers, 7–10 days after DNA electrotransfer seems to be necessary to recover a normal physiological state.

For all these reasons, optimizing protocols to limit damage is critical for accurate physiological, biochemical, and molecular measurements. For physiological and therapeutic applications, a compromise between efficient plasmid transfer and minimal cell toxicity has to be found. In physiological applications, if muscle damage occurs, the subsequent degeneration/regeneration of muscle fibers may lead to unexpected events, which should not occur in normal muscle fibers and which may modify study results. For therapeutic applications such as myopathies, muscle wasting disorders, and other pathophysiological conditions resulting in fragile muscles, the electrotransfer protocols must avoid further muscle damage.

3.3. Plasmid Biodistribution

To determine optimal conditions which maximise efficiency while reducing muscle damage, different protocols have been used to improve the access of plasmids to targeted cells. Several reports state that improved plasmid distribution in the skeletal muscle leads to an increase in DNA expression. Improved plasmid distribution could be achieved by preinjecting of a sucrose solution, which created spaces between muscle fibers (29), or by pretreating with hyaluronidase (30), an enzyme which breaks down components of the extracellular matrix containing hyaluronan and collagen (31). Moreover, improved gene transfer with hyaluronidase pretreatment allowed the use of lower voltages, resulting in a reduction of muscle damage (32). Molnar et al. (33) showed an increased muscular transfection

efficiency of 150–370% after pretreatment with hyaluronidase and electrotransfer (175 V/cm) in different strains of mice. In pathological conditions with fragile muscles such pretreatment can be particularly useful to avoid further muscle injuries: this protocol has allowed levels of transfection up to 60% with a reporter gene in the *mdx* model of Duchenne Muscular Dystrophy, with no detrimental change in muscle structure (34).

4. Treating Systemic Disease

The ability of the skeletal muscle to produce posttranslationally modified functional proteins opens up the possibility of using it as an endocrine tissue for secretion of proteins into the bloodstream and, therefore, its use in a large number of applications. Moreover, the stable expression which can be obtained by DNA electrotransfer makes it very appealing for the treatment of numerous pathologies. We report here some of the important therapeutic applications demonstrated in animal models in recent years.

4.1. Monogenic Diseases

Monogenic diseases were thought to be an obvious target for gene therapy approaches (and therefore were the first targeted diseases), provided the unique defective gene is known. Among these diseases, hemophilia is a good model, as even a small amount of the missing clotting factor (factor VIII for hemophilia A or factor IX for hemophilia B) would be sufficient to reverse the clinical phenotype of the disease. Raising the level of human factor IX to 2% of the normal one would be enough to be of therapeutic value in hemophilia B patients, and a 2% increase has been reached for several weeks in SCID mice (19) and 0.5–1% in dogs (35). However, immune responses against the human transgenic proteins were observed in both of these studies, restricting the length of the study. Human factor IX synthesized by myotubes is biologically active, although not all posttranslational modifications are complete (e.g., phosphorylation), suggesting that the muscle may not have absolute capacity for posttranslational protein modifications as shown by Arruda et al. (36). Recently, production by the muscle of murine factor VIII in hemophilic mice was successful, as all mice submitted to hemostatic challenge 6 days after treatment survived (37). In order to provide a long-term expression, an Epstein-Barr-virus (EBV)-based episomal plasmid encoding FVIII was studied (38). FVIII was thus detected in the serum for at least 90 days.

Erythropoietin (EPO) is another good candidate for gene therapy applications, since a small amount will produce a physiological effect: raising the hematocrit. Numerous studies report efficient EPO secretion after muscle plasmid electrotransfer. Therapeutic EPO levels have been achieved after a single injection of 1 μg of plasmid, with stable and long-lasting EPO production (20, 39). The efficiency of this

approach has been shown in β-thalassemic mice, in which not only the hematocrit level was increased, but also a phenotypic correction of erythrocytes was obtained with a 100% increase in the lifespan of erythrocytes (40). EPO electrotransfer was also efficient in uremic rats, producing erythropoiesis for more than 15 weeks and inducing a significant decrease in platelet counts. However, the typical side effect of EPO in uremic disease, hypertension, was also observed (41). Tetracycline regulatory systems have also been used to control EPO expression (18, 42, 43). An increase in plasma EPO levels was obtained by changing the EPO signal peptide by the hTAP (human tissue plasminogen activator) signal peptide, leading to an increased hematocrit level in cynomolgus monkeys (44).

4.2. Organ/Tissue-Specific Diseases

Since the muscle is able to secrete proteins after electrotransfer treatment, it can also be used for treating organ-specific diseases, distant from the electrotransferred muscle. Dilated cardiomyopathy, one of the major causes of severe heart failure, was studied in a hamster inherited model by intramuscular electrotransfer of a plasmid encoding hepatocyte growth factor (HGF) (45). This treatment led to attenuated cardiac function deterioration, reduced cardiac hypertrophy and fibrosis, and higher myocardial capillary density, which altogether might improve deleterious changes related to the disease.

HGF is also a mitogenic and antiapoptotic factor for alveolar and bronchial epithelial cells, but its very short half-life prevents its use as recombinant protein. Intramuscular electrotransfer of an HGF encoding plasmid was thus studied in a mouse bleomycin-induced lung fibrosis model and HGF protein was secreted for 4 weeks (46). HGF systemic secretion resulted in reduced fibrogenesis and apoptosis and improved survival rate.

Neurotrophin-3 (NT-3) production following intramuscular plasmid electrotransfer or adenovirus injection was compared in a mouse model of sensory neuropathy (47). Both delivery systems led to the production of NT-3 at levels promoting survival of the large fiber sensory neurons. In another study, a single electrotransfer of a cardiotrophin-1-encoding plasmid in neonate progressive motor neuropathy resulted in improvement of all electromyographic parameters, protection of myelinated axons and phrenic nerves, and a 25% increase in the mean lifespan (48).

4.3. Metabolic Disorders

Studies previously described about EPO gene delivery in β-thalassemic models are also useful for the treatment of severe anemia associated with chronic renal failure. In this context, studies on two rat models have shown increased hematocrit levels without inducing severe hypertension (49, 50).

Because of the difficulties of GH (growth hormone) proteotherapy and the short half-life of GHRH (GH releasing hormone) making it unsuitable for recombinant therapy for chronic diseases, electrotransfer of GHRH-encoding plasmids seemed a valuable alternative. This strategy was developed in pigs with a muscle-specific promoter (51), where long-term expression, linear growth, and improved fat metabolism have been obtained (52).

To better understand the regulatory mechanisms of leptin on food intake and energy metabolism in vivo, leptin gene transfer was investigated using intramuscular electrotransfer or hydrodynamic-based gene delivery (53). Both systems reduced food intake and fat levels in mice, although better results were obtained with the hydrodynamic method.

4.4. Immune-System-Related Diseases

The field of immune-system-related diseases is one of the most important applications of gene therapy, mainly for viral vectors, but also for the electroporation method.

4.4.1. Deciphering the Role of Proteins

Intramuscular electrotransfer has been used as a powerful tool to confirm the pro- or antiangiogenic properties of cytokines in peripheral ischemia models, in which both hypoxia and inflammation are involved. In this way, interleukin (IL)-10 and IL-18 were proven antiangiogenic (54, 55) whereas FGF and tumor necrosis factor alpha (TNF-α) soluble receptors were found to stimulate angiogenesis (56, 57). The use of plasmid electrotransfer to induce overexpression of VEGFB has allowed molecular events associated with VEGFB effects to be deciphered (58).

4.4.2. Diabetes

Skeletal muscle can be an efficient platform for the ectopic secretion of insulin (59). In a mouse model of severe diabetes, electrotransfer of a proinsulin-encoding plasmid led to reduced blood glucose and increased survival (60). Similar results were obtained after DNA electrotransfer of a plasmid encoding modified preproinsulin genes in a streptozotocin-induced model of autoimmune diabetes, where reduced blood glucose was observed for 5 weeks (61). In the same model, electrotransfer of a calcitonin-gene-related peptide (CGRP) significantly decreased morbidity due to diabetes, ameliorated hyperglycemia and insulin deficiency, and inhibited lymphocyte infiltration into the islets, thereby protecting β cells against autoimmune destruction (62).

4.4.3. Cardiovascular Diseases

IL-10 is one of the most studied transgenic proteins following electrotransfer, and, along with its usefulness against ischemia, overexpression of IL-10 has also been shown to reduce acute rejection risk in rat cardiac allografts (63) and to increase survival rates in autoimmune or viral models of myocarditis (64, 65). The difficulty in obtaining a high enough serum IL-10 level was overcome by creating a fusion protein between IL-10 and an Fc fragment, which resulted in a 100-fold higher serum peak (66). However, the level of the secreted protein depended on the mouse strain used (67).

IL-10 has also been proved effective in preventing the development of lesions in a mouse atherosclerosis model (68). The critical role of IL-18/IL-18BP was also demonstrated in apoE knockout mice, in which the ability of IL-18 to reduce the development and progression of atherosclerotic plaques was proven (69). In the same model, electrotransfer of a plasmid encoding plasma platelet-activating factor acetylhydrolase (PAF-AH) protected against the progression of atherosclerosis by inhibiting phospholipid accumulation in the aortic wall (70).

4.4.4. Rheumatoid Arthritis

Rheumatoid arthritis is a very good example to illustrate the advantages and challenges of DNA electrotransfer. It is a chronic inflammatory autoimmune disease in which the main observed characteristics are joint destruction and systemic inflammation. DNA electrotransfer of anti-inflammatory cytokines has shown promising results: IL-10 was efficient in both preventive and curative treatment of collagen-induced arthritis in mice (71). Electrotransfer of an IL-4-encoding plasmid could prevent the incidence of severe arthritis for at least 2 weeks and led to reduced IL-1β levels in the ankle joints (72), and IL-1 receptor antagonist treatment efficiently lowered both joint swelling and incidence of arthritis (73). However, elevated levels of circulating cytokines are very difficult to obtain.

But the most efficient current treatments are based on the use of anti-TNF-α molecules, the best results being obtained with recombinant proteins acting as soluble receptors for TNF-α (74). At least three recombinant proteins are currently used in clinics (Etanercept, Infliximab, and Adalimumab); however, serious adverse effects have been observed following these treatments. Infections, malignancy, heart failure, multiple sclerosis, and autoimmune responses have been reported, although it is not proven yet that this is a class effect (74). These side effects might be due to the high doses of recombinant proteins injected. DNA electrotransfer might allow a satisfying response at low doses and avoid a deleterious peak of protein following injection.

Different forms of TNF-α soluble receptors were studied, and TNF soluble receptors I or II associated with the Fc fragment of IgG appeared to be the most interesting form for rheumatoid arthritis (75, 76). Similar results, as with the

recombinant protein Etanercept, were obtained on the clinical score in a mouse model of collagen-induced arthritis with electrotransfer of a TNFR1-Fc encoding plasmid (which encodes a protein very similar to Etanercept), with circulating levels of secreted proteins as low as 5 ng/mL (75). Although there is high morbidity in rheumatoid arthritis patients, it is considered as a nonlethal disease. Therefore, from an ethical point of view, it is more difficult to start clinical trials of gene therapy, as the safety issue is still a matter of concern.

4.4.5. Eye Diseases

Another great advantage of DNA electrotransfer is the possibility to deliver a local treatment. The eye is an isolated organ difficult to reach via systemic administration. Eye diseases are then treated with intra- or periocular injections. Repeated injections in the eye are not suitable because of the risk of adverse effects, mainly infections. Recently, the electrotransfer method was developed for the local treatment of inflammatory eye disease by targeting the ciliary muscle, which is a particular smooth muscle with some characteristics of striated skeletal muscle. This method led to production and secretion in the ocular media of therapeutic levels of TNF-α soluble receptor, thus preventing clinical and histological signs in a rat uveitis model (77). Such a method would allow a single treatment at the best, or multiple spaced treatments at the worst, though reducing the risks.

4.5. Cancer

Direct intratumoral plasmid electrotransfer is a well-developed strategy for local production of therapeutic proteins. However, the efficacy of gene transfer into tumor cells in vivo is generally low, and intramuscular electrotransfer can also be efficiently used for distal tumor treatment, mainly with cytokine-encoding plasmids. In 2001, Lucas and Heller optimized electrical conditions for transferring an IL-12-encoding plasmid into mouse gastrocnemius and showed systemic IL-12 secretion for at least 21 days, as well as IFN-γ production (78). Secreted IL-12 induced regression of 50% of large subcutaneous tumors and significantly prolonged the lifespan of affected mice. In vivo electroporation of the IL-12 gene was also effective in suppressing subcutaneous and lung metastatic tumors, tumors of colon, adenocarcinomas, and melanomas (79). However, it has recently been shown that systemic treatment by IL-12 following intramuscular electrotransfer cannot eradicate established tumors since only NK-cell-dependent and not T-cell-dependent antitumor effects are observed following intramuscular administration (80). Other interesting results were the inhibition of tumor growth in various models with plasmids encoding IFN-α (81, 82), metalloproteinase-4 inhibitor for the treatment of kidney-derived cancers (83), endostatin (84), or an extracellular FGF-2 binding polypeptide encoding plasmid (85).

This distal approach may be very powerful when used in combination with a local treatment, for those tissues inaccessible to local administration, or in combination with another therapeutic strategy, e.g. chemotherapy. In the latter case, a synergistic effect may be obtained, since these methods use different mechanisms to kill cancer cells. As an example, we have shown that intratumoral electrotransfer of an antisense of MBD2, an enzyme involved in DNA methylation, results in a significant inhibition of tumor growth in a human tumor model grafted in nude mice (86). When combined with a single bleomycin injection, this treatment led to a 30% total remission of xenografted tumors (86).

4.6. *Monoclonal Antibodies*

Monoclonal antibodies have a wide range of clinical applications in the fields of autoimmune disease, cancer, and infectious disease (see (87) for a review). Injections of recombinant proteins are the easiest way to treat patients, but this method is not always useful for long-term treatments. However, skeletal muscle can secrete functional monoclonal antibodies, provided that the DNA sequences of the two chains are transfected. Electrotransfer of two plasmids encoding the light and heavy chain in the tibialis cranialis muscle of mice led to an elevated level of functional circulating antibody (88). This result shows that antibody production by the skeletal muscle is possible and could have laboratory applications or lead to preclinical studies.

5. Conclusion: Advantages and Challenges

In vivo electrotransfer is a nonviral technique for reasonably efficient gene transfer, particularly in the skeletal muscle. Furthermore, the muscle is able to produce functional posttranslationally modified proteins. This fact allows the use of the muscle as an organ for the production of secreted proteins targeting systemic diseases. DNA electrotransfer has the main advantages of being fast, easy to perform, usable in a wide range of tissues, and cheap. From the earlier-mentioned examples, it is now established that therapeutic levels of circulating proteins can be reached in animal models (good examples are EPO, coagulation factors, and TNF-α soluble receptors). As no immune response is induced against DNA, this treatment can also be repeated as often as desired, if necessary. But treatment may not have to be repeated very often as long-term expression in the muscle has commonly been reported. DNA electrotransfer seems to have tremendous advantages over recombinant protein therapy: it should avoid the possible adverse effects of an acute injection of an elevated dose of protein, since steady-state equilibrium in the blood concentration is reached for a long time. This fact could be particularly valuable in the case of small proteins for which clearance by kidneys is very fast (i.e. epoetin alpha, recombinant EPO, half-life is 4 to 6h), or for receptors which are rapidly used so that the blood half-life after

injection is short (Etanercept has to be injected twice a week at 25-mg doses.). Finally, as it was described in the case of the eye, DNA electrotransfer allows for a very local treatment. The local concentration of anti-TNF protein in the eye is enough to produce a physiological effect, while a very low (or undetectable) level of this protein is found in the general circulation, thereby limiting possible adverse effects.

Also, there are still challenges for the clinical use of DNA electrotransfer. First, as the levels of expression obtained with DNA electroporation might not be very high when compared with those with viruses, in vivo electroporation would not be a competitor, but an alternative to viral techniques for a specific use. Second, its exact mechanism is not yet elucidated, and improvement can be expected in its understanding from further studies. Particularly, parameters and DNA biodistribution have to be further investigated in order to optimize this technique and to limit its local toxicity. Third, the muscle is able to produce functional glycosylated proteins, but muscle cells may not have absolute capacity for posttranslational protein modifications as shown by Arruda et al. This disability has to be further investigated and has to be checked for each protein. Fourth, as most of the possible applications concern non lethal diseases, the safety of such a treatment is crucial, and there are ethical restraints to starting clinical trials that should carefully be discussed. Particularly, it will be necessary to develop an efficient regulation mechanism for gene expression, as it should be possible to stop the treatment at any desired time if a clinical problem or adverse effects are observed (as is the case for the anti-TNF recombinant proteins, for example). Such a regulation system is not yet available.

A clinical trial of DNA electrotransfer into skeletal muscle is currently on—a protocol of immunization with a plasmid DNA encoding a prostate tumor antigen (89), and hopefully others can be expected soon, as the clinical potential of DNA electrotransfer seems very encouraging.

References

1. Neumann, E., Schaefer-Ridder, M., Wang, Y., and Hofschneider, P.H. (1982) Gene transfer into mouse lyoma cells by electroporation in high electric fields. *EMBO J.* **1**, 841–845.
2. Belehradek, J., Jr., Orlowski, S., Ramirez, L.H., Pron, G., Poddevin, B., and Mir, L.M. (1994) Electropermeabilization of cells in tissues assessed by the qualitative and quantitative electroloading of bleomycin. *Biochim. Biophys. Acta.* **1190**, 155–163.
3. Mir, L.M. and Orlowski, S. (1999) Mechanisms of electrochemotherapy. *Adv. Drug Deliv. Rev.* **35**, 107–118.
4. Sersa, G., Stabuc, B., Cemazar, M., Miklavcic, D., and Rudolf, Z. (2000) Electrochemotherapy with cisplatin: the systemic antitumour effectiveness of cisplatin can be potentiated locally by the application of electric pulses in the treatment of malignant melanoma skin metastases. *Melanoma Res.* **10**, 381–385.
5. Rebersek, M., Cufer, T., Cemazar, M., Kranjc, S., and Sersa, G. (2004) Electrochemotherapy with cisplatin of cutaneous tumor lesions in breast cancer. *Anticancer Drugs.* **15**, 593–597.
6. Rols, M.P., Tamzali, Y., and Teissie, J. (2002) Electrochemotherapy of horses. A preliminary clinical report. *Bioelectrochemistry.* **55**, 101–105.
7. Heller, L.C., Ugen, K., and Heller, R. (2005) Electroporation for targeted gene transfer. *Expert Opin. Drug Deliv.* **2**, 255–268.

8. Li, S. (2004) Electroporation gene therapy: new developments in vivo and in vitro. *Curr. Gene Ther.* **4**, 309–316.

9. Trollet, C., Bigey, P., and Scherman, D. (2005) Electrotransfection: an overview. In: Schleef, M. (ed.). *DNA-pharmaceuticals.* Wiley-VCH, Weinheim, Chap. 11, pp. 189–218.

10. Prud'homme, G.J., Glinka, Y., Khan, A.S., and Draghia-Akli, R. (2006) Electroporation-enhanced nonviral gene transfer for the prevention or treatment of immunological, endocrine and neoplastic diseases. *Curr. Gene Ther.* **6**, 243–273.

11. Faurie, C., Phez, E., Golzio, M., et al. (2004) Effect of electric field vectoriality on electrically mediated gene delivery in mammalian cells. *Biochim. Biophys. Acta.* **1665**, 92–100.

12. Mir, L.M., Bureau, M.F., Rangara, R., Schwartz, B., and Scherman, D. (1998) Long-term, high level in vivo gene expression after electric pulse-mediated gene transfer into skeletal muscle. *C.R. Acad. Sci. III.* **321**, 893–899.

13. Vicat, J.M., Boisseau, S., Jourdes, P., et al. (2000) Muscle transfection by electroporation with high-voltage and short-pulse currents provides high-level and long-lasting gene expression. *Hum. Gene Ther.* **11**, 909–916.

14. Mir, L.M., Moller, P.H., Andre, F., and Gehl, J. (2005) Electric pulse-mediated gene delivery to various animal tissues. *Adv. Genet.* **54**, 83–114.

15. Mir, L.M., Bureau, M.F., Gehl, J., et al. (1999) High-efficiency gene transfer into skeletal muscle mediated by electric pulses. *Proc. Natl. Acad. Sci. U.S.A.* **96**, 4262–4267.

16. Lu, Q.L., Bou-Gharios, G., and Partridge, T.A. (2003) Non-viral gene delivery in skeletal muscle: a protein factory. *Gene Ther.* **10**, 131–142.

17. Goldspink, G. (2003) Skeletal muscle as an artificial endocrine tissue. *Best Pract. Res. Clin. Endocrinol. Metab.* **17**, 211–222.

18. Trollet, C., Ibanez-Ruiz, M., Bloquel, C., Valin, G., Scherman, D., and Bigey, P. (2004) Regulation of gene expression using a conditional RNA antisense strategy. *J. Genome Sci. Tech.* **3**, 1–13.

19. Bettan, M., Emmanuel, F., Darteil, R., et al. (2000) High-level protein secretion into blood circulation after electric pulse-mediated gene transfer into skeletal muscle. *Mol. Ther.* **2**, 204–210.

20. Kreiss, P., Bettan, M., Crouzet, J., and Scherman, D. (1999) Erythropoietin secretion and physiological effect in mouse after intramuscular plasmid DNA electrotransfer. *J. Gene Med.* **1**, 245–250.

21. Honigman, A., Zeira, E., Ohana, P., et al. (2001) Imaging transgene expression in live animals. *Mol. Ther.* **4**, 239–249.

22. Peng, B., Zhao, Y., Lu, H., Pang, W., and Xu, Y. (2005) In vivo plasmid DNA electroporation resulted in transfection of satellite cells and lasting transgene expression in regenerated muscle fibers. *Biochem. Biophys. Res. Commun.* **338**, 1490–1498.

23. Gehl, J. and Mir, L.M. (1999) Determination of optimal parameters for in vivo gene transfer by electroporation, using a rapid in vivo test for cell permeabilization. *Biochem. Biophys. Res. Commun.* **261**, 377–380.

24. Mathiesen, I. (1999) Electropermeabilization of skeletal muscle enhances gene transfer in vivo. *Gene Ther.* **6**, 508–514.

25. Leroy-Willig, A., Bureau, M.F., Scherman, D., and Carlier, P.G. (2005) In vivo NMR imaging evaluation of efficiency and toxicity of gene electrotransfer in rat muscle. *Gene Ther.* **12**, 1434–1443.

26. Hartikka, J., Sukhu, L., Buchner, C., et al. (2001) Electroporation-facilitated delivery of plasmid DNA in skeletal muscle: plasmid dependence of muscle damage and effect of poloxamer 188. *Mol. Ther.* **4**, 407–415.

27. Durieux, A.C., Bonnefoy, R., Busso, T., and Freyssenet, D. (2004) In vivo gene electrotransfer into skeletal muscle: effects of plasmid DNA on the occurrence and extent of muscle damage. *J. Gene Med.* **6**, 809–816.

28. Bertrand, A., Ngo-Muller, V., Hentzen, D., Concordet, J.P., Daegelen, D., and Tuil, D. (2003) Muscle electrotransfer as a tool for studying muscle fiber-specific and nerve-dependent activity of promoters. *Am. J. Physiol. Cell Physiol.* **285**, C1071–C1081.

29. Davis, H.L., Whalen, R.G., and Demeneix, B.A. (1993) Direct gene transfer into skeletal muscle in vivo: factors affecting efficiency of transfer and stability of expression. *Hum. Gene Ther.* **4**, 151–159.

30. Mennuni, C., Calvaruso, F., Zampaglione, I., et al. (2002) Hyaluronidase increases electro-gene transfer efficiency in skeletal muscle. *Hum. Gene Ther.* **13**, 355–365.

31. Favre, D., Cherel, Y., Provost, N., et al. (2000) Hyaluronidase enhances recombinant adeno-associated virus (rAAV)-mediated gene transfer in the rat skeletal muscle. *Gene Ther.* **7**, 1417–1420.

32. McMahon, J.M., Signori, E., Wells, K.E., Fazio, V.M., and Wells, D.J. (2001) Optimisation of electrotransfer of plasmid into skeletal muscle by pretreatment with hyaluronidase—increased expression with reduced muscle damage. *Gene Ther.* **8**, 1264–1270.

33. Molnar, M.J., Gilbert, R., Lu, Y., et al. (2004) Factors influencing the efficacy, longevity, and safety of electroporation-assisted plasmid-based gene transfer into mouse muscles. *Mol. Ther.* **10**, 447–455.

34. Ferrer, A., Foster, H., Wells, K.E., et al. (2004) Long-term expression of full-length human dystrophin in transgenic *mdx* mice expressing internally deleted human dystrophins. *Gene Ther.* **11**, 884–893.

35. Fewell, J.G., MacLaughlin, F., Mehta, V., et al. (2001) Gene therapy for the treatment of hemophilia B using PINC-formulated plasmid delivered to muscle with electroporation. *Mol. Ther.* **3**, 574–583.

36. Arruda, V.R., Hagstrom, J.N., Deitch, J., et al. (2001) Posttranslational modifications of recombinant myotube-synthesized human factor IX. *Blood.* **97**, 130–138.

37. Long, Y.C., Jaichandran, S., Ho, L.P., Tien, S.L., Tan, S.Y., and Kon, O.L. (2005) FVIII gene delivery by muscle electroporation corrects murine hemophilia A *J. Gene Med.* **7**, 494–505.

38. Mei, W.H., Qian, G.Q., Zhang, X.Q., Zhang, P., and Lu, J. (2006) Sustained expression of Epstein-Barr virus episomal vector mediated factor VIII in vivo following muscle electropo-ration. *Haemophilia.* **12**, 271–279.

39. Maruyama, H., Sugawa, M., Moriguchi, Y., et al. (2000) Continuous erythropoietin delivery by muscle-targeted gene transfer using in vivo electroporation. *Hum. Gene Ther.* **11**, 429–437.

40. Payen, E., Bettan, M., Rouyer-Fessard, P., Beuzard, Y., and Scherman, D. (2001) Improvement of mouse beta-thalassemia by electrotransfer of erythropoietin cDNA. *Exp. Hematol.* **29**, 295–300.

41. Maruyama, H., Ataka, K., Gejyo, F., et al. (2001) Long-term production of erythropoietin after electroporation-mediated transfer of plasmid DNA into the muscles of normal and uremic rats. *Gene Ther.* **8**, 461–468.

42. Dalle, B., Henri, A., Rouyer-Fessard, P., et al. (2001) Dimeric erythropoietin fusion protein with enhanced erythropoietic activity in vitro and in vivo. *Blood.* **97**, 3776–3782.

43. Rizzuto, G., Cappelletti, M., Maione, D., et al. (1999) Efficient and regulated erythropoietin production by naked DNA injection and muscle electroporation. *Proc. Natl. Acad. Sci. U.S.A.* **96**, 6417–6422.

44. Fattori, E., Cappelletti, M., Zampaglione, I., et al. (2005) Gene electro-transfer of an improved erythropoietin plasmid in mice and non-human primates. *J. Gene Med.* **7**, 228–236.

45. Komamura, K., Tatsumi, R., Miyazaki, J., et al. (2004) Treatment of dilated cardiomyopathy with electroporation of hepatocyte growth factor gene into skeletal muscle. *Hypertension.* **44**, 365–371.

46. Umeda, Y., Marui, T., Matsuno, Y., et al. (2004) Skeletal muscle targeting in vivo electroporation-mediated HGF gene therapy of bleomycin-induced pulmonary fibrosis in mice. *Lab. Invest.* **84**, 836–844.

47. Pradat, P.F., Kennel, P., Naimi-Sadaoui, S., et al. (2002) Viral and non-viral gene therapy par-tially prevents experimental cisplatin-induced neuropathy. *Gene Ther.* **9**, 1333–1337.

48. Lesbordes, J.C., Bordet, T., Haase, G., et al. (2002) In vivo electrotransfer of the cardio-trophin-1 gene into skeletal muscle slows down progression of motor neuron degeneration in *pmn* mice. *Hum. Mol. Genet.* **11**, 1615–1625.

49. Ataka, K., Maruyama, H., Neichi, T., Miyazaki, J., and Gejyo, F. (2003) Effects of erythropoietin-gene electrotransfer in rats with adenine-induced renal failure. *Am. J. Nephrol.* **23**, 315–323.

50. Rizzuto, G., Cappelletti, M., Mennuni, C., et al. (2000) Gene electrotransfer results in a high-level transduction of rat skeletal muscle and corrects anemia of renal failure. *Hum. Gene Ther.* **11**, 1891–1900.

51. Draghia-Akli, R., Ellis, K.M., Hill, L.A., Malone, P.B., and Fiorotto, M.L. (2003) High-efficiency growth hormone-releasing hormone plasmid vector administration into skeletal muscle mediated by electroporation in pigs. *FASEB J.* **17**, 526–528.

52. Draghia-Akli, R. and Fiorotto, M.L. (2004) A new plasmid-mediated approach to supplement somatotropin production in pigs. *J. Anim. Sci.* **82** (E-Suppl.), E264–E269.

53. Xiang, L., Murai, A., Sugahara, K., Yasui, A., and Muramatsu, T. (2003) Effects of leptin gene expression in mice in vivo by electroporation and hydrodynamics-based gene delivery. *Biochem. Biophys. Res. Commun.* **307**, 440–445.

54. Mallat, Z., Silvestre, J.S., Le Ricousse-Roussanne, S., et al. (2002) Interleukin-18/interleukin-18 binding protein signaling modulates ischemia-induced neovascularization in mice hindlimb. *Circ. Res.* **91**, 441–448.

55. Silvestre, J.S., Mallat, Z., Duriez, M., et al. (2000) Antiangiogenic effect of interleukin-10 in ischemia-induced angiogenesis in mice hindlimb. *Circ. Res.* **87**, 448–452.

56. Nishikage, S., Koyama, H., Miyata, T., Ishii, S., Hamada, H., and Shigematsu, H. (2004) In vivo electroporation enhances plasmid-based gene transfer of basic fibroblast growth factor for the treatment of ischemic limb. *J. Surg. Res.* **120**, 37–46.

57. Sugano, M., Hata, T., Tsuchida, K., et al. (2004) Local delivery of soluble TNF-alpha receptor 1 gene reduces infarct size following ischemia/reperfusion injury in rats. *Mol. Cell Biochem.* **266**, 127–132.

58. Silvestre, J.S., Tamarat, R., Ebrahimian, T.G., et al. (2003) Vascular endothelial growth factor-B promotes in vivo angiogenesis. *Circ. Res.* **93**, 114–123.

59. Kon, O.L., Sivakumar, S., Teoh, K.L., Lok, S.H., and Long, Y.C. (1999) Naked plasmid-mediated gene transfer to skeletal muscle ameliorates diabetes mellitus. *J. Gene Med.* **1**, 186–194.

60. Martinenghi, S., Cusella De Angelis, G., Biressi, S., et al. (2002) Human insulin production and amelioration of diabetes in mice by electrotransfer-enhanced plasmid DNA gene transfer to the skeletal muscle. *Gene Ther.* **9**, 1429–1437.

61. Croze, F. and Prud'homme, G.J. (2003) Gene therapy of streptozotocin-induced diabetes by intramuscular delivery of modified preproinsulin genes. *J. Gene Med.* **5**, 425–437.

62. Sun, W., Wang, L., Zhang, Z., Chen, M., and Wang, X. (2003) Intramuscular transfer of naked calcitonin gene-related peptide gene prevents autoimmune diabetes induced by multiple low-dose streptozotocin in C57BL mice. *Eur. J. Immunol.* **33**, 233–242.

63. Tavakoli, R., Gazdhar, A., Pierog, J., et al. (2006) Electroporation-mediated interleukin-10 overexpression in skeletal muscle reduces acute rejection in rat cardiac allografts. *J. Gene Med.* **8**, 242–248.

64. Nakano, A., Matsumori, A., Kawamoto, S., et al. (2001) Cytokine gene therapy for myocarditis by in vivo electroporation. *Hum. Gene Ther.* **12**, 1289–1297.

65. Watanabe, K., Nakazawa, M., Fuse, K., et al. (2001) Protection against autoimmune myocarditis by gene transfer of interleukin-10 by electroporation. *Circulation.* **104**, 1098–1100.

66. Adachi, O., Nakano, A., Sato, O., et al. (2002) Gene transfer of Fc-fusion cytokine by in vivo electroporation: application to gene therapy for viral myocarditis. *Gene Ther.* **9**, 577–583.

67. Jiang, J., Yamato, E., and Miyazaki, J. (2003) Sustained expression of Fc-fusion cytokine following in vivo electroporation and mouse strain differences in expression levels. *J. Biochem. (Tokyo).* **133**, 423–427.

68. Mallat, Z., Besnard, S., Duriez, M., et al. (1999) Protective role of interleukin-10 in atherosclerosis. *Circ. Res.* **85**, e17–e24.

69. Mallat, Z., Corbaz, A., Scoazec, A., et al. (2001) Interleukin-18/interleukin-18 binding protein signaling modulates atherosclerotic lesion development and stability. *Circ. Res.* **89**, E41–E45.

70. Hase, M., Tanaka, M., Yokota, M., and Yamada, Y. (2002) Reduction in the extent of atherosclerosis in apolipoprotein E-deficient mice induced by electroporation-mediated transfer of

the human plasma platelet-activating factor acetylhydrolase gene into skeletal muscle. *Prostaglandins Other Lipid Mediat.* **70**, 107–118.

71. Saidenberg-Kermanac'h, N., Bessis, N., Deleuze, V., et al. (2003) Efficacy of interleukin-10 gene electrotransfer into skeletal muscle in mice with collagen-induced arthritis. *J. Gene Med.* **5**, 164–171.

72. Ho, S.H., Hahn, W., Lee, H.J., et al. (2004) Protection against collagen-induced arthritis by electrotransfer of an expression plasmid for the interleukin-4. *Biochem. Biophys. Res. Commun.* **321**, 759–766.

73. Jeong, J.G., Kim, J.M., Ho, S.H., Hahn, W., Yu, S.S., and Kim, S. (2004) Electrotransfer of human IL-1Ra into skeletal muscles reduces the incidence of murine collagen-induced arthritis. *J. Gene Med.* **6**, 1125–1133.

74. Olsen, N.J. and Stein, C.M. (2004) New drugs for rheumatoid arthritis. *N. Engl. J. Med.* **350**, 2167–2179.

75. Bloquel, C., Bessis, N., Boissier, M.C., Scherman, D., and Bigey, P. (2004) Gene therapy of collagen-induced arthritis by electrotransfer of human tumor necrosis factor-alpha soluble receptor I variants. *Hum. Gene Ther.* **15**, 189–201.

76. Kim, J.M., Ho, S.H., Hahn, W., et al. (2003) Electro-gene therapy of collagen-induced arthritis by using an expression plasmid for the soluble p75 tumor necrosis factor receptor-Fc fusion protein. *Gene Ther.* **10**, 1216–1224.

77. Bloquel, C., Bejjani, R., Bigey, P., et al. (2006) Plasmid electrotransfer of eye ciliary muscle: principles and therapeutic efficacy using hTNF-alpha soluble receptor in uveitis. *FASEB J.* **20**, 389–391.

78. Lucas, M.L. and Heller, R. (2001) Immunomodulation by electrically enhanced delivery of plasmid DNA encoding IL-12 to murine skeletal muscle. *Mol. Ther.* **3**, 47–53.

79. Lee, S.C., Wu, C.J., Wu, P.Y., Huang, Y.L., Wu, C.W., and Tao, M.H. (2003) Inhibition of established subcutaneous and metastatic murine tumors by intramuscular electroporation of the interleukin-12 gene. *J. Biomed. Sci.* **10**, 73–86.

80. Li, S., Zhang, L., Torrero, M., Cannon, M., and Barret, R. (2005) Administration route- and immune cell activation-dependent tumor eradication by IL12 electrotransfer. *Mol. Ther.* **12**, 942–949.

81. Li, S., Zhang, X., Xia, X. et al. (2001) Intramuscular electroporation delivery of IFN-alpha gene therapy for inhibition of tumor growth located at a distant site. *Gene Ther.* **8**, 400–407.

82. Zhang, G.H., Tan, X.F., Shen, D., et al. (2003) Gene expression and antitumor effect following im electroporation delivery of human interferon alpha 2 gene. *Acta. Pharmacol. Sin.* **24**, 891–896.

83. Celiker, M.Y., Wang, M., Atsidaftos, E., et al. (2001) Inhibition of Wilms' tumor growth by intramuscular administration of tissue inhibitor of metalloproteinases-4 plasmid DNA. *Oncogene.* **20**, 4337–4343.

84. Cichon, T., Jamrozy, L., Glogowska, J., Missol-Kolka, E., and Szala, S. (2002) Electrotransfer of gene encoding endostatin into normal and neoplastic mouse tissues: inhibition of primary tumor growth and metastatic spread. *Cancer Gene Ther.* **9**, 771–777.

85. Bossard, C., Van den Berghe, L., Laurell, H., et al. (2004) Antiangiogenic properties of fibstatin, an extracellular FGF-2-binding polypeptide. *Cancer Res.* **64**, 7507–7512.

86. Ivanov, M.A., Lamrihi, B., Szyf, M., Scherman, D., and Bigey, P. (2003) Enhanced antitumor activity of a combination of MBD2-antisense electrotransfer gene therapy and bleomycin electrochemotherapy. *J. Gene Med.* **5**, 893–899.

87. Pelegrin, M., Gros, L., Dreja, H., and Piechaczyk, M. (2004) Monoclonal antibody-based genetic immunotherapy. *Curr. Gene Ther.* **4**, 347–356.

88. Perez, N., Bigey, P., Scherman, D., Danos, O., Piechaczyk, M., and Pelegrin, M. (2004) Regulatable systemic production of monoclonal antibodies by in vivo muscle electroporation. *Genet. Vaccines Ther.* **2**, 2.

89. Christian H. Ottensmeier et al. (2006) 4th international workshop on DNA vaccines, Trest. Complete information about this clinical trial can be found on the Internet site: http://www.inovio.com/

Chapter 15
Delivery of DNA into Skeletal Muscle in Large Animals

Patricia A. Brown, Amir S. Khan, and Ruxandra Draghia-Akli

Abstract Increased transgene expression after plasmid transfer to the skeletal muscle is obtained with electroporation in many species, but optimal conditions for individual species and muscle group are not well defined. Using a muscle-specific plasmid driving the expression of a secreted embryonic alkaline phosphatase (SEAP) reporter gene, we have optimized the electroporation conditions in a large mammal model, i.e. pig. The parameters optimized include electric field intensity, number of pulses, lag time between plasmid injection and electroporation, and plasmid delivery volume. Constant current pulses, between 0.4 and 0.6 A, applied 80 s after the injection of 0.5 mg SEAP-expressing plasmid in a total formulation volume of 2 mL produced the highest expression in semimembranosus muscle in pigs. These results could be extrapolated for a different muscle group in pigs, the biceps femoris, and may be an evaluation starting point for large muscle in veterinary species or humans (*see* **Note 1**).

Keywords: muscle, electroporation, pigs, secreted embryonic alkaline phosphatase

1. Introduction

Previous investigators have used electroporation devices that are designed to maintain constant voltage (1). Because of inherent variations in tissue resistance, a predetermined voltage pulse may cause an unregulated increase in the current flowing through a muscle during the duration of the pulse and a loss of the perfect square wave function (2). By contrast, a constant-current source actually maintains a square wave function with constant current through the muscle tissue (2). The unregulated current generated with conventional electroporation devices may generate an amount of heat in tissues that can easily kill cells (3, 4). For example, a typical 50-ms pulse with an average current of 0.5 A across a typical load impedance of 25 Ω can theoretically raise the temperature in tissue by 7.5°C, which is enough to kill cells (5). Electrical shock trauma tends to produce a very complex pattern of injury because of the multiple modes of frequency-dependent tissue-field interactions, such as heating, permeabilization of cell membranes, and direct electroconformational denaturation of

S. Li (ed.), *Electroporation Protocols: Preclinical and Clinical Gene Medicine.*
From *Methods in Molecular Biology, Vol. 423.*
© Humana Press 2008

macromolecules such as proteins (5). By contrast, the power dissipation is lower in a constant-current system with a decentralized pattern of pulse generation (maximum meridians of an electric field delineating an area instead of intersecting into a point) which prevents heating of a tissue. Thus, constant-current electroporation may reduce tissue damage and contribute to the overall success of the procedure. We have previously demonstrated that the constant-current electroporation procedure is effective for intramuscular plasmid delivery in many large animal species and does not cause permanent damage to adjacent cells (6–9). Furthermore, improvements in the electroporation conditions could increase the efficacy of plasmid transfer and lower the total amount of plasmid needed to generate targeted levels of biologically active proteins.

In this study, we delivered a muscle-specific plasmid encoding for secreted embryonic alkaline phosphatase (SEAP) to the skeletal muscle of pigs and measured SEAP levels in the serum. Although the SEAP protein is immunogenic in most adult mammals, including pigs, the immune-mediated clearance of the protein does not occur until 10–14 days after plasmid delivery. Thus, the levels of SEAP expression over a 2-week period can be analyzed and interpreted as a reliable measure of gene expression following intramuscular plasmid transfer (*see* **Note 2**). Electroporation conditions are stated for each experiment. In all cases, except where noted, we used a constant current, 0.4–1.0 A, 3 pulses, 52 ms/pulse, and 1 s between pulses (*see* **Note 3**).

2. Materials

2.1. Plasmid DNA

1. pSP-SEAP (5,019 bp) is a muscle-specific plasmid (Fig. 15.1) expressing SEAP, a reporter protein.
2. The promoter is SPc5-12, a strong, muscle-specific, synthetic promoter (10), and the 3′ ends of the SEAP transcripts are defined by the SV40 late poly(A) signal.
3. The plasmid was constructed by inserting a 394-bp *Acc*65I-*Hin*dIII fragment, containing the 334-bp SPc5-12 promoter sequences, between the *Acc*65I and *Hin*dIII sites of pSEAP-2 Basic Vector (Clontech Laboratories, Inc., Palo Alto, CA).
4. Plasmid preparations were diluted in sterile water and formulated 1% weight per weight with poly-L-glutamate sodium salt (mol. wt. = 10.5 kDa average).

2.2. In Vivo Experiments

1. Young hybrid pigs of mixed genders, 3–6 weeks old, weighing between 15 and 40 kg. Animals were group-housed in pens with ad libitum access to 24% protein

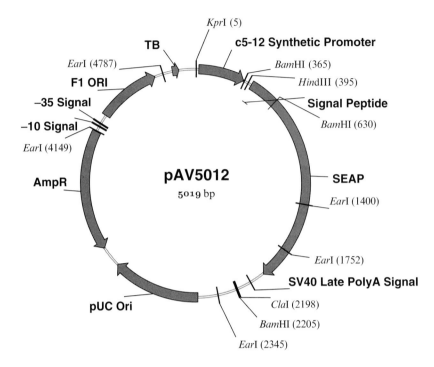

Fig. 15.1 pSP-SEAP is a muscle-specific plasmid for SEAP. The promoter is SPc5-12, a strong muscle-specific synthetic promoter, and the 3′ ends are defined by the SV40 late poly(A) signal

diet (Producers Cooperative Association, Bryan, TX) and water and were maintained in accordance with NIH, USDA, and Animal Welfare Act guidelines.
2. 1-cm^3 syringes and 21-gauge needles.
3. Plasmid pSP-SEAP.
4. CELLECTRA™ Adaptive Constant Current Electroporation Device (VGX Pharmaceuticals, Immune Therapeutics Division, The Woodlands, Texas) (Fig. 15.2).
5. Five needle-electrode sterile single-use arrays.

2.3. Collection and Processing of Blood Samples

1. Microtainer serum separator tubes.
2. 3-cm^3 syringes and 21-gauge needles.
3. Centrifuge.
4. Transfer pipettes.
5. Ice packs and insulated transportation container.

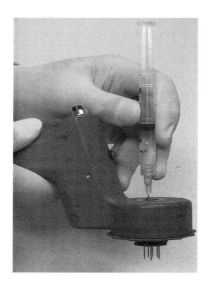

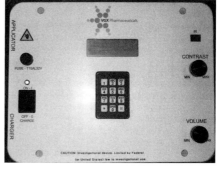

Fig. 15.2 The CELLECTRA™ device is battery-operated and portable; so it can be used in either laboratory or farm conditions. It is software-driven and stores several hundred EP pulse patterns that can be downloaded to any PC. The ergonomically designed applicator can be used in animals of any size, from rabbits, piglets, and primates to sows, cows, and horses

2.4. Secreted Embryonic Alkaline Phosphatase Assay

1. Phospha-light chemiluminescent reporter assay kit (Applied Biosystems, Bedford, MA), per manufacturer's instructions.
2. LUMIstar Galaxy luminometer (BMG Labtechnologies, Offenberg, Germany).
3. Control serum (for diluting samples).
4. Pig serum samples.

3. Methods

In this study, we delivered a muscle-specific plasmid encoding for secreted embryonic alkaline phosphatase (SEAP) to the skeletal muscle of pigs and measured SEAP levels in serum in order to identify optimal electroporation parameters. SEAP has been previously used *in vitro* (11) and *in vivo* in mice (12), dogs, and pigs (6) as a reporter to investigate the conditions required for expression of secreted proteins. The levels of SEAP expression over a 2-week

period can be analyzed and interpreted as a reliable measure of gene expression following intramuscular plasmid transfer.

3.1. Plasmid Preparation

Plasmid preparations were diluted in sterile water and formulated 1% weight per weight with HPLC-purified poly-L-glutamate sodium salt (mol. wt. = 10.5 kDa average). This formulation was shown to stabilize the plasmid *ex vivo* (the plasmid does not need to be refrigerated during the experiment, even if the work is performed in farm conditions at ambient temperature), prevent damage of the tissue during electroporation, and increase expression (6, 13).

3.2. Intramuscular Injection and Electroporation of pSP-SEAP

3.2.1. Preparation and Assembly of Electroporator

1. Attach charged battery pack to the connector closest to the power switch.
2. Attach the applicator assembly to the larger connector.
3. Turn on (switch is sloped away from the user) the electroporator.
4. Display reads "VGX Pharmaceuticals CELLECTRA Version 3.1."
5. Enter password (if entered incorrectly, simply re-enter it).
6. Current-operating parameters appear and may read as follows: Current = 0.5 A, Firing Delay = 80 s, Impedance Chk = EN.

3.2.2. Change Parameters

1. Using the numeric key pad, key in the number of the parameter to be changed (for instance, 1 = Current, 2 = Firing Delay, etc.).
2. The parameter value will start flashing.
3. Key in the value required.
4. The display will then show the corrected value.
5. When all parameters are at the required setting, the machine is ready for the injection and electroporation sequence.

3.2.3. Select IEP Site

An efficient injection and electroporation procedure (IEP) must ensure that the plasmid is delivered into skeletal muscle and not into the fat, fascia, or a blood vessel.

All needle electrodes of the EP array must be placed in the same muscle and should circumscribe the plasmid injection site. If the electrodes of the array are in different muscles, the EP process will be inefficient. If the plasmid injection is made into fascia or ligament, the plasmid will be ineffective. If EP needles are placed into fascia or a ligament, the resistance of the tissue during electroporation may be higher than normal and not uniform within the electric field, often leading to burns in the tissue or on the surface of the skin. Do not inject into an area that presents with skin laceration or obvious pathological condition. If this is the case, select a different site.

1. Site for IEP in pigs is situated in the internal (medial) inguinal area, such that the injection and electroporation needle-electrodes would penetrate and stay within the belly of the semimembranosus muscle. The inguinal area is heavily vascularized, and so, when choosing the IEP site, care should be taken to avoid large superficial vessels (Fig. 15.3). The actual IEP site will vary from animal to animal.
2. Manually restrain the animals (*see* **Note 4**).
3. Clean and dry the IEP site with a disposable towel, as infection at the injection site should be completely avoided.
4. If wet, the site should be thoroughly dried to avoid electric short circuit at the surface of the skin and possible skin burns.

3.2.4. Insertion of Needle-Electrode Array and Injection Needle

1. Prepare and activate the plasmid dose unit.
2. Place a disposable electrode disk in the receptacle of the applicator and lock into position.
3. Insert the needle of the plasmid dose unit into the center of the electrode array.

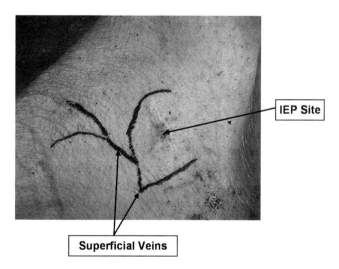

Fig. 15.3 The IEP site can be discerned from the superficial veins and other landmarks of the semimembranosus muscle on the inner thigh of the sow

4. Ensure injection needle is approximately the same length as the needle electrodes in the array.
5. Place the array over the selected IEP site and penetrate the tissue with one forceful motion, pushing down to ensure the electrode pins and injection needle are completely inserted into the muscle and the electrode lock disk is pressed tightly against the surface of the skin, slightly depressing the skin at the IEP site.
6. Check that the injection needle is not in a blood vessel by gently pulling back the piston of the syringe.
7. If blood flows back into the syringe, the tip of the injection needle is in a blood vessel and the device should be removed and reinserted into a different site.
8. The procedure should be repeated each time prior to injection of plasmid.

3.2.5. Injection and Electroporation Sequence

1. Press the applicator button once.
2. Using the key pad enter the unique study number of the animal and press the number key.
3. Press the applicator button again (you will hear a series of clicks) to initiate the impedance check to ensure that the needle electrodes are inserted into muscle.
4. If the indicator light on the applicator is red, remove, reposition, and reinsert all needles (press any button to clear display screen).
5. If circuit connection is correct and all needle electrodes are inserted into the muscle, the indicator light on the applicator will turn green and the display will read "Waiting for trigger press to start electroporation."
6. Inject the plasmid slowly into the muscle (*see* **Note 5**).
7. The plasmid should be delivered such that it stays within the area delineated by the needle-electrode arrays.
8. Push the applicator button one more time to initiate the countdown to EP.
9. Keep finger on the syringe to prevent plasmid backflow into the syringe.
10. The display will show the amperage, pulse, and impedance along with the time remaining until electroporation.
11. The EP counts down from either 4 or 80 s, and then, following audible beeps during the last 4 or 5 s of the countdown, the device delivers the EP. With each pulse, the muscle will contract—the degree of contraction depends on the ampere setting (higher amperes will incur a higher response).

3.3. Blood Collection

1. Approximately 2–3 mL of blood should be collected (by jugular vein puncture) into blood collection tubes on study days 0 (immediately before treatment), 3, 7, and 10.
2. Blood should be allowed to clot at room temperature for 10–15 min.

3. The blood samples are centrifuged at 3,000xg for 10 min, and the collected serum is aliquoted into at least 2 prelabeled tubes per sample.
4. All serum samples need to be immediately placed on ice for transfer back to the laboratory and then frozen at −80°C until further analysis.

3.4. Secreted Embryonic Alkaline Phosphatase Assay

1. Serum samples are thawed and 50 μL is assayed for SEAP activity using the Phospha-Light Chemiluminescent Reporter Assay Kit.
2. The assay kit is run as per manufacturer's instructions.
3. The lower limit of detection for the assay is 3 pg/mL.
4. More concentrated serum samples (usually samples from study days 7 and 10) are diluted 1:10 in control serum before assaying for SEAP activity (*see* **Note 6**).
5. All samples were read using LUMIstar Galaxy luminometer (BMG Labtechnologies, Offenburg, Germany).

3.5. Statistical Analysis

Data were analyzed using Microsoft Excel Statistics package. Specific values are obtained by comparison using t test or one-way ANOVA. A value of $p < 0.05$ is set as the level of statistical significance (*see* **Note 7**).

4. Notes

1. Muscle groups vary in resistance; therefore, it is recommended to perform a pilot study to compare muscles in selected animal species using a plasmid such as muscle-specific plasmid encoding for SEAP prior to use of specific transgenes.
2. Once animals have been injected with SEAP plasmid, retreatment will not be feasible because of the strong immune response against the human SEAP protein.
3. We have performed several pilot experiments in order to optimize the electric field intensity (Fig. 15.4) and lag time between injection and electroporation (Fig. 15.5) in pigs and other large animal models. It is recommended that users conduct similar pilot experiments in order to optimize the conditions of electroporation for their individual model.
4. Larger pigs may require an anesthetic to avoid injury to the operator or animal or both. Consult a licensed veterinarian for an anesthesia protocol recommendation.
5. Inject plasmid slowly to prevent leaking of solution.
6. A dilution of the samples collected at day 7 and day 10 may be necessary to maintain the SEAP serum values within the assay limits of detection. A correction factor is added during the analysis of the SEAP data in order to accurately compare the results.
7. Data are corrected for each animal's body weight, as a correction for the blood volume. Thus, at each time point serum SEAP levels, e.g., 140 pg/mL, are corrected for the animal's body weight, e.g., 10 kg, to achieve a final SEAP expression value of 14 pg/mL/kg. If this correction is not performed, animals that are larger, e.g. 20 kg, will demonstrate artificially lower SEAP unless body weight is taken into account.

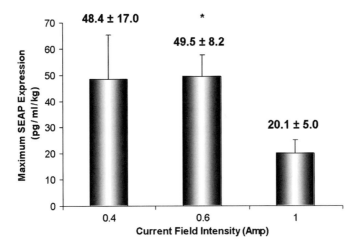

Fig. 15.4 SEAP expression is measured as a function of the electric field intensity, and will decrease at higher ampere settings ($^*p < 0.05$, compared with that at 0.6 and 1 A)

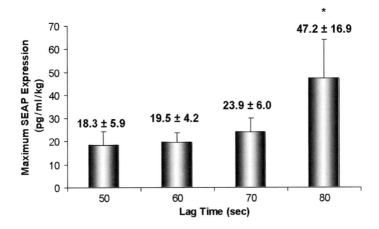

Fig. 15.5 SEAP expression is measured as a function of the lag time between injection and EP. Expression is significantly increased at 80 s ($^*p < 0.05$, compared with that at 80 s and both 50 and 60 s)

References

1. Satkauskas, S., Andre, F., Bureau, M.F., Scherman, D., Miklavcic, D., and Mir, L.M. (2005) Electrophoretic component of electric pulses determines the efficacy of in vivo DNA electrotransfer. *Hum. Gene Ther.* **16**, 1194–1201.
2. Draghia-Akli, R. and Smith, L.C. (2003) Electrokinetic enhancement of plasmid DNA delivery *in vivo*. In: Templeton, N.S. and Lasic, D.D. (eds.). *Gene and cell therapy.* Marcel Derker, New York, NY, pp. 245–263.

3. Martin, G.T., Pliquett, U.F., and Weaver, J.C. (2002) Theoretical analysis of localized heating in human skin subjected to high voltage pulses. *Bioelectrochemistry.* **57**, 55–64.

4. Pliquett, U.F., Martin, G.T., and Weaver, J.C. (2002) Kinetics of the temperature rise within human stratum corneum during electroporation and pulsed high-voltage iontophoresis. *Bioelectrochemistry.* **57**, 65–72.

5. Lee, R.C., Zhang, D., and Hannig, J. (2000) Biophysical injury mechanisms in electrical shock trauma. *Annu. Rev. Biomed. Eng.* **2**, 477–509.

6. Draghia-Akli, R., Khan, A.S., Cummings, K.K., Parghi, D., Carpenter, R.H., and Brown, P.A. (2002) Electrical enhancement of formulated plasmid delivery in animals. *Technol. Cancer Res. Treat.* **1**, 365–371.

7. Khan, A.S., Smith, L.C., Abruzzese, R.V., et al. (2003) Optimization of electroporation parameters for the intramuscular delivery of plasmids in pigs. *DNA Cell Biol.* **22**, 807–814.

8. Brown, P.A., Davis, W.C., and Draghia-Akli, R. (2004) Immune enhancing effects of growth hormone releasing hormone delivered by plasmid injection and electroporation. *Mol. Ther.* **10**, 644–651.

9. Tone, C.M., Cardoza, D.M., Carpenter, R.H., and Draghia-Akli, R. (2004) Long-term effects of plasmid-mediated growth hormone releasing hormone in dogs. *Cancer Gene Ther.* **11**, 389–396.

10. Li, X., Eastman, E.M., Schwartz, R.J., and Draghia-Akli, R. (1999) Synthetic muscle promoters: activities exceeding naturally occurring regulatory sequences. *Nat. Biotechnol.* **17**, 241–245.

11. Durocher, Y., Perret, S., Thibaudeau, E., et al. (2000) A reporter gene assay for high-throughput screening of G-protein-coupled receptors stably or transiently expressed in HEK293 EBNA cells grown in suspension culture. *Anal. Biochem.* **284**, 316–326.

12. Nicol, F., Wong, M., MacLaughlin, F.C., et al. (2002) Poly-L-glutamate, an anionic polymer, enhances transgene expression for plasmids delivered by intramuscular injection with in vivo electroporation. *Gene Ther.* **9**, 1351–1358.

13. Hebel, H.L., Attra, H.E., Khan, A.S., and Draghia-Akli, R. (2006) Successful parallel development and integration of a plasmid-based biologic, container/closure system and electrokinetic delivery device. *Vaccine.* **24**, 4607–4614.

Chapter 16
Delivery of DNA into Skin via Electroporation

Babu M. Medi and Jagdish Singh

Abstract Delivery of DNA into skin is an attractive method, because skin is the most accessible somatic tissue for gene transfer and can be monitored conveniently. Skin is especially suitable for immunization using plasmid-DNA-based vaccines; however, a low level of transfection is the major limitation to the use of DNA-based therapeutics. Several chemical and physical methods are being investigated to improve the transfection of target cells with plasmid DNA. Electroporation is a physical method of gene transfer by applying electric pulses to the target cells. Most of the electroporation studies involve insertion of electrode needles into the tissues. In this chapter, we discuss that the DNA delivery into skin can be greatly enhanced by topical electroporation of the DNA injection site in rabbits using a tweezer electrode. Furthermore, the immune responses following a DNA vaccine delivery by using electroporation have been explored. Electroporation shows great potential for enhancing the DNA delivery into the skin.

Keywords: electroporation, skin, DNA delivery, DNA vaccine, reporter plasmid, plasmid DNA, gene delivery

1. Introduction

Skin is the largest organ of the human body and functions as a barrier to exogenous harmful influences while maintaining homeostasis. Skin is an attractive site for DNA delivery because it is easily accessible and easy to monitor. In the case of gene-based therapy, perhaps the most fundamental hurdle is the appropriate DNA delivery to the target cells. To be therapeutically useful, the DNA should be delivered into the target cells efficiently with minimal toxic effects (1). In spite of the research focusing on the development of gene delivery tools spanning more than a decade, the efficient delivery of genes to target cells and, in reality, to appropriate subcellular compartments, still remains largely elusive.

Recently, DNA delivery using electroporation has been shown to be useful for gene transfer. Electroporation involves the application of controlled, short, and

high-voltage electric pulses to reversibly permeabilize the target cell/tissue for macromolecules such as genes and proteins (2, 3). Initially developed for gene transfer, electroporation is now used for the delivery of a variety of molecules such as ions, drugs, dyes, tracers, antibodies, oligonucleotides, RNA, and DNA (4, 5). Also, electroporation has become a standard method of in vitro cell transfection and cell loading (6). Progress during the past few years clearly demonstrates the great potential of this technology (7, 8). Electroporation has been evaluated in animals and humans for the delivery of chemotherapeutic agents with high efficiency. Furthermore, it has been employed in studies involving delivery of plasmid DNA in vivo to different types of tissues with improved transfection efficiency (9–13). Most of these studies involve insertion of electrode needles into the tissue, after plasmid DNA injection, for electroporation. In this chapter, the DNA delivery to skin using topical electroporation after intradermal DNA injection is discussed.

2. Materials

2.1. Plasmid DNA and Other Reagents

1. Plasmid DNA coding β-galactosidase, gWiz™ β-gal (Aldevron LLC, Fargo, ND).
2. Plasmid DNA coding for Hepatitis B surface antigen (HBsAg), pCMV-S, and HBsAg protein (Aldevron LLC).
3. Micro BCA™ protein assay kit, 3,3′,5,5′-tetramethylbenzidine (TMB), and hydrogen peroxide (Pierce Biotechnology, Rockford, IL).
4. X-gal (4-chloro-5-bromo-3-indolyl β-D-galactopyranoside).
5. Assay kit (including lysis buffer) for β-galactosidase (Promega Corp., Madison, WI).
6. Other chemicals were obtained from Sigma-Aldrich (St. Louis, MO).
7. Ultrapure water obtained with a Barnstead Nanopure Infinity® water system (Barnstead, Boston, MA) having resistivity of ≥18 MΩ cm was used to prepare all solutions and buffers.

2.2. In Vivo Animal Experiments

1. New Zealand White (NZW) rabbits (*Oryctolagus cuniculus*), 10 weeks old and about 2.0–3.0-kg body weight, were used in the study. The animals were housed and cared for in accordance with the *Guide for the Care and Use of Laboratory Animals* (National Academy Press, 1996). All the animal experiments were performed according to the protocols approved by Institutional Animal Care and Use Committee of North Dakota State University (NDSU).

2. Pentobarbital sodium (Nembutal®) (Abbott Laboratories, North Chicago, IL).
3. Microsyringe with a 28-gauge needle (Hamilton Company, Reno, NV).
4. Square wave electroporator (CUY21 EDIT version) and stainless steel tweezer electrode (10 mm × 5 mm, CUY 663B) (NEPA Gene Co., Chiba, Japan).

3. Methods

New Zealand White (NZW) rabbits (*O. cuniculus*) were used for the DNA delivery to skin with this electroporation technique. The effects of electroporation pulse amplitude and pulse length were studied using a reporter plasmid DNA coding β-galactosidase and a DNA vaccine (pCMV-S) for Hepatitis B virus (HBV).

3.1. Preparation of Animal

1. Remove the hair on the back of the rabbits by using an electric clipper carefully, with great care to avoid damage to the skin, 24 h prior to the beginning of the study (*see* **Note 1**).
2. At the beginning of the study, anesthetize the rabbits using pentobarbital sodium (40 mg/kg; Nembutal®) injected intraperitoneally. It may take 15–30 min after injection for the Nembutal to take effect.
3. Clean the skin with 70% (v/v) isopropyl alcohol and clearly mark an area of the skin on the back of the rabbit with a permanent marker (*see* **Note 2**).

3.2. DNA Delivery Using Electroporation

1. Turn on the electroporator and set the electroporation parameters (pulse voltage, pulse length, number of pulses, and time between each pulse).
2. Once the rabbit is under anesthesia, inject 50 µg (20 µL of 2.5 mg/mL) of the plasmid DNA encoding β-galactosidase, gWiz™ β-gal, in phosphate-buffered saline (PBS, pH 7.4) intradermally using a microsyringe with a 28-gauge needle in the marked area of the rabbit (*see* **Note 3**). For delivery of DNA vaccine, follow the same procedure using HBsAg DNA vaccine.
3. Immediately after injection, clamp the site of plasmid injection with a stainless steel tweezer electrode by raising the skin around the site of injection (*see* **Note 4**). An electrode clamped on to the rabbit skin is shown in Fig. 16.1.
4. Press the application button to perform the electroporation (*see* **Note 5**).
5. Move the rabbits back into their cages and allow them to recover.

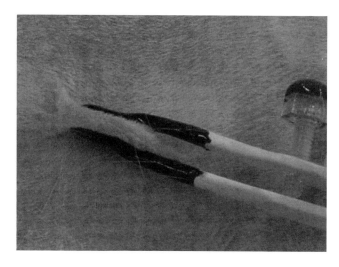

Fig. 16.1 The clamping of tweezer electrode on the rabbit skin for delivery of DNA using electroporation. The figure in the inset shows the drawing of the tweezer electrode

3.3. Sample Collection for β-Galactosidase Expression Studies

1. Euthanize the animals using pentobarbital sodium (100 mg/kg) injected intraperitoneally.
2. Remove the skin samples from the site of DNA delivery using a surgical scalpel and scissors and place them into appropriate tubes in liquid nitrogen and then store at −70°C until further use (*see* **Note 6**).

3.4. Quantification of Reporter Gene (β-Galactosidase) Expression

1. Thaw the skin samples and mince the samples with surgical scissors or surgical blade (petri dish may be useful for mincing). Collect the minced samples into an appropriate tared tube and record the weight of the sample.
2. Add reporter lysis buffer to the minced sample and homogenize using a hand-held homogenizer (*see* **Note 7**). while keeping the samples on ice.
3. Centrifuge the tissue homogenate at 17,624*g* and 4°C for 15 min and collect the clear supernatant for further analysis (*see* **Note 8**).
4. Use the above sample to measure total protein content with the Micro BCA™ protein assay reagent kit following manufacturer's protocol. Sample dilution may be necessary.
5. Based on the protein content results, calculate the protein content per unit weight of the skin sample.

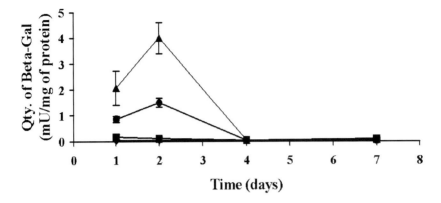

Fig. 16.2 Effect of electroporation pulse voltage on β-galactosidase DNA delivery to skin. *Key:* (♦) passive (injection without pulses), (■) 100 V, (●) 200 V, and (▲) 300 V. Five pulses of the specified voltage and 10-ms pulse length were applied with 1-s interval between each pulse. The values shown are mean ± SD ($n = 3$). (Reproduced from (13), with permission from Elsevier Science)

6. Using the same sample, measure the β-galactosidase (reporter molecule) by using the β-galactosidase enzyme assay method following the manufacturer's protocol. Sample dilution may be necessary.
7. Calculate the enzyme activity in the samples and express it as milliunits per milligram of protein (*see* **Note 9**). The effect of electroporation pulse voltage on β-galactosidase DNA delivery to the skin is shown in Fig. 16.2.

3.5. In Situ *Histochemical Staining of Skin for β-Galactosidase*

1. Collect the skin samples as described in sect. 3.3.
2. Cut the skin samples into 1–2-mm strips on a microscopic glass slide using a surgical blade and forceps
3. Fix the samples in freshly prepared ice-cold 2% (v/v) formaldehyde, 0.2% glutaraldehyde, and 2 mM $MgCl_2$ in PBS (pH 7.4) overnight at 4°C.
4. Wash the tissue samples thrice within 2 h with PBS containing 2 mM $MgCl_2$, 0.1% sodium deoxycholate, and 0.02% NP-40.
5. Stain the tissue samples in the dark at 37°C overnight with X-gal (1 mg/mL) in PBS containing 5 mM potassium ferricyanide, 5 mM potassium ferrocyanide, 2 mM $MgCl_2$, 0.02% NP-40, and 0.1% sodium deoxycholate at pH 8.
6. Fix the stained tissues in 10% formalin for 2 h, section the samples into 10-μm sections, and counterstain with hematoxylin and eosin (*see* **Note 10**).
7. Observe the stained tissue sections under a light microscope (Meiji Microscope, Osaka, Japan) with 40× magnification and take pictures using Polaroid® Microcam camera (*see* **Note 11**).

3.6. Measurement of Specific Anti-HBsAg IgG Antibody Levels

1. Collect blood samples from the mid-ear artery of rabbits at different time points after immunization and separate the serum.
2. The anti-HBsAg IgG levels in the individual rabbit sera (diluted 1:1000) were measured, in triplicate, using enzyme-linked immunosorbent assay (ELISA).
3. For ELISA, coat the wells of the 96-well Nunc-Immuno MaxiSorp microplate (Nalge Nunc International, Rochester, NY) with 100 μL of 10 μg HBsAg/mL in carbonate coating buffer overnight at 4°C.
4. After washing thrice with PBS-Tween, block the wells with 3% BSA in coating buffer for 2 h at room temperature.
5. Add the diluted sera to the wells and incubate at room temperature for 2 h.
6. Remove sera and wash the plate with PBS-Tween and incubate with 1:10,000 dilution of goat anti-rabbit IgG conjugated with horseradish peroxidase.
7. After washing the conjugate solution, incubate the plate with TMB and hydrogen peroxide for 30 min and stop the reaction with 2M H_2SO_4 and read the optical density at 450 nm in a plate reader.
8. Report the optical densities of different groups of animals that are being compared (*see* **Note 12**).

4. Notes

1. Check the rabbit skin for any abnormalities or skin rashes, and if any are present, avoid using the animal.
2. Swab the area of the skin to be injected with DNA using a 70% isopropyl alcohol pad and allow it to dry before injecting and electroporating the site.
3. For intradermal injection of 20 μL plasmid DNA a microsyringe was used with a 28-gauge needle. The needle was inserted at 10°–15° angle to the skin and the DNA was injected. After intradermal injection a small bleb is visible on the skin.
4. Adjust the distance between the electrode plates with the adjustable screw on the electrode before clamping. Keep the distance between the electrode plates constant for comparison studies. For clamping the electrode, raise the skin in such a way that the site of the skin injected with DNA is on one side of the skin fold and place the electrode properly to cover the skin area completely with the cathode on the side of the injection. During electroporation, the electrode must be held properly at the site, as the animal might move.
5. Clean the electrode plates after each electroporation by using 70% isopropyl alcohol and allow them to dry.
6. If hair has grown at the DNA delivery site (usually encountered after 48 h) on the back of the rabbits, remove them using an electric clipper, taking care to avoid damage to the skin. Also do not cut the fatty tissue under the skin.
7. The skin sample should be cut into small pieces using surgical scissors before homogenizing the samples with a handheld Tissue Tearor™ homogenizer at 17,624g. A 5-mL disposable tube works great for use with 1 mL lysis buffer.
8. The supernatant from the homogenate can be stored at −70°C if not used immediately.
9. The protein content of the skin samples should be normalized based on the weight of the skin samples used for homogenization.

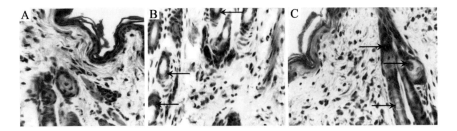

Fig. 16.3 In situ histochemical staining with X-Gal for local expression of β-galactosidase in skin. Key: (**A**) control skin, (**B**) five pulses of 200 V and 10-ms pulse length with 1-s interval between each pulse, and (**C**) five pulses of 300 V and 10-ms pulse length with 1-s interval between each pulse. Blue stain indicates the expression of β-galactosidase (as shown by *arrows*). (Reproduced from (13), with permission from Elsevier Science)

10. The samples were actually sent to diagnostic laboratories for sectioning and staining; however, the stained samples can be processed using a microtome.
11. The slides with the stained tissue samples can be observed under any light microscope and the blue-colored spots represent β-galactosidase expression (Fig. 16.3). Pictures can be taken using a digital camera attached to the microscope.
12. The end point titers can be reported if a serial dilution of sera is tested in ELISA.

References

1. Verma, I.M. and Somia, N. (1997) Gene therapy—promises, problems and prospects. *Nature.* **389**, 239–242.
2. Gehl, J. (2003) Electroporation: theory and methods, perspectives for drug delivery, gene therapy and research. *Acta. Physiol. Scand.* **177**, 437–447.
3. Medi, B.M. and Singh, J. (2003) Electronically facilitated transdermal delivery of human parathyroid hormone (1–34). *Int. J. Pharm.* **263**, 25–33.
4. Jubin, R. (2004) Optimizing electroporation conditions for intracellular delivery of morpholino antisense oligonucleotides directed against the Hepatitis C virus internal ribosome entry site. *Methods Mol. Med.* **106**, 309–322.
5. Rao, M., Baraban, J.H., Rajaii, F., and Sockanathan, S. (2004) In vivo comparative study of RNAi methodologies by in ovo electroporation in the chick embryo. *Dev. Dyn.* **231**, 592–600.
6. Hui, S.W. (1995) Effects of pulse length and strength on electroporation efficiency. *Methods Mol. Biol.* **55**, 29–40.
7. Li, S. (2004) Electroporation gene therapy: new developments in vivo and in vitro. *Curr. Gene Ther.* **4**, 309–316.
8. Wells, D.J. (2004) Gene therapy progress and prospects: electroporation and other physical methods. *Gene Ther.* **11**, 1363–1369.
9. Zhang, L., Li, L., Hoffman, G.A., and Hoffman, R.M. (1996) Depth-targeted efficient gene delivery and expression in the skin by pulsed electric fields: an approach to gene therapy of skin aging and other diseases. *Biochem. Biophys. Res. Commun.* **220**, 633–636.
10. Glasspool-Malone, J., Somiari, S., Drabick, J.J., and Malone, R.W. (2000) Efficient nonviral cutaneous transfection. *Mol. Ther.* **2**, 140–146.

11. Matsumoto, T., Komori, K., Shoji, T., et al. (2001) Successful and optimized in vivo gene transfer to rabbit carotid artery mediated by electronic pulse. *Gene Ther.* **8**, 1174–1179.
12. Liu, F. and Huang, L. (2002). Electric gene transfer to the liver following systemic administration of plasmid DNA. *Gene Ther.* **9**, 1116–1119.
13. Medi, B.M., Hoselton, S., Marepalli, R.B., and Singh, J. (2005) Skin targeted DNA vaccine delivery using electroporation in rabbits. I: Efficacy. *Int. J. Pharm.* **294**, 53–63.

Chapter 17
Electroporation-Mediated Gene Delivery to the Lungs

Rui Zhou, James E. Norton, and David A. Dean

Abstract Electroporation is a safe, efficient, and inexpensive method to transfer naked plasmid DNA into various tissues. For electroporation-mediated gene transfer to the mouse lung, a plasmid solution is delivered to the lungs via the trachea. Immediately after plasmid delivery, eight square wave pulses are delivered by two pregelled electrodes placed on each side of the chest. The optimal field strength is 200 V/cm, with a pulse duration of 10 ms each and a 1 s interval between pulses. High-level gene expression can be achieved within 24 h in all cell types in the lung, with very little inflammation and no apparent trauma.

Keywords: electroporation, nonviral gene therapy, plasmid, lungs, whole lung tissue extraction, bronchoalveolar lavage fluid (BALF), intubation

1. Introduction

Electroporation has proven to be a highly effective technique for the in vivo delivery of genes to a number of solid tissues. In most of the methods, DNA is injected into the target tissue and electrodes are placed directly on or in the tissue for application of the electric field (1). While this works well for solid tissues, there are many tissues and organs that are not amenable to such an approach. The lung is one such target (2). Although multiple techniques for gene delivery to the lung have been developed, including the use of adenoviruses, adeno-associated viruses, lipoplex, and polyethyleneimine, all have drawbacks, including inefficiency of gene transfer, immunological responses, inflammation, nonspecificity of cell targeting, and low levels of gene expression (3, 4).

Our lab and others have recently shown that electroporation can also be used to deliver genes efficiently to the lungs of mice and rats (5–8). The technique is simple, quick, and yields high levels of gene transfer and expression. The method results in gene transfer and expression to all cell types in the lung, including airway and alveolar epithelial cells, endothelial cells, and airway and vascular smooth-muscle cells. Most importantly, almost no detectable inflammatory response is

S. Li (ed.), *Electroporation Protocols: Preclinical and Clinical Gene Medicine.*
From *Methods in Molecular Biology, Vol. 423.*
© Humana Press 2008

observed and no trauma is apparent, making this a safe and effective procedure. In this chapter, we will describe the procedure for use in mice and several methods for analysis of tissues following the procedure.

2. Materials

2.1. Electroporation

1. BTX ECM830 Electroporator (Genetronics, San Diego, CA).
2. Pregelled disposable surface electrode (Medtronic, Skovlunde, Denmark) (Figs. 17.1 and 17.2).
3. Alligator clip cables (Figs. 17.1 and 17.2).
4. Angle adjustable surgical board with a piece of 3-0 suture taped across the top of the unhinged end of the table (Fig. 17.1).

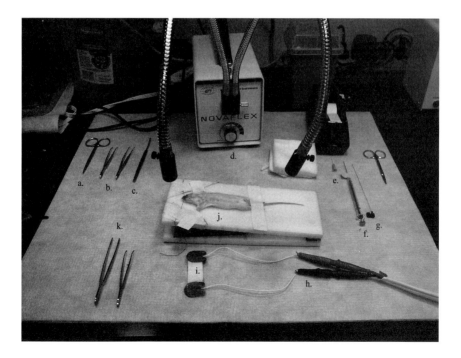

Fig. 17.1 Electroporation equipment. (**a**) Surgical scissors, (**b**) eyedressing forceps, (**c**) surgical blade handle, (**d**) surgical lamp, (**e**) angiocatheter, (**f**) Hamilton syringe, (**g**) guide wire from Arrow Radial Artery Catheter Kit, (**h**) alligator clip cable, (**i**) Medtronic pregelled electrode pads, (**j**) angled surgery table, (**k**) forceps

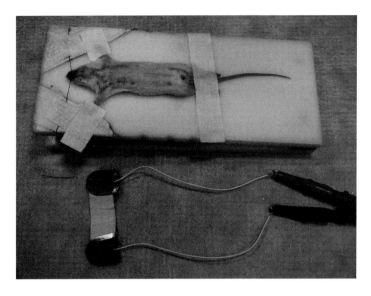

Fig. 17.2 Animal positioning on surgical table and electrode attachment to alligator cables

5. Autoclave or surgical tape (autoclave tape is preferred for its stronger adhesive).
6. Sodium pentobarbital.
7. U 100 insulin syringe 0.5 cm³, 28 GA, 0.5 in.
8. Plasmid DNA: Dilute purified plasmid DNA to 1.0 µg/µL in 10 mM tris, pH 8.0, 1 mM EDTA, and 140 mM NaCl (TEN Buffer) and mix thoroughly but gently by flicking (do not vortex) (*see* **Note 1**).
9. Surgical lubricant (e.g., Surgilube or KY Jelly).
10. Squirt bottle of 70% ethanol.
11. Gauze pads.
12. Ruler.

2.2. Intubation (Fig. 17.1, 17.3)

1. 18–20 GA catheter cut down to ~1.9 cm in length, with the end cut at a 60° angle.
2. 0.13-mm thick spatula (Fine Science Tools 10094–13).
3. Eye dressing tweezers.
4. Guide wire from a Radial Artery Catheterization Set (Arrow International, Reading, PA).
5. 25 or 50 µL Hamilton syringe with tape at the base of the needle reducing the exposed needle length to 3.8 cm.
6. Novaflex surgical light.

2.3. Aspiration

1. Flat forceps (Fig. 17.1).
2. P-200 pipetteman.
3. Novaflex surgical light (Fig. 17.1).

2.4. Tracheal Injection

1. Sterile gauze.
2. Sterile surgical instrument kit.

    (a) Two pairs of eye dressing forceps.
    (b) One pair of sharp surgical scissors.
    (c) One pair of needle holders.

3. Bench coat.
4. 4-0 Dexon II Suture.
5. Triple antibiotic ointment.
6. Betadine antiseptic.

2.5. Sample Collection

1. Surgical instrument kit (see sect. 2.4).
2. Sterile 1× PBS.

2.5.1. Whole Lung Tissue Extract

1. Tissue pulverizer (Biospec Products, Inc.).
2. Hammer.
3. Microcentrifuge tubes.
4. Liquid nitrogen.
5. Bench-top microcentrifuge.
6. 5× Promega lysis buffer (Promega).
7. 1 M dithiothreitol.

2.5.2. Cryoembedding

1. Tissue Tek O.C.T. Compound (Sakura Finetek Inc.).
2. 30% sucrose.
3. Syringes (1 mL and 10 mL).
4. Injection needle (25 GA).

5. 20 GA angiocatheter: Cut down to 5 mm long, with a tapered end.
6. Petridish.
7. Cryomold.
8. Dry ice.

2.5.3. Paraffin Embedding

1. Two 2 in. pieces of 3-0 suture.
2. 20 GA angiocatheter, marked with a pen about half way up the catheter.
3. Petridish.
4. 4% paraformaldehyde in 1× PBS.
5. 1 mL syringe.
6. Small animal surgery table.

2.5.4. Bronchoalveolar Lavage Fluid (BALF) Collection

1. Ice and ice bucket.
2. Microcentrifuge tubes.
3. Bench-top microcentrifuge.
4. 20 GA angiocatheter.
5. 1 mL syringe.
6. 4-0 silk suture, 2–3 in. long.

3. Methods

3.1. Electroporation Preparation

1. Cover the bench where electroporation will take place with a piece of bench coat.
2. Cut two pregelled electrodes into individual pads and tape together about 1 in. apart (wide enough to span the width of the animal's chest to position the electrodes on opposite sides) (Figs. 17.1 and 17.2).
3. Clip the plastic plugs off the electrodes and carefully strip the insulation from the wires.
4. Program parameters for BTX ECM830 Electroporator as follows:

   (a) Mode: Low voltage.
   (b) Voltage: Set to deliver 200 V/cm chest width at delivery point (to be determined in sect. 3.4).
   (c) Pulse Length: 10 ms.
   (d) Number of Pulses: 8.
   (e) Interval: 1 s.

3.2. Animal Preparation

1. Sedate mouse with 50 mg/kg sodium pentobarbital by intraperitoneal injection with U100 insulin syringe.
2. Once the animal is completely anesthetized, place it on surgical table in a supine position. Secure the animal head to the surgical table by placing the 3-0 suture across the top mandible behind the incisors. Stretch the animal to full length along the table and secure the animal's hindquarters with a piece of autoclave or surgical tape across the base of the tail (Fig. 17.2) (*see* **Note 2**).

3.3. Techniques for Introducing Plasmid to Airway (*see* Note 3)

3.3.1. Intubation

1. Elevate the animal's upper body by positioning the surgical table at about a 60° angle. Position the surgical lamp to illuminate inside the mouth, esophagus, and tracheal opening of the mouse.
2. Slide the catheter onto the A-line guide wire and place them on the table on the side of your dominant hand.
3. Using the tweezers in your dominant hand, gently grip the tongue and pull it out of the mouth. Using the weak hand, insert the spatula blade into the back of the animal's throat, and gently push the back of the tongue up toward the bottom of the mouth and out of the way of the esophageal opening allowing visualization of the epiglottis and tracheal opening.
4. When the epiglottis has been visualized, set the tweezers down and pick up the A-line guide wire with the catheter on it. Gently slide the A-line guide wire into the back of the mouse's throat using the incisors and the roof of the mouth to steady the wire. Carefully insert the guide wire between the vocal chords and into the tracheal opening about 0.5 cm.
5. Slide the catheter down the A-line wire into the trachea. Some resistance may be felt at the vocal chords. Gently twist and push the catheter past the vocal chords, inserting the entire catheter into the mouth of the mouse up to the base of the catheter. Remove the guide wire.
6. Verify that the catheter is in the trachea and not the esophagus by momentarily blocking the open end of the catheter with your thumb, being careful not to force the catheter base any deeper into the mouth. If the catheter is in the trachea, occluding the opening will block the mouse's airway causing the breathing rate to slow and become deeper as the mouse tries to catch its breath.
7. Once the position of the catheter in the trachea has been confirmed, DNA can be injected into the lungs with the Hamilton syringe. Starting with the syringe plunger at the 5 μL position, draw up 25 μL of 1 μg/μL DNA in TEN Buffer.
8. Carefully slide the needle of the Hamilton syringe into the catheter that is inserted in the trachea of the mouse until the tape on the syringe needle is at the

opening to the catheter (this should position the end of the needle at the opening of the catheter just above the carina). With one smooth, continuous motion, depress the plunger to inject 25 µL of DNA solution into the lungs. Remove the syringe and prop the surgical table up to the 85° angle allowing the DNA to drain further into the lungs and the animal to clear the catheterized airway.

9. Once the animal is breathing regularly, repeat step 8 to deliver a final total volume of 50 µL of DNA (*see* **Note 4**).

10. Once the mouse is breathing regularly again, you may proceed to the electroporation procedure. Once the DNA has been inserted in the airway, electroporate immediately for highest efficiency (*see* **Note 5**).

3.3.2. Aspiration

1. Elevate the unhinged end of the surgical table and the animal's head to about a 60° angle. Position the surgical lamp to illuminate the mouth, esophagus, and tracheal opening of the mouse.

2. Draw up 50 µL of DNA solution into a 200 µL pipette tip (*see* **Note 4**). Using the forceps gently pull the tongue out and hold it in position momentarily with pipette tip. While holding tongue in the extended position, reposition one blade of the forceps to the back of the throat pushing the tongue up against the bottom of the mouth to visualize the tracheal opening and vocal chords. Do not pinch the tongue and jaw with the forceps, and use only one blade of the forceps like a spatula to depress the tongue against the floor of the mouth.

3. While holding the tongue with the forceps, insert the pipette tip with 50 µL of DNA into the back of the throat to the tracheal opening. Inject the entire DNA into the back of the throat covering the tracheal opening.

4. Hold the tongue in the extended position with the forceps blade to prevent the mouse from swallowing the DNA. Keep the tongue restrained until the animal has aspirated the DNA solution into the lungs and is breathing regularly. Proceed to electroporation as soon as regular breathing is restored.

3.3.3. Tracheal Injection

1. Sterilize the field on the anterior of the neck with a 3–5-min betadine scrub.

2. Using a pair of forceps, gently pull the loose skin on the anterior of the neck away from the muscle layer and cut a small piece of about 0.5 cm² of skin from under the mandible to about the top of the sternum with a pair of surgical scissors to expose the layer of muscles surrounding the trachea.

3. Carefully expose the trachea by teasing the dorsal muscles apart using the small eye dressing forceps.

4. Position the surgical table at a 70° angle and insert a 30GA ½ needle attached to a 1 mL syringe filled with DNA solution at a 45° angle between the cartilaginous tracheal rings located about two-thirds of the way down from the lower mandible.

5. Slowly inject 50 μL of DNA solution into trachea in ~15–25 μL doses. Allow the animal to inhale DNA solution and clear airway between doses (*see* **Note 4**).
6. As soon as the animal has cleared its airway and is breathing with a regular pattern, proceed to electroporation.
7. After electroporation is completed and animal is breathing regularly on its own, close the open wound with two to three sutures of Dexon II.
8. Coat stitches and wound with triple antibiotic ointment and place animal in a prone position on a heating pad to recover.

3.4. Electroporation (*see* **Note 6**)

Following plasmid DNA delivery, electroporation should be applied as soon as possible to minimize DNA degradation.

1. Peel plastic covering off electrode surface and moisten pads with about a dime-sized dollop of surgical lubricant to facilitate transmission of current across the thoracic cavity.
2. Tape the fore limbs out to the side to expose the chest area for placement of electrode pads (Figs. 17.1, 17.2, and 17.4).
3. Place moistened electrode pads on opposite sides of the chest, under the forelimbs, holding them firmly against the body to ensure electrode contact. Be careful to ensure that surgilube gel does not make contact between the pads creating a short circuit (caution: make sure that no metal or liquids are touching the electrodes, wires, or cables).
4. Measure the gap distance between the electrodes and use this to set the voltage parameter on the electroporator so that a field strength of 200 V/cm can be achieved.
5. While holding the electrode pads against the mouse, discharge the current from the pulse generator.
6. After discharging the voltage through the animal, immediately remove the electrodes, free front forelimbs from under tape, and gently wipe the surgilube from the animal's chest to help stimulate breathing (*see* **Note 5**).
7. After the mouse is breathing on its own with a regular pattern (typically within 30 s), wipe the excess surgilube from the electrode pads and recover the pad surface with the plastic covers (electrodes can be reused about 10 times before current resistance levels begin to reduce voltage output).
8. If DNA was delivered via tracheal injection, close the incision at this point and proceed.
9. Remove the mouse from surgical table and lay prone on a covered, warm heating pad to allow recovery from sedation. Monitor temperature and maintain at approximately 37°C to prevent hyper or hypothermia.
10. Once the mouse has regained consciousness, return it to its cage.

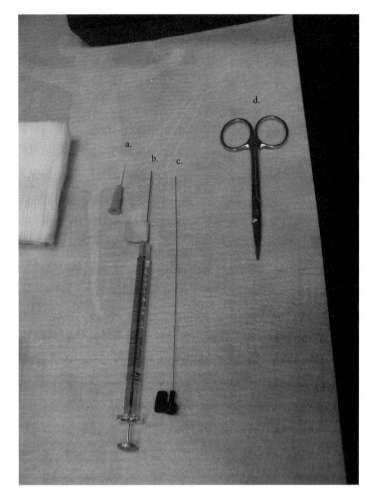

Fig. 17.3 Mouse Intubation Kit. (**a**) 18 GA catheter, (**b**) 50 μL Hamilton Syringe, (**c**) guide wire from an Arrow Radial Artery Catheterization Kit, (**d**) surgical scissors

3.5. Sample Collection

Depending on the promoter used, gene expression can be detected within 24 h of electroporation and can last 3–5 days (i.e., with the CMV promoter (5, 6)) or up to several months (i.e., the UbC promoter (2, 8)). Thus, at the appropriate time after gene transfer, the expression of the transferred genes can be evaluated at the protein or RNA level. The most common approaches are to measure RNA or protein levels and activity in tissue extracts or bronchial alveolar lavage fluids, or by immunohistochemistry in tissue sections.

Fig. 17.4 Electrode placement for in vivo electroporation of the lung

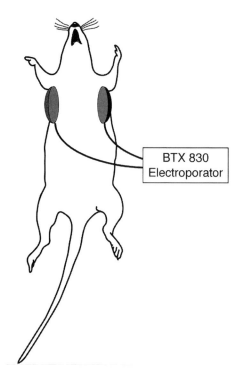

3.5.1. Whole Lung Tissue Extract

3.5.1.1. Tissue Collection and Cryopulverization

1. Prechill the tissue pulverizer in −80°C freezer.
2. Euthanize the mouse by intraperitoneal injection of sodium pentobarbital (100 mg/kg). As soon as the mouse is unresponsive to whisker pull and toe pinch, proceed.
3. Secure the animal on the surgical table as shown in Fig. 17.2.
4. Carefully pull up the neck skin and cut open about 2 cm.
5. Tear the dorsal muscle apart using the small eye dressing forceps to expose the trachea.
6. Lift the abdominal skin and muscle using a pair of forceps and excise the skin and muscle to open abdominal cavity.
7. Gently push intestines to one side to expose the kidney. Cut the renal artery. Use gauze to absorb blood.
8. Hold the sternum and insert the tip of a pair of closed scissors through the diaphragm carefully and spread open the scissors. Be very careful not to cut the lung.
9. Cut longitudinally on both sides of the chest to expose the heart and the lung.
10. Hold the flap up. Continue to cut open toward anterior until the trachea, lung, and heart are completely exposed.
11. Hold the trachea with forceps and gently pull it up.

12. Cut the trachea and carefully separate the trachea from esophagus.
13. Continue to lift the trachea up and detach the lung from the chest cavity.
14. Excise the lung and heart. Rinse in PBS.
15. Cut out the lung lobes at extralobar pulmonary arteries. Snap-freeze the lung lobes in liquid nitrogen. Snap-freeze the trachea if gene expression in the trachea needs to be measured. Pulverization of the frozen trachea can be done using a glass tissue grinder bathed in liquid nitrogen.
16. Place the tissue pulverizer in a stainless steel tray and fill it with liquid nitrogen.
17. Transfer the frozen lung tissue into the well of mortar.
18. Place the pestle into the mortar and smash the frozen lung by hitting the pestle using a hammer several times.
19. Remove the pestle. The lung tissue should appear as a fine pink powder. If there are large particles, pour a little more liquid nitrogen into the pulverizer and repeat the pulverization.
20. Transfer the tissue powder into a microfuge tube that has been prechilled in liquid nitrogen. The sample is ready for protein and RNA extraction.

3.5.1.2. Protein extraction

1. Prepare lysis buffer. Dilute the 5× Promega lysis buffer to 1× (1:5) and add dithiothreitol to 1 mM (1:1,000).
2. Weigh 100 mg of the tissue powder into another microfuge tube.
3. Add 0.5 mL lysis buffer. Vortex briefly to mix.
4. Freeze the mixture in liquid nitrogen for 5 min.
5. Thaw the mixture in a room temperature water bath. The lid of the microfuge tube could pop open violently in the water bath. This problem can be prevented by loosening the lid before putting it into the water bath.
6. Repeat freeze–thaw twice.
7. After the final thawing, centrifuge the mixture at a maximum speed in a microcentrifuge for 10 min at 4°C.
8. Transfer the supernatant into a clean microfuge tube.
9. The supernatant is ready for protein and gene expression assays.

3.5.1.3. RNA Extraction

RNA extraction can be performed using Qiagen RNeasy Plus RNA isolation kit, as described by the manufacturer or by trizol extraction.

3.5.2. Cryoembedding

1. Mix equal volumes of O.C.T. and 30% sucrose together and stir gently. Because the liquid is very viscous, prepare well in advance so that the air bubbles resulting from the mixing can be removed.

2. Euthanize the mouse by intraperitoneal injection of sodium pentobarbital (100 mg/kg). As soon as the mouse is unresponsive to whisker pull and toe pinch, proceed.

3. Secure the animal on the surgical table as shown in Fig. 17.2.

4. Lift the abdominal skin and muscle using a pair of forceps and excise the skin and muscle to open abdominal cavity.

5. Gently push intestines to one side to expose the kidney. Cut the renal artery. Use gauze to absorb blood.

6. Hold the sternum and insert the tip of a pair of closed scissors through the diaphragm carefully and spread open the scissors. Be very careful not to cut the lung.

7. Cut longitudinally on both sides of the chest to expose the heart and the lung.

8. Hold the flap up. Continue to cut open toward the anterior until the trachea, lung, and heart are completely exposed.

9. Inject ice-cold PBS into the right ventricle slowly to perfuse the lung until the lung turns white.

10. Carefully cut open the neck skin.

11. Tear the dorsal muscle apart using the small eye dressing forceps to expose the trachea.

12. Carefully separate the trachea from the esophagus.

13. Cut a small incision in the trachea.

14. Place a 4-0 silk suture underneath the trachea at a position below the incision.

15. Carefully insert the catheter into the incision and gently push it down into the trachea. Twist the catheter while pushing it down if necessary.

16. Tie the suture around the trachea with the inserted catheter to secure the catheter in position. To prevent the catheter from slipping out, make a loose knot first, and then hold the catheter while tying the suture.

17. Inject 1 mL O.C.T.–sucrose mixture slowly into the lung via the catheter in the trachea.

18. Inflate the lung until it is visually expanded. Tie the trachea below the incision using a 4-0 silk suture such that the O.C.T.–sucrose is sealed in the lung.

19. Hold the trachea and take out the whole heart–lung complex.

20. Cut out the lung lobes in a petridish and embed them into a cryomold filled with O.C.T.

21. Place the mold on dry ice to freeze the tissue. If the tissue floats up, use forceps to push it down gently.

3.5.3. Paraffin Embedding

1. Deliver enough sodium pentobarbital (100 mg/kg) intraperitoneally to ensure euthanasia.

2. As soon as mouse is nonresponsive to whisker pulls and toe pinches, place animal in supine position on surgical table and tape appendages down with autoclave tape (Figs. 17.1 and 17.2).

3. Gently pull the skin around the throat away from the body and cut an extended skin flap from sternum just below the lower mandible with scissors.

4. Carefully separate the trachea from the dorsal muscle and slide a 3-0 suture under the trachea, being careful not to crush or tear the trachea.

5. Snip the trachea open with surgical scissors about one-third of the way down from the mandible, being careful not to cut the trachea in half.

6. Carefully push the catheter into the open trachea between the cartilage rings until the half way mark of the catheter is at the lower edge of the mandible.

7. Tie the suture around the trachea and catheter to hold the catheter in place.

8. Pull the skin of the abdomen away from the body using forceps and carefully open the abdomen by cutting the extended skin away with scissors. Locate the kidneys by gently pushing the stomach and intestines to one side. Exsanguinate the animal by cutting the renal artery and raise animal up to a 90°angle to drain blood. Pack the abdomen with gauze to absorb the blood.

9. When the blood loss has slowed, but while the heart is still pumping, carefully cut a lateral midline incision across the top of both sides of the chest just below the ribs to open the abdomen further.

10. Pull the liver down to expose the diaphragm and carefully puncture the diaphragm by inserting the closed tip of the scissors just under the sternum and opening them slowly to spread the diaphragm.

11. Cut the sternum away by making two longitudinal cuts from posterior to anterior along the ribs on both sides of the sternum using scissors. Make a lateral cut across the top to remove the sternum flap, being careful not to puncture the lungs or trachea.

12. Open the rest of the chest, being careful not to puncture the lungs or separate them from the trachea.

13. Carefully detach the lungs, trachea, and heart with catheter attached from the body cavity being careful not to rip or puncture them.

14. Place the organs in one half of a petridish and wash thoroughly with 1× PBS removing as much blood and fluid as possible.

15. Insert a 1 mL syringe filled with 1 mL of ice-cold 4% paraformaldehyde into catheter and slowly inject paraformaldehyde into the lungs causing them to slowly inflate.

16. When lungs are full and with the syringe still attached and plunger depressed, carefully tie off the end of the trachea between the lungs and the catheter to keep the lungs inflated with paraformaldehyde.

17. Place paraformaldehyde-filled lungs in a 50 mL conical tube with 10 volumes (10–20 mL) of 4% paraformaldehyde and store at 4°C for at least 12 h before processing for paraffin embedding.

3.5.4. BALF Collection

1. Prechill sterile PBS on ice.

2. Euthanize the mouse by intraperitoneal injection of sodium pentobarbital (100 mg/kg).

3. Secure the animal on the surgical table as shown in Fig. 17.2.

4. Gently pull the skin on the neck away from the muscle layer and cut a small piece of the skin about 1–2 cm long.

5. Tear the dorsal muscle apart using the small eye dressing forceps to expose the trachea.

6. Carefully separate the trachea from the esophagus.

7. Cut a small incision in the trachea using a pair of small scissors.

8. Place a 4-0 silk suture underneath the trachea at a position below the incision.

9. Carefully insert the catheter into the incision and gently push it down into the trachea. Twist the catheter if necessary.

10. Tie the suture around the trachea with the inserted catheter to secure the catheter in position. To prevent the catheter from slipping out, make a loose knot first, and then use the two thumb knuckles to hold the catheter while tying the suture.

11. Inject 0.9 mL sterile ice-cold PBS into the lung via the catheter.

12. Gently pull the PBS back into syringe until air bubbles come out from the lung.

13. Push the PBS back into the lungs.

14. Repeat two more times.

15. Transfer the BALF (~0.6 mL) into a microfuge tube and keep it on ice for further analysis.

4. Notes

1. Plasmids can be purified by CsCl banding or a number of commercially available purification kits, such as those from Qiagen.

2. We have found a relatively steep learning curve to exist with this procedure. Expect significant mortality with the initial applications of this procedure (first 5–10 animals). We believe that most mortality is caused by an occlusion of the animal's airway by fluid or foam coughed up during the electroporation process.

3. We routinely use any one of the following three methods for DNA or fluid delivery to the lungs. All yield equivalent levels of gene expression and can be used interchangeably.

4. We have found that introducing DNA volumes greater than 50 μL significantly increases mortality, likely due to suffocation. Volumes greater than 100 μL usually result in death after electroporation.

5. When using the intubation method to introduce DNA, we have decreased mortality to nearly 0% by ventilating the animal for 10–30 s on a MiniVent Type 845 (Harvard Apparatus, March-Hugstetten, Germany). To ventilate the animal, leave the angiocatheter in the trachea during the electroporation procedure, taping it down to the surgical table to prevent dislocation. After freeing the animal's forelimbs and wiping off lubricant, carefully attach the catheter to a running minivent (150 strokes/min and 120 μL stroke volume). Ventilate the animal for 10–20 s to clear any fluid from the trachea. Detach the ventilator from the catheter, being careful not to remove the catheter from the trachea and verify normal breathing. If the animal continues to struggle to establish a normal breathing pattern, reattach to ventilator for an additional 30–45 s. Once the animal is capable of breathing on its own, unaided by the ventilator, remove the catheter from the trachea and place the animal on the heating pad to recover from anesthesia.

6. Electroporation has also been used successfully for gene transfer to rat lungs using very similar procedures and parameters. The only differences would be volumes of DNA delivered (less

than 500 μL of DNA over a 2 s period), a larger angiocatheter for intubation (16 GA), and slightly larger electrodes (~3 × 4 cm²).

References

1. Somiari, S., Glasspool-Malone, J., Drabick, J.J., et al. (2000) Theory and in vivo application of electroporative gene delivery. *Mol. Ther.* **2**, 178–187.
2. Dean, D.A. (2003) Electroporation of the vasculature and the lung. *DNA Cell Biol.* **22**, 797–806.
3. West, J. and Rodman, D.M. (2001) Gene therapy for pulmonary diseases. *Chest.* **119**, 613–617.
4. Weiss, D. (2002) Delivery of gene transfer vectors to lung: obstacles and the role of adjunct techniques for airway administration. *Mol. Ther.* **6**, 148–152.
5. Dean, D.A., Machado-Aranda, D., Blair-Parks, K., Yeldandi, A.V., and Young, J.L. (2003) Electroporation as a method for high-level non-viral gene transfer to the lung. *Gene Ther.* **10**, 608–1615.
6. Machado-Aranda, D., Adir, Y., Young, J.L., et al. (2005) Gene transfer of the Na⁺, K⁺-ATPase β1 subunit using electroporation increases lung liquid clearance in rats. *Am. J. Respir. Crit. Care Med.* **171**, 204–211.
7. Jones, M.R., Simms, B.T., Lupa, M.M., Kogan, M.S., and Mizgerd, J.P. (2005) Lung NF-kappaB activation and neutrophil recruitment require IL-1 and TNF receptor signaling during pneumococcal pneumonia. *J. Immunol.* **175**, 7530–7535.
8. Gazdhar, G., Bilici, M., Pierog, J., et al. (2006) In vivo electroporation and ubiquitin promoter—a protocol for sustained gene expression in the lung. *J. Gene Med.* **8**, 910–918.

Chapter 18
Delivery of DNA into Bladder via Electroporation

Masaki Yoshida, Hitoshi Iwashita, Masayuki Otani, Koichi Masunaga, and Akito Inadome

Abstract The possibility of in vivo gene transfer into the rat bladder by electroporation (EP) was evaluated. The bladder was exposed through an abdominal midline incision in 8-week-old male rats. Plasmid DNA of marker genes, green fluorescent protein (GFP) and luciferase, and the neuronal nitric oxide synthase (nNOS) gene were then injected into the subserosal space of the bladder and EP was applied. At 72 h after gene transfer, GFP and luciferase were assayed in the isolated bladder, and immunohistochemical staining was used to detect nNOS. NOx released from isolated bladder strips was also assessed using microdialysis procedure. From the luciferase assay, 45 V, 1 Hz, 50 ms, and 8 pulses were selected as the optimum conditions for EP. Bladder specimens with GFP genes injected by EP showed numerous bright sites of GFP expression in the smooth-muscle layer. In rats with the nNOS gene injected by EP, there was marked nNOS immunoreactivity, and NOx released from bladder strips was significantly greater than that in the control groups.

These results suggest that EP is a useful technique for in vivo gene transfer into rat bladder smooth muscles, and that the nNOS gene transferred by this procedure functionally expresses and contributes to NO production.

Keywords: in vivo electroporation, nitric oxide, gene transfer, bladder smooth muscle, rat

1. Introduction

A safe and highly effective gene transfer method is absolutely necessary for the success of gene therapy. Current in vivo gene transfer methods include viral vectors, cation liposomes, and injections of purified DNA. Generally, the viral vector systems can provide efficient gene transfer, but there are still some disadvantages, such as the potential for toxicity associated with chronic overexpression or insertional mutagenesis in the retroviral and the adenovirus systems, the possibility of nonspecific inflammatory response, and antivector cellular immunity (1–3).

S. Li (ed.), *Electroporation Protocols: Preclinical and Clinical Gene Medicine.*
From *Methods in Molecular Biology, Vol. 423.*
© Humana Press 2008

Electroporation (EP) is a well-established laboratory technique owing to the transfer of plasmid DNA in vitro and is one of the most efficient nonviral methods for introducing exogenous molecules into cells by high-voltage electric pulses (4–6). Several studies demonstrated that in vivo EP procedures could transfer DNA into skin, muscle, liver, and tumor cells (7–12); however, there is little information on the application of in vivo EP procedures to the urinary tract (13, 14).

In this chapter, we introduce in vivo gene transfer by synthesizing a neuronal NOS (nNOS) expression vector and transferring it into the rat bladder using in vivo EP. The functional activity of the transferred nNOS gene was also evaluated.

2. Materials

2.1. Synthesis of nNOS Expression Vector

This vector is synthesized as follows. The nNOS expression vector (pcDNA3.1-mouse nNOS) is constructed by ligating the 4.8-kb Nhe1 and EcoR1 fragment of mouse nNOS cDNA from RDB1386 (MBNOS71) into the unique site between Nhe1 and EcoR1 site of the pcDNA3.1(+) (Invitrogen, Carlsbad, CA). The direction of the ligated fragment is confirmed by enzyme digestion (Fig. 18.1). All plasmids are extracted from *Escherichia coli* and purified by using the Plasmid Mega Kit.

The plasmid DNA is adjusted to a concentration of 1 mg/mL in PBS. pcDNA nNOS is then transfected into DT40 (RCB1464) and nNOS expression is confirmed by Western blotting, as described previously (15, 16). Immunoblotting analysis is conducted using an anti-nNOS antibody as the primary antibody and a horseradish peroxidase–conjugated rabbit anti-mouse IgG (1:2,000) as the secondary antibody. The bound antibody is detected using the ECL kit.

2.2. In Vivo Gene Transfer by EP and Expression Analyses Using Luciferase and Green Fluorescent Protein Assays

1. All the experiments on gene transfer were done with 8-week-old male Sprague-Dawley rats weighing 200–250 g. In all procedures, animal care was in accordance with guidelines of Kumamoto University.
2. Sodium pentobarbital.
3. Gentamicin sulfate.
4. 4–0 absorption threads.
5. Phosphate-buffered solution (PBS): 30.8 mM NaCl, 120.7 mM KCl, 8.1 mM Na_2HPO_4, 1.46 mM KH_2PO_4, and 10 mM $MgCl_2$ in distilled water.
6. Electroporator (CUY-21; Tokiwa Science, Tsukushino, Japan).

Mouse Neuronal NOS cDNA

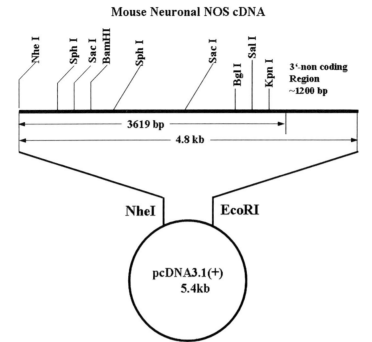

Fig. 18.1 Vector map of the nNOS expression vector. Mouse nNOS cDNA, derived from RDB1386 (MB-NOS71) into pcDNA3.1(+), was inserted

7. Electrode (Tokiwa Science, Tsukushino, Japan): Two-square parallel electrodes made of gilded stainless steel mounted on insulated tweezers with a plastic screw, by which the distance between the two electrodes can be fixed.
8. Expression vectors for luciferase gene (plasmid pGL3-control, Promega Corporation, Madison, WI).
9. Mutant green fluorescent protein (GFP) gene (plasmid pEGFP-C1) (Clontech Laboratories, Inc., Palo Alto, CA).
10. Luciferase-lysis buffer.
11. Luciferase assay substrate (Promega Corporation, Madison, WI).
12. Luminometer (Luminocounter 700, NITI-ON, Tokyo, Japan).
13. Cryostat (CM1800, Leica, Tokyo, Japan).
14. Confocal microscope (Fluoroview, Olympus, Tokyo, Japan).

2.3. Immunohistochemical Staining for NOS Expression

1. Cold acetone.
2. PBS.
3. BSA (1%, Sigma-Aldrich) in PBS. Prepare the solution and store at 4°C.

4. Primary anti-Nos1 antibody (sc-648 Santa Cruz Biotechnology, Inc.). Make 1:100 dilution in 1% BSA in PBS and place in a humidity chamber overnight at 4°C.

5. Secondary antibody: Biotin-conjugated rabbit anti-rat Ig antibody (Santa Cruz Biotechnology, Inc). Make 1:100 dilution in 1% BSA in PBS for 30 min at ambient temperature.

6. Methanol (70%) containing 0.1% H_2O_2.

7. Enzyme substrate: Diaminobenzidine (DAB), stored at −20°C.

8. Vectastain ABC kit (Vector Labs, Burlingame, CA), stored at 4°C.

9. Tris-HCl (0.05%), pH 7.6, containing 0.075% H_2O_2.

2.4. Measurement of NO Release from Isolated Bladder Strips by Microdialysis Procedure

1. Krebs-Henseleit (K-H) solution: 117.7 mM NaCl, 4.69 mM KCl, 2.16 mM $CaCl_2$, 1.20 mM $MgSO_4$, 24.39 mM $NaHCO_3$, 1.20 mM KH_2PO_4, and 9.99 mM glucose.

2. Microdialysis probe (outer diameter, 220 μm; inner diameter, 200 μm; length, 10 mm; cellulose membrane, molecular cut-off, 50 kDa).

3. Ringer solution: 147 mM NaCl, 4.0 mM KCl, 2.3 mM $CaCl_2$.

4. Microsyringe pump (EP-60, Eicom, Kyoto, Japan).

5. Organ baths (20 mL; AME-400C, JTTohsi, Tokyo, Japan).

6. Force–displacement transducer (TB-611T, Nihon Khoden, Tokyo, Japan).

7. NOx analyzing system using a syringe loading sample injector (Model 7725; Eicom, Kyoto, Japan).

3. Methods

3.1. Injection of DNA into Rat Bladder and EP Procedure

1. Anesthetize rats by an intraperitoneal injection of pentobarbital sodium (30 mg/kg) using a 26 GA needle.

2. Expose the bladder by a lower abdominal midline incision (~3 cm).

3. Withdraw intravesical urine with a 27 G needle. Then inject (viewing under a microscope and using a 27 G needle) 50 μL of potassium phosphate–buffered solution containing 50 μg of each plasmid DNA into the subserosal regions of the anterior and posterior bladder wall (*see* **Note 1**).

4. Put the bladder between two-square parallel electrodes and pulse twice from an electroporator (*see* **Note 2**).

5. After the gene transfer, apply gentamicin sulfate to the bladder surroundings, and suture the abdominal layer with 4–0 absorbable suture.

3.2. Expression Analyses of In Vivo Gene Transfer by EP Using Luciferase

3.2.1. Assay Luciferase Activity to Determine the Optimum Conditions

Assay luciferase activity in various conditions (*see* **Note 3**). Seventy-two hours after the gene transfer, dissect the bladder under pentobarbital anesthesia (30 mg/kg) and immediately freeze it at –80°C. Optimal electric pulse conditions for EP in the rat bladder were determined by luciferase expression (Fig. 18.2). At 8 pulses, 50 ms/pulse, and 1 Hz, enhancement of expression is detected from 30 V and reduced at 50 V and above (*see* **Note 4**).

3.2.2. Assay Luciferase Activity

1. Thaw, homogenize, and incubate the bladder in 2 mL of luciferase lysis buffer for 30 min at room temperature.

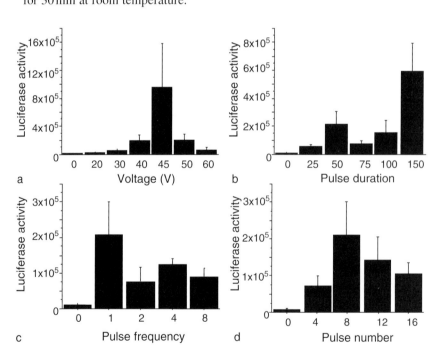

Fig. 18.2 Luciferase activities in various conditions for each variable during electroporation. (**a**) Effect of voltage (eight pulses; 50 ms/pulse, 1 Hz). The value for 45 V is significantly ($p = 0.007$) higher than that of the control. (**b**) Effect of pulse length (8 pulses, 45 V, 1 Hz). The values for 150 and 50 ms are significantly ($p = 0.001$ and 0.048, respectively) higher than that of the control. (**c**) Effect of the frequency of pulse delivery (8 pulses, 45 V, 50 ms/pulse). The value for 1 Hz is significantly ($p = 0.008$) higher than that of the control. (**d**) Effect of pulse number (45 V, 50 ms/pulse, 1 Hz). The value for 8 times is significantly ($p = 0.007$) higher than that of the control. The spacing of the electrodes was constant at 2 mm. Each bar represents the mean (SEM) of five experiments

2. Centrifuge the sample at 12,000g for 30 min.
3. Mix 100 μL of the luciferase assay substrate with 20 μL of the supernatant from the sample.
4. Measure luciferase activity using a luminometer as reported previously (17).

3.3. Expression Analyses of In Vivo Gene Transfer by EP Using GFP Assays (see Note 5)

1. Seventy-two hours after gene transfer, fix the bladder with 4% paraformaldehyde overnight.
2. Wash with PBS.
3. Cut the bladder into 5-μm slices using a cryostat at −20°C.
4. Visualize GFP expression with a confocal microscope at excitation and emission wavelengths of 460 and 490 nm, respectively (see Note 6).

3.4. Transfer n-NOS Expression Vector into Rat Bladder, and Immunohistochemical Staining for Examining the Expression of nNOS in the Rat Bladder (18)

1. Perform nNOS gene transfer under the optimal condition, as described previously.
2. Seventy-two hours after gene transfer, cut the isolated rat bladder into 6-μm slices using a cryostat at −20°C.
3. Treat the section in 70% methanol containing 0.1% H_2O_2 for 30 min to block endogenous peroxidase activity.
4. Incubate the section with 2.5% normal rabbit serum for 20 min to block nonspecific staining.
5. After rinses in PBS, incubate the section with nNOS primary antibody at 4°C overnight.
6. Incubate the section consecutively with biotinylated secondary antibody for 30 min, elite ABC reagent for 30 min, and 0.05% diaminobenzidine in 0.05% tris-HCl, pH 7.6, containing 0.075% H_2O_2 for 10 min.
7. After washing in cold water, seal the sections with coverslips (see Note 7).

3.5. Measurement of NO_2^-/NO_3^- (NOx) Release from Isolated Rat Bladder Strips (19)

1. Isolate the bladder from rats under pentobarbital anesthesia.
2. Dissect smooth-muscle strip ($3 \times 10\,mm^2$) from the bladder (see Note 8).
3. Suspend each strip in a 20-mL organ bath filled with Krebs-Henseleit solution at 37°C (see Note 9).
4. Insert a microdialysis probe into the smooth-muscle strips (see Note 10).

5. Each strip is stretched to the optimum resting tension (1.0 g).
6. After 2 h of equilibration contractions, perfuse the microdialysis tube with Ringer solution at a constant flow rate of 2 μL/min, using a microsyringe pump.
7. After 2 h of equilibration, collect the dialysate for 10 min (20 μL) in polyethylene tubes at room temperature.
8. To quantify NOx in the dialysis fraction, directly inject 10 μL dialysate into an automated NO analysis system using a syringe-loading sample injector (*see* **Note 11**).
9. Determine the level of NOx in standard solution and sample as a mixture of NO_2^-/NO_3^- (NOx) release (*see* **Note 12**).

4. Notes

1. Insertion of the needle is just beneath the bladder outer layer. Deep insertion may cause the penetration of the bladder wall.
2. The space between the electrodes was stable at 2 mm, which is the suitable distance to grasp the rat bladder.
3. We are using the four variables: voltage (0–60 V), pulse length (0–150 ms), pulse number (0–16 times), and the frequency of pulse delivery (0–8 Hz). The pulse values that gave the highest gene expression are used for the remaining gene transfer experiments with GFP and functional nNOS genes.
4. Under the optimal conditions, there is no tissue damage in the rat bladder, and luciferase expression was ~300-fold higher compared with the control group.
5. GFP analysis is performed under the optimal electric pulse conditions obtained from luciferase assay.

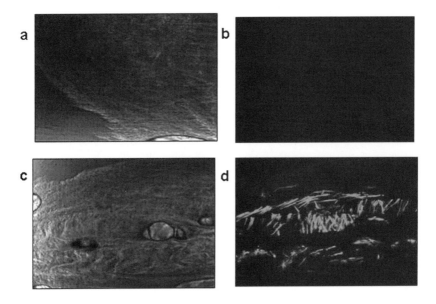

Fig. 18.3 GFP fluorescence images in the rat bladder after 100 mg GFP gene injection without (**a**, **b**) or with (**c**, **d**) in vivo electroporation. Each image consists of the phase-contrast images (**a** and **c**) and GFP fluorescence images (**b** and **d**), all at 100×

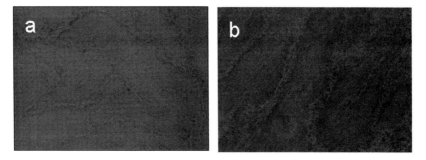

Fig. 18.4 Immunohistochemical staining of nNOS gene with anti-nNOS protein (nNOS-1) in rat bladder. (**a**) Bladder with nNOS gene injection only (control, 200×). (**b**) Bladder with nNOS gene injection and electroporation (200×). Bladder smooth muscles in (**b**) were clearly stained by anti-nNOS protein antibody compared with those in the control group

6. Phase-contrast images of the same slices were also recorded. A bladder with only plasmid DNA injection was used as a control. In the rats with EP, there were numerous bright green fluorescent signals along the bundles of smooth muscles in the bladder wall (Fig. 18.3).
7. In the rat with nNOS DNA transfer, the bladder smooth-muscle staining had a more intense reaction to anti-nNOS protein antibody than did the control group (Fig. 18.4).
8. All rats are divided into three groups (only nNOS injection, only EP, and nNOS injection with EP).
9. The solution is bubbled with 95% O_2 and 5% CO_2, giving a pH of 7.4, and attached to two L-shaped metal specimen holders by tying both ends of the preparations with silk ligatures. One end of each strip is connected to a force–displacement transducer.
10. Insertion of the microdialysis probe is carefully performed just beneath the urothelium. The whole portion of the dialysis membrane should be placed inside the bladder tissue.
11. The assay is based on Griess reaction.
12. The value is expressed in terms of the amount of NOx in $10\,\mu L$ dialysate fraction, divided by the weight of the strip (pmol/g wet weight of bladder). The release was not inhibited by pretreatment with tetrodotoxin ($1.0\,\mu M$), but was inhibited by pretreatment with $100\,\mu M$ N^{ω}-nitro-L-arginine (a NOS inhibitor). After nNOS gene injection with EP the NOx release is significantly higher than that in groups treated with EP only or gene injection only.

References

1. Verma, I.M. and Somia, N. (1997) Gene therapy—promises, problems and prospects. *Nature.* **389**, 239–242.
2. van der Eb, M.M., Cramer, S.J., Vergouwe, Y., et al. (1998) Severe hepatic dysfunction after adenovirus-mediated transfer of the herpes simplex virus thymidine kinase gene and ganciclovir administration. *Gene Ther.* **5**, 451–458.
3. McMenamin, M.M., Byrnes, A.P., Charlton, H.M., Coffin, R.S., and Latchman, D.S. (1998) A γ34.5 mutant of herpes simplex 1 causes severe inflammation in the brain. *Neuroscience.* **83**, 1225–1237.
4. Neumann, E., Schaefer-Ridder, M., Wang, Y., and Hofschneider, P.H. (1982) Gene transfer into mouse lyoma cells by electroporation in high electric fields. *EMBO J.* **1**, 841–845.
5. Fromm, M., Taylor, L.P., and Walbot, V. (1985) Expression of genes transferred into monocot and dicot plant cells by electroporation. *Proc. Natl. Acad. Sci. USA.* **82**, 5824–5828.

6. Reiss, M., Jastreboff, M.M., Bertino, J.R., and Narayanan, R. (1986) DNA-mediated gene transfer into epidermal cells using electroporation. *Biochem. Biophys. Res. Commun.* **137**, 244–249.

7. Titomirov, A.V., Sukharev, S., and Kistanova, E. (1991) In vivo electroporation and stable transformation of skin cells of newborn mice by plasmid DNA. *Biochim. Biophys. Acta.* **1088**, 131–134.

8. Zhang, L., Li, L., Hoffmann, G.A., and Hoffman, R.M. (1996) Depth targeted efficient gene delivery and expression in the skin by pulsed electric fields. An approach to gene therapy of skin aging and other diseases. *Biochem. Biophys. Res. Commun.* **220**, 633–636.

9. Aihara, H. and Miyazaki, J. (1998) Gene transfer into muscle by electroporation in vivo. *Nat. Biotechnol.* **16**, 867–870.

10. Mir, L.M., Bureau, M.F., Gehl, J., et al. (1999) High-efficiency gene transfer into skeletal muscle mediated by electric pulses. *Proc. Natl. Acad. Sci. USA.* **96**, 4262–4267.

11. Heller, R., Jaroszeski, M., Atkin, A., et al. (1996) In vivo gene electroinjection and expression in rat liver. *FEBS Lett.* **389**, 225–228.

12. Goto, T., Nishi, T., Tamura, T., et al. (2000) Highly efficient electrogene therapy of solid tumor by using an expression plasmid for the herpes simplex virus thymidine kinase gene. *Proc. Natl. Acad. Sci. USA.* **97**, 354–359.

13. Iwashita, H., Yoshida, M., Nishi, M., and Ueda, S. (2004) In vivo transfer of a neuronal nitric oxide synthase expression vector into the rat bladder by electroporation. *BJU Int.* **93**, 1098–1103.

14. Otani, M., Yoshida, M., Iwashita, H., et al. (2004) Electroporation-mediated muscarinic M3 receptor gene transfer in to rat urinary bladder. *Int. J. Urol.* **11**, 1001–1008.

15. Yu, W.J., Juang, S.W., Chin, W., Chi, T.C., Wu, T.J., and Cheng, J.T. (1999) Decrease of nitric oxide synthase in the cerebrocortex of streptozotocin-induced diabetic rats. *Neurosci. Lett.* **272**, 99–102.

16. Wei, C.L., Khoo, H.E., Lee, K.H. and Hon, W.M. (2002) Differential expression and localization of nitric oxide synthases in cirrhotic livers of bile duct-ligated rats. *Nitric Oxide.* **7**, 91–102.

17. Yoshizato, K., Nishi, T., Goto, T., et al. (2000) Gene delivery with optimized electroporation parameters shows potential for treatment of gliomas. *Int. J. Oncol.* **16**, 899–905.

18. Garcia-Pascual, A., Costa, G., Labadia A., Persson K., and Triquero D (1996) Characterization of nitric oxide synthase activity in sheep urinary tract: functional implications. *Br. J. Pharmacol.* **118**, 905–914.

19. Seshita H., Yoshida, M., Takahashi, W., et al. (2002) Prejunctional α-adrenoceptors regulate nitrergic neurotransmission in the rabbit urethra. *Eur. J. Pharmacol.* **400**, 271–278.

Chapter 19
Analysis of Gene Function in the Retina

Takahiko Matsuda and Constance L. Cepko

Abstract The Retina is a good model system for studies of neural development and disease because of its simplicity and accessibility. To analyze gene function rapidly and conveniently, we developed an electroporation technique in mice and rats for use in vivo and in vitro. The efficiency of electroporation into the neonatal retina is quite good, and transgene expression persists for more than a month. With this technique, various types of DNA constructs, including RNA interference (RNAi) vectors, are readily introduced into the retina without DNA size limitation. In addition, more than two different DNA constructs can be introduced into the retina at once, with very high cotransfection efficiency.

In vivo and in vitro electroporation will provide a powerful method to analyze the molecular mechanisms of retinal development and disease.

Keywords: retina, development, progenitors, photoreceptor cells, ganglion cells, electroporation, subretinal space, vitreous chamber, explant, Green Fluorescent Protein (GFP)

1. Introduction

The retina has been an excellent model system for studies of central nervous system (CNS) development and disease, because of its relatively simple structure and accessibility for gene delivery. To deliver genes into the rodent retina, several types of viral vectors, including murine oncoretrovirus (1, 2), lentivirus (3), adenovirus (4–6), and adeno-associated virus (7), have been used. However, there are disadvantages inherent in the use of such vectors. First, it is time-consuming to prepare high titer virus stocks to achieve efficient gene transfer. Second, viral vectors have a size limitation for insert DNA. Third, in general, such vectors do not readily allow introduction of more than two genes into the same cells. Fourth, biosafety is a concern for some viral vectors having broad host ranges.

To bypass these problems, we developed a system to deliver plasmid DNA directly into neonatal mouse and rat retinas by in vivo electroporation ((8, 20),

Figs. 19.1–19.4). This method is faster and, in some cases, safer than viral gene transfer methods. The efficiency of electroporation into the developing postnatal retina is quite good, and transgene expression persists for more than a month. Moreover, in vivo electroporation has several advantages. First, various types of DNA constructs, including RNA interference (RNAi) vectors as well

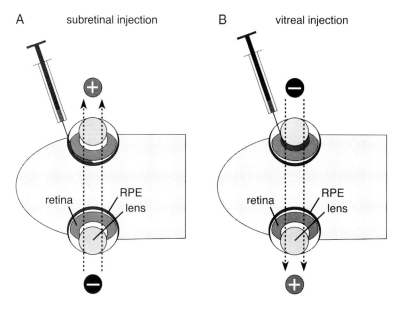

Fig. 19.1 Strategy for in vivo electroporation. (**A**) Electroporation from the scleral (retinal pigment epithelium) side of the retina. (**B**) Electroporation from the vitreal side of the retina

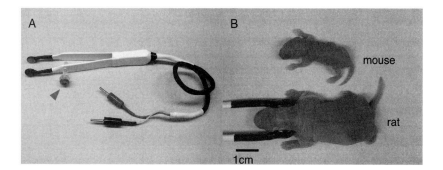

Fig. 19.2 Electrodes and procedure for in vivo electroporation. Tweezer-type electrodes (**A**) are placed to hold the head of newborn (P0) rat or mouse (**B**). *Arrowhead* indicates the plus side of the electrodes

bright field   GFP   overlay

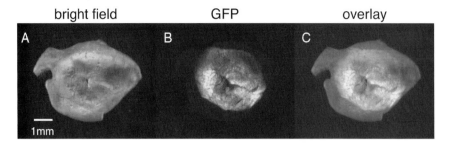

Fig. 19.3 In vivo electroporated retina—whole mount. Whole mount preparation of rat retina in vivo electroporated with pCAG-GFP (GFP expression vector driven by the CAG promoter) at P0 and harvested at P21. Pictures were taken from the scleral side. Bright-field (**A**), GFP (**B**), and merged (**C**) images are shown. (Reprinted from (8), © 2004, National Academy of Science, USA)

as conventional gene expression vectors, are readily introduced into the retina without DNA size limitation. Second, more than two different DNA constructs can be introduced into the retina at once. We found that at least five plasmids can be co-electroporated without a significant reduction in co-electroporation frequency.

We are currently applying this technique to various types of studies, including gain-of-function, loss-of-function, and promoter analyses in the developing retina (8–10, 20).

1.1. Basic Strategy

Figure 19.1A shows the basic strategy for in vivo electroporation into newborn mouse and rat pups. DNA constructs are injected into the subretinal space between the retina and retinal pigment epithelium (RPE). Then, electrodes are placed on the heads of pups (Fig. 19.2), and electric pulses are applied to the eyes in the direction shown in Fig. 19.1A. The DNA constructs are transduced from the scleral side of the retina, where undifferentiated mitotic and newly postmitotic cells exist (*see* **Note 1**). In addition to the strategy shown in Fig. 19.1A, it is also theoretically possible to transfect DNAs from the vitreous side of the retina by injecting DNAs into the vitreous chamber, and by applying electric pulses in the direction opposite to that shown in Fig. 19.1A (Fig. 19.1B). Indeed, other groups reported that DNA constructs could be transduced to ganglion cells, which line the surface of the retina facing the vitreous body, by in vivo electroporation using this strategy (11, 12). However, our data show that efficiency of transfection from the vitreal side (ganglion cells) of the neonatal retina, as well as into the adult retina, is much lower than that from the scleral side (progenitor/precursor cells) of the neonatal retina (Fig. 19.7).

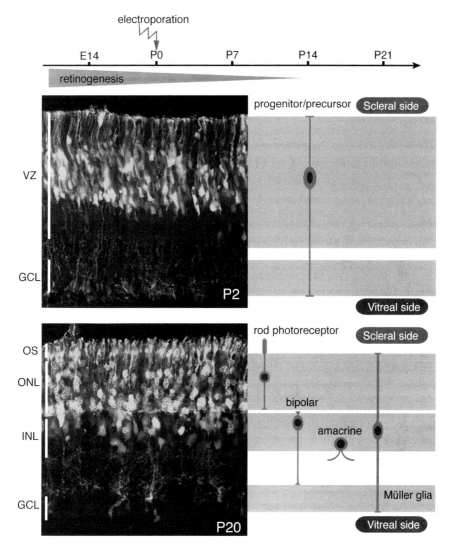

Fig. 19.4 In vivo electroporated retina—section. Rat retinas were in vivo electroporated with pCAG-GFP at P0 and harvested at P2 (*upper* panel) or P20 (*lower* panel). At P2, most of the GFP-positive cells have the morphology of progenitor/precursor cells, suggesting that DNAs are preferentially transfected to progenitor/precursor cells. Retinogenesis is completed within the first 2 weeks after birth. At P20, GFP is observed in four differentiated cell types, including rod photoreceptors, bipolar cells, amacrine cells, and Müller glial cells. *VZ* ventricular zone, *GCL* ganglion cell layer, *OS* outer segment, *ONL* outer nuclear layer, *INL* inner nuclear layer

1.2. In Vivo Electroporation vs. In Vitro Electroporation

In organ cultures of embryonic or neonatal retina, progenitor cells differentiate into neurons and glia and form three layers, mimicking normal development. Taking advantage of this, we also developed a system to electroporate DNAs into isolated

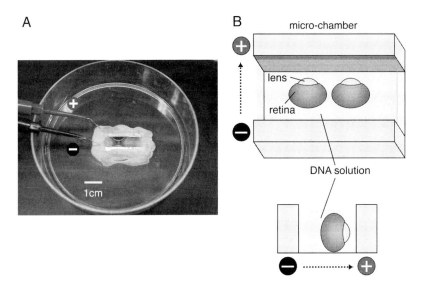

Fig. 19.5 Microchamber for in vitro electroporation. (**A**) Commercially available microchamber. (**B**) Orientation of the retina in the chamber. Maximum transfection efficiency can be obtained when the scleral side is facing the negative electrode

retinas (in vitro electroporation) using a micro electroporation chamber ((8), Figs. 19.5–19.7). Electroporated retinas are cultured for a few days to weeks.

Compared with in vivo electroporation, in vitro electroporation has several advantages. First, in vitro electroporation is easier and less skill-dependent than in vivo electroporation. All retinas subjected to electroporation become Green Fluorescent Protein (GFP) or Red Fluorescent Protein (RFP)-positive when GFP (RFP)-expression vectors are used (*see* **Note 2**). Second, it is relatively easy to handle a large number of retinas in a day. Third, in vitro electroporation can be easily applied not only to postnatal retina, but also to embryonic retina, to which in vivo electroporation (*in utero* electroporation) is hard to apply. Fourth, real-time monitoring of GFP (RFP)-transduced cells is possible under a fluorescence microscope. However, in vitro electroporation has several disadvantages inherent to organ culture. First, the morphology of cultured retina is frequently poor, and photoreceptor outer segments are poorly formed. Second, it is hard to culture retinas for a long period. In our experience, retinas tend to become unhealthy when cultured for more than two weeks (*see* **Note 3**).

1.3. Limitations of Electroporation

Although in vivo and in vitro electroporations are very powerful techniques, they have several limitations. One should take account of these points. First, unlike retroviral vectors that integrate into the host genome and stably express foreign genes for a long time, gene expression from DNA constructs introduced by in vivo electroporation is not so stable. Although we confirmed that the GFP expression is

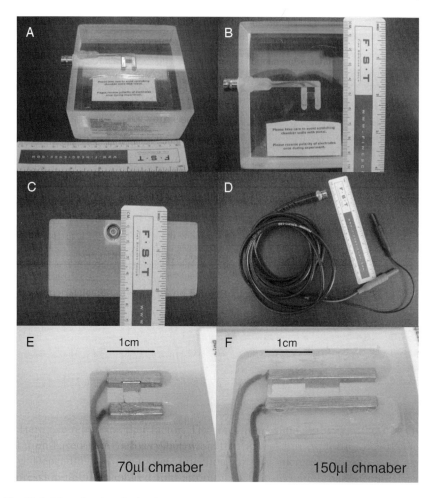

Fig. 19.6 Microchambers for in vitro electroporation—custom made. (**A, B**) The electroporation microchamber made in the Machine Shop at Harvard Medical School. (**C**) Side view of the chamber showing a BNC type female connector. (**D**) Cables with a BNC type male connector. (**E**) High-magnification view of a 70-μL chamber. (**F**) High-magnification view of a 150-μL chamber. Pictures were kindly provided by Dr. Douglas Kim

visible for at least 50 days after electroporation, it is unlikely that the gene expression persists for more than several months (*see* **Note 4**). Second, undifferentiated progenitor/precursor cells are highly transfectable, while transfection efficiency of differentiated neurons, including photoreceptors and ganglion cells located at the surface of the retina, is very low for unknown reasons. Thus, electroporation of photoreceptor cells in the adult retina is not practical. Finally, the DNA constructs electroporated to the postnatal retinal progenitor/precursor cells are inherited by late born retinal cell types (rod photoreceptor, bipolar, amacrine, and Müller glial cells) but not by early born cell types (cone photoreceptor, horizontal and

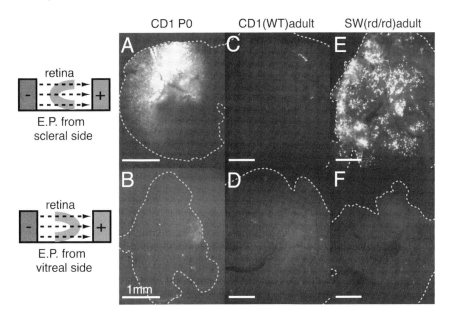

Fig. 19.7 In vitro electroporated retinal explant—whole mount. Mouse retinas of P0 CD1 (**A**, **B**), adult CD1 (**C**, **D**), or adult Swiss Webster having a retinal degeneration mutation (**E**, **F**) were in vitro electroporated with pCAG-GFP from the scleral side (A, C, E) or from the vitreal side (B, D, F), and cultured for 5 days. Pictures a, c, and e were taken from the scleral side, and b, d, and f were taken from the vitreal side. Note that only the scleral side of developing retina or of degenerated retina is highly transfectable. In panel e, most of GFP-positive cells are Müller glial cells. (Reprinted from (8), © 2004, National Academy of Science, USA)

ganglion cells) whose progenitor/precursor cells exist only in the embryonic retina (Fig. 19.4). To deliver genes into the early born cell types by electroporation, one needs to target the embryonic retina. Although subretinal injection into the E13/E14 mouse retina is technically possible (13, 14), it is not as easy as that into the postnatal retina.

2. Materials

2.1. Animals

All the animal experiments were approved by the Institutional Animal Care and Use Committee at Harvard University.

1. Timed or untimed pregnant CD1 mice (*see* **Notes 5,6**).
2. Timed or untimed pregnant Sprague-Dawley rats.

2.2. Plasmid DNA

Both ubiquitous and retinal cell type specific promoters can be used. As a ubiquitous promoter, we are using the CAG (chicken β-actin promoter with cytomegalovirus (CMV) enhancer) promoter (15) or human ubiquitin C promoter (16) (*see* **Note 7**). As a reporter, we are mainly using GFP (EGFP) or RFP (DsRed2) purchased from Clontech. Expression vectors used in our laboratory, as well as their detailed sequence information, are available through Addgene (http://www.addgene.org/connie_cepko).

1. Plasmid DNAs are prepared using Qiagen Plasmid Maxi kit (Qiagen). For in vivo injection, plasmids are ethanol precipitated and suspended in phosphate-buffered saline (PBS, Invitrogen) (final, 2.0–6.0 μg/μL). Approximately 10 μL is needed to inject into 10–20 pups. For in vitro injection, plasmids are ethanol precipitated and suspended in Hanks' balanced salt solution (HBSS, Invitrogen) (final, 0.5–2.0 μg/μL). Approximately 100 μL is needed to perform one electroporation. DNA can be stored at −20°C (*see* **Note 8**).
2. 1% Fast Green FCS (J.T. Baker) in H_2O: filter through a 0.45-μm filter and store at room temperature. Add 1/10 volume of 1% Fast Green to DNA solution as a tracer (final, 0.1%).

2.3. Electroporator

1. Square pulse electroporator CUY21 (Nepagene, Japan) or ECM830 (BTX): Both models work well. CUY21 has a function to display the current measurements after electroporation. ECM830 is compact and less expensive, but does not display the current measurements.
2. Foot pedal switch (BTX, model 1250FS): This is optional but facilitates the electroporation procedure, because both hands are occupied holding an animal and electrodes during electroporation.

2.4. In Vivo Electroporation

1. Tweezer-type electrodes (BTX, model 520, 7 mm diameter). Model 522 (10 mm diameter) works as well.
2. Injection syringe with a 33-gauge blunt end needle (Hamilton, no. 0159666) or injection syringe with a 32-gauge blunt end needle (Hamilton, no. 87931) (*see* **Note 9**).
3. Disposable 30-gauge ½ needle (Becton Dickinson, no. 5106).
4. Cotton swab.
5. 70% ethanol.
6. PBS.

7. Slide warmer, heating pad, or heat lamp to warm anesthetized pups.
8. Dissecting microscope.

2.5. In Vitro Electroporation and Explant Culture

1. Microchamber for electroporation: An electroporation chamber is commercially available from Nepagene (model CUY520P5, 3 mm × 8 mm × 5 mm, ~100 μL volume; Fig. 19.5). We made the electroporation chambers with the help of the Machine Shop at Harvard Medical School Neurobiology Department (Fig. 19.6). These chambers (3 mm × 4 mm × 5.5 mm, ~70 μL volume; and 3 mm × 8 mm × 5.5 mm, ~150 μL volume) were made on acrylic blocks (10 cm × 10 cm × 5 cm) with pure gold plate bars (Alfa Aesar, 3 mm × 12 mm × 3 mm for the 70-μL chamber or 3 mm × 25 mm × 3 mm for the 150-μL chamber) as electrodes, and a BNC-type female connector (Allied Electronics no. 885-5369). These chambers were designed by Dr. Douglas Kim.
2. Culture medium: Dulbecco's modified Eagle medium (DMEM) containing 10% fetal calf serum and antibiotics (penicillin (100 U/mL) and streptomycin (100 mg/mL) (*see* **Note 10**).
3. 6-well tissue culture dish.
4. Nucleopore polycarbonate membrane (Whatman, no. 110606, 25 mm diameter, 0.2-μm pore).
5. Dissecting instruments: fine scissors (e.g., F.S.T. no. 14085-08) and fine forceps (e.g., Dumont no. 5) for dissection. Curved forceps (Dumont no. 7) for transfer of dissected retina.
6. HBSS.
7. 10- or 6-cm Petri dish.

2.6. Analysis of Electroporated Retina

2.6.1. Sectioning

1. 4% (w/v) paraformaldehyde in PBS: dissolve 8.0 g of paraformaldehyde (J.T. Baker) in 40 mL of H_2O in a 50-mL centrifuge tube. Add 20 μL of 10 M NaOH, and heat at 65°C, occasionally inverting the tube until it dissolves. Filter through a 0.45-μm filter. Stable at 4°C for up to a week. Before use, make 4% solution in PBS.
2. 30% (w/v) sucrose/PBS: filter through a 0.45-μm filter and store at room temperature.
3. O.T.C. compound.
4. Embedding mold (Ted Pella, no. 106, 12 cavities).
5. Liquid nitrogen or dry ice.

6. Superfrost Plus slide glass (Fisher Scientific).
7. Cryostat.

2.6.2. Dissociation into Single Cells

1. Papain (Worthington, no. LS003126, milky solution). Store at 4°C. Stable for more than a year.
2. 50 mM cysteine in H_2O: filter through a 0.45-μm filter and store at 4°C.
3. 10 mM EDTA in H_2O. Store at room temperature.
4. 60 mM 2-mercaptoethanol in H_2O. Store at 4°C.
5. Poly-D-lysine (10 mg/mL) in H_2O. Store at 4°C. Before use, dilute in H_2O at 1:100 (final, 0.1 mg/mL).
6. HBSS or PBS.
7. DMEM with 10% fetal calf serum.
8. DNaseI (Roche, no. 776785), 10 units/μL.
9. 8-well slide glass (Cell-Line/ERIE Scientific Co. 8 Rect. $11 \times 13\,mm^2$).

2.6.3. Immunostaining

1. Goat serum (Gemini-Bioproducts): Heat-inactivate at 56°C for 30 min, and store at −20°C.
2. PBS.
3. PBST: PBS containing 0.1% (v/v) Triton X-100.
4. Primary antibodies (**Table 19.1**).
5. Secondary antibodies: Cy2-, Cy3-, or Cy5-conjugated anti-mouse, anti-rabbit, or anti-rat IgG. Reconstitute in 50% glycerol and store in dark at −20°C. Use at 1:500.
6. 4', 6-diamidino-2-phenylindole (DAPI) (1 mg/mL; Roche) in H_2O. Store in dark at 4°C. Use at 1:2000.
7. Slide staining dish and holder.
8. Gel/Mount (Biomedia).
9. Cover glass.

3. Methods

3.1. In Vivo Electroporation

3.1.1. DNA Injection

For in vivo electroporation of the retina, DNA injection is the most critical step to achieve good transfection. We are using two injection methods (Figs. 19.8–19.10). One is to inject DNA without opening the eyelid (Method 1). This is a "blind

Table 19.1 Retinal cell type specific antibodies

Antibody	Host	Source	Retinal cell type	Dilution
Anti-rhodopsin (Rho4D2)	Mouse	Dr. R.S. Molday (17)	Rod	1:100
Anti-Gt1α	Rabbit	SantaCruz (no. sc-389)	Rod	1:500
Anti-blue opsin	Rabbit	Chemicon (no. AB5407)	Blue cone	1:500
Anti-green opsin	Rabbit	Chemicon (no. AB5405)	Green cone	1:500
Anti-Gt2α	Rabbit	SantaCruz (no. sc-390)	Cone	1:100
Anti-PKCα	Mouse	Oncogene (no. OP74)	Rod bipolar	1:100
Anti-Chx10	Rabbit	Our laboratory	Bipolar (+progenitor)	1:500
Anti-glutamine synthetase	Mouse	Chemicon (no. MAB302)	Müller glia	1:500
Anti-syntaxin 1a (HPC-1)[a]	Mouse	SantaCruz (no. sc-12736)	Amacrine	1:200
Anti-calbindin D28K[b]	Mouse	Sigma (no. C9848)	Horizontal	1:200
Anti-Thy-1.1[c]	Mouse	SantaCruz (no. sc-19614)	Ganglion	1:200
Anti-Thy-1.2[d]	Rat	PharMingen (no. 553000)	Ganglion	1:200

[a]Syntaxin 1a is also weakly expressed in horizontal cells
[b]Calbindin is also expressed in a small subset of amacrine cells
[c]This antibody works for rat retina, but may not react with most mouse strains (e.g., C57/BL6) expressing Thy-1.2 instead of Thy-1.1
[d]This antibody works for most of the mouse strains expressing Thy-1.2, but does not work for rat retina

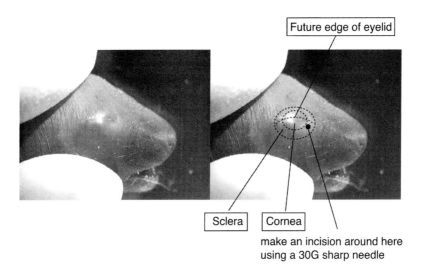

Future edge of eyelid

Sclera

Cornea

make an incision around here using a 30G sharp needle

Fig. 19.8 Head of newborn rat

injection," but causes less damage to the animals. This method works well particularly for rat pups. The other one is to inject DNA after cutting the future edge of the eyelid (Method 2). In this method, the tip of an injection needle can be seen through the lens (Fig. 19.10G), and it is easy to check whether DNA was correctly

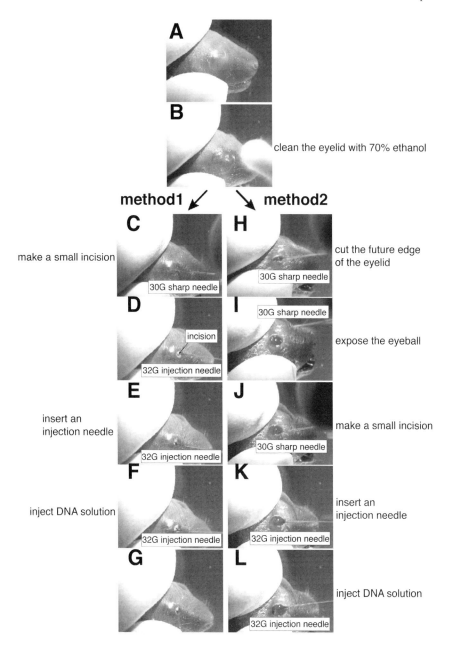

Fig. 19.9 Procedure for subretinal injection—pictures. Two injection methods, injection without opening the eyelid (**A**, **B**, **C–G**) and injection after opening the eyelid (**A**, **B**, **H–L**), are shown. See text for detailed description of the procedure

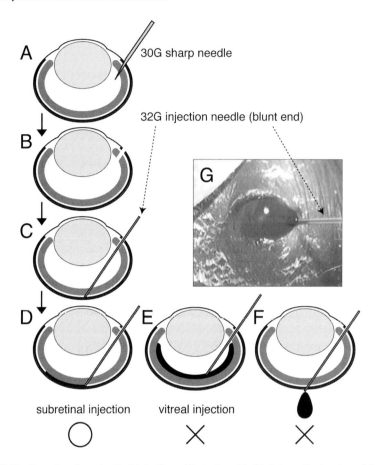

Fig. 19.10 Procedure for subretinal injection—illustration. (**A–D**) Schematic drawing of the sub-retinal injection procedure. See text for detailed description the procedure. (**E, F**) Examples of bad injection. (**G**) Injection needle can be seen through the lens if the eyelid is opened before injection

injected into the subretinal space. DNA injection is done under a dissecting microscope.

3.1.1.1. Method 1 (Injection Without Opening the Eyelid)

1. Anesthetize newborn pups on ice for several minutes until they stop moving. It takes several minutes (*see* **Note 11**).
2. Clean the eyelid with 70% ethanol using a cotton swab (Fig. 19.9B). By wetting the eyelid, the shape of the eyeball and the future edge of the eyelid can be easily recognized.
3. Make a small incision in the sclera near the cornea through the skin using the tip of a sharp 30-gauge needle (Figs. 19.9C and 19.10A, B).

4. Carefully insert an injection needle (Hamilton syringe, 32 or 33 gauge) into the eyeball through the incision until you feel a slight resistance (Figs. 19.9D, E and 19.10C (*see* **Note 12**)). When you feel resistance, the tip of the injection needle is located at the subretinal space between RPE and retina.

5. Slowly inject DNA solution containing 0.1% Fast Green into the subretinal space (Figs. 19.9F and 19.10D (*see* **Note 13**)). For rat newborn pups, we inject 0.3–0.5 µL of DNA, and for mouse newborn pups, we inject 0.2–0.3 µL of DNA. If DNA is injected correctly, the eyeball becomes green (Fig. 19.9G).

3.1.1.2. Method 2 (Injection After Opening the Eyelid)

1. Anesthetize newborn pups on ice for several minutes.
2. Clean the eyelid with 70% ethanol using a cotton swab (Fig. 19.9B).
3. Carefully cut the future edge of the eyelid using the tip of a sharp 30-gauge needle (Fig. 19.9H (*see* **Notes 14,15**)).
4. Expose the eyeball by pulling down the skin (Fig. 19.9I).
5. Make a small incision in the sclera near the cornea using the tip of a sharp 30-gauge needle (Figs. 19.8, 19.9J, and 19.10A).
6. Insert an injection needle (Hamilton syringe, 32 or 33 gauge) into the eyeball through the incision until you feel resistance (Figs. 19.9K and 19.10C). The inserted needle can be seen through the lens (Fig. 19.10G).
7. Slowly inject DNA into the subretinal space (Figs. 19.9L and 19.10D). If you correctly inject DNA into the subretinal space, the dye spreads within a relatively small area (usually not in the entire retina), and you can see "green" and "nongreen" areas in the retina by slightly rotating the injected animal. If you inject DNA into the vitreous chamber, the dye spreads more rapidly and uniformly in the eyeball (Fig. 19.10E). If you inject DNA outside the eyeball, the eyeball does not become green (Fig. 19.10F).

3.1.2. Electroporation

1. Soak the tweezer-type electrodes in PBS (optional). This step is to increase the contact between the pup and the electrodes.
2. Place the tweezer-type electrodes to hold the head of the pup, and slightly squeeze them (Fig. 19.2B). The positive electrode, marked by a plastic screw (Fig. 19.2A, arrowhead), should be at the DNA-injected side, if you transfect DNA from the subretinal space into the retina.
3. Apply five square pulses of 50-ms duration with 950-ms intervals using a pulse generator. For newborn rat pups, we apply 100-V pulses, and for newborn mouse pups, we apply 80-V pulses. We usually apply electric pulses right after DNA injection. The measured current is 0.10–0.20 mA (*see* **Notes 16,17**).
4. Warm the operated pups (e.g., on a 37°C slide warmer) until they recover from anesthetic, and then return them to their mother (*see* **Notes 18,19**).

3.2. In Vitro Electroporation

In the neonatal or fetal retina, the scleral side is highly transfectable, whereas the vitreal side is not (Fig. 19.7). Therefore, good transfection can be achieved when the scleral side is facing the negative electrode in an electroporation chamber (Fig. 19.5B). Multiple retinas can be electroporated at once, and the DNA solution can be used several times, although transfection efficiency gradually decreases. All the procedures are done at room temperature. We are not using a tissue culture hood for dissection and electroporation.

1. Put polycarbonate filters on culture medium (2 mL/well) in 6-well dishes, and keep the plates in a CO_2 incubator. Do not sink the filters into the medium.
2. Dissect eyeballs in HBSS in a Petri dish under a dissecting microscope, and carefully take out the retina with lens. RPE is usually removed with sclera (*see* **Note 20**).
3. Collect the retinas in a new Petri dish. The dissected retina can be kept in HBSS for 1 h at room temperature.
4. Transfer the retina(s) to a micro electroporation chamber filled with a DNA solution.
5. Set the position and orientation of the retina(s) in the chamber. When the vitreal side (lens) is facing the positive electrode, transfection efficiency becomes maximum (Figs. 19.5B and 19.7)
6. Apply five square pulses (30 V) of 50-ms duration with 950-ms intervals using a pulse generator. Air bubbles are generated only from the negative electrode (*see* **Note 21**).
7. When electroporation is repeated using the same DNA solution, gently stir the DNA solution by pipetting several times. Do not make air bubbles. If necessary, DNA solution can be recovered and stored at −20°C after removing cell debris by centrifugation for 3 min at 13,000 rpm.
8. Transfer the electroporated retina(s) into HBSS in a Petri dish to wash out the DNA.
9. Remove the lens from the retina, and flatten the retina. If necessary, carefully make several incisions to facilitate flattening.
10. Using curved forceps, carefully transfer the retina onto polycarbonate filters in 6-well dishes with the scleral side down.
11. Apply 20 μL of the culture medium onto the retina.
12. Culture the retina at 37°C in a CO_2 incubator.
13. Change half of the culture medium every 3 days.

3.3. Analysis of Electroporated Retina

Electroporated retinas are harvested 2 days to several weeks after electroporation and dissected under a fluorescent dissecting microscope (Leica, MZFL III) to select

GFP (RFP)-positive retinas (*see* **Note 22**). Dissected retinas are analyzed by making sections to examine the morphology of GFP (RFP)-positive cells. Alternatively, retinas may be dissociated into single cells. These single cells can be used for immunocytochemistry or for FACS sorting of GFP (RFP)-positive cells to analyze gene expression profiles (Reverse Transcription-PCR (RT-PCR) and microarray analyses).

3.3.1 Sectioning

1. Dissect eyeballs in HBSS or PBS in a Petri dish, and carefully take out the retina with lens (*see* **Note 23**).
2. Fix the harvested retina with 4% paraformaldehyde in PBS for 30–60 min at room temperature (*see* **Note 24**).
3. Incubate the retina in 30% sucrose in PBS for several hours to overnight at 4°C (*see* **Note 25**).
4. Embed the retinas in O.C.T. compound using an embedding mold.
5. Snap-freeze the retinas in the O.C.T. compound with liquid nitrogen or dry ice. Frozen retina can be stored at −80°C for more than a year.
6. Cut cryosections (20 μm thickness) on a cryostat.
7. Air-dry the slides at room temperature for 30 min. Dried slides can be stored in a tightly sealed slide box with several pieces of dry silica gel (Drierite) at −20°C for several months.

3.3.2. Dissociation into Single Cells

1. Prepare a papain activation solution by mixing H_2O (7 mL) with 50 mM cysteine (1 mL), 10 mM EDTA (1 mL), and 60 mM 2-mercaptoethanol (100 μL).
2. Activate papain by diluting 5 μL (× tube numbers) of papain in 200 μL (× tube numbers) of the activation solution. When the color of this solution changes from white to clear, the papain solution is ready to use. Usually it takes a few minutes.
3. Put retina(s) into a 1.5-mL microtube containing 200 μL of HBSS.
4. Add 200 μL of the activated papain solution into the tube.
5. Incubate at 37°C for ~10 min. During incubation, tap the tubes several times by finger to promote digestion (*see* **Note 26**).
6. Add 600 μL of DMEM/10%FCS to stop the digestion (*see* **Note 27**).
7. Add 50 u (5 μL) DNase, and incubate at 37°C for ~5 min.
8. Suspend the dissociated retinal cells by gently pipetting up and down using a P1000 pipetman (10–15 times).
9. Centrifuge the tubes for 30 s at 4,000 rpm.
10. Remove the supernatant (but not completely), and then break the cell pellet by finger tapping (*see* **Note 28**).

11. Add 600 μL of DMEM/10%FCS, and suspend by gently pipetting up and down using a P1000 pipetman (5–10 times). Dissociated retinal cells can be kept on ice for 1 h.
12. Coat 8-well glass slides with poly-D-lysine (0.1 mg/mL) for more than 30 min at room temperature.
13. Aspirate the poly-D-lysine solution, and air-dry the slides at room temperature. Coated slides can be stored at room temperature.
14. Plate the dissociated retinal cells (50 μL/well) on coated 8-well slides.
15. Incubate the slides at 37°C for 40–60 min in a CO_2 incubator.
16. Aspirate the medium, and fix the cells with 4% paraformaldehyde in PBS at room temperature for 5 min.
17. Wash the slides twice with PBS.
18. Aspirate PBS and air-dry the slides at room temperature for 30 min. Dried slides can be stored in a tightly sealed slide box with several pieces of dry silica gel (Drierite) at −20°C for several months.

3.3.3. Immunostaining

1. Wash slides in PBS for 5 min at room temperature.
2. Block the slides with 10% goat serum in PBST for 1 h at room temperature.
3. Remove the blocking solution, and apply primary antibody diluted in the blocking solution.
4. Incubate overnight at 4°C in the dark.
5. Remove the antibody solution, and wash the slides in PBS thrice for 5 min each at room temperature.
6. Apply secondary antibody diluted in the blocking solution containing 0.0005% DAPI.
7. Incubate for several hours (more than 2 h) at room temperature in the dark.
8. Wash the slides in PBS thrice for 5 min each at room temperature.
9. Air-dry the slides at room temperature for 30 min.
10. Mount the slides using Gel/Mount.

Acknowledgments We thank Dr. Douglas Kim for providing the pictures. This work was supported by NIH EYO 8064 and 9676 and Howard Hughes Medical Institute.

4. Notes

1. It is possible to transfect DNA into the RPE in the adult mouse and rat by injecting DNA into the subretinal space, and by applying electric pulses in the direction opposite to that shown in Fig. 19.1A (12, 18).
2. After in vitro electroporation with GFP (RFP)-expression vectors, 5–20% of cells become GFP (RFP)-positive in the retina, depending upon the DNA concentration, type of promoters in the expression vectors, and orientation of the retina in the electroporation microchamber.

3. Another group reported that neonatal mouse retinal explants could be maintained for more than 4 weeks using improved culture conditions (19).

4. This problem might be partly overcome by the use of ΦC31 integrase, which stably integrates the plasmid DNA containing bacterial attB site into the mammalian genome (18).

5. We examined several outbred mouse strains maintained by Charles River and Taconic, and found that CD1 mice from Charles River have normal retinal morphology. Most other outbread mouse strains, including Swiss Webster (Taconic, Charles River), ICR (Taconic), and Black Swiss (Taconic), have a retinal degeneration (*rd1*) mutation and are not suitable for the study of retinal development.

6. Inbred mouse strains, such as C57BL/6, can also be used. However, inbred mice sometimes do not take care of their pups after the pups are subjected to surgery. Careful monitoring is needed when inbred mouse pups are returned to their mothers after surgery.

7. We compared several ubiquitous promoters, including CAG, ubiquitin C, cytomegalovirus (CMV), and human elongation factor 1α (EF-1α) promoters, in the developing rat retina and found that the CAG and ubiquitin C promoters work very well. The ubiquitin promoter is stronger than the CAG promoter in the developing rat retina. The CMV and the EF-1α promoters also work but appear to be silenced in photoreceptor cells (8).

8. Repeated freeze-thawing of DNA solution sometimes generates an insoluble precipitate, which is probably an aggregate of DNA. To prevent an injection needle from clogging, such an aggregate has to be removed by centrifugation at 13,000 rpm for 3 min.

9. We are using a blunt end needle (point style 3) instead of a sharp beveled needle. Use of a blunt end needle facilitates subretinal injection. Pulled sharp glass needles also can be used, but practice is needed for successful injection.

10. We tested several culture media, and found that the serum-free medium (Neurobasal medium (Invitrogen) with 1× B-27 serum-free supplement (Invitrogen)) also works well. In the serum-free culture, however, the number of surviving ganglion cells is lower than that in the serum-containing medium. Other cell types are largely not affected.

11. If newborn pups are left on ice too long (more than 10 min), some pups do not recover.

12. Note that even a blunt end needle can easily penetrate the sclera of newborn pups when the needle is pushed strongly. Careful and slow needle insertion is a key point.

13. If the same injection syringe is used for different DNA solutions, wash the syringe by filling and ejecting water several times until the dye disappears from the syringe.

14. You may cut the entire edge of the eyelid (Fig. 19.8). But if you correctly make a small cut in the edge line of the eyelid (just above the lens), and pull down the skin, the eyelid will open like a "zipper" with little damage.

15. After cutting eyelids of ~10 pups, needles appear to become dull. Change the needle to a new one, or clean the needle tips with a kimwipe and 70% ethanol.

16. We usually transfect DNA into only one eye per animal. It is possible to electroporate DNA into both eyes, but there must be more than 1-h interval between electroporations in the same animals. This is particularly true when newborn mouse pups are used. Shorter interval electroporations in the same animal occasionally damage the operated pups.

17. When electric pulses are correctly applied, slight twitching of the muscles can be observed.

18. Almost all operated pups survive and are apparently healthy after electroporation. In our hands, an average of more than 80% rat retinas and more than 50% mouse retinas express GFP, when GFP expression vector is used. In a good transfection, GFP expression is observed in a wide area of the retina (Fig. 19.3).

19. When a heat lamp is used to warm the pups, the temperature has to be carefully monitored. If the distance between the lamp and the pups is too close, the pups can die because of high temperature.

20. The lens may be removed from the retina at this step. But if the lens is attached to the retina, scleral side and vitreal side can be easily distinguished. In the flattened retina, it is often hard to distinguish these two sides. In addition, the lens prevents the retina from

directly touching the electrode if the orientation of the retina in the chamber is as shown in Fig. 19.5B.

21. When the retina touches the electrode during electroporation, the contacted area may be only slightly (but not seriously) damaged.

22. When the CAG promoter is used to express GFP, GFP can be visible 24 h after electroporation, but the expression is low at this time point. Maximum expression is observed a few days after electroporation.

23. Lens may be removed at this step. But if the lens is removed before fixation, the retina tends to curl up during the following procedures. Good shape (cup-like structure) is maintained when the lens is removed from the retina after fixation.

24. Fluorescence of GFP (RFP) diminishes by fixation but recovers during the cryoprotection procedure. Overnight fixation with 4% paraformaldehyde significantly reduces the brightness of GFP (RFP) and is not recommended.

25. Long cryoprotection (more than 12 h) increases the level of autofluorescence, particularly in the photoreceptor outer segments.

26. Do not suspend the retina by pipetting at this step. Pipetting at this step significantly decreases the viability of the dissociated retinal cells.

27. If serum is not desired in the experiments, 10% FCS can be substituted with bovine serum albumin (1 mg/mL; BSA).

28. If medium is added without finger tapping, resuspension of the cell pellet becomes hard.

References

1. Price, J., Turner, D., and Cepko, C. (1987) Lineage analysis in the vertebrate nervous system by retrovirus-mediated gene transfer. *Proc. Natl. Acad. Sci. U.S.A.* **84**, 156–160.

2. Turner, D.L. and Cepko, C.L. (1987) A common progenitor for neurons and glia persists in rat retina late in development. *Nature.* **28**, 131–136.

3. Miyoshi, H., Takahashi, M., Gage, F.H., and Verma, I.M. (1997) Stable and efficient gene transfer into the retina using an HIV-based lentiviral vector. *Proc. Natl. Acad. Sci. U.S.A.* **94**, 10319–10323.

4. Bennett, J., Wilson, J., Sun, D., Forbes, B., and Maguire, A. (1994) Adenovirus vector-mediated in vivo gene transfer into adult murine retina. *Invest. Ophthalmol. Vis. Sci.* **35**, 2535–2542.

5. Li, T., Adamian, M., Roof, D.J., et al. (1994) In vivo transfer of a reporter gene to the retina mediated by an adenoviral vector. *Invest. Ophthalmol. Vis. Sci.* **35**, 2543–2549.

6. Jomary, C., Piper, T.A., Dickson, G., et al. (1994) Adenovirus-mediated gene transfer to murine retinal cells in vitro and in vivo. *FEBS Lett.* **347**, 117–122.

7. Ali, R.R., Reichel, M.B., Thrasher, A.J., et al. (1996) Gene transfer into the mouse retina mediated by an adeno-associated viral vector. *Hum. Mol. Genet.* **5**, 591–594.

8. Matsuda, T. and Cepko, C.L. (2004) Electroporation and RNA interference in the rodent retina in vivo and in vitro. *Proc. Natl. Acad. Sci. U.S.A.* **101**, 16–22.

9. Yang, T.L. and Cepko, C.L. (2004) A role for ligand-gated ion channels in rod photoreceptor development. *Neuron.* **41**, 867–879.

10. Yang, T.L., Matsuda, T., and Cepko, C.L. (2005) The noncoding RNA taurine upregulated gene 1 is required for differentiation of the murine retina. *Curr. Biol.* **15**, 501–512.

11. Dezawa, M., Takano, M., Negishi, H., Mo, X., Oshitari, T., and Sawada, H. (2002) Gene transfer into retinal ganglion cells by in vivo electroporation: a new approach. *Micron.* **33**, 1–6.

12. Kachi, S., Oshima, Y., Esumi, N., et al. (2005) Nonviral ocular gene transfer. *Gene Ther.* 12, 843–851.

13. Turner, D.L., Snyder, E.Y., and Cepko, C.L. (1990) Lineage-independent determination of cell type in the embryonic mouse retina. *Neuron.* **4**, 833–845.
14. Surace, E.M., Auricchio, A., Reich, S.J., et al. (2003) Delivery of adeno-associated virus vectors to the fetal retina: impact of viral capsid proteins on retinal neuronal progenitor transduction. *J. Virol.* **77**, 7957–7963.
15. Niwa, H., Yamamura, K., and Miyazaki, J. (1991) Efficient selection for high-expression transfectants with a novel eukaryotic vector. *Gene.* **108**, 193–199.
16. Schorpp, M., Jager, R., Schellander, K., et al. (1996) The human ubiquitin C promoter directs high ubiquitous expression of transgenes in mice. *Nucleic Acids Res.* **24**, 1787–1788.
17. Molday, R.S. (1989) Monoclonal antibodies to rhodopsin and other proteins of rod outer segments. *Prog. Ret. Res.* **8**, 173–209.
18. Chalberg, T.W., Genise, H.L., Vollrath, D., and Calos, M.P. (2005) phiC31 integrase confers genomic integration and long-term transgene expression in rat retina. *Invest. Ophthalmol. Vis. Sci.* **46**, 2140–2146.
19. Caffe, A.R., Ahuja, P., Holmqvist, B. et al. (2001) Mouse retina explants after long-term culture in serum free medium. *J. Chem. Neuroanat.* **22**, 263–273.
20. Matsuda, T., and Cepko, C.L. (2007) Controlled expression of transgenes introduced by in vivo electroporation. *Proc. Natl. Acad. Sci. U.S.A.* **84**, 156–160.

Chapter 20
Optical In Vivo Imaging of Electrically Mediated Delivery of siRNA into Muscle for Gene Function Analysis

Muriel Golzio and Justin Teissié

Abstract Short interfering RNAs (siRNAs) represent new potential therapeutic tools owing to their capacity to induce strong, sequence-specific gene silencing in cells. However, this development requires new, safe, and efficient in vivo siRNA delivery methods. In this study, we reported that gene silencing was efficiently obtained in vivo in an adult mammal (mouse) with chemically synthesized siRNA after its electrical delivery in muscles. The associated gene silencing was followed on the same animal and lasted more than 11 days. Gene silencing was obtained in muscles not only on young adult mice but also on much older animals. No tissue damage was detected under our electrical conditions. Therefore, this method should provide an efficient approach for gene function analysis by a localized delivery of siRNAs.

Keywords: electropulsation, electroporation, siRNA, gene silencing, whole body imaging, GFP

1. Introduction

Since the discovery of RNA interference (1), the identification of the short interfering RNAs (siRNAs) involved in this process and their use for sequence-specific gene silencing has offered a new approach for molecular therapeutics by taking advantages of the progress in genomics (2, 3). SiRNAs appear as a very promising new therapeutic agent, but, besides the problem of safe and efficient delivery (4), an unanswered problem is to know how long its effect lasts after a single dose delivery (5). SiRNA gene silencing could be obtained in vivo on reporter as well as endogenous genes. The limit was the delivery method, where efficiency and targeting were requested. Chemical methods brought a systemic delivery that needs a long time of contact, because the injected siRNA is the target of RNAses. A rapid transfer method is, therefore, requested to keep intact the nucleic acids before its introduction in the cytoplasm of the target tissue cells. Another problem with the chemical additives is they may be responsible for the interferon-like response

associated to siRNA. The demonstration in 1998 of drug and plasmid electrotransfer and gene expression in tumors (6, 7) led to the proposal that in vivo electroporation was a promising tool for exogenous agents' delivery (8). Electroporation is obtained by applying a voltage pulse between two electrodes in direct contact with the tissue. Electrically mediated gene transfer had been shown to be effective on many tissues: liver (9), skin (10), muscle (11, 12), and heart (13). Delivery is targeted to the tissue volume where the field pulse is applied, i.e. under the control of the electrode localization and geometry.

Efficient electrodelivery is controlled on one hand by the way the solution of the nucleic acid is injected in the tissue and on the other hand by the electrical parameters. It is known that expression of the electrotransfered gene is under the control of the field strength (voltage and geometry of electrodes), on the pulse duration (long pulses appear to be needed as an electrophoretic drag must push the nucleic acid against the cell surface), and on the vectoriality of the field (electrophoretic accumulation on different part of the cell). Problems may be present such as tissue burning (when the electric conditions are too drastic) or reactive oxygen species generation due to the cellular stress. During the last 6 years, optimization of electrogenetransfer in muscle has been obtained. Two strategies are present with similar efficiencies. In both cases, plate electrodes are brought in contact with the shaved skin and the field is delivered across the skin. This treatment is noninvasive and gives a fairly homogeneous field distribution. The difference is in the sequence of pulses. In the most popular method, a train of 8 pulses lasting 20 ms with a voltage to electrode distance of 200 V/cm is delivered with a 1 Hz frequency. Recently, it was proposed to apply a high voltage pulse (100 µs, 1,300 V/cm) followed by a long LV pulse (400 ms, 80 V/cm).

Recent developments in optical imaging provide continuous monitoring of gene delivery and expression in living animals (14). As imaging methods are noninvasive, they allow for long-term studies on a single animal without sacrificing the animal. Indeed, reporter gene activity can be accurately followed on the same animal as a function of time with no adverse effects either on the reporter gene product or on the animal itself. In this study, we investigated the effectiveness of electroporation for the localized delivery of siRNA in adult mice by using the thigh muscle as a model target. The resulting inhibitory effect was monitored in living animals over several weeks by whole body fluorescence imaging of the green fluorescent protein (GFP).

2. Materials

2.1. Nucleic Acids

1. Plasmid DNA: The pEGFP-C1 plasmid (Clontech, Palo Alto, CA) expresses a mammalian-enhanced version of the green fluorescent protein (eGFP).
2. efgp22 oligo DNAs: The sense and antisense egfp22 siRNA sequences are 5′-r(GCA AGC UGA CCC UGA AGU UCA U) and 5′ r(GAA CUU CAG GGU

CAG CUU GCC G), respectively. These oligos are directed against GFP mRNA and designed according to (15).

3. The P76 oligo DNAs: The sense and antisense P76 siRNA sequences are 5′ r(GCG GAG UGG CCU GCA GGU A)dTT and 5′ r(UAC CUG CAG GCC ACU CCG C)dTT, respectively. These oligos are directed against an unrelated human P76 mRNA and shows no significant homology to mouse transcripts. It serves as a control for specificity of the siRNA construct. These oligos were obtained from Qiagen Xeragon (Germantown, MD).

4. Annealing buffer: 100 mM KCl, 30 mM Hepes KOH, pH 7.4, 2 mM magnesium acetate.

2.2. In Vivo Experiments

1. Mice: The female C57Bl/6 mice were 9–10 weeks old at the beginning of the experiments, weighing 20–25 g and were considered as young mice. The Balb/c mice were 8 months old at the beginning of the experiments and were therefore considered as old mice. They were both maintained at constant room temperature with a 12 h light cycle in a conventional animal colony. Before the experiments, the C57Bl/6 mice were subjected to an adaptation period of at least 10 days.

2. Cream (Veet) (*see* **Note 1**).

3. Hamilton syringe with a 26 GA needle (Hamilton, Bonaduz, Switzerland).

4. Isoflurane.

5. Anesthetic machine with air compressor (TEM, Bordeaux, France) and isoflurane vaporizer (Xenogene, Alameda, CA).

6. Electroporator PS 10 CNRS (Jouan, St Herblain, France). This is a square wave pulse generator (constant voltage output during the preset pulse duration) (*see* **Note 2**). All parameters were monitored on line with an oscilloscope (Metrix, Annecy, France). An electronic switch cutting the pulse as soon as its intensity was 5 A brought safety against current surge.

7. Plate parallel electrodes (length, 1 cm; width, 0.6 mm) (IGEA, Carpi, Italy) (*see* **Note 3**).

8. Conducting paste (Eko-gel, Egna, Italy).

2.3. In Vivo Visualization of Gene Expression or Gene Silencing

1. A fluorescence stereo-microscope. The fluorescence excitation was obtained with a Mercury Arc lamp, GFP, or the G filter sets.

2. Cooled CCD Camera Coolsnap fx (Roper Scientific, Evry, France)

3. The MetaVue software (Universal, USA) drives the camera from a Dell computer and allowed quantitative analysis of the GFP fluorescence level.

3. Methods

In this study, electrotransfer of plasmid DNA encoding the GFP reporter gene was used to show the efficiency of in vivo electro-administration of specific siRNA after intramuscular injection in young and old adult mice. We compared treatment groups using egfp22 siRNA, electric field alone, and nonrelevant p76 siRNA. It is important to use the same amount of DNA or siRNA among groups as well as the same volumes for injection to obtain reliable results. The first step was to determine the kinetics of GFP gene expression using different electric field parameters and then to determine whether the injection of egfp22 siRNA affects the GFP fluorescence expression. The expression of the fluorescent reporter gene was determined by in vivo fluorescence stereomicroscopy, and a simple set-up was used to quantify the fluorescence on the digitized images.

3.1. In Vivo Electropulsation

1. Two days before the treatment, shave one of the legs of the mice with the cream (*see* **Note 1**).
2. Anesthetize mice by isoflurane inhalation (*see* **Note 4**).
3. Inject 10 μg of eGFP plasmid (0.5 μg/μL) in PBS mixed with 2 μg of siRNA (0.4 μg/μL in annealing buffer) slowly (about 15 s) with a Hamilton syringe through a 26 GA needle into tibialis muscle in mice. In the control conditions, add 5 μL of PBS to 20 μL of plasmid solution to keep the injection conditions similar (a final volume of 25 μL was injected).
4. Add a conducting paste to obtain a good electric contact between the skin and the flat electrodes (*see* **Note 5**).
5. Place the parallel plate electrodes on two sides of the leg 30 s after injection (*see* **Note 6**).

A train of eight 120 V square wave pulses was delivered with a 200 V/cm voltage to electrode width, 20 ms pulse duration, and 1 Hz pulse frequency (Fig. 20.1A). These experimental conditions were shown to be the most efficient when a train of repetitive pulses was applied to muscles.

3.2. Whole Body Imaging

The electrically mediated GFP gene transfer in the mouse muscle was detected directly on the anesthetized animal through its skin by digitized stereomicroscopy. The GFP fluorescence from the muscle was quantitatively evaluated at different days and thereafter with weekly intervals until the GFP fluorescence was no longer detectable (Fig. 20.1B).

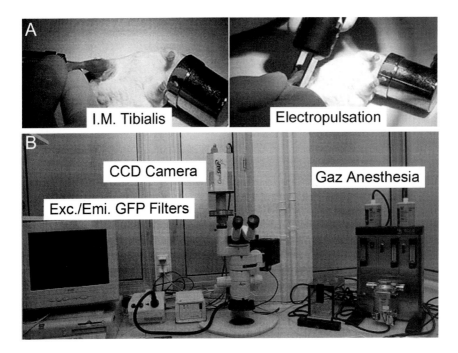

Fig. 20.1 Experimental set-up. **A**. *Left*, Intramuscular injection in the tibialis of an anesthetized mouse. The nose of the animal is set in the isoflurane mask. A volume of 25 μL was injected slowly into the tibialis muscle (*see* **Note 6**). *Right*, The flat electrodes set-up, being positioned on the two sides of the leg, allowed to easily apply a train of eight square wave pulses. All electrical parameters are preset (120 V, 20 ms pulse duration, and a 1 Hz frequency of pulses) (*see* **Notes 7,8**). **B**. Digitized stereomicroscopy imaging set-up. The image is obtained on the PC monitor illustrated on the *left*

3.2.1. Fluorescence Data Acquisition

1. Anesthetize the mouse.
2. Hold the leg and place it under the stereo fluorescence microscope. The whole muscle is observed as a 12 bits 1.3 M pixels image with a cooled CCD camera. Capture the image under the visible light (*see* **Note 9**) prior to capturing the fluorescence.
3. Obtain the fluorescence image using a Mercury Arc lamp. The exposure time was set at 1 s with no binning. Two different fluorescence images were acquired by selecting the GFP and the G filter sets.

3.2.2. Fluorescence Data Analysis

1. A direct light image, a fluorescent image with GFP filter set, and a fluorescent image with G (red) filter set were taken on each leg. From the visible light image, the tibialis cranialis muscle was located and gated to give the region of interest. On the image taken with the GFP filter set, the mean fluorescence in the gated area (whole muscle) was

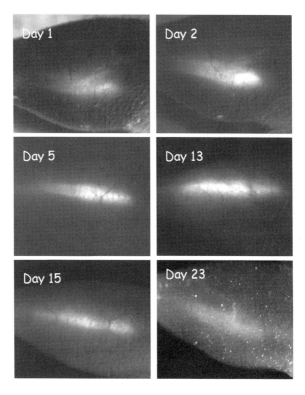

Fig. 20.2 Time lapse follow up of GFP expression in 8-month-old Balb/c mouse leg muscle. GFP expression resulting from the plasmid alone electrotransfer was observed on days 1, 2, 5, 13, 15, and 23

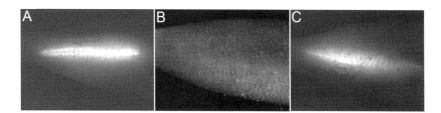

Fig. 20.3 RNA interference in mouse leg muscle (each image is 1.3 cm wide) as illustrated by the representative images of the GFP fluorescence from the mouse leg. **A**. Control experiment resulting from the plasmid alone electrotransfer was observed on day 15. **B**. GFP expression silencing in the leg when the plasmid was co-transferred with specific egfp22 siRNA. **C**. GFP expression when unrelated siRNA (P76) was co-transferred with the plasmid

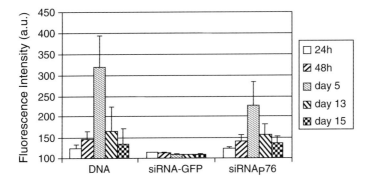

Fig. 20.4 Changes in the mean fluorescence emission with time was quantified by fluorescence imaging from day 1 to 15. Control experiment results from the plasmid alone. Specific silencing was observed when plasmid DNA was co-transferred with specific egfp22 siRNA. No silencing was observed with unrelated siRNA (P76) co-transferred with the plasmid DNA. Bars represents standard deviations ($n = 4$)

quantitatively estimated (measure/region measurement). The image taken with the G filter set depicted the autofluorescence from the skin and hair follicles. In our experiments, the background fluorescence was sufficiently low enough not to interfere with our quantification when GFP emission was present (Figs. 20.2 and 20.3) (*see* **Note 10**).

2. To quantify the relative knockdown effect induced by the postintroduction of siRNA (Fig. 20.4), use the respective intensity of each muscle on day 2 as an animal-specific internal control. Relative fluorescence on day x is the mean of the ratio of the fluorescence on day x to this "internal control."

3.3. Statistical Analysis

Treat four different animals for each condition and treat only one leg for each animal to avoid cross reaction between the successive treatments. Compare fluorescent level differences between the different conditions through an unpaired *t*-test using the prism software (version 4.02, Graphpad).

3.4. Conclusions

The described intramuscular injection of siRNA was safe for the animal. A muscle contraction was observed when the electric field was applied. No local burns, edema, or loss of limb functions were observed.

Intramuscular injection of GFP plasmid DNA induced GFP expression in the injected muscles of C57Bl/c mice for more than 23 days, peaking at day 11.

Co-delivery of siRNA against GFP completely silenced GFP expression during the first 11 days and partially silenced the GFP expression thereafter but remained statistically significant. The unrelated siRNA 9P76 siRNA could not silence GFP expression at the same experimental condition. A similar trend was observed in 8-month old Balb/c mice. Delaying injection of siRNA by 48 h resulted in a weaker silencing activity in young mice.

Acknowledgments This work was supported by grants from the CNRS CEA "Imagerie du petit animal" program, the Region Midi Pyrenees (Therapie génique et cellulaire), and the Association française contre les Myopathies.

4. Notes

1. The cream should be used two days before the fluorescence imaging because some components fluoresce under blue excitation. This cream should be used carefully, as it may cause some irritations in the skin of the leg. Rinse the cream with lots of water.
2. Square wave pulse generators are needed. As the tissue conductance is affected by the field-induced cell membrane permeabilization, the time constant of a capacitor discharge pulse generator is changing during the pulse. This leads to loss in control in the duration of the effective pulse.
3. Carefully clean the surface of the electrodes at the end of the experiments to avoid rusting due to the electrochemical reactions associated with the pulses.
4. Isoflurane inhalation is safe; mice recover very fast after the electrical treatment. It can be used every day for in vivo imaging, with no pain for the observed animal.
5. Conductive paste is very important to ensure a good electrical contact with the skin. One should pay attention that the paste is not continuous between the two electrodes as the field will pass through the paste and not through the muscle.
6. A too rapid injection could give false-positive muscle fibers. This injection needle must be parallel to the fibers.
7. The pulse delivery must be controlled on line either on an oscilloscope or after digitization on a laptop. Because of the high current, one may observe that the voltage is slightly decreasing at the end of the millisecond pulse.
8. Because of power limitation of the pulse generator in most cases, it is difficult to work a train frequency larger than 1 Hz. This is a technical reason. There is no evidence that it is the optimized setting from a biophysical point of view.
9. One person should be responsible for holding the leg of the mouse in the same position under the stereomicroscope to avoid lack of reproducibility in the exposition to the excitation beam.
10. Color fluorescence imaging is needed when the tissue autofluorescence is high. Discrimination between the yellowish autofluorescence and the green emission of GFP is obtained.

References

1. Fire, A., Xu S., Montgomery, M.K., Kostas, S.A., Driver, S.E., and Mello, C.C. (1998) Potent and specific genetic interference by double-stranded RNA in Caenorhabditis elegans. *Nature*. **391**, 806–811.

2. McManus, M.T. and Sharp, P.A. (2002) Gene silencing in mammals by small interfering RNAs. *Nat. Rev. Genet.* **3**, 737–747.

3. Sioud, M. (2004) Therapeutic siRNAs. *Trends Pharmacol. Sci.* **25**, 22–28.

4. Zamore, P.D. and Aronin, N. (2003) siRNAs knock down hepatitis. *Nat. Med.* **9**, 266–267.

5. Herweijer, H. and Wolff, J.A. (2003) Progress and prospects: naked DNA gene transfer and therapy. *Gene Ther.* **10**, 453–458.

6. Mir, L.M., Glass, L.F., Sersa, G., et al. (1998) Effective treatment of cutaneous and subcutaneous malignant tumours by electrochemotherapy. *Br. J. Cancer.* **77**, 2336–2342.

7. Rols, M.P., Delteil, C., Golzio, M., Dumond, P., Cros, S., and Teissie, J. (1998) In vivo electrically mediated protein and gene transfer in murine melanoma. *Nat. Biotechnol.* **16**, 168–171.

8. Potts, R.O. and Chizmadzhev, Y.A. (1998) Opening doors for exogenous agents. *Nat. Biotechnol.* **16**, 135.

9. Heller, R., Jaraszeski, M., Atkin, A., et al. (1996) In vivo gene electroinjection and expression in rat liver. *FEBS Lett.* **389**, 225–228.

10. Glasspool-Malone, J., Somian, S., Drabick, J.J., and Malone, R.W. (2000) Efficient nonviral cutaneous transfection. *Mol. Ther.* **2**, 140–146.

11. Mathiesen, I. (1999) Electropermeabilization of skeletal muscle enhances gene transfer in vivo. *Gene Ther.* **6**, 508–514.

12. Aihara, H. and Miyazaki, J.-I. (1998) Gene transfer into muscle by electroporation in vivo. *Nat. Biotechnol.* **16**, 867–870.

13. Harrison, R.L., Byrne, B.J., and Tung, L. (1998) Electroporation-mediated gene transfer in cardiac tissue. *FEBS Lett.* **435**, 1–5.

14. Yang, M., Baranov, E., Li, X.M., et al. (2001) Whole-body and intravital optical imaging of angiogenesis in orthotopically implanted tumors. *Proc. Natl. Acad. Sci. USA.* **98**, 2616–2621.

15. Caplen, N.J., Parrish, S., Imani, F., Fire, A., and Morgan, R.A. (2001) Specific inhibition of gene expression by small double-stranded RNAs in invertebrate and vertebrate systems. *Proc. Natl. Acad. Sci. USA.* **98**, 9742–9747.

Chapter 21
Electroporation of Adult Zebrafish

N. Madhusudhana Rao, K. Murali Rambabu, and S. Harinarayana Rao

Abstract We generated transient transgenic zebrafish by applying electrical pulses subsequent to injection of DNA into muscle tissue of 3–6-month old adult zebrafish. Electroporation parameters, such as number of pulses, voltage, and amount of plasmid DNA, were optimized and found that 6 pulses of 40 V/cm at 15 µg/fish increased the luciferase expression by 10-fold compared with those in controls. By measuring the expression of luciferase, in vivo by electroporation in adult zebrafish and in vitro using fish cell line (*Xiphophorus xiphidium* A2 cells), the strength of three promoters (CMV, human EF-1α, and Xenopus EF-1α) was compared. Subsequent to electroporation after injecting DNA in the mid region of zebrafish, expression of green fluorescent protein was found far away from the site of injection in the head and the tail sections. Thus, electroporation in adult zebrafish provides a rapid way of testing the behavior of gene sequences in the whole organism.

Keywords: zebrafish, electroporation, DNA injection, GFP, luciferase

1. Introduction

Application of electrical pulses to cells or tissue causes the formation of transient pores in the cellular membranes, leading to increased cell permeability for polar solutes and electrophoresis of the DNA into the cell (1, 2).This phenomenon was exploited to deliver nucleic acids into the cells, resulting in a tool to investigate regulatory properties of gene sequences and also to alleviate disease state by inhibiting "gain-of-function" mutations or correct "loss-of-function" mutations (3). Electrotransfer of DNA into muscle tissue has become a very popular method of gene delivery because of easy access of the muscle tissue, long life span of the muscle cell, abundant blood supply, and its suitability for the production of proteins as systemic therapeutic agents (2).

Zebrafish has proven to be a useful model organism to understand vertebrate developmental biology because of the following traits: easy husbandry, large supply

S. Li (ed.), *Electroporation Protocols*: *Preclinical and Clinical Gene Medicine*.
From *Methods in Molecular Biology, Vol. 423.*

of eggs, and transparent early stages of development (4). The small size of the zebrafish allows the use of forceps electrodes to apply pulses to bring about in vivo gene transfer. Optimization of electroporation parameters, mortality, comparison of promoter strengths, and tissue distribution of expression in zebrafish were investigated.

2. Materials

1. Cell line: A2 cell line (from late embryonic tissue of *Xiphophorus xiphidium* was a kind gift from Dr. N. Sivakumar of University of Hyderabad). Morphologically, the A2 cells appear cuboidal and are of epithelial origin. Cells were cultured in Dulbecco Modified Eagle's Medium supplemented with 15% fetal bovine serum, penicillin (50 μg/mL), streptomycin (60 μg/mL), and kanamycin (100 μg/mL).

2. Molecular Biology: pCMV-Luc was a gift from Dr. Robert Debs, University of California at San Diego. Dr. Suresh Kumar of National University of Singapore provided plasmids pESG and pBOS-H2b/GFP. Restriction enzymes Hind III, Xba I, EcoR1, and Not1 were procured from New England Biolabs.

3. Culturing zebrafish: Original stock of zebrafish was obtained from a licensed supplier. Equipment for maintaining the tanks, aerators, thermostats, fish feed (earth worms), and reproduction tanks were either purchased or fabricated locally.

4. DNA injection and electroporation: Borosilicate capillaries with Kwik-Fil feature (OD/ID, mm – 1.0/0.75, WPI, UK), needle puller (Narasighe model PC-10), tweezer type electrodes (Tweezertrodes, 0.7 cm wide and 11.5 cm long, Model 520, BTX), and electroporator (Electrosquare porator ECM 830, BTX, San Diego, CA)

5. Histology and fluorescence imaging: Locally available high quality glass slides are coated with gelatin (0.5% w/v in double distilled water (DDW)). Photographs are taken using Contax 16Mt camera attached to a fluorescence microscope (Zeiss Axiovision).

6. Tissue lysis buffer: 8.25 mM Tris-H_3PO_4, pH 7, 2 mM DTT, 2 mM 1,2-diamino-cyclohexane N,N,N',N'-tetracetic acid, 10% glycerol, and 1% Triton X-100.

7. Cell lysis buffer: 250 mM Tris-HCl, pH 8.0, containing 0.5% NP40.

8. Culture medium: Dulbecco's Modified Eagle's medium containing 15% FBS, penicillin (50 μg/mL), streptomycin (60 μg/mL), and kanamycin (100 μg/mL).

3. Methods

The involvement of a large portion of the body in electroporation for gene delivery, as described here for zebrafish, has not been attempted earlier. The application of electroporation of zebrafish eggs for transgene production has been established (5). Transgene expression in adult zebrafish has been demonstrated upon muscle injection (6, 7) and by electroporating the zebrafish fins (8). A portion of the muscle,

usually the thigh muscle in mice, was used for electroporation (3). Since electroporation is known to cause tissue damage due to leakage of cell contents, careful optimization of the electroporation parameters is required. The objective of optimization is to achieve maximum gene expression with the least toxicity/death to the zebrafish. Compared with naked DNA injection, the application of electrical pulses at the site of injection enhances transgene expression several fold. Since electrical pulses are applied across the body, involving several organs of the fish, it is important to assess the extent of expression and also sensitivity of various tissues present between the electrodes.

The efficiency of transgene expression after electroporation depends on the intensity, shape, duration, and frequency of the pulse(s). In several contexts, pulse sequences such as high voltage followed by low voltage pulses were shown to be efficient. Protocols for transient expressions of transgenes are important to understand the roles of gene sequences. In vitro data on putative roles of gene sequences often needs substantiation of in vivo situations. The several fold enhancement of transgene expression upon electroporation in the adult organisms allows unambiguous assessment of the behavior of gene sequences.

3.1. General Methods for Culturing Zebrafish

1. Two months old zebrafish were purchased from the local licensed fish supplier and were kept in observation tanks for one week before transferring them to experimental tanks. This period also serves for acclimatization of fish to the more controlled conditions of the laboratory.
2. The fish tanks are of 40-L capacity [(2.5 L × 1 W × 1.5 H) m^3] and filled with tap water. The tanks are housed in rooms maintained at 25°C by thermostat-controlled heaters. Add commercial sea salts and minerals to the stored water (50 mg of Instant Ocean per liter of water). Allow chlorine in the water to evaporate by exposing the water to air for couple of days after the addition of dechlorination drops (*see* **Notes 1–5**). Usually 50–75 fish can be maintained in each tank. Remove two-thirds of water every alternate day. This helps the removal of debris. The storage water is maintained free from contaminants by the addition of antifungal solution (Methylene blue). Change the tanks every week and clean with potassium permanganate solution. Feed the fish twice a day with live earthworms. Commercial fish pellets are also suitable. Discard diseased fish, which are found occasionally because of fungal infections. Early stages of infection in fish can be treated with RIDAL, an antifungal solution, for 15 min and then transferred into main tanks.

3.2. Plasmid Preparation

1. The choice of promoters is based on the need to compare a strong promoter (CMV) with other eukaryotic promoters. Three plasmids were constructed with

each containing either CMV (pCMV-Luc), human elongation factor (EF 1α promoter, pBOS-Luc), or Xenopus EF-1α (pESG-Luc) promoter, with luciferase as the reporter gene. The choice of promoter and plasmid backbone can be replaced depending on the experiment.

2. Plasmids pCMV/GFP, pCMV/Luc, pESG/Luc, and pBOS/Luc were isolated by CsCl density gradient method. Digest the fragment from pGL-3/Luc (Promege) with Hind III and XbaI to obtain the luciferase gene (9).

3. Double digest pESGFP plasmid with Hind III and Xba I restriction enzymes and ligate the purified backbone with the luciferase insert. We have constructed the pBOS/Luc in a similar fashion except that we have used EcoRI and NotI enzymes for taking the luciferase and inserting in the pBOS to generate pBOS/Luc. Transform the plasmids in DH5α host and purify the plasmids using Qiagen plasmid purification kits. Subsequent protocols such as gel electrophoreses, gel extraction, plasmid purifications, etc. were carried out as per standard procedures (7).

3.3. Microinjection Needle Preparation

1. Several methods of injecting the DNA solution into the fish muscle were carefully tested. Because of the small size and delicate nature of the fish, we examined metal needles (36GA), drawn from capillaries manufactured by several companies for efficiency of liquid delivery (*see* **Note 6**). The efficiency of injection can be tested by injecting a deeply colored solution (any blue dye, e.g. methylene blue and blue dextran) and monitoring the leakage (*see* **Note 7**). Metal needles were not suitable since the length of the bevel causes leakage of contents.

2. After several trials, we found glass needles drawn from borosilicate capillaries $(1.0 \times 90\,mm^2$ size) having the Kwik-fil feature more reliable. Draw the needles using a needle puller (Narishige Model PC-10). Needles attached to a rubber bulb with a rubber tube can be used to draw and expel the liquid (Fig. 21.1) (*see* **Note 8**). Several needles can be pulled and calibrated by drawing a concentrated colored solution to a mark and estimating the volume by measuring the adsorption of the liquid in a spectrophotometer and comparing with the standard graph prepared using commercial microsyringes.

3.4. Injection of DNA into the Fish

1. Use 3–5-month old zebrafish for electroporation.
2. Hold the fish gently between the fingers or in the palm as shown in Fig. 21.1.
3. Clean the injection site with 70% ethanol. Preload the needles with the DNA solution (5–20 μL of pCMV-luc) and keep them ready for injection. Once the fish is immobile, the needle can be inserted at about a 45° angle to a depth of 2–3 mm in the mid region of the fish. Carefully expel the solution from the needle by pressing

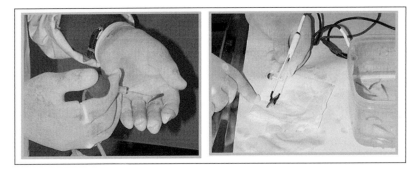

Fig. 21.1 Picture showing the handling of zebrafish for injection (*left*) and electroporation

the bulb gently. Withdraw the needle at a slightly different angle after waiting for a few seconds to allow the absorption of the liquid (*see* **Notes 9–11**).

3.5. Electroporation

1. Immediately after the injection, place the fish in a narrow wooden groove padded with cotton. The groove is such that the fish would be held immobile and at the same time the sides of the fish are accessible for placing the electrodes (*see* **Note 12**).
2. Deliver the electric pulses with the set parameters to the fish (*see* **Note 13**).
3. Immediately place the fish in the water containing gentamycin (50 μg/mL) to prevent any bacterial infections.
4. After electroporation, keep the fish in tanks at 28°C and maintain the treated ones like normal fish, fed daily twice with worm pellet (*see* **Note 14**).
5. Illustration of highest level of transient gene expression at 6 pulses of 40 V/cm and 15 μg of DNA (Fig. 21.2) (*see* **Note 15**).

3.6. Estimation of Luciferase Activity

1. Monitor the fish for morbidity and death for the next 48 h. Then, dissect the fish with a scalpel to separate the head, midsection, and tail region. Freeze the tissue immediately in liquid nitrogen.
2. Add 1 mL of tissue lysis buffer to the dried tissue and homogenize using a small Teflon coated homogenizer. Remove the debris by brief centrifugation (13,000 rpm at 4°C for 5 min). Monitor luciferase activity as per the instructions provided in the kit supplied by Promega. Record the light counts using a luminometer (LUMAC Biocounter M2000). Dispense 50 μL of the Luciferase

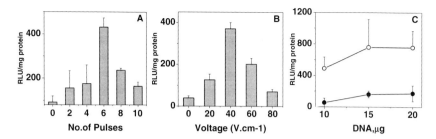

Fig. 21.2 Influence of various parameters of electroporation on reporter gene activities. Electrical pulses were applied following intramuscular injection of plasmid DNA. **A.** Pulse number; **B.** voltage strength; **C.** amount of plasmid DNA present in fixed volume with (open circles) and without (closed circles) electroporation. Relative luminescence units were normalized for the amount of protein. Each data point is an average of values obtained from 6–8 fish and each experiment was repeated three times. All values of reporter gene activities on electroporation were significantly ($p < 0.01$) different from the control values (with permission from BioMedCentral)

assay reagent into luminometer tubes or microcentrifuge tubes (1.5 mL). Initiate the reading by adding 5–10 µL of the lysate. The luminometer was programmed to measure the counts after a 3-s delay followed by a 15-s measurement. Use 5 µL of the cell lysate for protein estimation by modified Lowry's method.

3.7. Protein Assay

1. The protein content in the lysate was estimated using the Bio-Rad Protein Assay kit for normalizing the luciferase activity. Prepare the assay reagent by diluting 1 volume of the dye stock with 4 volumes of DDW. The solution should appear brown and have a pH of 1.1. It is stable for weeks in a dark bottle at 4°C. Protein standards supplied by the manufacturer were used.
2. Make up the volume of 2 µL of the lysate to 20 µL with DDW. To this solution, add 1 mL of Biorad protein assay solution. Mix and incubate at room temperature for 5 min. Absorbance of the samples is taken within an hour. Measure the absorbance at 595 nm and calculate the protein value from the standard curve prepared on the same day (*see* **Note 16**).

3.8. Evaluation of Promoter Strength In Vivo (Zebrafish) and In Vitro (A2 Cell Lines)

1. The utility of in vivo electroporation would be beneficial if transgene expressions obtained in vivo match with in vitro results. For this purpose we chose

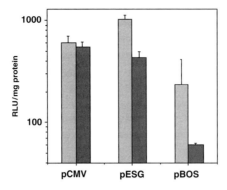

Fig. 21.3 Comparison of strengths of three promoters in vivo (zebrafish, light gray) and in vitro (in A2 cell lines, dark gray). The relative luminometer counts obtained with A2 cell lines were divided by 100 for comparison with the in vivo values obtained with zebrafish (with permission from BioMedCentral)

      three promoters viz. CMV (pCMV-Luc), human EF-1α (pBOS-Luc), or Xenopus EF-1α (pESG-Luc) promoter. In all these constructs, luciferase was the reporter gene (*see* **Note 17**). The promoter strength was tested in A2 cells, a cell line derived from late embryonic tissue from *Xiphophorus xiphidium* and also in adult zebrafish by the method described earlier.

2. Maintain A2 cells at 28°C in DMEM medium in a 5% CO_2 environment. Propagate by transferring into fresh medium every 48 h. Plate the cells at a density of 16,000–20,000 cells per well in a 96-well plate one day prior to transfection. This would result in ~70% confluency.

3. Lipofectamine, an efficient cationic liposomal preparation, was used to complex the DNA and deliver the DNA inside the cells. To form a transfection complex, incubate Lipofectamine (1–3 µL) and plasmid (0.3 µg/well) together for 30 min in serum-free DMEM. Add these complexes to cells for about 3 h. Replace the medium with DMEM medium containing 10% serum.

4. Incubate cells for 24 h before estimating the reporter gene activity. After 24 h, remove the medium, wash cells with PBS, and lyse with 50 µL lysis buffer for 10 min at room temperature. Use 5 µL of cell lysate for protein estimation by modified Lowry's method.

5. Assay reporter gene activity with these three plasmids in A2 cell line at various charge ratios of lipid to DNA. With all three constructs, the maximum reporter gene expression was obtained at a charge ratio (+/−) of 1:1 and reporter gene activity decreased either on increase or decrease of charge ratio from 1:1.

6. Estimate luciferase activity in the tissue as described in sect. 3.6.

7. Figure 21.3 shows the data obtained with the expression of three different promoters of variable strength both in vivo (electroporation) and in vitro (by lipid-mediated transfection)

3.9. Expression of Green Fluorescent Protein (GFP) in Fish After Electroporation

1. Electroporate zebrafish by the protocol described earlier except that the plasmid in this study was pCMVGFP. Inject zebrafish with 10 µg of pCMVGFP and electroporate by applying 6 pulses at 40 V/cm.
2. Sacrifice the fish after 2 days. Tissues around the injection site, head, and tail regions can be taken. Inject in mid-dorsal region approximately at 60% of the total length from the head. The sections for the head region and tail region are taken at least 1 cm away on either side from the point of injection (Fig. 21.4). Snap freeze the tissues in liquid nitrogen and maintain at this temperature till cryosectioning.
3. Embed the frozen tissue in O.C.T. embedding medium (Tissue-Tek® Embedding Supplies; Sakura Finney) for cryosectioning. Carefully transfer sections of 7 µm onto gelatin (0.5%) coated slides. Store the slides at −20°C until examination. Both phase-contrast and fluorescence images (490 nm) of sections

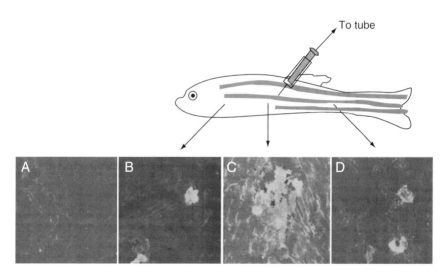

Fig. 21.4 Effect of electroporation on GFP expression in zebrafish. Fluorescence images from control (**A**) and electroporated samples (**B, C,** and **D**). **A** is an image taken from the tissue at the site of injection. **B, C,** and **D** represent the images obtained from head, midsection, and tail region, respectively. Fluorescence from tissue sections from head and tail of control fish is similar to **A** (hence not shown). At least 20 fish were electroporated in different days with pCMVGFP. We observe some variation in extent of GFP expression, but the expression was always maximum in the midregion. The presented data is from one fish (with permission from BioMedCentral)

are obtained using a Zeiss Axiovison fluorescent microscope with a camera (Contax 16MT).

4. The expression levels of GFP in head, midsection, and tail region are shown in Fig. 21.4. The sections are also stained using 4′,6-diamidino-2-phenylindole (DAPI), a simple nuclear staining fluorescent dye. Dilute DAPI (1 mg/mL) solution 1:1,000 with phosphate buffered saline and add to the section. Wash off the stain after 10 min with fresh PBS.

4. Notes

1. Maintenance of temperature at 28°C in the fish tanks is essential to achieve expected growth of zebrafish.
2. A light–dark cycle (L:D) of 12 h × 12 h was used in the room where fish tanks are maintained.
3. Dechlorination of the water is very important since the municipal water supply contains variable amounts of chlorine.
4. Tanks were filled with tap water 48 h before introducing the fish. Aged tap water is known to have reduced chlorine content.
5. Standard zebrafish husbandry practices are employed. Fish were fed twice a day. For more specific information, the following site has many useful tips: http://zfin.org/zf_info/zfbook/ zfbk.html.
6. Metal needles of any gauge were not found suitable since the extent of tissue puncture is unacceptable.
7. Careful monitoring of the leakage of solution after withdrawing the needle is very critical for complete delivery of the DNA solutions. This step, in our opinion, is the most important one for reproducible results. Needles are not reused. Needles are drawn to a very fine tapering end but, at the same time, firm enough to withstand the resistance from the tissue.
8. The design of tubing and the attachment to connect the needle can be replaced with something more convenient. We normally held the fish in our left hand and the rubber bulb in the right hand. To keep the right hand free sometimes we put the tube in the mouth and blew air to expel the liquid.
9. DNA solutions of more than 20 μL were found leaky with our needles. We often found withdrawing of the needle crucial since the DNA solution might leak back along the incision made by the needle. We have practiced the injection and withdrawal by using a colored solution.
10. The zebrafish would be under stress once taken out of water, and hence the DNA injection and application of electrical pulses needs to be completed as rapidly as possible. We completed our operation in less than a minute.
11. After electroporation, the fish were found to be sluggish in their mobility for a day and then recovered their normal mobility.
12. Restraining the zebrafish at the time of applying pulses is essential.
13. Voltages more than 80 V/cm were found to increase death of the fish substantially. The mortality was observed to be 6, 15, 25, and 30% at 20, 40, 60, and 80 V/cm respectively.
14. Expression of a transgene was observed after one week of electroporation whenever tested.
15. Electroporation conditions of 6 pulses of 40 V/cm were also found to be efficient in another fish, Indian carp.
16. Performing control with the cell lysis buffers in protein estimation is essential since lysis buffer gives significant blue color, which must be corrected.
17. For isolation of plasmids, we have used plasmid purification and gel extraction kits marketed by Qiagen. The protocols are followed as described in the manufacturer's brochure.

References

1. Bigey, P., Bureau, M.F., and Scherman, D. (2002) In vivo plasmid DNA electrotransfer. *Curr. Opin. Biotechnol.* **13,** 443–447.
2. Herweijer, H. and Wolff, J.A. (2003) Progress and prospects: naked DNA gene transfer and therapy**.** *Gene Ther.* **10,** 453–458.
3. Fattori, E., La Monica, N., Ciliberto, G., and Toniatti, C. (2002) Electro-gene-transfer: a new approach for muscle gene delivery**.** *Somat. Cell Mol. Genet.* **27,** 75–83.
4. Neumann, C.J. (2002) Vertebrate development: a view from the zebrafish. *Semin. Cell Dev. Biol.* **13,** 469.
5. Powers, D.A., Hereford, L., Cole, T., et al. (1992) Electroporation: a method for transferring genes into the gametes of zebrafish (*Brachydanio rerio*), channel catfish (*Ictalurus punctatus*), and common carp (*Cyprinus carpio*). *Mol. Mar. Biol. Biotechnol.* **1,** 301–308.
6. Sudha, P.M., Low, S., Kwang, J., and Gong, Z. (2001) Multiple tissue transformation in adult zebrafish by gene gun bombardment and muscular injection of naked DNA. *Mar. Biotechnol. (NY)* **3,** 119–125.
7. Tan, J.H. and Chan, W.K. (1997) Efficient gene transfer into zebrafish skeletal muscle by intramuscular injection of plasmid DNA**.** *Mol. Mar. Biol. Biotechnol.* **6,** 98–109.
8. Tawk, M., Tuil, D., Torrente, Y., Vriz, S., and Paulin, D. (2002) High-efficiency gene transfer into adult fish: a new tool to study fin regeneration. *Genesis.* **32,** 27–31.
9. Sambrook, J., Fritsch, E.F., and Maniatis, T. (eds.) (1989) *Molecular cloning. A laboratory manual*, part II. Cold Spring Harbor Laboratory Press, New York, NY.

Part IV
Treatment of Cancer via Electroporation Gene Therapy

Chapter 22
Flow Electroporation with Pulsed Electric Fields for Purging Tumor Cells

Abie Craiu and David Scadden

Abstract Electroporation has been used in biological laboratories for many years to transiently porate cell membranes and permit plasmid or protein transfection. It has been shown that the application of pulsed electric fields (PEFs) of defined strength will kill off larger cells and select for viable small cells, in samples containing heterogeneous cells. This permits the selective killing of several blood and bone marrow–resident tumor cells. PEF technology is being applied to tumor purging of progenitor-cell transfusions, in support of high-dose chemotherapy, for the treatment of cancers such as lymphoma and multiple myeloma. Autologous stem-cell transplantation, in the setting of hematologic malignancies such as lymphoma, improves disease-free survival if the graft has undergone tumor purging. Progenitor cells are preserved or enriched. To overcome issues of electrical resistance, purging fidelity, and large sample volume, a flowing chamber PEF apparatus was designed and constructed for large-scale purging of clinical quantities of progenitor-cell transfusions. The specifics of this technique are described here. Treatment of greater than 10^9 cells is achieved in 30 min, under optimized flow conditions designed to overcome surface area or resistance issues and to optimize exposure of cells to electric fields. Efficient, large volume tumor purging of greater than 3 logs, for mixtures of tumor cells and mononuclear cells, is routinely achieved under defined conditions.

Keywords: pulsed electric fields, electroporation, tumor cells, tumor purging, progenitor cells, flowing system, lymphoma, myeloma

1. Introduction

The cellular lipid membrane is nonconducting, whereas the cytosol conducts electricity. When an external electric field is applied to cells in solution, charge separation across the cell membrane results at the poles of the cells aligned with the electric field. The voltage developed across roughly spherical cells scales proportionally with cell diameter and electric field strength (Fig. 22.1) (1–6). When the potential exceeds a critical value (V_{mc}) permanent membrane pore formation occurs, resulting

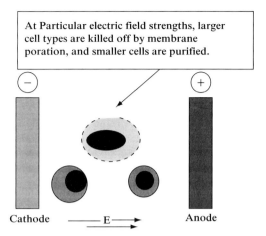

At Particular electric field strengths, larger cell types are killed off by membrane poration, and smaller cells are purified.

Cathode ——— E ——→ Anode

Fig. 22.1 PEF size selection principle. For a fixed externally imposed electric field, cells larger than a critical size will be porated and, thereby, killed. The remaining viable cells will include stem cells and progenitor cells

in loss of semipermeable function and rapid cell death (4, 5) (*see* **Note 1**). Under defined electric field strengths and conditions, this attribute permits selective eradication of larger cells in a mixture of cells. For instance, it was shown that tumor cells are selectively killed with pulsed electric field (PEF) strengths between 1.35 and 1.40 kV/cm in the PEF apparatus (7–8). At the same electric field strength, stem cells and progenitor cells, which are generally quiescent and among the smallest cells in the hematopoietic system, are preserved or enriched.

The intended clinical application for PEF described here is for tumor purging in association with autologous progenitor-cell transfusions. Hematologic cancers, such as lymphomas, myeloma, and leukemia, are often treated with high-dose chemotherapy to kill rapidly proliferating cancer cells. An essential component of supportive care following treatment is the replenishment of hematopoietic stem cells using mobilized peripheral blood cell or bone marrow mononuclear cell transfusions (9–12). Autologous transplants are often contaminated with cancer cells, which are a potential cause of tumor relapse following intensive chemotherapy (12–14). The effective purging of cancer cells from transplant infusions is now recognized as an important goal and is currently being explored using a number of technologies.

Clinically mobilized peripheral blood or bone marrow specimens withdrawn for transplant purposes generally contain more than 10^9 cells. Even at a high cell density, an electroporation chamber of larger than 10-mL volume would be required for efficient purging. However, because of the relationship between cell buffer resistivity, desired electric field strength, and chamber dimensions, it would not be possible to devise a single static electroporation chamber to efficiently purge 100 mL of dense cells (8). Therefore, there is a practical limitation to single electrode

chamber volume, which limits the volume of cells to be treated (*see* **Note 2**). The design and construction of a flowing PEF chamber effectively overrides the issue of chamber dimension limitations by permitting continual flow of cells through a chamber of fixed dimensions appropriate for generating an electric field of desired strength and attributes (8).

In addition, there are regions within the edges of electroporation chambers in which cells are not exposed to electric fields, and gravitational cell settling also affects purging efficacy. For this reason, SRL designed a curved flow channel for flowing PEF, in which nonlaminar flow mixes cells in transit permitting optimal exposure of all cells passing through the flowing PEF chamber to PEFs and tumor-cell inactivation. Whereas these considerations are of some importance in the field of maximizing electroporation potential, they are of crucial importance for maximizing the tumor purging potential of PEF. Accurate tumor purging of at least 3 logs is desired since it is possible that just one or very few unpurged cancer cells could be sufficient to cause cancer relapse if introduced into a patient during transfusion.

The materials section 2 and methods section 3 are based on the acquisition of the Cytopulse pulse driver described later, or its equivalent. These sections also assume pulse driver and oscilloscope set up and connections will be performed according to manufacturer's instructions. In addition, SRL's PEF flowing chamber unit or its equivalent is required for flow purging according to the specific methods mentioned here.

2. Materials

2.1. Cell Buffers and Reagents

1. Phosphate-buffered saline (PBS).
2. PEF pulsing buffer: 90% isotonic (282 mM) dextrose, 10% PBS, pH 7.2, sterile filtered.
3. RPMI-1640, supplemented with pennicillin, streptomycin and glutamine and 2% fetal bovine serum.
4. Trypan blue (0.4%).
5. Ficoll-hypaque (Amersham Pharmacia, Piscataway, NJ).

2.2. Electroporation System

A closed-system, sterile device incorporating the various components described here has been designed at Science Research laboratory Inc. (SRL, Somerville, MA). The components described here can be used to prepare a breadboard device with a Cytopulse electric pulse driver system and PEF chambers that can be fabricated and purchased from SRL. Ideally, components 1–6 can be assembled and ready at a workstation.

1. An electric pulse driver system (Cytopulse P-2000, Columbia, MD), capable of delivering bipolar, rectangular-shaped voltage pulses to loads greater than 8 Ω, at electric field strengths up to 2 kV and repetition rates as high as 10 Hz. The electric field pulse length is variable over the range 4–40 μs. Other flow-through electroporation systems, which do not employ PEF, have not been assessed for purging.
2. Syringe pump (standard Pump 22 infusion) (Harvard Apparatus, Holliston, MA).
3. Oscilloscope (Tektronix, Beaverton, OR).
4. Circulating water bath, for maintaining PEF electrode temperature, e.g. Brinkmann Instruments MGW Lauda RC6 (Westbury, NY).
5. Control computer.
6. Cytopulse operation control program CD (Cytopulse, Rockville, MD).
7. Four-way valve for purging trapped air from the upstream syringe pump lines (McMaster Carr, New Brunswick, NJ).
8. 1/8″ internal diameter silicone tubing, male and female luer locks (McMaster Carr, New Brunswick, NJ).
9. PEF flow chamber: Fabrication and ordering from SRL (Somerville, MA); composed of water-cooled copper plates, Iso-67 graphite plates, and polycarbonate flow plate.

2.3. Cell Staining and Analysis

1. Nuclear exclusion dyes: 7-aminoactinomycin-D (7-AAD) and To-Pro-3™ or 4′,6-diamidino-2-phenylindole, dihydrochloride (DAP-I).
2. Fluorochrome-conjugated monoclonal antibodies: anti-CD20, anti-CD5, anti-CD10, anti-CD38, anti-CD138, anti-CD45, and anti-CD34, conjugated to fluorochrome fluorescein isothiocyanate (FITC), phycoerythrin (PE), allophycocyanin, or peridinin chlorophyll protein (PerCP).
3. Flow cytometer.

3. Methods

3.1. Preparation of Tumor-Contaminated Cell Specimens for Purging

1. The primary cell specimens to be treated by this method are derived from mobilized peripheral blood or bone marrow of patients prior to treatment with high-dose chemotherapy. Rapidly thaw frozen specimens in a 37°C water bath and immediately transfer contents into twice the volume RPMI in a tissue culture hood.
2. Spin down cells in a tabletop swingbucket centrifuge at 150 g. Wash and resuspend pellets. Repeat the step twice.

3. Resuspend at greater than 10^7 cells/mL in RPMI/2% FBS in 15- or 50-mL tubes. Layer over Ficoll-hypaque and spin in desktop centrifuge at $200\,g$ (no brake).
4. Collect cells from interface, wash 3 times, and spin down 3 times with PBS.
5. After PEF equipment is ready (see next section), resuspend cells at 10^7/mL in pulsing buffer and begin PEF purging.

3.2. Preparation of Flowing PEF Chamber and Apparatus

See Fig. 22.2 for schematic of procedure (provided by Dr. Henry Eppich) and equipment and Fig. 22.3 for photograph of flow plate. The PEF treatment chamber and flow plates are fabricated at SRL.

1. The leads from the pulse driver are inserted, clipped, or sealed onto the anode and cathode of the flow chamber. Follow Cytopulse manufacturer's instructions. The oscilloscope should also be set up to detect actual voltage and current, according to the manufacturer's instructions.
2. The Cytopulse control program should be downloaded onto the computer at the workstation, and the USB port connection from the pulse driver to the computer should be attached. The Cytopulse program can be used to enter electric field strength, pulse length, pulse frequency, and duration of pulses during operation.
3. The flow chamber and flow plates should be autoclaved for sterilization and assembled according to manufacturer's instructions. The inlet and outlet points of the flow plate will be exposed.
4. Switch on cooling system for electroporation flow plates (mediated through water-cooled channels in electrodes). All procedures hereon are performed under sterile conditions.
5. Cut and attach $4 \times 20\,cm^2$ lengths of silicone tubing to the four-way valve using luer locks. The free ends of the four lengths of tubing will later be attached using luer locks, as follows:

   (a) to syringe A (buffer alone) in syringe pump,
   (b) to the flow plate inlet, using an 18-gauge needle,
   (c) to syringe B (cells in buffer) in syringe pump,
   (d) open (air bleed).

6. Fill a 60 mL syringe (A) with PEF pulsing buffer and place it in syringe pump A. Attach it to tubing from four-way valve. Turn valve to permit flow through valve open positions (a) and (b) above.
7. Pump buffer through flow chamber at 7.5 mL/min. Wait till buffer reaches silicone tubing attached to tubing at exit end of flow plate. Switch off syringe pump.
8. Check safety prior to switching on and operating pulse driver (*see* **Note 3**). Follow safety instructions for Cytopulse equipment.
9. Start up the Cytopulse control program according to manufacturer's instructions. Enter values of 0.52 kV electric field strength (corresponding to 1.4 kV/cm) (*see* **Note 4**), 20 μs pulse length, and 0.3 s interpulse intervals. Optimal tumor purging

A

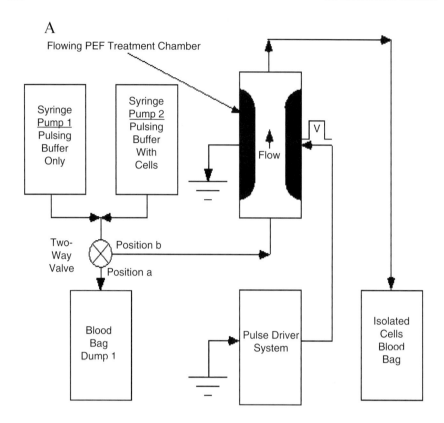

Fig. 22.2 A schematic of the flowing PEF process

Fig. 22.3 PEF polycarbonate flow-plate resting on electrode component. Entry and egress of cell suspensions is mediated through a hypodermic needle inserted through silicone rubber into the curved flow channel. The flow plate is sandwiched between two electrode plates during PEF

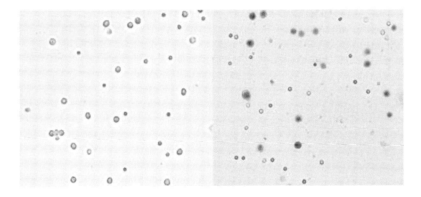

Fig. 22.4 Selective killing of primary myeloma cells by PEF. Primary bone marrow containing myeloma cells were PEF-treated at 1.4 kV/cm, and cells were analyzed immediately afterwards with trypan blue staining and microscopy. Untreated cells (*left*) show a variety of cells of different sizes, including intact larger cells. After PEF at 1.4 kV/cm (*right*) the large cells are dead and trypan blue positive. Smaller cells, including lymphocytes and progenitor cells, remain intact after PEF

of lymphoma and myeloma contaminants, with preservation of progenitor cells, was achieved at these values using the equipment described. Corresponding modifications would need to be implemented for PEF treatment using flow channels of different dimensions or for different buffers (*see* **Note 2**).

10. Test pulse in pulsing buffer. Begin pulsing in PEF pulsing buffer. The presence of PEF buffer in the chamber permits waveform monitoring on an oscilloscope prior to PEF-treating the specimen. The input voltage can therefore be adjusted up or down over a few pulsing sessions till the waveform is indicated at 0.48 kV (*see* **Notes 2 and 4**).

11. The oscilloscope also permits monitoring of waveform shape, which should be rectangular, with very little tailing off (*see* **Note 5**). Always switch off the control program pulses and pulse driver when not in use.

3.3. PEF Purging Procedure

1. Tubing from the flow plate outlet is attached into a sterile collection bottle containing 50 mL RPMI.
2. Syringe B is filled with the tumor-containing cell suspension. It is attached by a luer lock to the tubing/four-way valve and inserted in the syringe pump.
3. Switch the valve to open positions (c) and (d) (see sect. 3.2.5), and pump at 7.5 mL/min for a few seconds until air is bled. Switch off pump.
4. Switch valve to open positions (b) and (c).
5. Enter Pulse on pulse driver control program. Then switch on syringe pump again at 4 mL/min. Monitor waveform during run. 50 mL will be purged in ~12.5 min.

6. When the cells have been plunged through the syringe, after about 12 min, switch the four-way valve to positions (a) and (b). Pump syringe A (containing pulsing buffer again) for 3 min to chase and PEF purge remaining cells in chamber.
7. Switch off pulse driver. Collect container with PEF-treated cells.

3.4. Handling and Analysis of Tumor-Purged Cell Preparations After PEF

1. Spin down cells in centrifuge immediately after PEF treatment is complete. Wash twice and resuspend in RPMI. Purging of tumor cells can be confirmed by trypan blue staining and cell counts on a hemacytometer, using an inverted microscope (see Fig. 22.4).
2. Live cells can be further purified by standard procedures.
3. Flow cytometry can be applied for accurate enumeration of progenitor cells and tumor cells after PEF treatment (*see* **Note 6**).
4. Cells can be frozen in dextrose- or DMSO-containing buffer for future clinical use.

4. Notes

1. The voltage developed across the diameter of spherical cells is as follows: $V_c = 3\, d\, E/2$, where V_c is the voltage developed across the diameter of the cell, d is the diameter of the cell, and E is the strength of the imposed electric field. The critical electric field strength for inactivation of spherical cells larger than d_c is given by the following equation: $E_c = 4\, V_{mc}/3\, d_c$. Therefore, electric field strength is the key parameter for setting the demarcation size between viable and PEF inactivated cells. Total electric field exposure time, t_{ex}, which is the sum of the on-time of all electric field pulses applied to the suspension of cells, also determines the reduction in numbers of the PEF affected cells.
2. The formula for calculating anticipated chamber electrical resistance, in a pulsing buffer of known resistivity, is as follows: $R = \rho l/A$. Here, R is resistance (Ω), ρ is buffer resistivity (in Ωcm), l is the length of the electrode gap (distance between electrodes), and A is surface area in contact with cell suspension exposed to electric field (17 cm²). The electric pulse generator imposes a minimal resistance load of 6 Ω. Therefore, for physiologic buffers such as PBS, with ρ of ~50 Ωcm, $A/l < 6.2$ (because $A/l = \rho/R$) for resistance load > 8 Ω. Using a dextrose-based buffer (90% isotonic dextrose (282 mM)/10% PBS) with known resistivity of 500 Ωcm, A/l can be 62, permitting a 31-cm² surface area with electrode gap of 0.5 cm and a total volume of 15.5 cm³. Practical pulse drivers for use with this unit need only generate up to 2 kV for efficient pulsing.
3. Safety: The operating electric field strengths for these procedures are sufficient to cause severe injury or death to the user if accidentally shocked. Follow standard safety procedures when dealing with controlled electrical equipment. Always switch off and unplug the system when not in use. The unit is enclosed in nonconducting acrylic when pulsing.
4. The optimal electric field strength for killing of lymphoma and myeloma cells, with maximal preservation of progenitor cells was achieved at 1.4 kV/cm. Since the electrode gap for the current flow plate is 0.32 cm, the desired electric field strength to enter is actually 0.52 kV, with some further accounting for the droop due to buffer resistivity (final electric field strength of 0.48 kV).

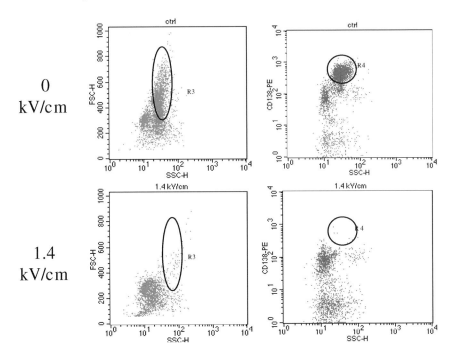

Fig. 22.5 PEF selectively purges primary myeloma cells from bone marrow. Flow cytometry analysis: After PEF treatment, cells were stained with anti-CD138-PerCP, anti-CD38-PE, anti-CD45-FITC, or anti-CD34-FITC (not shown), and the nuclear exclusion dye 7-AAD. Forward and side scatter profiles, demonstrating depletion of the population of large myeloma cells, by PEF are shown on the *left*. Cells gated as 7-AAD-negative, CD45 low, and CD38 high were analyzed for the presence of CD138++ high side scatter myeloma cells. The myeloma cells (CD138++) are noticeably absent in the dot plot on the right for cells pulsed at 1.4 kV/cm. The limit of detection based on input bone marrow cell number and percent myeloma cells within bone marrow specimens is ~3 logs (or 0.1% of starting number). Cells staining positive for the CD38 high CD45 low CD138 ++ phenotype constituted 2–7% of total bone marrow mononuclear cells from multiple myeloma samples analyzed. Small cells, including lymphocytes, are preserved

5. The rectangular pulses generated allow extremely efficient and accurate delivery of defined electric field pulses and minimize heat loss. These operating conditions are not possible with pulse drivers that deliver exponentially-decaying pulses.
6. Please see Fig. 22.5 for sample flow cytometry data demonstrating purging of primary myeloma cells by PEF. A synopsis of staining and flow methods is provided in the legend. A small aliquot of unpurged tumor cells can also serve as a control for staining. Cells are stained with antibodies defining unpurged progenitor cells (e.g. CD34), lymphocytes (e.g. CD3 and CD19), and tumor cells, including myeloma cells (generally CD138++ and CD38) or different lymphoma cell subtypes (e.g. CD5, CD10, CD15, or CD20). It is important to include at least one nuclear exclusion dye (7-AAD, To-Pro-3, or DAP-I) with each staining or analysis panel to distinguish live cells from dead cells for flow analysis. Staining and flow cytometry specifics are beyond the scope of this review.

References

1. Coulson, C.A. (ed.) (1951) *Electricity*. Oliver and Boyd, London.
2. Sale, A.J.H. and Hamilton, W.A. (1968) Effects of high electric fields on microorganisms: III. Lysis of erythrocytes and protoplasts. *Biochim. Biophys. Acta.* **163**, 37–43.
3. Sixou, S. and Teissie, J. (1990) Specific electropermeabilization of leucocytes in a blood sample and application to large volumes of cells. *Biochim. Biophys. Acta.* **1028**, 154–160.
4. Tsong, T.Y. (1996) Electrically stimulated membrane breakdown, In: Lynch, P.T., Davey, M.R. (eds). *Electrical manipulation of cells*. Chapman & Hall, NY, pp. 15–37.
5. Hulsheger, Potel, H.J. and Niemann. E.G. (1981) Killing of bacteria with electric pulses of high field strength. *Radiat. Environ. Biophys.* **20**, 53–65.
6. Phez, E., Faurie, C., Golzio, M., Teissie, J., and Rols, M.P. (2005) New insights in the visualization of membrane permeabilization and DNA/membrane interaction of cells submitted to electric pulses. *Biochim. Biophys. Acta.* **1724,** 248–254.
7. Eppich, H.M., Foxall, R., Gaynor, K., et al. (2000) Pulsed electric fields for selection of hematopoietic cells and depletion of tumor cell contaminants. *Nat. Biotech.* **18**, 882–887.
8. Craiu, A., Saito, Y., Limon, A., et al. (2005) Flowing cells through pulsed electric fields efficiently purges stem cell preparations of contaminating myeloma cells while preserving stem cell function. *Blood.* **105**, 2235–2238.
9. Melillo, L., Cascavilla, N., Lerma, E., Corsetti, M. T., and Carella, A. M. (2005) The significance of minimal residual disease in stem cell grafts and the role of purging: is it better to purge in vivo or in vitro? *Acta Haematol.* **114,** 206–213.
10. van Besien, K., Loberiza F.R., Jr., Bajorunaite, J,. Armitage, J.O., and Bashey, A. (2003) Comparison of autologous and allogeneic hematopoietic stem cell transplantation for follicular lymphoma. *Blood.* **102**, 3521–3529.
11. Bierman, P.J., Sweetenham, J.W., Loberiza F.R., Jr., Taghipour, G., Lazarus, H.M; The Lymphoma Working Committee of the International Bone Marrow Transplant Registry and the European Group for Blood and Marrow Transplantation. (2003) Syngeneic hematopoietic stem-cell transplantation for non-Hodgkin's lymphoma: a comparison with allogeneic and autologous transplantation. *J. Clin. Oncol.* **21**, 3744–3753.
12. Kumar, S., Lacy, M.Q., Dispenzieri, A., et al. (2004) High-dose therapy and autologous stem cell transplantation for multiple myeloma poorly responsive to initial therapy. *Bone Marrow Transplant.* **34**, 161–167.
13. Imai, Y., Chou, T., Tobinai, K., Tanosaki, R., Morishima, Y; CliniMACS Study Group (2005). Isolation and transplantation of highly purified autologous peripheral CD34 + progenitor cells: purging efficacy, hematopoietic reconstitution in non-Hodgkin's lymphoma (NHL): results of Japanese phase II study. *Bone Marrow Transplant.* **35**, 479–487.
14. Negrin, R.S., Atkinson, K., Leemhuis, T., Nhanania, E., and Juttner, C. (2000) Transplantation of highly purified CD34 + Thy-1 + hematopoietic stem cells in patients with metastatic breast cancer. *Biol. Blood Marrow Transplant.* **6**, 262–271.

Chapter 23
Delivery of DNA into Tumors

Shulin Li

Abstract Delivery of plasmid DNA encoding therapeutic genes into tumors is
one of the main applications of electroporation. This chapter summarizes various
investigators' electroporation parameters for intratumoral gene delivery. In addition
to electroporation parameters, injection volume is also critical for achieving a high
level of gene expression via electroporation. In this study, we attempt to provide
a strategy for determining the optimal injection volume for intratumoral injection
via electroporation. Unlike muscle tissues, the optimal volume for gene delivery
into tumors via electroporation may vary greatly based on the tumor size and the
electroporation parameters. More efforts in defining the optimal injection volume
should be made to further advance intratumoral electroporation gene therapy for
treating tumors.

Keywords: electroporation, reporter gene, intratumoral delivery, injection volume

1. Introduction

Electroporation-mediated gene delivery increases the level of gene expression by
1–3 logs compared with injection of naked plasmid DNA without electric pulses
(1–4). This increase is primarily dependent on the selected electroporation parameters
and the target tissues (5, 6). Regardless of the tissues, an array of electroporation
parameter sets have been claimed to achieve a high level of gene expression and a
significant therapeutic efficacy. The electroporation parameters used by different investigators for performing intratumoral gene delivery are summarized in Table 23.1. This
array of electric parameters can be roughly classified into low, medium, and high
voltage-based parameters. Low and medium voltages, at a range of 25–250 and 350–
500 V/cm, respectively, have to be used with long-duration pulses to achieve a
maximum level of gene expression (see Table 23.1). On the other hand, high voltages, at a range of 600–1600 V/cm, need short duration pulses to achieve a high
level of gene expression or therapeutic effect (see Table 23.1). Either combination
seems to be effective in gene delivery.

S. Li (ed.), *Electroporation Protocols: Preclinical and Clinical Gene Medicine.*
From *Methods in Molecular Biology, Vol. 423.*
© Humana Press 2008

Table 23.1 Three types of Electoporation settings for intratumoral gene delivery

Authors	Parameters	Year	References
Setting I. Low voltages combined with long pulse duration			
Goto T, et al.	66 V/cm, 8 pulses, 50 ms	2000	(10)
Kishida T, et al.	50 V/0.5 cm, 6 pulses, 100 ms	2001	(11)
Lohr F, et al.	100 V/0.6 cm, 6 pulses, 50 ms	2001	(12)
Shibata MA, et al.	100 V/0.5cm, 8 pulses, 20 ms	2002	(13)
Kishida T, et al.	125 V/cm, 6 pulses, 50 ms	2003	(14)
Setting II. Mid- to high voltages combined with mid-long pulse duration			
Li S. et al.	400-500 V/cm, 2 pulses, 20-25 ms	2002	(15)
Heller LC, et al	800 V/cm, 10 pulses, 5 ms	2002	(16)
Matsubara H, et al.	1000 V/cm, 8 pulses, 100 ms	2001	(17)
Setting III. Short pulses combined with mid- to high- voltages			
Nishi T, et al.	600 V/cm, 8 pulses, 95-99 μs	1996	(18)
Matsubara H, et al.	1000 V/cm, 8 pulses, 100 μs	2001	(19)
Mikata K, et al.	1000 V/cm, 8 pulses, 99 μs	2002	(20)
Niu G, et al.	1500 V/cm, 14 pulses, 100 μs	1999	(21)
Lucas ML, et al.	1500 V/cm, 6 pulses, 99 μs	2002	(22)

Here, we report another factor: The injection volume greatly affects electroporation-based gene delivery but was often ignored in exploring this technology. Depending on the tissue, injection volume increases the level of gene expression up to a hundred fold at optimal electroporation parameters. Tumor tissue is more sensitive to the injection volume than other tissues in electroporation-based gene delivery. In our model, low voltage-based electroporation yields a higher level of gene expression than high voltage-based gene delivery at a fixed injection volume. The optimal volume may vary between tumors with a diameter between 3 and 4 mm and greater than 1 cm. A larger volume is needed for the large tumors.

2. Materials

2.1. Plasmid DNA

1. The luciferase gene construct used for the in vivo study was obtained from Valentis, Inc. (Burlingame, CA) and contains DNA fragments encoding firefly

luciferase. The luciferase encoding gene is driven by a CMV promoter and terminated by an independent bovine growth hormone polyadenylation signal.

2. The plasmid DNA was manufactured using the QIAGEN Endo-Free Prep kit. Working solutions were prepared minutes before injection by diluting 10 μg DNA in the indicated volume containing 0.45% saline for each experiment (*see* **Note 1**).

3. Endofree plasmid Maxi kit (Qiagen, Valencia, CA).

2.2. In vivo Experiments

1. Six- to eight-week-old C3H mice, weighing 18–20 g, from the in-house animal breeding facility were used for this study and were maintained under National Institutes of Health guidelines, approved by the Institutional Animal Care and Use Committee of Louisiana State University.

2. Insulin syringes.

3. IsoFlo (Isoflurane).

4. Anaesthetic machine with oxygen cylinder and isoflurane vaporizer (Abbott Laboratories, North Chicago, IL).

5. SCCVII cells, a spontaneously arising murine squamous cell carcinoma that has been well characterized in C3H/HeJ mice, were used to generate syngeneic transplant tumors (7).

6. Electroporator BTS EC830 (Inovio, Inc., San Diego, CA).

7. Caliper electrode (Inovio, Inc., San Diego, CA).

2.3. Cell Culture

1. SCCVII cells. The cells were maintained in Dulbecco's Modified Eagle's Medium (DMEM) supplemented with penicillin, streptomycin, glutamine, and 10% fetal bovine serum (FBS).

2. 0.5% trypsin–EDTA solution.

2.4. Collection and Processing of Mouse Samples

1. Aluminum foil.

2. Silica beads.

3. 5× cell lysis buffer and luciferase assay substrate (Promega, Madison, WI).

4. Mini-Bead Beater (BioSpec Products, Bartlesville, OK).

5. Liquid nitrogen dewar.

6. BCA protein assay kit (Pierce, Rockford, IL).

3. Methods

3.1. Preparation of SCCVII Cells for Generating Transplant Tumors

1. Harvest SCCVII cells with trypsin–EDTA when approaching confluence.
2. Resuspend cells in PBS and count under the microscope.
3. Dilute cells to 2×10^5 cells/30 μL PBS, which is the number of cells to be administered to each mouse.
4. Anesthetize mice with isoflurane.
5. Shave one spot on the back with electric razor and subcutaneously inoculate with 2×10^5 SCCVII cells in a volume of 30 μL using an insulin syringe.
6. Tumors are measured with a caliper to determine the volume.
7. Tumor volumes were determined using the following equation: $V = \pi/8\,(a \times b^2)$, where V = tumor volume, a = maximum tumor diameter, and b = diameter at $90°$ to "a" (8).

3.2. Determine the Optimal Injection Volume for Intratumoral Electroporation

1. When tumors reach 60–80 mm³, plasmid DNA in various volumes were intratumorally administered with a syringe followed by electric pulses using a caliper electrode and the previously optimized electroporation parameters, 450 V/cm and two 25 ms pulses (9).
2. Twenty-four hours after injection of luciferase encoding plasmid DNA via electroporation, luciferase activity was analyzed using luminometer (*see* **Note 2**). DNA injection volumes of 50 or 100 μL, compared with other low or high volumes, induce high levels of gene expression (Fig. 23.1); however, this result is dependent on the electroporation parameters and tumor volumes (*see* **Notes 3,4**).

3.3. Determine Luciferase Activity from Tumor Tissues

1. Place tumors wrapped in foil in liquid nitrogen for a few minutes.
2. Remove sample from liquid nitrogen, and crush tumor tissues in foil with a hammer and further homogenize tumor tissues in a bead beater tube containing 3–5 silica beads using a bead beater.
3. Spin the homogenized tissues in the same tube at maximum speed for 10 min.
4. Draw 20 μL supernatant and mix with 80 μL luciferase substrate for measuring the luciferase activity using the luminometer.

Fig. 23.1 The optimal injection volume for intratumoral electroporation to achieve the maximum level of gene expression. Ten micrograms of plasmid DNA encoding reporter gene luciferase was administered into tumors via electroporation and the luciferase activity was determined. The values at each injection volume represent the ratio between tumor volumes and injection volumes. Two panels represent two independent experiments

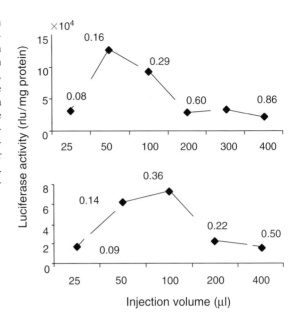

Fig. 23.2 Comparison of luciferase activities at a fixed injection volume under different electroporation fields. Three sets of representative electroporation parameters were used: low, medium, and high electric fields ($n = 5$). Low electric field, 100 V/cm with two pulses at 50 ms duration; medium electric field, 450 V/cm with two pulses at 25 ms duration; high electric field, 1,600 V/cm with 6 pulses at 100 µs duration

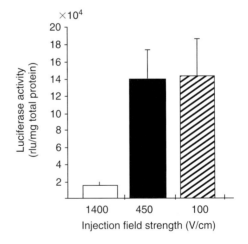

4. Notes

1. Isoflurane inhalation is safe, mice recover very fast, and it can be used several times.
2. Tumors should be collected within 96 h after intratumoral electroporation for easy detection of luciferase activity. Exceeding 96 h greatly reduces the detected signals due to either gene silencing or plasmid DNA degradation.

Fig. 23.3 Injection volume for large tumors (*n* = 5)

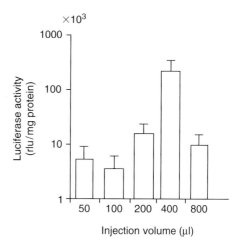

3. Because tumor sizes vary among the animals, the absolute injection volume is difficult to translate into actual practice. To avoid this problem, we determine the ratio of tumor volume versus injection volume, referred to as tumor–injection volume ratio (TIVR) in this study. High level of gene expression is associated with a ratio between 0.14 and 0.36. A ratio less than 0.1 or more than 0.5 will yield a low level of gene expression.

4. Because various optimal electroporation conditions have been used for gene delivery, it is important to know whether this TIVR is dependent on the electric parameters chosen; therefore, high voltage (1,600 V/cm, six 100 μs pulses), low voltage (100 V/cm, two 50 ms pulses), and our medium voltage-based parameters were used in the same experimental setting. Low voltage-mediated gene delivery yields a similar level of gene expression as medium voltage-mediated gene delivery. High voltage-based gene delivery yields a seven-fold less expression than low or medium voltage-based gene delivery (Fig. 23.2). We hypothesize that high voltage-based gene delivery requires a different injection volume to maximize the level of gene expression. To test this hypothesis, 10 μg plasmid DNA, which were dissolved into different volumes of saline, were injected into tumors. Injection of a 400 μL volume yields the highest level of reporter gene expression, resulting in a 38-fold higher level of gene expression compared with injection of a 50 μL volume with the same amount of plasmid DNA (Fig. 23.3). This injection volume will yield an optimal TIVR of 1.4 because the tumor volume in this experiment is 550 mm³ on average for each group of mice. Further decreases or increases of this ratio do not improve the transfection efficiency in tumors. This TIVR ratio for high voltage-mediated gene delivery is bigger than the one for medium voltage-mediated gene delivery.

References

1. Aihara, H. and Miyazaki, J. (1998) Gene transfer into muscle by electroporation in vivo. *Nat. Biotechnol.* **16**, 867–870.
2. MacLaughlin, F.C., Li, S., Fewell, J., Rolland, A., and Smith, L.C. (2000) Plasmid gene delivery: advantages and limitations. In: Gregoriadis, G., McCormack, B. (eds.) *Targeting of drugs: Strategies for gene constructs and delivery*, IOS Press, The Netherlands, **323**, 81–91.

3. Mir, L.M., Bureau, M.F., Gehl, J., et al. (1999) High-efficiency gene transfer into skeletal muscle mediated by electric pulses. *Proc. Natl. Acad. Sci. U.S.A.* **96**, 4262–4267.
4. Li, S., Zhang, X., Xia, X., et al. (2001) Intramuscular electroporation delivery of IFN-alpha gene therapy for inhibition of tumor growth located at a distant site. *Gene Ther.* **8**, 400–407.
5. Li, S. and Benninger, M. (2002) Applications of muscle electroporation gene therapy. *Curr. Gene Ther.* **2**, 101–105.
6. Li, S. (2004) Electroporation gene therapy: new developments in vivo and in vitro. *Curr. Gene Ther.* **4**, 309–316.
7. O'Malley, B.W., Jr., Cope, K.A., Johnson, C.S., and Schwartz, M.R. (1997) A new immunocompetent murine model for oral cancer. *Arch. Otolaryngol. Head Neck Surg.* **123**, 20–24.
8. Puisieux, I., Odin, L., Poujol, D., et al. (1998) Canarypox virus-mediated interleukin 12 gene transfer into murine mammary adenocarcinoma induces tumor suppression and long-term antitumoral immunity. *Hum. Gene Ther.* **9**, 2481–2492.
9. Li, S., Xia, X., Zhang, X., and Suen, J. (2002) Regression of tumors by IFN-alpha electroporation gene therapy and analysis of the responsible genes by cDNA array. *Gene Ther.* **9**, 390–937.
10. Goto, T., Nishi, T., Tamura, T., Dev, S.B., Takeshima, H., Kochi, M., Yoshizato, K., Kuratsu, J., Sakata, T., Hofmann, G.A. *et al.* (2000) Highly efficient electro-gene therapy of solid tumor by using an expression plasmid for the herpes simplex virus thymidine kinase gene. *Proc Natl Acad Sci USA.* **97**, 354–359.
11. Kishida, T., Asada, H., Satoh, E., Tanaka, S., Shinya, M., Hirai, H., Iwai, M., Tahara, H., Imanishi, J. and Mazda, O. (2001) In vivo electroporation-mediated transfer of interleukin-12 and interleukin-18 genes induces significant antitumor effects against melanoma in mice. *Gene Ther.* **8**, 1234–1240.
12. Lohr, F., Lo, D.Y., Zaharoff, D.A., Hu, K., Zhang, X., Li, Y., Zhao, Y., Dewhirst, M.W., Yuan, F. and Li, C.Y. (2001) Effective tumor therapy with plasmid-encoded cytokines combined with in vivo electroporation. *Cancer Res.* **61**, 3281–3284.
13. Shibata, M.A., Morimoto, J. and Otsuki, Y. (2002) Suppression of murine mammary carcinoma growth and metastasis by HSVtk/GCV gene therapy using in vivo electroporation. *Cancer Gene Ther.* **9**, 16–27.
14. Kishida, T., Asada, H., Itokawa, Y., Yasutomi, K., Shin-Ya, M., Gojo, S., Cui, F.D., Ueda, Y., Yamagishi, H., Imanishi, J. *et al.* (2003) Electrochemo-gene therapy of cancer: intratumoral delivery of interleukin-12 gene and bleomycin synergistically induced therapeutic immunity and suppressed subcutaneous and metastatic melanomas in mice. *Mol Ther.* **8**, 738–745.
15. Li, S., Zhang, X. and Xia, X. (2002) Regression of tumor growth and induction of long-term antitumor memory by interleukin 12 electro-gene therapy. *J Natl Cancer Inst.* **94**, 762-768.
16. Heller, L.C. and Coppola, D. (2002) Electrically mediated delivery of vector plasmid DNA elicits an antitumor effect. *Gene Ther.* **9**, 1321–1325.
17. Matsubara, H., Gunji, Y., Maeda, T., Tasaki, K., Koide, Y., Asano, T., Ochiai, T., Sakiyama, S. and Tagawa, M. (2001) Electroporation-mediated transfer of cytokine genes into human esophageal tumors produces anti-tumor effects in mice. *Anticancer Res.* **21**, 2501–2503.
18. Nishi, T., Yoshizato, K., Yamashiro, S., Takeshima, H., Sato, K., Hamada, K., Kitamura, I., Yoshimura, T., Saya, H., Kuratsu, J. *et al.* (1996) High-efficiency in vivo gene transfer using intraarterial plasmid DNA injection following in vivo electroporation. *Cancer Res.* **56**, 1050–1055.
19. Matsubara, H., Maeda, T., Gunji, Y., Koide, Y., Asano, T., Ochiai, T., Sakiyama, S. and Tagawa, M. (2001) Combinatory anti-tumor effects of electroporation-mediated chemotherapy and wild-type p53 gene transfer to human esophageal cancer cells. *Int J Oncol.* **18**, 825–829.
20. Mikata, K., Uemura, H., Ohuchi, H., Ohta, S., Nagashima, Y. and Kubota, Y. (2002) Inhibition of growth of human prostate cancer xenograft by transfection of p53 gene: gene transfer by electroporation. *Mol Cancer Ther.* **1**, 247–252.

21. Niu, G., Heller, R., Catlett-Falcone, R., Coppola, D., Jaroszeski, M., Dalton, W., Jove, R. and Yu, H. (1999) Gene therapy with dominant-negative Stat3 suppresses growth of the murine melanoma B16 tumor in vivo. *Cancer Res.* **59,** 5059–5063.
22. Lucas, M.L., Heller, L., Coppola, D. and Heller, R. (2002) IL-12 plasmid delivery by in vivo electroporation for the successful treatment of established subcutaneous B16.F10 melanoma. *Mol Ther.* **5,** 668–675.

Chapter 24
Intratumoral Bleomycin and IL-12 Electrochemogenetherapy for Treating Head and Neck Tumors in Dogs

Jeffry Cutrera, Marina Torrero, Keijiro Shiomitsu, Neal Mauldin, and Shulin Li

Abstract Bleomycin and Interleukin 12 have been used clinically to treat tumors; however, the co-administration of Bleomycin and Interleukin 12 followed by electroporation has not been tested clinically. In this study, dogs with spontaneous head and neck tumors were treated with one co-administration of Bleomycin and Interleukin 12 plasmid DNA followed by electroporation. The regression of the recurrent papillary tumor and the adjacent metastatic bone tumor was analyzed by multiple CT scans. The papillary tumor was completely eradicated in less than 2 weeks, and the bone tumor was not visible 23 weeks after the administration.

Keywords: electroporation, dog, canine, bleomycin, IL-12, gene therapy, spontaneous tumor treatment

1. Introduction

The anti-tumor chemotherapeutic agent bleomycin (BLM) is a G-, M-, and S-phase cell-cycle specific drug (1). In the presence of O_2 and a metal ion cofactor such as iron (2) or copper, BLM forms an intermediate metal complex (3) which leads to both single- and double-strand scission in the DNA (2). This scission results in the inhibition of DNA, RNA, and protein synthesis (4). Electroporation allows the BLM to enter the cytosol and greatly increases its cytotoxicity (5). The pro-inflammatory cytokine Interleukin 12 (IL-12), which is produced by dendritic cells and phagocytes, produces several immunological responses, such as inducing the production of interferon-γ and tumor necrosis factor-α and stimulating the differentiation of T helper 1 cells (6). Already, the antitumor effects of IL-12 protein have been shown by intravenous (7) and subcutaneous administration (8). Also, preclinical studies show that in vivo electroporation of the IL-12 gene induces tumor regression and antitumor memory (9). A preclinical murine study has shown that the co-administration of BLM and IL-12 gene eradicated 100% of squamous-cell carcinoma VII tumors and prevented tumor recurrence in 80% of the mice (10).

S. Li (ed.), *Electroporation Protocols: Preclinical and Clinical Gene Medicine.*
From *Methods in Molecular Biology, Vol. 423.*
© Humana Press 2008

In this study, dogs that were histopathologically diagnosed with head and neck tumors underwent a CT scan to determine the best administration site and were then treated with 150 μg of IL-12 gene and 0.5 units of BLM per cm^2 of tumor followed by electroporation. A CT scan taken 7 weeks after the administration showed complete eradication of the spontaneous tumor, and a CT scan taken 23 weeks after the administration showed a complete regression of the spontaneous tumor as well as the adjacent bone tumor (when present).

2. Materials

2.1. General Anesthesia

1. 22 GA BD Angiocath catheter (Becton, Dickinson and Company, Franklin Lakes, NJ).
2. Jorvet 3.5 mm J-615B endotracheal tube (Jorgensen Laboratories, Inc., Loveland, CO).
3. Propofol.
4. Isoflurane (IsoFlo).
5. Capnocheck II Capnograph/Oximeter (Smith Medical PM, Inc., Wausheka, WI).

2.2. Histopathological Diagnosis of Malignant Oral Tumors

2.2.1. Complete Blood Count/Chemisty Profile

1. 22 GA (0.7 mm × 25 mm) needles.
2. 6 mL syringes.
3. 3 mL BD vacutainer containing the anticoagulant K_2EDTA.
4. 4 mL BD vacutainer with SST Gel and Clot Activator (Becton, Dickinson and Company, Franklin Lakes, NJ).
5. ADVIA 120 Hematology System (Bayer Diagnostics, Tarrytown, NJ).
6. AU600 Chemistry Analyzer (Olympus, Melville, NY).

2.2.2. Urinalysis

1. 22 GA (0.7 mm × 38.1 mm) needles.
2. 12 mL syringes.
3. 10 mL BD vacutainer 364979.
4. 3 mL BD vacutainer.
5. Multistix 10SG (Bayer Diagnostics, Tarrytown, NJ).

2.2.3. Radiograph

1. Polydoros SX 80 (Siemens Medical Solutions, Cranbury, NJ).

2.2.4. Biopsy

1. Seamless Premier Uni-Punch Disposable Biopsy Punch (Premier Medical Products Company, Plymouth Meeting, PA).
2. 10% Formaldehyde.

2.3. CT Scan

1. PQ5000 CT scanner.
2. Omnipaque, the iodinated contrast medium (Amersham Health, Inc., Princeton, NJ).

2.4. Intratumoral Injection and Electroporation for Treating Tumors in Dogs

1. IL-12 gene construct (Valentis, Inc., Burlingame, CA). The construct contains DNA fragments which encode for both the p35 and p40 subunits in the same plasmid vector. Both subunits are driven by independent CMV promoters and terminated by independent bovine growth hormone polyadenylation signals (11).
2. Bleomycin (GensiaSicor Pharmaceuticals, Irvine, CA).
3. Electroporator BTS EC830 (Inovio, San Diego, CA).
4. Caliper electrode (Inovio, San Diego, CA).

3. Methods

Dogs with a histopathologic diagnosis of malignant head and neck tumors were used. BLM is metabolized mainly in the liver and kidneys (1), and 45–70% is excreted through urine in the first 24 h (4); therefore, dogs with a history of high renal values (1.5 times normal) or hepatic values (2.0 times normal) determined by blood chemistry profiles were excluded. After treatment, radiographs and CT scans should be performed to monitor the effects of the treatment and the dog's health.

3.1. General Anesthesia

1. Insert catheter into cephalic vein (*see* **Note 1**).
2. Anesthetize using propofol induction into the catheter at a dose of 6 mg/kg.
3. Insert endotracheal tube.
4. Maintain anesthesia using IsoFlo (1.0–3.0%) inhalation through the endotracheal tube.
5. The Capnocheck II Capnograph/Oximeter is used to monitor the dog during anesthesia.

3.2. Biochemical and Histopathological Analysis of Tumor-Bearing Dogs

3.2.1. Complete Blood Count/Chemistry Profile

1. Insert needle attached to syringe into jugular vein.
2. Acquire 6 mL of blood in the syringe.
3. Inject 3 mL of blood into the K_2EDTA vacutainer.
4. Inject 3 mL of blood into the SST gel and Clot Activator vacutainer.
5. Invert both vacutainers 5 times.
6. Place the K_2EDTA vacutainer in the ADVIA 120 Hematology System to measure the number of red blood cells, number of white blood cells, amount of hemoglobin, hematocrit (fraction of blood composed of red blood cells), platelet count, and the mean corpuscular volume (size of red blood cells).
7. Let the SST gel and Clot Activator vacutainer sit for 30 min.
8. Centrifuge at full speed for 15 min.
9. Place SST gel and Clot Activator vacutainer in AU600 Chemistry Analyzer to measure the levels of chemicals released from tissues.

3.2.2. Urinalysis (*see* **Note 2**)

1. Insert needle through the abdominal wall and into the bladder.
2. Acquire 12 mL of urine.
3. Inject 10 mL of urine into the 10 mL vacutainer.
4. Inject the remaining urine into the 3 mL vacutainer for storage.
5. Remove rubber stopper from 10 mL vacutainer.
6. Insert a Multistix 10SG into the tube and compare the color from all blocks according to chart on Multistix 10 SG bottle.

3.2.3. Radiograph

1. Take right lateral, left lateral, and dorsal ventral radiograph views.

3.2.4. Biopsy

1. Administer general anesthesia as previously described.
2. Insert biopsy punch into tumor.
3. Twist biopsy punch until section of tumor is cut away.
4. Remove specimen from punch.
5. Immediately place specimen in 10% formaldehyde.
6. Specimens should be delivered to a pathologist for microscopic examination and diagnosis.

3.3. CT Scan

1. Administer general anesthesia as previously described.
2. Place dog in sternal recumbency, head-first into the gantry of PQ5000.
3. Obtain initial scout images and subsequent 2 mm contiguous transverse images.
4. Administer intravenous bolus (0.5 mg/kg) of iodinated contrast medium, Omnipaque.
5. Acquire contrast images using same imaging technique.
6. Window width and level are manipulated before and after contrast administration to best visualize the tumor measurement and evaluation.

3.4. Intratumoral Injection and Electroporation

1. Use the CT images to determine the size of tumor and best injection site.
2. Administer general anesthesia as previously described.
3. Inject 150 µg of IL-12 gene construct and 0.5 units of BLM per cm^2 of tumor (*see* **Note 3**).
4. Immediately after injection, place caliper electrode tight around tumor and apply 450 V/cm for 2 pulses of 25 ms (*see* **Note 4**).

4. Notes

1. It is critical to have a veterinarian who works side by side with an investigator for this procedure.
2. It is critical to perform this analysis for selecting proper dogs on account of the toxicity of BLM.
3. Using this IL-12 and BLM dose, the recurrent 3–4 cm in diameter tumor was successfully eradicated from the mouth as illustrated by both the CT scan and visual inspection (Figs. 24.1 and 24.2); however, this ratio could be changed for maximizing tumor eradication.
4. For oral cavity, the caliper electrode is safe; however, for deep oral cavity, a needle electrode has to be used. In these circumstances, precautions, such as using a thick wall inhalation tube, should be implemented since the inserted inhalation tube contains oxygen, and an electric spark could cause an explosion.

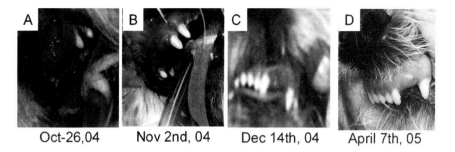

Oct-26,04   Nov 2nd, 04   Dec 14th, 04   April 7th, 05

Fig. 24.1 Regression of the recurrent papillary tumors by co-administration of BLM and IL-12 plasmid DNA via electroporation. **A.** Recurrent papillary tumor prior to treatment. **B.** Reduction of recurrent papillary tumor volume 7 days posttreatment. **C.** Eradication of the visible papillary tumor 7 weeks after treatment. **D.** Confirmation that the papillary tumor is eradicated 5.5 months after treatment

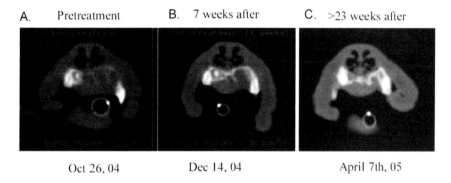

A. Pretreatment   **B.** 7 weeks after   **C.** >23 weeks after

Oct 26, 04   Dec 14, 04   April 7th, 05

Fig. 24.2 Eradication of the invasive bone tumor by bystander effect. **A.** CT scan showing $4\,cm^3$ invasive bone tumor prior to treatment. **B.** CT scan showing regression of invasive bone tumor volume to $2.2\,cm^3$ 7 weeks after treatment. **C.** CT scan showing eradication of the metastatic bone tumor 23 weeks after treatment

References

1. Dorr, R.T. and Von-Huff, D.D. (1994) Drug monographs. In: Dorr, R.T. and Von-Huff, D.D., (eds.). Cancer chemotherapy handbook, Appleton and Lange, Norwalk, CT, pp. 227–236.
2. Berger R.M., Projan S.J., Horwitz S.B., and Peisach J. (1986) The DNA cleavage mechanism of iron-bleomycin. J. Biol. Chem. 261, 15955–15959.
3. Perry M. (ed.) (2001) The chemotherapy source book. Lippincott Williams & Wilkins, Philadelphia, PA.
4. Chabner B.A. and Longo D.L. (eds.) (2001) *Cancer chemotherapy and biotherapy.* Lippincott, Williams, and Wilkins, Philadelphia, PA.
5. Poddevin B., Orlowski S., Belehradek, J., Jr., and Mir L.M. (1991) Very high toxicity of bleomycin introduced into the cytosol of cells in culture. *Biochem. Pharmacol.* **42**, S67–S75.
6. Trinchieri G. (2003) Interleukin-12 and the regulation of innate resistance and adaptive immunity. *Nat. Rev. Immunol.* **3**, 133–146.

7. Robertson M.J., Cameron C., Atkins M.B., et al. (1999) Immunological effects of Interleukin 12 administered by bolus intravenous injection to patients with cancer. *Clin. Cancer Res.* **5**, 9–16.
8. Motzer R.J., Rakhit A., Schwartz L.H., et al. (1998) Phase I trial of subcutaneous recombinant human Interleukin-12 in patients with advance renal cell carcinoma. *Clin. Cancer Res.* **4**, 1183–1191.
9. Li S., Zhang X. and Xia X. (2002) Regression of tumor growth and induction of long-term antitumor memory by Interleukin 12 electrogenetherapy. *J. Natl. Cancer Inst.* **94**, 668–675.
10. Torrero M.N., Henk W.G., and Li S. (2006) Regression of high-grade malignancy in mice by bleomycin and Interleukin-12 electrochemogenetherapy. *Clin. Cancer Res.* **12**, 257–263.
11. Coleman M., Muller S., Quezada A., et al. (1998) Nonviral interferon alpha gene therapy inhibits growth of established tumors by eliciting a systemic response. *Hum. Gene Ther.* **9**, 2223–2230.

Chapter 25
Systemic IL-12 Gene Therapy for Treating Malignancy via Intramuscular Electroporation

Shiguo Zhu and Shulin Li

Abstract Interleukin 12 (IL-12) is effective in treating systemic microscopic malignancies by inducing T helper 1 (T_H1) response, inhibiting angiogenesis, and triggering secondary cytokine production. Unfortunately, daily systemic administration of an acute dose of IL-12 protein is very costly and severely toxic. Here, a simple, economic, and less toxic approach, intramuscular administration of IL-12 gene, is provided for treating tumors in three tumor models. The results indicate that intramuscular administration of IL-12 encoding plasmid DNA via electroporation is a promising technology for treating systemic residual malignancies (less than 3–5 mm in diameter), as illustrated by the inhibition of tumor growth and lung metastases as well as the extension of survival rate. This approach is not effective in treating tumors larger than 3–5 mm in diameter.

Keywords: electoporation, gene transfer, muscle, cancer, gene therapy, IL-12, tumor immunity, tumor model

1. Introduction

Surgery, radiotherapy, and chemotherapy are the current major approaches for treating malignant tumors, and gene therapy may be the fourth method to complement the current approaches. The challenge for gene therapy is to develop an efficient and safe delivery system (1). Several types of cancer, such as melanoma and head and neck cancer, are amenable for direct intra-tumoral administration; however, most late stage patients carrying metastatic tumors can only be reached by systemic gene administration (2). Regardless of local or systemic gene transfer, most gene therapy studies focus on virus vectors, and almost two-thirds of cancer gene therapy trials are based on viral vectors (3). Virus-mediated gene transfer may achieve high effective transfer, but safety is always a concern (4–9).

Intramuscular administration of plasmid DNA via electroporation is one of the most promising strategies for treatment of systemic malignant diseases for several reasons (10–14). First, direct injection of plasmid DNA into skeletal

muscle is easy, simple, inexpensive, and safe. Second, gene expression can be increased dramatically (100–1,000 folds). Third, gene expression can be sustained and specified. Fourth, therapeutic protein generated from the injected gene in muscle tissue can be secreted into the blood yielding a systemic antitumor effect.

IL-12 is a proinflammatory cytokine that plays an essential role in the interaction between innate resistance and adaptive immunity (15, 16). Treatment with IL-12 has been shown to have a marked antitumor effect on several different tumor models by inhibiting the establishment of tumors or by inducing tumor regression (10, 17–19). IL-12 exerts its antitumor activity by inducing the production of secondary cytokines, like IFN-γ, augmenting the responses of T_H1 cells and cytotoxic T lymphocytes (CTL), and inhibiting angiogenesis (20–25).

In this protocol, the techniques and methods for using intramuscular IL-12 gene delivery via electroporation is described. To illustrate the IL-12 effects, the tumor growth, lung metastases, survival rate, and induction of IFN-γ are analyzed after administration of IL-12 gene.

2. Materials

2.1. Plasmid DNA and In Vivo Electroporation Buffer

1. The constructs of murine IL-12 plasmid DNA and control plasmid DNA (Valentis, Inc., Burlingame, CA, USA).
2. Plasmid maxi preparation kit (Qiagen, Inc., Valencia, CA)
3. Biophotometer and bench microcentrifuge.
4. Half-strength saline: dilute saline (0.9% sodium chloride, Abbott laboratories, Abbott Park, IL) into equal volume sterilized water, and store at room temperature.

2.2. Cell Lines and Cell Culture

1. Cell lines: 4T1, a murine mammary tumor cell line and B16F10, a murine melanoma cell line, were obtained from ATCC. Murine squamous cell carcinoma cell line, SCCVII, was obtained from Dr. Candice Johnson's laboratory at Roswell Park Cancer Research Institute (Buffalo, New York).
2. Complete medium: Dulbecco's Modified Eagle's Medium (DMEM) supplemented with 10% fetal bovine serum (FBS), and 1% antibiotic-antimycotic (penicillin G sodium 10,000 units/mL, streptomycin sulfate 10,000 units/mL, and amphoterin B 25 μg/mL in 0.9% saline).
3. 0.05% Trypsin-EDTA and 0.4% trypan blue staining.

2.3. Experimental Animals

1. Six-week-old BALB/C, C3H, and C57BL/6 mice.
2. Maintain mice under National Institutes of Health guidelines that were approved by the Institutional Animal Care and Use Committee of Louisiana State University.
3. Five mice per cage and per treatment.
4. ELISA kit (R&D Systems, Minneapolis, MN), microplate reader (Molecular Devices, Menlo Park, CA), and Promega Reporter gene assay system (Promega, Madison, WI)
5. Bouin's Fixative (Labchem Inc., Pittsburgh, PA)

2.4. Electroporation Apparatus and Other Necessary Reagents

1. IsoFlo (Isoflurane).
2. Insulin syringes.
3. 70% ethanol: Dilute 70 mL of 95% ethanol into 25 mL of sterilized water.
4. Power supply (ECM 830) and caliper electrodes (BTX instrument, San Diego, CA).

3. Methods

3.1. Plasmid DNA Preparation and Quantitation

1. Prepare plasmid DNA following Qiagen Endotoxic Free Plasmid Maxi-preparation protocol.
2. Measure DNA concentration. Measure DNA absorption at 260 nm using Biophotometer (the OD 260/280 ratios of these plasmid preparations ranged from 1.5 to 1.7) and then run an aliquot of DNA in 1% of agarose gel electrophoresis with standard DNA to confirm DNA concentration. The percentage of supercoiled DNA to total DNA preparations ranged from 80 to 90%.
3. Dilute to 5 μg plasmid DNA per 30 μL using half strength saline prior to injection (*see* **Notes 1–3**).

3.2. Tumor Model

1. Maintain 4T1, SCCVII, and B16F10 cells in DMEM supplemented with 10% FBS and 1% antibiotic-antimycotic.

2. Trypsin cells using 0.05% Trypsin-EDTA and count cell number under microscope after staining cells with 0.4% trypan blue.
3. Dilute cells to 1×10^5 cells/30 μL for 4T1 cells and 2×10^5 cells/30 μL for SCCVII and B16F10 cells in PBS.
4. Anesthetize mice using Isoflurane, then shave the hair from the back of the mouse. Spray 70% ethanol prior to subcutaneously (s.c.) inoculating 1×10^5 4T1 cells into each BALB/C mouse, 2×10^5 SCCVII cells into each C3H mouse, and 2×10^5 B16F10 cells into each C57BL/6 mouse.
5. Check tumor growth 3 days after inoculation and start to treat when tumors reach 2–4 mm in diameter.

3.3. Electroporation of Plasmid DNA into Skeletal Muscle

1. Plug electrode to power supplies, turn on electroporator and adjust to the following parameters: 35 V/mm, 20 ms pulse duration, and 2 pulses with a 100 ms interval between the pulses.
2. Anesthetize all mice in a covered box by Isoflurane.
3. Randomly pick one mouse from the anesthetized mice and spray 70% ethanol onto each hind limb.
4. Intramuscularly inject 30 μL of plasmid DNA solution (5 μg/30 μL) into each hind limb tibialis muscle (*see* **Notes 4–5**).
5. Put two metal heads on each side of injected muscle and tightly clamp the limb immediately after injection, then apply square wave pulses to the injected muscles.
6. Repeat for the second muscle in the same mouse (*see* **Notes 6–7**).
7. Clip a small hole in the left ear on each mouse for tracking individual tumor growth.
8. Place the treated animal into the cage and clearly mark the cage in detail, including treatment method, DNA dose, gene name, and date.
9. Perform the second treatment 10 days later (*see* **Notes 8–9**).

3.4. Determine the Expression of IL-12 and IFN-γ

1. Draw 50 μL of blood from each mouse 3 days after IL-12 intramuscular electroporation.
2. Clot on ice for 30 min, then gently centrifuge at 3,000g for 10 min at 4°C using a bench centrifuge. Transfer yellow serum to a new tube with pipette and store at −80°C.
3. Assay IL-12 and IFN-γ protein with an ELISA kit, and determine concentration using a precision microplate reader.
4. Calculate the IL-12 and IFN-γ protein as μg/mL serum (Figs. 25.1A, B).

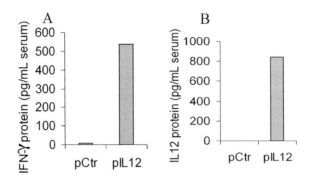

Fig. 25.1 IL-12 and IFN-γ protein in serum on day 3 after intramuscular IL-12 electroporation. 5 µg plasmid DNA was injected into each hind-limb tibialis muscle of 4T1 tumor-bearing mice via electroporation. Three days after IL-12 treatment, sacrifice mice and collect blood. **A.** IL-12 protein; **B.** IFN-γ protein

3.5. Tumor Measurement and Calculation

1. Monitor the size of tumors using a caliper every 3 days, and sacrifice mice with CO_2 asphyxiation when the tumor reaches 2 cm in diameter (*see* **Note 10**).
2. Input all the data into Excel software, and calculate tumor volume with the formula $V = \pi/8(ab^2)$, where 'V' is the tumor volume, 'a' is the maximum tumor diameter, and 'b' is the diameter at 90° to "a".
3. Draw the tumor growth curve using Excel software (Figs. 25.2A–C).
4. A two-sided student t test is used to compare the means of individual treatments. $p < 0.05$ is considered statistically significant.

3.6. Survival Assay

1. Treat five mice per group.
2. Monitor the size of tumors using the caliper every 3 days.
3. Count mortality when mice die either naturally or when tumors exceed 2 cm in diameter.
4. Input all data into STATISTICS 7 software and draw cumulative proportion surviving graph (Figs. 25.3A, B).

3.7. Lung Metastases

1. Sacrifice mice using CO_2 asphyxiation 30 days after IL-12 treatment via intramuscular electroporation.

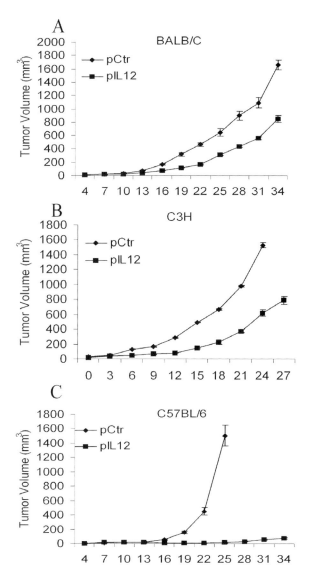

Fig. 25.2 Inhibition of tumor growth by intramuscular IL-12 electroporation. 5 μg, 1 μg, and 2.5 μg of plasmid DNA were respectively injected into each hind-limb tibialis muscle of 4T1 tumor-bearing BALB/C mice (**A**), SCCVII tumor-bearing C3H mice (**B**), and B16F10 tumor-bearing C57BL/6 mice (**C**) via electroporation. Error bars represent means standard errors (SE). pCtr and pIL-12 represent control and IL-12-encoding plasmid DNA, respectively

2. Remove lungs from mice immediately after injecting Bouin's Fixative via trunk into lungs, place lungs into 50 mL Falcon tube containing Bouin's fixative and store at 4°C (*see* **Note 11**).
3. Cut lungs into five pieces, and count the number of metastatic nodules using microscopy (Fig. 25.4).

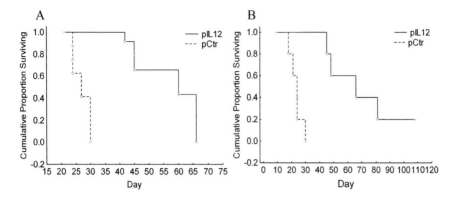

Fig. 25.3 Increase in survival rate after intramuscular IL-12 electroporation. 5 or 2.5 µg plasmid DNA was injected into each hind-limb tibialis muscle of 4T1 tumor-bearing BALB/C mice (**A**) or B16F10 tumor-bearing C57BL/6 mice (**B**) via electroporation. pCtr and pIL-12 represent control and IL-12-encoding plasmid DNA, respectively

Fig. 25.4 Inhibition of lung metastases after intramuscular IL-12 electroporation. 5 µg of plasmid DNA was injected into each hind-limb tibialis muscle of 4T1 tumor-bearing BALB/ C mice via electroporation. On Day 30 after treatment, count the number of lung metastases using microscopy. pCtr and pIL-12 represent control and IL-12-encoding plasmid DNA, respectively

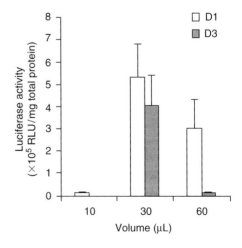

Fig. 25.5 Dependence of luciferase expression on injection volumes. 5 µg of luciferase expression plasmid DNA is diluted in different volumes and injected into hind-limb tibialis muscles. On days 1 and 3 after treatment, mice were killed and injected muscles were removed. Luciferase was examined using Promega Reporter assay system

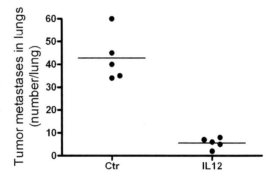

Table 25.1 Intramuscular electroporation parameters

Author	Year	Electric parameters	Reference
Aihara and Miyazaki	1998	200 V/cm, 3 pulses, 50 ms	(11)
Mir et al.	1999	200 V/cm, 8 pulses, 20 ms, 2 Hz	(12)
Lawson et al.	2000	200 V/cm, 3 pulses, 50 ms	(27)
Nakano et al.	2001	6 pulses, 50 ms, 1 pulse/s	(28)
Mcmahon et al.	2001	100–200 V/cm, 10–20 ms, 1 pulse, 1 Hz	(29)
Vilquin et al.	2001	200 V/cm, 8 pulses, 20 ms, 2 Hz	(30)
Lucas and Heller	2001	100 V/cm, 20 ms	(31)
Nicol et al.	2002	375 V/cm, 2 pulses, 25 ms	(32)
Lee et al.	2002	125 V/cm, 4 pulses × 2, 50 ms, 1 Hz	(26)
Lesbordes et al.	2002	500 V/cm, 8 pulses, 20 ms, 1 Hz	(33)
Martinenghi et al.	2002	200 V/cm, 9 pulses, 1 ms, 10 Hz	(34)
Celiker et al.	2002	200 V/cm, 3 pulses, 50 ms	(35)
Satkauskas et al.	2002	800 V/cm 100 µs; 80 V/cm, 100 ms	(36)
Takahashi et al.	2003	100 V/cm, 3 pulses, 50 ms	(37)
Murakami et al.	2003	100 V/cm, 6 pulses, 50 ms, 1 Hz	(38)
Gollins et al.	2003	175 V/cm, 10 pulses, 20 ms 1 Hz	(39)
Molnar et al.	2004	175–1,800 V/cm, 8 pulses, 0.2–20 ms	(40)
Quaglino et al.	2004	375 V/cm, 8 pulses, 25 ms	(41)
Bureau et al.	2004	800 V/cm, 0.1 ms, 10 s, 80 V/cm, 4 pulse, 83 ms	(42)
Rabinovsky and Draghia-Akli	2004	150 V/cm, 3 pulses, 20 ms	(43)
Brown et al.	2004	80–120 V/cm, 5 pulses, 52 ms	(44)
Otten et al.	2005	200 V/cm, 6 pulses, 25 ms	(45)
Tjelle et al.	2005	250 mA, 5 pulses, 20 ms	(46)
Peng et al.	2005	200 V/cm, 6 pulses, 50 ms	(47)
Schertzer et al.	2005	50–200 V/cm, 3 pulses, 20 ms, 1 Hz	(48)

Acknowledgments This work was supported by National Cancer Institute/NIH grant RO1CA120895 and National Institute of Dental and Craniofacial Research/NIH grant R21DE14682. We thank Jeffry Cutrera for assistance in the preparation of this manuscript.

4. Notes

1. It is very important to run an agarose gel for plasmid DNA quantitation. If plasmid DNA is partially degraded, it should not be used in animal treatment. Degradation is indicated by a smeared band using agarose gel analysis.
2. Usually, saline is used to dissolve DNA, however, Lee MJ et al. report that half strength saline is an optimal vehicle for in vivo electroporation of naked DNA in skeletal muscle (26).
3. When DNA is diluted to half saline strength, the original plasmid DNA concentration should be more than 1 µg/µL. If DNA concentration is lower than 1 µg/µL, then concentrate the DNA by speed vacuum.
4. Injection volume is an important factor for intramuscular plasmid DNA electroporation. Usually, 30 µL per tibialis in mice can generate a significant high level of gene expression (Fig. 25.5). This volume should be modified when injecting into a large muscle.
5. Five microgram DNA per limb is enough to generate an efficient antitumor response. Increasing the amount of plasmid DNA will not enhance the antitumor effect.
6. Electroporation of two limbs for each treatment can obtain consistent result compared with one-limb treatment.

7. The electrotransfer parameter is the key factor for an effective intramuscular gene delivery. Usually, we use 35 V/mm, 20 ms pulse duration, and 2 pulses with a 100 ms interval between the pulses. Because of the variation of the electroporation device and experiment condition, different electric parameters may be used to achieve the best results. Table 25.1 shows some intramuscular electroporation parameters that have been used by others.

8. Start to treat mice when tumors reach 2–4 mm in diameter. If it is larger than 4–5 mm, intramuscular electroporation is not effective. It usually takes 3–4 days for 4T1 tumors, 4–5 days for SCCVII tumors, and 5–6 days for B16F10 tumors to reach 2–4 mm.

9. Prior to treatment, mice should be anesthetized. Injectable anesthesia and gas anesthesia are usually adopted, and the mainstay of general anesthesia is gas anesthesia. We use the safest and most effective, Isoflurane, for gas anesthesia, which requires specialized equipment, such as oxygen, endotracheal tube, and vaporizer.

10. Monitor tumor growth of different experiments by one person to reduce variation.

11. For accurate counting of metastatic tumor nodules, expand lungs using Bouin's fixative to increase contrast between tumors and lung tissues. To expand lungs, some investigators inject fixative solution into the lungs via trunks, then tie the trunks and isolate the expanded lungs; however, it is not necessary. Just remove the lungs and directly inject fixative solution into lungs via the trunk, then immediately submerge the expanded lungs into a tube containing the fixative solution.

References

1. Anderson, W.F. (1998) Human gene therapy. *Nature*. **392**, 25–30.
2. Ogris, M. and Wagner, E. (2002) Targeting tumors with non-viral gene delivery systems. *Drug Discov. Today*. **7**, 479–485.
3. Scollay, R. (2001) Gene therapy: a brief overview of the past, present, and future. *Ann. N. Y. Acad. Sci.* **953**, 26–30.
4. Kay, M.A. and Nakai, H. (2003) Looking into the safety of AAV vectors. *Nature*. **424**, 251.
5. Li, Z., Dullmann, J., Schiedlmeier, B., et al. (2002) Murine leukemia induced by retroviral gene marking. *Science*. **296**, 497.
6. Chuah, M.K., Collen, D., and VandenDriessche, T. (2003) Biosafety of adenoviral vectors. *Curr. Gene Ther.* **3**, 527–543.
7. Davis, M.E. (2002) Non-viral gene delivery systems. *Curr. Opin. Biotechnol.* **13**, 128–131.
8. Debyser, Z. (2003) Biosafety of lentiviral vectors. *Curr. Gene Ther.* **3**, 517–525.
9. Check, E. (2003) Harmful potential of viral vectors fuels doubts over gene therapy. *Nature*. **423**, 573–574.
10. Li, S., Zhang, X., and Xia, X. (2002) Regression of tumor growth and induction of long-term antitumor memory by interleukin 12 electro-gene therapy. *J. Natl. Cancer Inst.* **94**, 762–768.
11. Aihara, H. and Miyazaki, J. (1998) Gene transfer into muscle by electroporation in vivo. *Nat. Biotechnol.* **16**, 867–870.
12. Mir, L.M., Bureau, M.F., Gehl, J., et al. (1999) High-efficiency gene transfer into skeletal muscle mediated by electric pulses. *Proc. Natl. Acad. Sci. U.S.A.* **96**, 4262–4267.
13. Li, S. (2004) Electroporation gene therapy: new developments in vivo and in vitro. *Curr. Gene Ther.* **4**, 309–316.
14. Li, S. and Benninger, M. (2002) Applications of muscle electroporation gene therapy. *Curr. Gene Ther.* **2**, 101–105.
15. Trinchieri, G. (1995) Interleukin-12: a proinflammatory cytokine with immunoregulatory functions that bridge innate resistance and antigen-specific adaptive immunity. *Annu. Rev. Immunol.* **13**, 251–276.

16. Trinchieri, G. (2003) Interleukin-12 and the regulation of innate resistance and adaptive immunity. *Nat. Rev. Immunol.* **3**, 133–146.

17. Brunda, M.J., Luistro, L., Warrier, R.R., et al. (1993) Antitumor and antimetastatic activity of interleukin 12 against murine tumors. *J. Exp. Med.* **178**, 1223–1230.

18. Torrero, M.N., Henk, W.G., and Li, S. (2006) Regression of high-grade malignancy in mice by bleomycin and interleukin-12 electrochemogenetherapy. *Clin. Cancer Res.* **12**, 257–263.

19. Nanni, P., Nicoletti, G., De Giovanni, C., et al. (2001) Combined allogeneic tumor cell vaccination and systemic interleukin 12 prevents mammary carcinogenesis in HER-2/neu transgenic mice. *J. Exp. Med.* **194**, 1195–1205.

20. Gately, M.K., Wolitzky, A.G., Quinn, P.M., and Chizzonite, R. (1992) Regulation of human cytolytic lymphocyte responses by interleukin-12. *Cell Immunol.* **143**, 127–142.

21. Manetti, R., Parronchi, P., Giudizi, M.G., et al. (1993) Natural killer cell stimulatory factor (interleukin 12 [IL-12]) induces T helper type 1 (Th1)-specific immune responses and inhibits the development of IL-4-producing Th cells. *J. Exp. Med.* **177**, 1199–1204.

22. Voest, E.E., Kenyon, B.M., O'Reilly, M.S., Truitt, G., D'Amato, R.J., and Folkman, J. (1995) Inhibition of angiogenesis in vivo by interleukin 12. *J. Natl. Cancer Inst.* **87**, 581–586.

23. Yao, L., Pike, S.E., Setsuda, J., et al. (2000) Effective targeting of tumor vasculature by the angiogenesis inhibitors vasostatin and interleukin-12. *Blood.* **96**, 1900–1905.

24. Hsieh, C.S., Macatonia, S.E., Tripp, C.S., Wolf, S.F., O'Garra, A., and Murphy, K.M. (1993) Development of TH1 CD4+ T cells through IL-12 produced by Listeria-induced macrophages. *Science.* **260**, 547–549.

25. Ohteki, T., Fukao, T., Suzue, K., et al. (1999) Interleukin 12-dependent interferon gamma production by CD8alpha + ymphoid dendritic cells. *J. Exp. Med.* **189**, 1981–1986.

26. Lee, M.J., Cho, S.S., Jang, H.S., et al. (2002) Optimal salt concentration of vehicle for plasmid DNA enhances gene transfer mediated by electroporation. *Exp. Mol. Med.* **34**, 265–272.

27. Lawson, B.R., Prud'homme, G.J., Chang, Y., et al. (2000) Treatment of murine lupus with cDNA encoding IFN-gammaR/Fc. *J. Clin. Invest.* **106**, 207–215.

28. Nakano, A., Matsumori, A., Kawamoto, S., et al. (2001) Cytokine gene therapy for myocarditis by in vivo electroporation. *Hum. Gene Ther.* **12**, 1289–1297.

29. McMahon, J.M., Signori, E., Wells, K.E., Fazio, V.M., and Wells, D.J. (2001) Optimisation of electrotransfer of plasmid into skeletal muscle by pretreatment with hyaluronidase—increased expression with reduced muscle damage. *Gene Ther.* **8**, 1264–1270.

30. Vilquin, J.T., Kennel, P.F., Paturneau-Jouas, M., et al. (2001) Electrotransfer of naked DNA in the skeletal muscles of animal models of muscular dystrophies. *Gene Ther.* **8**, 1097–1107.

31. Lucas, M.L. and Heller, R. (2001) Immunomodulation by electrically enhanced delivery of plasmid DNA encoding IL-12 to murine skeletal muscle. *Mol. Ther.* **3**, 47–53.

32. Nicol, F., Wong, M., MacLaughlin, F.C., et al. (2002) Poly-L-glutamate, an anionic polymer, enhances transgene expression for plasmids delivered by intramuscular injection with in vivo electroporation. *Gene Ther.* **9**, 1351–1358.

33. Lesbordes, J.C., Bordet, T., Haase, G., et al. (2002) In vivo electrotransfer of the cardiotrophin-1 gene into skeletal muscle slows down progression of motor neuron degeneration in pmn mice. *Hum. Mol. Genet.* **11**, 1615–1625.

34. Martinenghi, S., Cusella De Angelis, G., Biressi, S., et al. (2002) Human insulin production and amelioration of diabetes in mice by electrotransfer-enhanced plasmid DNA gene transfer to the skeletal muscle. *Gene Ther.* **9**, 1429–1437.

35. Celiker, M.Y., Ramamurthy, N., Xu, J.W., et al. (2002) Inhibition of adjuvant-induced arthritis by systemic tissue inhibitor of metalloproteinases 4 gene delivery. *Arthritis Rheum.* **46**, 3361–3368.

36. Satkauskas, S., Bureau, M.F., Puc, M., et al. (2002) Mechanisms of in vivo DNA electrotransfer: respective contributions of cell electropermeabilization and DNA electrophoresis. *Mol. Ther.* **5**, 133–140.

37. Takahashi, T., Ishida, K., Itoh, K., et al. (2003) IGF-I gene transfer by electroporation promotes regeneration in a muscle injury model. *Gene Ther.* **10**, 612–620.

38. Murakami, T., Nishi, T., Kimura, E., et al. (2003) Full-length dystrophin cDNA transfer into skeletal muscle of adult mdx mice by electroporation. *Muscle Nerve.* **27**, 237–241.

39. Gollins, H., McMahon, J., Wells, K.E., and Wells, D.J. (2003) High-efficiency plasmid gene transfer into dystrophic muscle. *Gene Ther.* **10**, 504–512.
40. Molnar, M.J., Gilbert, R., Lu, Y., et al. (2004) Factors influencing the efficacy, longevity, and safety of electroporation-assisted plasmid-based gene transfer into mouse muscles. *Mol. Ther.* **10**, 447–455.
41. Quaglino, E., Iezzi, M., Mastini, C., et al. (2004) Electroporated DNA vaccine clears away multifocal mammary carcinomas in her-2/neu transgenic mice. *Cancer Res.* **64**, 2858–2864.
42. Bureau, M.F., Naimi, S., Torero Ibad, R., et al. (2004) Intramuscular plasmid DNA electrotransfer: biodistribution and degradation. *Biochim. Biophys. Acta.* **1676**, 138–148.
43. Rabinovsky, E.D. and Draghia-Akli, R. (2004) Insulin-like growth factor I plasmid therapy promotes in vivo angiogenesis. *Mol. Ther.* **9**, 46–55.
44. Brown, P.A., Davis, W.C., and Draghia-Akli, R. (2004) Immune-enhancing effects of growth hormone-releasing hormone delivered by plasmid injection and electroporation. *Mol. Ther.* **10**, 644–651.
45. Otten, G.R., Schaefer, M., Doe, B., et al. (2006) Potent immunogenicity of an HIV-1 gag-pol fusion DNA vaccine delivered by in vivo electroporation. *Vaccine.* **24**, 4503–4509.
46. Tjelle, T.E., Salte, R., Mathiesen, I., and Kjeken, R. (2006) A novel electroporation device for gene delivery in large animals and humans. *Vaccine.* **24**, 4667–4670.
47. Peng, B., Zhao, Y., Lu, H., Pang, W., and Xu, Y. (2005) In vivo plasmid DNA electroporation resulted in transfection of satellite cells and lasting transgene expression in regenerated muscle fibers. *Biochem. Biophys. Res. Commun.* **338**, 1490–1498.
48. Schertzer, J.D., Plant, D.R., and Lynch, G.S. (2005) Optimizing plasmid-based gene transfer for investigating skeletal muscle structure and function. *Mol. Ther.* **13**, 795–803.

Chapter 26
Treatment of SCCVII Tumors with Systemic Chemotherapy and Interleukin-12 Gene Therapy Combination

Marina Torrero and Shulin Li

Abstract Cyclophosphamide and Interleukin-12 (IL-12) have been successfully used in clinical trials for treating malignancies. In this study, we explore the coadministration of cyclophosphamide and IL-12 plasmid DNA followed by electroporation for treating SCCVII in mice. Cyclophosphamide, IL-12 plasmid DNA, or a combination of both was injected intramuscularly in mice bearing SCCVII tumors. The tumor growth, survival, cytokine expression, cytotoxic T lymphocyte activity, and vascular density were analyzed. Coadministration of cyclophosphamide and IL-12 plasmid DNA via electroporation delays tumor growth and increases survival in mice. This combination therapy has great potential to be translated to a clinical setting for treating malignancies.

Keywords: Interleukin-12, SCCVII, electroporation, gene therapy, cyclophosphamide

1. Introduction

Cyclophosphamide (CTX) can stimulate delayed-type hypersensitivity, and it induces a T_H1 antitumor response in combination with interleukin-12 (IL-12) (1, 2). Previous studies in mice have shown that combination therapy using IL-12 and CTX was superior to each molecule alone for tumor regression in the murine MB-49 bladder carcinoma, B16 melanoma (3), and MCA207 sarcoma (1, 4). More recently, treatment with IL-12, followed by CTX plus IL-12, has successfully cured established SCCVII tumors in mice (5). These groups delivered intraperitoneally (i.p.) 200–500 ng of recombinant IL-12 and 100 mg/kg CTX per mouse; however, the combination of CTX with IL-12 gene therapy has not been reported, and intramuscular (i.m.) gene delivery followed by electroporation is a simple approach for this treatment. Several investigators contributed to the optimization of intramuscular gene electrotransfer (6–8). In this study, we treated mice bearing SCCVII tumors with low doses of CTX (5 mg/kg) in combination with IL-12 plasmid DNA (10 μg per mouse) i.m. followed by electroporation. Tumor growth and vessel density, for monitoring angiogenesis, were measured in mice. In addition, CTL and cytokine

expression were measured to monitor the immune response associated with this combination therapy.

2. Materials

2.1. Plasmid DNA and Antibiotics

1. The IL-12 gene construct used for the in vivo study was obtained from Valentis (Burlingame, CA) and contains DNA fragments encoding both p35 and p40 subunits in the same backbone. The two subunits are driven by two independent CMV promoters and terminated by two independent bovine growth hormone polyadenylation signals (9).
2. The control plasmid DNA consists of a deletion of the IL-12 gene from the IL-12 construct. The plasmid DNA was manufactured using the QIAGEN Endo-Free Prep kit. Working solutions were prepared a few minutes before injection by diluting 10 µg DNA in 60 µL half-strength saline.
3. Cyclophosphamide (CTX, Sigma-Aldrich, St Louis, MO) was dissolved in water at 100 mg/mL and stored in aliquots at −80°C. Working solution should be prepared a few minutes before injection by diluting 100 µg CTX in 60 µL 0.5× saline.
4. Endotoxin-free plasmid purification kit (Qiagen).
5. 0.5× saline (75 mM sodium chloride).

2.2. In Vivo Experiments

1. Six- to eight-week-old C3H mice, weighing 18–20 g, from the in-house animal breeding facility were used for this study and were maintained under National Institutes of Health guidelines, approved by the Institutional Animal Care and Use Committee of Louisiana State University.
2. Insulin syringes.
3. IsoFlo (Isoflurane).
4. Anesthetic machine with oxygen cylinder and isoflurane vaporizer (Abbott Laboratories, North Chicago, IL).
5. SCCVII cells, a spontaneously arising murine squamous cell carcinoma that has been well characterized in C3H/HeJ mice (10, 11), were used to generate syngeneic transplant tumors. This cell line was obtained from Dr. Thomas Carey at the University of Michigan (Ann Arbor, MI).
6. Electroporator BTS EC830 (Inovio, San Diego, CA).
7. Caliper electrode (Inovio).

2.3. Cell Culture

1. SCCVII cells: The cells were maintained in Dulbecco's modified Eagle's medium (DMEM) supplemented with penicillin, streptomycin, glutamine, and 10% fetal bovine serum (FBS).
2. Red blood lysis solution (Gentra Systems, Minneapolis, MN).
3. 10× Trypsin-EDTA solution: 2% trypsin and 50 mM EDTA in Hanks' Balanced Salt Solution (HBSS) buffer.
4. Mitomycin C: 50 µg/mL.
5. Cell strainer: Containing pores of 70 µm diameter.
6. CyToxiLux kit (OncoImmunin, Gaithersburg, MD): CTL activity assay kit.
7. Biosafety cabinet: For cell culture.

2.4. Collection and Processing Mouse Samples

1. Aluminum foil squares.
2. Silica beads.
3. 5× cell lysis buffer (Promega, Madison, WI).
4. Mini-Bead Beater (BioSpec Products, Bartlesville, OK).
5. Liquid nitrogen dewar.
6. Murine p70 IL-12 and interferon-γ (IFN-γ) quantitative enzyme-linked immunosorbent assay (ELISA) kits (R&D Systems, Minneapolis, MN).
7. BCA protein assay kit.

2.5. Immunostaining

1. Cold acetone.
2. Phosphate-buffered saline (PBS).
3. 1% bovine serum albumin (BSA) in PBS: Prepare the solution and store at 4°C.
4. Primary antibody: rat anti-mouse CD31. Make 1:400 dilution in 1% BSA in PBS before using.
5. Secondary antibody: biotinylated goat anti-rat Ig. Make 1:400 dilution in 1% BSA in PBS before using.
6. 30% hydrogen peroxide.
7. Enzyme substrate: diaminobenzidine (DAB). Store at −20°C.
8. Vectastain ABC kit (Vector Labs, Burlingame, CA): Store at 4°C.
9. 100%, 95%, and 75% ethyl alcohol (EtOH).
10. Xylene.

3. Methods

In this study, SCCVII-tumor-bearing syngeneic C3H mice were used to show the therapeutic effects of CTX and IL-12 plasmid DNA coadministration followed by electroporation. We compared treatment groups using CTX plus IL-12 plasmid DNA, IL-12 plasmid DNA, CTX, and control plasmid DNA. It is important to use the same amount of DNA or antibiotic among groups, as well as the same volumes for injection, to obtain reliable results. The first step was to determine whether the combination of CTX and IL-12 plasmid DNA affects the expression of the administered IL-12 plasmid DNA in the mouse. The expression of IL-12 in tumors was determined using ELISA. Tumor growth was measured with a caliper from the day we started treatment to the end of the experiment. We set up a different experiment to collect samples (spleens and tumors) for CTL activity and immunostaining.

3.1. Preparation of SCCVII Cells for Generating SCCVII Tumors

1. Harvest SCCVII cells with trypsin/EDTA when approaching confluence.
2. Resuspend cells in PBS and count under the microscope.
3. Dilute cells to 2×10^5 cells/30 μL PBS, which is the number of cells to be administered to each mouse.
4. Anesthetize mice by providing isoflurane (*see* **Note 1**).
5. Shave a spot on the back with electric razor and subcutaneously inoculate with 2×10^5 SCCVII cells in a volume of 30 μL using an insulin syringe (*see* **Note 2**).

3.2. Intramuscular Injection and Electroporation for Treating SCCVII-Tumor-Bearing Mice

Four days after tumor inoculation, when tumors are 2–4 mm, we deliver the IL-12 plasmid DNA, CTX, IL-12 plasmid DNA plus CTX, or control plasmid DNA to the left and right hind tibialis muscles followed by electroporation. In each experiment, five animals for each treatment or control group were used. Three administrations of 10 μg DNA per mouse, 100 μg of CTX per mouse, or both were performed once a week. Electrotransfer parameters for intramuscular injection were optimized previously (12, 13).

1. Anesthetize mice by isoflurane inhalation.
2. Rub hind legs with a 70% alcohol pad to visualize the tibialis muscles.
3. Using an insulin syringe, slowly (to avoid leakage or muscle burst) inject each hind tibialis muscle with 5 μg of DNA, 50 μg of CTX, or both in a volume of 30 μL.
4. Hold the muscle tight with a caliper covering the area that was injected, measure gap, and apply two 20-ms pulses of 35 V/mm (*see* **Note 3**).

5. Monitor tumor growth every 3 days with a caliper as soon as the treatment was initiated (*see* **Note 4**). Calculate tumor volume with the following formula: $V = \pi/8(a \times b^2)$, where V is volume, a is maximum tumor diameter, and b is diameter at 90° to a. An example is shown in Fig. 26.1.

6. Monitor mouse survival from the beginning of tumor inoculation to the end of the experiment. An example is shown in Fig. 26.2 (*see* **Note 5**).

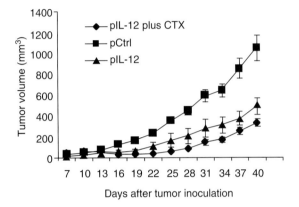

Fig. 26.1 Inhibition of tumor growth by intramuscular coadministration of cyclophosphamide (CTX) and IL-12 plasmid DNA. Three coadministrations were performed once a week, starting 4 days after inoculation of SCCVII tumor cells. Ten micrograms of control plasmid DNA (pCtrl), IL-12 plasmid DNA (pIL-12), or a combination of pIL-12 and 100 μg of CTX were injected into C3H tumor-bearing mice via electroporation at 35 V/mm and 20 ms for two pulses. Five mice were used for each treatment group

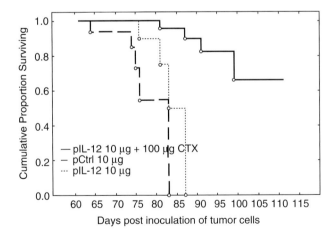

Fig. 26.2 Extension of tumor-bearing mice survival by coadministration of cyclophosphamide and IL-12 plasmid DNA. See legend of Fig. 26.1 for treatment details and electroporation condition. Kaplan–Meier survival curves from different treatment groups were drawn using Statistica software

3.3. Analysis of Cytokine Expression

1. Remove and wrap tumors in aluminum foil and immediately drop tumors into liquid nitrogen to preserve tumor tissues and prevent protein degradation (*see* **Note 6**).
2. Smash tumors with a hammer and dissolve the crushed tissue powder in 1× cell lysis buffer before the samples are thawed.
3. Homogenize the tumor tissue with a mini-bead beater in the presence of 2-mm silica beads for 2 min.
4. Centrifuge samples for 3 min at maximum speed using Eppendorf microcentrifuge.
5. Analyze the expression of IL-12 and IFN-γ by using 50 μL of the tumor extract and ELISA kits following the manufacturer's protocol.
6. Determine total protein of each sample using the BCA protein assay kit to normalize the level of cytokine expression in tumors. An example is shown in Fig. 26.3.

3.4. Analysis of CTL Activity

T cell cytotoxicity was evaluated using a CyToxiLux kit (OncoImmunin), a single-cell-based fluorogenic cytotoxicity assay (14).

3.4.1. Preparation of Stimulator Cells

1. Prepare SCCVII cell suspension in 2 mL complete RPMI media, and wash the cells with PBS to remove trypsin-EDTA.

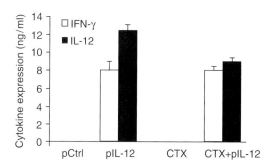

Fig. 26.3 Analysis of IL-12 and IFN-γ expression in tumors after intramuscular coadministration of cyclophosphamide (CTX) and IL-12 plasmid DNA via electroporation. Ten micrograms of control DNA (pCtrl), IL-12 plasmid DNA (pIL-12), 100 μg of CTX, or a combination of pIL-12 and CTX were injected into C3H tumor-bearing mice via electroporation at 35 V/mm and 20 ms for two pulses. Tumors (*n* = 5) were removed on day 3 after treatment and tumor extracts were isolated as described in sect. 3.3. The level of cytokines was determined using ELISA kits

2. Add 0.5 mg of mitomycin C.
3. Incubate 1 h at 37°C in a tube protected from light. Prepare effector cells.
4. Wash thrice with PBS to completely remove the mytomicin C.
5. Adjust the cell concentration to 5×10^6/mL.

3.4.2. Effector Cell Preparation

1. Kill mice using CO_2 asphyxiation.
2. Dip the mouse in 70% EtOH. Remove spleen with sterile tools to avoid contamination and place it in 50-mL tube containing RPMI 1640 media.
3. Under the biosafety hood, transfer the spleen into 60-mm tissue culture dish. Mince spleens using the core of a 5-mL syringe in a 70-μm cell strainer.
4. Pour 10 mL RPMI 1640 media supplemented with penicillin, streptomycin, and glutamine through the cell strainer to recover cells.
5. Pipette to resuspend the splenocytes and transfer into 50-mL Falcon conical tube.
6. Centrifuge for 10 min at 250 g.
7. Remove the supernatant using a vacuum, resuspend the pellet in 10 mL red blood cell lysis solution, and incubate for 5 min on ice.
8. Centrifuge and wash the cells with RPMI media.
9. Resuspend in complete RPMI media (10% FBS) and adjust the cell concentration to 12.5×10^6/mL.

3.4.3. Stimulation of Effector Cells

1. Add 2 mL of effector cells and 0.2 mL of stimulator cells (ratio 25:1) to each well in a 6-well plate. Set up 3 wells per group and incubate cells for 3 days at 37°C.

3.4.4 Preparation of the Stimulated Effector Cells for CTL Activity Assay

1. Harvest nonadherent effector cells and centrifuge for 10 min at 250 g.
2. Adjust cell concentration to 3×10^7/mL with serum-free medium.

3.4.5. Preparation of Target Cells for CTL Activity Assay

1. Trypsinize SCCVII cells, add complete media, and centrifuge.
2. Wash target cells in PBS and resuspend at 3×10^5/mL RPMI.
3. Transfer an aliquot of the reconstituted target cell fluorescence label (TFL) from CytotoxiLux kit into the target cells (1:3,000) to label the target cells with red fluorescence.

4. Incubate at 37°C for 1 h.
5. Wash cells thrice with PBS.
6. Adjust the cells to 3×10^5/mL in serum-free RPMI media.

3.4.6. Determination of the CTL Activity by Coincubation of Target and Effector Cells

1. Mix effector and target cells in each microfuge tube at E:T ratios of 100:1, 50:1, and 25:1.
2. Set up enough tubes for treatment groups, control (target cells only), and CD8 and NK antibodies for depletion (negative controls).
3. Incubate 8–16 h for apoptosis in humidified incubator.
4. Centrifuge at low speed and resuspend cells very gently in 50 µL of wash buffer (*see* **Note 7**).
5. Count the number of left-over live red fluorescence target cells under a fluorescence microscope. Start with the control group and calculate the total number of red fluorescence cells per field (*see* **Note 8**).
6. Calculate CTL activity using the equation CTL activity = 100 × [total number of target cells/field – number of alive red cells in each group (mean)]/(total number of target cells/field). An example is shown in Fig. 26.4.

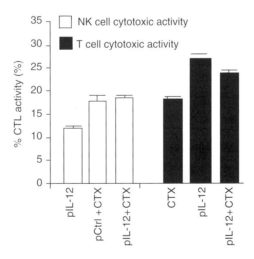

Fig. 26.4 Cytotoxic T lymphocyte (CTL) activity mediated by coadministration of cyclophosphamide and IL-12. See legend of Fig. 26.3 for treatment groups and dose. The treatment was performed once a week. Mice (*n* = 3) were killed 1 week after the second treatment and spleens were removed. Spleen cells were used for analysis of CTL activity against SCCVII tumor cells. To determine the NK cell activity, CD8 T cells were depleted (*open bars*). To determine the CTL activity, NK cells were depleted with antibodies (*filled bars*)

3.5. Immunostaining Analysis for Vessel Density

1. Bring frozen tumor sections to room temperature for 15 min.
2. Fix slides in cold acetone for 5 min on ice.
3. Rinse in PBS thrice for 5 min each.
4. Incubate in freshly prepared 0.2% H_2O_2 (30 μL of 30% in 50 mL PBS) for 2–3 min.
5. Rinse in PBS thrice for 5 min each.
6. Block with 1% BSA in PBS (200 μL per section) for 1 h at room temperature.
7. Drain slides and apply 50 μL/section primary antibody (1:200–1:500 dilution in 1% BSA in PBS), covering the tissue sections with Parafilm (*see* **Note 9**).
8. Incubate overnight at 4°C in humidified box (add water to a tip box, lay slide on tip rack, and cover).
9. Bring slides to room temperature and wash thrice in PBS for 5 min each.
10. Drain slides and apply 50 μL/section secondary antibody (1:400 dilution in 1% BSA in PBS), covering the tissue sections with Parafilm and incubate at room temperature for 45–60 min (*see* **Note 9**).
11. Prepare ABC reagent: one drop of A and one drop of B in 1–2 mL PBS. Let stand at room temperature for 30 min.
12. Rinse slides in PBS thrice for 5 min each.
13. Add 100 μL ABC reagent/section and incubate at room temperature for 30–45 min.
14. Prepare DAB solution: mix 20 mg DAB in 50 mL PBS, cover the tube with aluminum foil to avoid light, and let stand at room temperature for 30 min.
15. Rinse slides in PBS thrice for 5 min each.
16. Add 30 μL of 30% H_2O_2 to 50 mL of DAB solution just before using, and mix well.
17. Incubate one slide in DAB for 1–15 min, then check for color development under the microscope. Stain rest of the slides for the same amount of time.
18. Rinse slides in water for 0.5 min.
19. Place slides in 75, 95, and 100% ethanol for 2 min each.
20. Place completely dried slides in xylene for 1–2 min. Add a drop of mounting media (Permount, Fisher) and cover with cover slip.
21. Determine the average number of vessels per field under a microscope at 40× magnification (*see* **Note 10**).

3.6. Statistical Analysis

A two-sided Student's *t* test was performed to compare tumor growth, the number of vessels, CTL activity, and the expression levels of cytokines between treatment groups. The survival analysis was done using χ^2 tests, followed by Gehan's Wilcoxon test to compare means of individual treatments. *p* values less than 0.05 were considered statistically significant.

4. Notes

1. Isoflurane inhalation is safe. So the mice recover very fast and it can be used several times. It is recommended to start with a low dose and observe the mice at the beginning of the experiment to decide the appropriate time and dose of anesthesia.
2. Inject the cells very slowly to avoid breaking the skin, and a bubble will appear in the area of injection.
3. Six-week-old mice usually have 5-mm gap in the thigh. It will require 175 V. The other two measurements (20 ms and 2 pulses) are standard and do not change.
4. One person should be responsible for measuring the tumors from the beginning to the end of the experiment to avoid variation among measurements.
5. The treated mice died from either primary-tumor-volume-dependent or tumor-volume-independent causes. Primary-tumor-volume-dependent death is defined as death occurring when tumors reach 2 cm in diameter, at which time mice are killed for humane reasons. Tumor-volume-independent death describes the natural death occurring before the primary tumor size reaches 2 cm in diameter.
6. Use heavy-duty aluminum foil to wrap the tumors. Make sure the tumors are completely covered with the aluminum foil. Do not wrap the tumors too tightly to avoid breaking the aluminum foil with the hammer.
7. This assay is based on counting live cells; therefore, it is critical to avoid cell lysis by pipetting or shaking the cells.
8. Count one group at a time and start with the highest ratio 100:1, which should be the highest percent of CTL activity. In addition, check the antibody depletion for negative control. Count all the groups the same day to avoid spontaneous cell lysis.
9. Use filter paper to absorb the PBS before adding the antibody. Make sure the antibody covers the tissue and avoid bubbles.
10. Start at a low magnification to observe the whole tissue. Identify a field that is homogeneous and not close to the edge, then set up the microscope at 40× and count the vessels.

References

1. Tsung, K., Meko, J.B., Tsung, Y.L., Peplinski, G.R., and Norton, J.A. (1998) Immune response against large tumors eradicated by treatment with cyclophosphamide and IL-12. *J. Immunol.* **160**, 1369–1377.
2. Kaufmann, S.H.E., Hahn, H., and Diamantstein, T. (1980) Relative susceptibilities of T cell subsets involved in delayed-type hypersensitivity to sheep red blood cells to the in vitro action of 4-hydroperoxycyclophosphamide. *J. Immunol.* **125**, 1104–1110.
3. Teicher, B.A., Ara, G., Buxton, D., Leonard, J., and Schaub, R.G. (1997) Optimal scheduling of IL-12 and chemotherapy in the murine MB-49 bladder carcinoma and B16 melanoma. *Clin. Cancer Res.* **3**, 1661–1667.
4. Karnbach, C., Daws, M.R., Niemi, E.R., and Nakamura, M.C. (2001) Immune rejection of a large sarcoma following cyclophosphamide and IL-12 treatment requires both NK and NK T cells and is associated with the induction of a novel NK T cell population. *J. Immunol.* **167**, 2569–2576.
5. Mandpe, A.H., Tsung, K., and Norton, J.A. (2003) Cure of an established nonimmunogenic tumor, SCCVII, with a novel IL-12 based immunotherapy regimen in C3H mice. *Arch. Otolaryngol. Head Neck Surg.* **129**, 786–792.
6. Titomirov, A.V., Sukharev, S., and Kistanova, E. (1991) In vivo electroporation and stable transformation of skin cells of newborn mice by plasmid DNA. *Biochim. Biophys. Acta.* **1088**, 131–134.

7. Aihara, H. and Miyazaki, J. (1998) Gene transfer into muscle by electroporation in vivo. *Nat. Biotechnol.* **16**, 867–870.

8. Mir, L.M., Bureau, M.F., Gehl, J. et al. (1999) High-efficiency gene transfer into skeletal muscle mediated by electric pulses. *Proc. Natl. Acad. Sci. U.S.A.* **96**, 4262–4267.

9. Coleman, M., Muller, S., Quezada, A. et al. (1998) Nonviral interferon alpha gene therapy inhibits growth of established tumors by eliciting a systemic immune response. *Hum. Gene Ther.* **9**, 2223–2230.

10. Grandis, J., Chang, M.J., Yu, W.D., and Johnson, C. (1999) Antitumor activity of IL-1α and cisplatin in a murine model system. *Arch. Otolaryngol. Head Neck Surg.* **121**, 197–200.

11. O'Malley, B., Cope, K., Johnson, C., and Schwartz, M. (1997) A new immunocompetent murine model for oral cancer. *Arch. Otolaryngol. Head Neck Surg.* **123**, 20–24.

12. Li, S., Zhang, X., Xia, X. et al. (2001) Intramuscular electroporation delivery of IFN-α gene therapy for inhibition of tumor growth located at a distant site. *Gene Ther.* **8**, 400–407.

13. Li, S., Xia, X., Zhang, X., and Suen, J. (2002) Regression of tumors by IFN-α electrotransfer gene therapy and analysis of the responsible genes by cDNA array. *Gene Ther.* **9**, 390–397.

14. Liu, L., Chahroudi, A., Silvestri, G. et al. (2002) Visualization and quantification of T cell-mediated cytotoxicity using cell-permeable fluorogenic caspase substrates. *Nat. Med.* **8**, 185–189.

Chapter 27

Electroporation for Drug and Gene Delivery in the Clinic: Doctors Go Electric

Julie Gehl

Abstract Electroporation is a unique system for drug and gene delivery, as it is possible to very specifically target certain tissues within the body with whatever drug, gene, isotope, or other product is desired in a specific situation. An increasing number of clinical trials are being launched, and sophistication of equipment and protocols continues. This chapter reviews present knowledge from clinical trials, describes important issues in the patient management when using electroporation, and outlines future perspectives of the technology.

Keywords: electroporation, clinical protocol, electrode, electrochemotherapy, gene electrotransfer

1. Introduction

After a long-winded road of discovery (1) from physicists laying down the theoretical fundament of electroporation, through numerous investigations with artificial lipid membranes to work with cells, to eventually animal experiments, electroporation came to use in patients in the early 1990s (2). From the beginning, interest has been focussed on the following three areas: (1) electroporation for drug delivery—particularly for chemotherapy, (2) electroporation for gene transfer, and (3) electroporation for delivery of other molecules such as isotopes.

For several reasons, the first trials have focussed on electroporation for the enhancement of delivery of chemotherapeutic drugs to tumors. The ability of electrochemotherapy (ECT) to markedly enhance efficacy of chemotherapeutic agents in the region encompassed by the electrodes is a fascinating way of selectively hitting tumor tissue. Furthermore, there is a large need for improved treatment strategies for cancer patients.

The experiences gathered with electrochemotherapy over the past decade, as well as the development of equipment and protocols for clinical use, greatly benefit further development of the clinical trials on gene transfer. The transition

S. Li (ed.), *Electroporation Protocols: Preclinical and Clinical Gene Medicine.*
From *Methods in Molecular Biology, Vol. 423.*
© Humana Press 2008

from electrochemotherapy to electro gene transfer protocols will also mark a transition from electroporation-based treatments being solely for cancer patients to these treatments being an option for patients with a wide variety of chronic illnesses. Electroporation enables the delivery of any molecule to any tissue, and it is a technology for which the perspectives are just beginning to be unveiled.

2. Patient Management in Electroporation Protocols

2.1. Pulse Generators

Considerable advances have been made in the fabrication of electroporation equipment for clinical use. In the case of pulse generators, several models are available or are in the process of becoming so. What can be noted is that these machines are square-wave generators, enabling independent control of pulse amplitude and length. For use in patients, the equipment must be designed with strict separation between circuits of the power supply and the treatment end. Verification of actual treatment is not only the desired standard but also very useful for the physician to know whether pulses have been administered satis-factorily. One generator (Cliniporator, IGEA, Italy) offers online registration of pulse voltage and current delivered. Other generators may be tested using an oscilloscope.

2.2. Electrodes

One very important aspect of electroporation is electrodes. Plate electrodes can be used to treat superficial tissues, either skin or tumors, while needle electrodes can be used to treat more difficult areas. Currently, several different models of needle electrodes are in production. A six-needle circular electrode was developed by Gilbert et al. (3), a hexagonal electrode by Mir et al., and a needle electrode with two opposing arrays of electrodes by Gehl et al. (4). These electrodes give rise to different field characteristics (4), and it is important to understand the relationship between electrode configuration, distance between electrodes, and applied voltage for optimal treatment. In treating cancer patients, the optimal solution is to have options in the choice of electrodes depending on the need of the particular patient and the tumor being treated. Plate electrodes can be useful for superficial tumors, while needle array electrodes and hexagonal electrodes may be useful for smaller tumors, particularly in sensitive areas, and larger tumors, respectively.

2.3. Protocols for Electroporation

Current electroporation protocols are based on preclinical work, and thus, most electrochemotherapy protocols use 6–8 high-voltage pulses to deliver small molecules and some gene transfer protocols use high-voltage pulses (5). Still, other protocols prefer a combination of high- and low-voltage pulses to facilitate an electrophoretic effect on the DNA (6, 7).

An observation is that pain from the electroporation procedures is proportional to absolute applied voltage (8, 9); therefore, the applied voltage needs to stay as low as possible while achieving efficient permeabilization and transfer. Authors report using 1,000 V/cm (applied voltage/electrode distance ratio) up to approximately 1,300 V/cm for the pulses used for electrochemotherapy (8, 10). One simple way of lowering

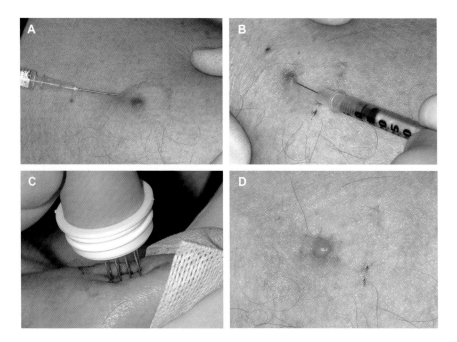

Fig. 27.1 Electro gene transfer to a tumor nodule of disseminated malignant melanoma. **A**: Lidocaine 2% is injected in a square surrounding the treatment area. **B**: 100 μg of DNA in 100 μL is injected into the tumor nodule. **C**: The needle array electrode is positioned in such a way that the nodule is covered. The array consists of two arrays with four needles each, connected to the positive or negative pole. If the nodule is larger than the electrode, several applications can be made. A pulse consisting of a high-voltage pulse of 800 V/cm (voltage to electrode distance ratio), followed after 1 s by a low-voltage pulse of 80 V/cm, of 400-ms duration. The Cliniporator (IGEA) was used. As can be seen, the doctor's left hand is lifting the treatment area away from the underlying muscle, using a gauze pad in order to get a good grip. This alleviates muscle contraction substantially; indeed in this case, the patient reported that the muscle contraction was barely noticeable. **D**: Immediately after the electrotransfer procedure. The entire procedure takes only a few minutes

the applied voltage is to decrease the gap between electrodes, e.g., 0.4 cm (Fig. 27.1). As for gene transfer protocols, the pulse amplitude varies from being similar to the pulses used for ECT (5) to a combination of one high-voltage pulse followed by a low-voltage pulse (6), to several pulses of longer duration and lower voltage (11).

2.4. Handling the Electroporation Procedure

Electroporation can easily be handled as an out-patient procedure, since only local anesthesia is needed to decrease pain associated with insertion of the needles and the electric pulses. Briefly, a "square" of local anesthesia (e.g., 2% lidocaine with epinephrine) is put around the treatment area, after which electrodes are positioned and the pulses applied (12). In more complicated cases where there are several tumors or the tumors are located in such a way that the local anesthetic cannot be applied satisfactorily, general anesthesia is recommended (12).

Contractions of underlying muscle are also associated with the application of pulses. The contractions are strongly related to the distance between the tips of the electrode and the surface of the muscle and can be alleviated by lifting up the lesion to be treated away from the muscle. Patients generally describe the muscle contraction as being surprising, sometimes unpleasant, but not painful. The muscle contraction is always regional and confined to the muscle just underneath the electrodes.

Also, a vascular effect of high-voltage pulses has been observed (13). In normal tissue, there is a brief and reflexory constriction of afferent arterioles (13), and, in tumor tissue with more fragile and tortuous vessels, long-term hypoperfusion can occur after electroporation (14). Possibly, this vascular effect is beneficial in electrochemotherapy, because it may mean that a higher concentration of drug is entrapped in the tumor due to lack of wash-out at the time of electroporation (13). Similarly, in gene therapy, transient hypoperfusion has been shown to enhance expression (15, 16).

3. Electrochemotherapy Protocols

3.1. Drugs, Doses, and Administration Route

Several studies have investigated the use of electroporation to enhance the efficacy of a large proportion of the drugs used clinically today (17–19) and for the treatment of various cancer types (18). A consistent finding is that lipo- or amphiphilic drugs traverse the cell membrane well without electroporation, while an enhancement in cytotoxicity is found with drugs that, under normal circumstances, do not pass the cell membrane easily. The most prominent example is bleomycin, which is a well-known and quite remarkable drug. Acting as an enzyme, one bleomycin

molecule can cause several DNA strand breaks and is highly toxic once inside the cell. Therefore, the enhancement of toxicity is several hundred fold, once the bleomycin accesses the cytosol. Gothelf et al. have reviewed ECT using bleomycin (20). Drug doses used in bleomycin-based electrochemotherapy have differed. Some groups have used intratumoral injection with relatively high doses (8), while other have settled for lower doses. Also, for i.v. administration, bleomycin is generally given in the doses used in standard treatment protocols (12). The results of the different regimens are seemingly comparable (8, 10, 20), but there may be more pronounced necrosis with the higher doses and a better chance to conserve normal tissue with the lower doses. Another option is cisplatinum (also known as cisplatin), which can be used for direct intratumoral injection. Its cytotoxicity is enhanced by approximately a factor 8 (21), and Sersa et al. have completed a comprehensive review of the use of this drug (22).

3.2. Posttreatment Considerations

Electrochemotherapy is generally a one time treatment because of its high efficacy; however, re-treatment is fully possible and can be considered in some cases, such as when there is regrowth of the tumor due to insufficient treatment of the margins in the first treatment. If the tumor penetrates the skin, a crust is formed 1–2 days after the treatment, and this crust will persist until the normal tissue underneath heals (often several weeks). In cases where the nodules are subcutaneous, it is generally possible to stick the needles through the skin and treat the tumor underneath while preserving the overlying normal skin, even while using i.v. bleomycin and uninsulated needles. So, there is a differential effect between the tumor tissue and the normal skin, which enables selective tumor destruction.

3.3. Standard Operating Procedures for Electrochemotherapy

A group of European electroporation expertise centers have performed a clinical study on electrochemotherapy (ESOPE) (10). Furthermore, a set of detailed standard operating procedures have been produced to enable doctors to safely deliver electrochemotherapy (12).

3.4. Electrochemotherapy Outcome

Thorough reviews on the outcome of electrochemotherapy using bleomycin and cisplatinum have been given by Gothelf et al. (20) and by Sersa et al. (22), respectively. The highly encouraging results are confirmed in the ESOPE study (10). The conclusions from these results are as follows: (1) electrochemotherapy has been efficient in every cancer histology type tested (e.g., breast, colon, bladder, renal

cell, malignant melanoma, basal cell carcinoma); (2) complete eradication of treated nodules occurs in approximately 75% of the cases, and at least a partial remission occurs in 85–90% of the treated patients; (3) i.v. or intratumoral administration of bleomycin and intratumoral administration of cisplatinum are equal treatment options; and (4) the toxicity profile of the treatment is very favorable, and patients may receive a one time treatment with very limited toxicity.

4. Electro Gene Transfer Protocols

Numerous studies on gene transfer on a wide variety of tissues in animal models have been published, with various indications (23). Most of the studies investigate the treatment of protein deficiencies and cancers using cytokines. These avenues are sought, as gene transfer to muscle may enable it to function as a protein factory, yielding long-term production of proteins, and gene transfer of cytokines to tumors may yield a local, as well as a systemic, antitumor response.

At this moment, no human electroporation-based gene transfer trials have been completed and published; however, several trials are underway, and some preliminary results are published in abstract form. At the University of South Florida, patients with advanced malignant melanoma are being treated by transfecting genes coding for interleukin-12 (IL-12) into melanoma metastases (American Society of Gene Therapy annual meeting 2006). Six high-voltage pulses are used with the Medpulser (Inovio, San Diego).

In a European study under the ESOPE group, patients are receiving electro gene transfer using a gene coding for β-galactosidase, with the aim of testing the efficacy and toxicity. Tumor tissue and normal skin are transfected using one high-voltage pulse, followed by a low-voltage electrophoretic pulse. The Cliniporator (IGEA) is used for these treatments.

4.1. Handling the Electro Gene Transfer Procedure

Considerable knowledge has been obtained from the clinical use of electrochemotherapy, as previously described in this chapter, and is also detailed by Mir et al. (12). To perform electro gene transfer, the following procedures are recommended.

4.2. Patient Information and Reception

1. As electro gene transfer is an investigational treatment, adequate permissions from the ethical committee, medicines agency, data registration permission, etc., must be obtained according to national regulations. Accordingly, written informed consent must be obtained prior to treatment.

2. Inform patient thoroughly about the procedure. Explain that some contraction of underlying musculature is to be expected, but the local anesthesia should protect against pain.
3. The treatment is done in an outpatient setting. Place the patient in a comfortable, lying position. Expose the region to be treated and cover the patient with a blanket to ensure the patient is warm and relaxed. No premedication is necessary.

4.3. Local Anesthesia and Disinfection

1. Disinfect the area to be treated using the standard procedure for your institution, e.g., Klorhexidine-alcohol with two independent applications, allowing the area to dry between applications.
2. Inject a 2% lidocaine solution with epinephrine in a square around the area to be treated, to ensure good coverage. In case of treating several areas, be sure to respect the maximum total dose of lidocaine allowed.

4.4. DNA Injection and Electric Pulsing

1. Inject 100 μg of DNA in 100 μL of solvent (good manufacturing practice (GMP) grade required) per area to be treated. Use an insulin syringe to precisely deliver the planned volume.
2. Electroporation equipment must be ready and a test run must have been performed prior to receiving the patient. Use the Cliniporator (IGEA), or a similar device, which delivers combinations of high-and low-voltage pulses and monitors the actual voltage and current delivered.
3. Choose appropriate electrodes, e.g., the array electrode (Fig. 27.1).
4. Insert the electrode through the skin. Hold the handle of the electrode with your right hand. Lift the area away from the underlying muscle using your left hand.
5. Activate electroporator through foot pedal.
6. Check that the voltage and current delivered were appropriate.
7. Remove electrodes. Bandaging of treated area is generally not necessary, but can be applied as a dry dressing.
8. After treatment the patient can go home.

5. Perspectives in Electroporation-Based Technologies

There are many perspectives in electroporation-based therapies. For electrochemotherapy, the technology is now so far ahead that it is anticipated to become a standard treatment offered in many cancer centers within the next few years. For gene

transfer, further work must be done to characterize optimal pulsing conditions for different tissues in patients, enhancing expression, and allowing control of expression. However, it is very encouraging to see several clinical trials already underway. Electro gene transfer is much safer than viral transfer for the patient, as well as for the working and general environments. Furthermore, the precise placement of electrodes enables precise delivery to certain tissues, enabling a much more specific approach than with other technologies. This may mean that electroporation-based gene transfer can be used in a variety of situations from vaccines to treatment of protein deficiency syndromes and to cancer treatment.

Finally, other molecules may also be transferred in the future. Several options are open, such as the transfer of isotopes or molecules for brachytherapy (e.g., iodine or boron for boron neutron capture), small interfering RNA (siRNA), locked nucleic acids (LNAs).

Electroporation allows you to deliver any molecule to any tissue. That is the reason doctors go electric.

Acknowledgments The ESOPE study was supported by the European Union's 5th FP, under the leadership of Lluis M. Mir. Dr. Poul F. Geertsen photographed the treatment session shown in Fig. 27.1.

References

1. Jaroszeski, M., Heller, R., and Gilbert, R. (2000) Electrochemotherapy, electrogenetherapy, and transdermal drug delivery. Humana, Totowa, NJ.
2. Belehradek, M., Domenge, C., Luboinski, B., et al. (1993) Electrochemotherapy, a new antitumor treatment. First clinical phase I-II trial. *Cancer.* **72**, 3694–3700.
3. Gilbert, R.A., Jaroszeski, M.J., and Heller, R. (1997) Novel electrode designs for electrochemotherapy. *Biochim. Biophys. Acta.* **1334**, 9–14.
4. Gehl, J., Sørensen, T.H., Nielsen, K., et al. (1999) In vivo electroporation of skeletal muscle: Threshold, efficacy and relation to electric field distribution. *Biochim. Biophys. Acta.* **1428**, 233–240.
5. Heller, R., Jaroszeski, M., Atkin, A., et al. (1996) In vivo gene electroinjection and expression in rat liver. *FEBS Lett.* **389**, 225–228.
6. Bureau, M.F., Gehl, J., Deleuze, V., Mir, L.M., and Scherman, D. (2000) Importance of association between permeabilization and electrophoretic forces for intramuscular DNA electrotransfer. *Biochim. Biophys. Acta.* **1474**, 353–359.
7. Mir, L.M., Bureau, M.F., Gehl, J. et al. (1999) High efficiency gene transfer into skeletal muscle mediated by electric pulses. *Proc. Natl. Acad. Sci. U.S.A.* **96**, 4262–4267.
8. Heller, R., Jaroszeski, M.J., Glass, L.F. et al. (1996) Phase I/II trial for the treatment of cutaneous and subcutaneous tumors using electrochemotherapy. *Cancer.* **77**, 964–971.
9. Gehl, J. and Geertsen, P. (2000) Efficient palliation of hemorrhaging malignant melanoma skin metastases by electrochemotherapy. *Melanoma Res.* **10**, 585–589.
10. Marty, M., Sersa, G., Garbay, J.R., Gehl, J., Collins, C.G., Snoj, M., et al. (2006) Electrochemotherapy – An easy, highly effective and safe treatment of cutaneous and subcutaneous metastases: Results of ESOPE (European Standard Operating Procedures of Electrochemotherapy) study. *Ejc Supplements.* **4**, 3–13.

11. Aihara, H. and Miyazaki, J.I. (1998) Gene transfer into muscle by electroporation in vivo. *Nat. Biotechnol.* **16**, 867–870.
12. Mir, L.M., Gehl, J., Sersa, G., Collins, C.G., Garbay, J.R., Billard, V., et al. (2006) Standard operating procedures of the electrochemotherapy: Instructions for the use of bleomycin or cis-plantin administered either systemically or locally and electric pulses delivered by the Cliniporator (TM) by means of invasive or non-invasive electrodes. *Ejc Supplements.* **4**, 14–25.
13. Gehl, J., Skovsgaard, T., and Mir, L.M. (2002) Vascular reactions to in vivo electroporation: characterization and consequences for drug and gene delivery. *Biochim. Biophys. Acta.* **1569**, 51–58.
14. Sersa, G., Cemazar, M., Parkins, C.S., and Chaplin, D.J. (1999) Tumour blood flow changes induced by application of electric pulses. *Eur. J. Cancer.* **35**, 672–677.
15. Tsurumi, Y., Takeshita, S., Chen, D., et al. (1996) Direct intramuscular gene transfer of naked DNA encoding vascular endothelial growth factor augments collateral development and tissue perfusion. *Circulation.* **94**, 3281–3290.
16. Takeshita, S., Isshiki, T., and Sato, T. (1996) Increased expression of direct gene transfer into skeletal muscles observed after acute ischemic injury in rats. *Lab. Invest.* **74**, 1061–1065.
17. Orlowski, S., Belehradek, J., Jr., Paoletti, C., and Mir, L.M. (1988) Transient electropermea-bilization of cells in culture. Increase of the cytotoxicity of anticancer drugs. *Biochem. Pharmacol.* **37**, 4727–4733.
18. Jaroszeski, M.J., Dang, V., Pottinger, C., Hickey, J., Gilbert, R., and Heller, R. (2000) Toxicity of anticancer agents mediated by electroporation in vitro. *Anticancer Drugs.* **11**, 201–208.
19. Gehl, J., Skovsgaard, T., and Mir, L.M. (1998) Enhancement of cytotoxicity by electroper-meabilization: an improved method for screening drugs. *Anticancer Drugs.* **9**, 319–325.
20. Gothelf, A., Mir, L.M., and Gehl, J. (2003) Electrochemotherapy: results of cancer treatment using enhanced delivery of bleomycin by electroporation. *Cancer Treat. Rev.* **29**, 371–387.
21. Sersa, G., Cemazar, M., and Miklavcic, D. (1995) Antitumor effectiveness of electrochemo-therapy with *cis*-diamminedichloroplatinum(II) in mice. *Cancer Res.* **55**, 3450–3455.
22. Sersa, G., Cemazar, M., and Rudolf, Z. (2003) Electrochemotherapy: advantages and draw-backs in treatment of cancer patients. *Cancer Ther.* **1**, 133–142.
23. Mir, L.M., Moller, P.H., Andre, F., and Gehl, J. (2005) Electric pulse-mediated gene delivery to various animal tissues. *Adv. Genet.* **54**, 83–114.

Chapter 28
IL-2 Plasmid Electroporation: From Preclinical Studies to Phase I Clinical Trial

Holly M. Horton, Peggy A. Lalor, and Alain P. Rolland

Abstract Electroporation (EP)-assisted intralesional delivery of Interleukin-2 (IL-2) plasmid (pDNA) has the potential to increase the local concentration of the expressed cytokine for an extended time in the injected tumors while minimizing its systemic concentration, in comparison with systemic delivery of the recombinant cytokine. Nonclinical Investigational New Drug application-enabling studies were performed in mice to evaluate the effect of intratumoral administration of murine IL 2 pDNA on local expression and systemic distribution of IL-2 transgene as well as the inhibition of established tumor growth. The safety of repeated administrations of a human IL-2 pDNA product candidate with EP was evaluated in rats. Following the nonclinical safety and efficacy studies, a human IL-2 pDNA product candidate intralesionally administered with EP to metastatic melanoma patients is currently being investigated in a phase I clinical trial.

Keywords: interleukin-2 (IL-2) plasmid, murine melanoma tumor model, in vivo electroporation, cationic lipid delivery system, repeated-dose safety, phase I clinical trial, metastatic melanoma

1. Introduction

Numerous studies have evaluated cytokine-encoded plasmid DNA (pDNA) delivered by conventional needle and syringe with electroporation (EP) for treatment of cancer in murine tumor models. A number of studies have been performed with pDNA encoding IL-12 or IL-2, and some of these studies have compared different routes of delivery as well as combinations of cytokine-encoded plasmids (Table 28.1). Hanna et al. (1) and Li et al. (2) demonstrated that intramuscular (i.m.) injection of IL-12 pDNA with EP reduced the growth of established squamous cell tumors (SCCVII) by 75% and 40%, respectively. In contrast, Lucas and Heller (3) found that EP-assisted intratumoral (i.t.), but not i.m., delivery of IL-12 pDNA resulted in

Table 28.1 Murine models of EP-assisted delivery of cytokine-encoded pDNA

Cytokine	Route	Tumor	EP conditions	Dose of pDNA	Primary tumor[a]	Metastases[b]	References
IL-12	i.t.	B16F10	1,500 V/cm, 6 pulses, 99 μs/pulse, 6 needle electrodes	50 μg, days 0 and 7 – tumors of 3–5 mm	Yes	NA	(12)
IL-12	i.m.	B16F10	100 V/cm, 12 pulses, 20 ms/pulse, 4 needle electrodes	50 μg, days 0 and 7 – tumors of 3–5 mm	No	NA	(12)
IL-12	i.t.	CT26 and Renca	66 V/cm, 8 pulses, 50 ms/pulse, 6 needle electrode	25 μg, days 0, 2, 3; and days 0 and 7 – tumors of 5 mm	Yes	NA	(4)
IL-12	i.m.	SCCVII	375 V/cm, 2 pulses, 20 ms/pulse, caliper electrodes	10 μg, days 7, 14, and 21 – tumors of 2–3 mm	Yes	NA	(1)
IL-12	i.t.	HCC	150 V, 10 pulses, 50 ms/pulse, 2 needle electrodes	100 μg, day 0 – tumors of 0.5 cm^3	Yes	Yes	(7)
IL-12	i.t.	SCCVII	450 V/cm, 2 pulses, 20 ms/pulse	10 μg, 2× per week	Yes	NA	(2)
IL-12	i.m.	SCCVII	350 V/cm, 2 pulses, 20 ms/pulse	10 μg, 2× per week	No	NA	(2)
IL-12 and IL-18	i.t.	B16F10	50 V, 6 pulses, 100 ms/pulse, 6 needle electrodes	5.5 pmol, days 0, 2, 10, and 12 – tumors of 75 mm^3	Yes	NA	(6)
IL-12 and bleomycin	i.t.	4T1 mammary	450 V/cm, 2 pulses, 20 ms/pulse,	30 μg, days 8 and 15	Yes	Yes	(5)
IFN-α	i.m.	SCCVII	375 V/cm, 2 pulses, 25 ms/pulse, caliper electrodes	10 μg, days 7, 14, and 21 – tumors of 15–30 mm^3	Yes	NA	(8)
IL-2	i.t.	B16F10	100 V, 6 pulses, 50 ms/pulse, caliper electrodes	50 μg, day 0 – tumors of 5–7 mm	Yes	NA	(9)
IL-2	i.t.	Esophageal	1,000 V/cm, 8 pulses, 100 ms/pulse, caliper electrodes	25 μg, day 0 – tumors of 8 mm	Yes	NA	(10)
IL-2 + GM-CSF	i.t.	IMEA.7R.1 hepatoma	1,300 V, 6 pulses, 99 μs/pulse, 6 needle electrode	50 μg, days 0, 1, and 2 – tumors of 4 mm	Yes	NA	(11)

IL interleukin, *i.t.* intratumoral, *NA*-Not Applicable, *i.m.* intramuscular, *TNF-α* tumor necrosis factor alpha.

[a] Inhibition in primary tumor growth.

[b] Reduction in establishment of metastases.

tumor regression in 47% of melanoma-bearing mice. Similarly, EP-assisted i.t. administration of IL-12 pDNA significantly inhibited the growth of CT26 colon carcinoma and Renca renal tumors (4).

EP-assisted IL-12 pDNA therapy has also been combined with other immunotherapies. Local delivery of bleomycin coadministered with EP-assisted i.t. injection of IL-12 pDNA significantly reduced the growth of 4T1 mammary carcinoma and squamous cell carcinoma (5). In a similar study, IL-18 and IL-12 pDNA injected i.t., followed by EP of established tumors, resulted in a significant suppression of B16F10 melanoma growth (6). In some studies, local administration of IL-12 pDNA with EP was shown to prevent the establishment of tumor metastases (5, 7). Overall, in most of the studies, i.t. injection of IL-12 pDNA, followed by EP, resulted in a significant antitumor effect on primary tumors and, in some cases, also reduced the establishment of metastases.

EP-assisted delivery of other cytokine pDNAs has also been evaluated in a number of tumor models. Li et al. (8) found that i.m. injection of interferon- α (IFN-α) pDNA, followed by EP, significantly reduced the growth of squamous cell carcinoma. Similarly, Lohr et al. (9) and Matsubara et al. (10) demonstrated that i.t. administration of IL-2, followed by EP, could markedly inhibit the growth of both B16F10 melanoma and human esophageal tumors in nude mice. IL-2 pDNA delivered i.t. in combination with GM-CSF pDNA, followed by EP, also led to a marked reduction in the growth of s.c. hepatic carcinomas (11).

The previously mentioned studies have clearly demonstrated antitumor efficacy in a wide variety of murine tumor models when EP-assisted i.t. delivery of cytokine-encoded pDNA has been used therapeutically. These studies have demonstrated both local and systemic anti-tumor effects. Some additional work has been done that begins to elucidate the mechanism of action of EP-assisted i.t. pDNA therapy. In several of the studies listed in Table 28.1, an increase in tumor-infiltrating NK and T cells was noted after the cytokine pDNA-EP therapy, suggesting an immune-mediated mechanism of action (6, 7). In addition, increased levels of the encoded cytokine detected in the tumor after local pDNA-EP therapy were often correlated with increased antitumor efficacy (2, 4, 11, 12). It should be noted, however, that although levels of the cytokine in tumor after i.t. injection were increased by 10- to 100-fold, its levels in serum remained low, suggesting that possible side effects occurring with systemic delivery of recombinant cytokine may be reduced with local cytokine pDNA therapy.

The following section describes some of the nonclinical efficacy and safety studies performed to evaluate a pDNA product candidate encoding human IL-2 (hIL-2) administered i.t. with EP and summarizes the ongoing phase I clinical trial of a lead therapeutic candidate for stage III and IV metastatic melanoma.

2. Materials

2.1. Murine IL-2 Plasmid—VR1111

The IL-2 pDNA used in the reported nonclinical studies, VR1111 (Vical, San Diego), was constructed by cloning murine IL-2 cDNA into the expression plasmid VR1055 (13). VR1055 contains the cytomegalovirus immediate early (CMV IE) gene promoter/enhancer, a polyadenylation sequence from the rabbit β-globin gene for transcriptional termination, and a bacterial kanamycin resistance gene. Prior to insertion into the backbone plasmid, the 5′ untranslated (UT) sequence of the murine IL-2 cDNA and the first two amino acids of the leader peptide were removed and replaced with the rat insulin II gene 5′ UT sequence and the coding region of the first five amino acids of the rat preproinsulin leader peptide. The modified murine IL-2 cDNA was then cloned into the multiple cloning site of VR1055 to generate the pDNA VR1111. The backbone pDNA, VR1055, served as the control pDNA for all studies.

2.2. Human IL-2 Plasmid—VCL-1102

Human IL-2 pDNA (VCL-1102) contains predicted open reading frames that code for the kanamycin-resistance protein and human IL-2. The IL-2 gene contains a 5′ untranslated region (UTR) and a signal peptide coding sequence (amino acids 1–5) derived from the rat insulin II gene fused in-frame to the sequence encoding human IL-2. The signal peptide of the newly synthesized protein is cleaved to produce the mature secreted form of human IL-2. Transcription of the IL-2 gene in mammalian cells is controlled by the human cytomegalovirus (hCMV) immediate early (hCMV-IE) gene promoter. The primary transcript includes 3′ UTR sequences and a polyadenylation [poly(A)] signal from the bovine growth hormone (BGH) transcriptional terminator region. The BGH poly(A) signal directs cleavage of the precursor RNA and polyadenylation of the mature transcript in mammalian cells.

2.3. Murine Cell Line

The Cloudman S91 M3 melanoma cell line was obtained from ATCC (Rockville, MD) and grown in F12-K medium supplemented with 15% horse serum, 2.5% (fetal bovine serum) FBS, and 0.1% β-mercaptoethanol.

3. Methods

All in vivo studies were performed in accordance with Institutional Animal Care and Use Committee approved protocols and in accordance with recommendations put forth in *The Guide for the Care and Use of Laboratory Animals* (14).

3.1. In Vivo EP Conditions

In all efficacy studies, mice with preestablished tumors were anesthetized for all treatment procedures by using inhalant isoflurane (*see* **Note 1**). A square-wave pulse generator (model T820, BTX, Holliston, MA) and modified caliper electrodes (model 384, BTX) were used to electroporate murine tumors following pDNA injection. The modified calipers consist of $15 \times 15 \times 3\,mm^3$ stainless steel electrode plates with added protruding stainless steel extensions measuring $7 \times 6 \times 3\,mm^3$ at the base to facilitate EP of 20–80-mm^3 tumors. Before EP, conductivity gel (Spectrode 360 electrode gel, Parker Labs, Fairfield, NJ) was applied to the surface of the plates. Immediately after pDNA injection, the plates were applied to the clipped skin on either side of the tumor and eight consecutive square-wave electrical pulses each of 1-ms duration were administered. The constant voltage delivered was adjusted for each tumor based on the size of the tumor treated. This allowed electroporation of each tumor with a comparable field strength generated between the electrodes of approximately 800 V/cm.

For the repeated-dose toxicity studies, Sprague-Dawley rats were anesthetized using isofluorane and injected with human IL-2 pDNA (VCL-1102). Each injected site was electroporated within several minutes after injection by using the MedPulser® Electroporation Therapy (EPT) System, (Inovio Biomedical Corporation, San Diego, CA). The system used for these studies was the same one intended to be used in the phase I clinical trial with some modifications. The device consists of the same type of electric pulse generator to be used in the clinic but fitted with a nondisposable applicator wand and a disposable 0.5-cm hexagonal stainless steel six-needle array. The six-needle array was inserted into the injected site such that the needles surrounded the injected site. Six square-wave electric pulses of 560 V lasting approximately 0.1 ms each were delivered one time to the injected area by using an automated preprogrammed pulsing sequence. This voltage generates a field strength of approximately 1,300 V/cm to the injected site. This device produces the same field strength as the device intended for use in the phase I clinical trial.

3.2. Nonclinical Studies in a Murine Tumor Model

To evaluate the intratumoral production and systemic distribution of IL-2 transgene, DBA/2N mice were injected s.c. with 5×10^5 Cloudman melanoma cells (*see* **Note 2**). On day 9 after tumor cell injection, mice bearing palpable tumors of 20–80 mm

(*see* **Note 3**) were randomly assigned to groups (n = 5 per timepoint) and were injected i.t. with 30 μg of IL-2 pDNA ± DMRIE/DOPE (D/D) cationic lipid delivery system (5:1 DNA:DMRIE mass ratio) in saline (*see* **Notes 3,4**). Each injected tumor was electroporated. On days 1, 2, and 8 post-pDNA injection, tumors and sera were harvested from euthanized mice and levels of murine IL-2 were determined for tumor lysates and sera by using a murine IL-2 ELISA (enzyme-linked immunosorbent assay; R&D Systems, Minneapolis, MN).

The results from this study show that EP-assisted delivery resulted in the highest levels of IL-2 in the tumor. Levels of IL-2 in tumors were 3- to 7-fold higher after i.t. injection of IL-2 pDNA (± D/D) with EP, compared with those after delivery of the IL-2 pDNA without EP (Table 28.2). Although not statistically significant, mice injected i.t. with IL-2 pDNA + D/D + EP had approximately 3-fold higher levels of IL-2 in the tumor when compared with mice injected with IL-2 pDNA + EP (without D/D). On day 1 post-pDNA injection, mice had an average of 1,445 pg of IL-2/mL after i.t. injection of IL-2 pDNA + D/D + EP versus 545 pg/mL after i.t. injection of IL-2 pDNA + EP. By the second day, mice had an average of 582 pg/mL after i.t. injection of IL-2 pDNA + D/D + EP versus 191 pg/mL after i.t. injection of IL-2 pDNA + EP. By day 8, levels of IL-2 in the tumor after i.t. injection of either IL-2 pDNA + D/D + EP or IL-2 pDNA + EP were similar (83 vs. 92 pg/mL). Thus, there was a trend toward increased IL-2 levels in the tumor at the early timepoints when injecting IL-2 pDNA formulated with cationic lipid, followed by EP, compared with IL-2 pDNA in saline, followed by EP.

Levels of murine IL-2 in serum after i.t. administration of IL-2 pDNA were 20- to 200-fold lower when compared with those in the tumor, suggesting primarily localized expression and distribution of IL-2 at the site of injection. Highest levels of IL-2 in the serum were observed after i.t. injection of IL-2 pDNA + D/D + EP (68 pg/mL on day 1). By day 2, the levels were lower (15 pg/mL), and by day 8, IL-2 was nondetectable. These results suggest that i.t. injection of IL-2 pDNA + EP results in primarily sustained local levels of IL-2 within the tumor and relatively low, short-lived serum levels, which may lessen the side effects commonly associated with systemic IL-2 protein therapy.

Table 28.2 Levels of IL-2 (picograms per milliliter) in tumors and sera after i.t. injection of IL-2 pDNA ± D/D ± electroporation

| | Days post-pDNA injection | | | | | |
| | 1 | | 2 | | 8 | |
	Tumors[a]	Serum[a]	Tumors	Serum	Tumors	Serum
Control pDNA	0	1	0	0	0	0
IL-2 pDNA	198	10	97	1	30	1
IL-2 pDNA + D/D	580	21	92	0	24	0
IL-2 pDNA + EP	545	31	191	0	92	0
IL-2 pDNA + D/D + EP	1445	68	582	15	83	0

pDNA plasmid DNA, *IL-2* interleukin-2, *EP* electroporation, *D/D* DMRIE/DOPE.
[a] Tumor lysates and sera were assayed for levels of IL-2 (picograms per milliliter) using a murine IL-2 ELISA (R&D system). Mean IL-2 levels (picograms per milliliter) are shown (n = 5/group). Control pDNA is the plasmid backbone lacking the IL-2 gene.

To evaluate efficacy, multiple experiments were performed. DBA/2N mice were injected once s.c. with 5×10^5 Cloudman S91 M3 melanoma cells. Nine to ten days later, mice bearing palpable tumors of approximately 20–80 mm^3 were randomly assigned to groups such that each group had the same mean tumor volume. Mice were then anesthetized with inhalant isoflurane and injected intratumorally with 30 μg of either control pDNA or IL-2 pDNA, followed by electroporation, every 4th day for a total of three treatments (days 9, 13, and 17). Tumors were measured in three dimensions (length × width × height) using calipers. Tumor volume was determined from the formula for volume of an ellipsoid (15): tumor volume (mm^3) = 0.52 (length × width × height). An example of the results of an antitumor efficacy study is shown in Fig. 28.1. Collectively, results from multiple studies show that EP-assisted delivery of IL-2 pDNA induces tumor regression more effectively than does administration of IL-2 pDNA without EP, irrespective of whether the plasmid was formulated or not with D/D. Mice treated by i.t. injection of IL-2 pDNA ± D/D, followed by EP, had a significant inhibition of tumor growth ($p < 0.05$) (Fig. 28.1A). By day 34 of the study, mice treated with IL-2 pDNA ± D/D + EP had a mean tumor volume of 350–450 mm^3, while mice treated with IL-2 ± D/D without EP averaged 1,300 mm^3 mean tumor volume (Fig. 28.1B). Mice treated with control pDNA + D/D + EP averaged 1,000 mm^3 mean tumor volume, indicating that D/D or EP alone did not appear to cause the antitumor effect. These results suggest that EP can enhance the antitumor efficacy of i.t.-administered IL-2 pDNA. Formulating the IL-2 pDNA with D/D lipid was not required, as EP delivery of IL-2 pDNA with or without D/D showed similar antitumor effects.

The levels of murine IL-2 seen in the tumors (Table 28.2) appear to correlate with inhibition of tumor growth (Fig. 28.1) for unformulated pDNA. Although the addition of D/D increased the i.t. levels of murine IL-2, it did not enhance the inhibition of tumor growth beyond that seen with the pDNA delivered with EP.

On the basis of results from these efficacy studies and other murine studies (16) demonstrating a distal effect of EP-assisted delivery of IL-2 pDNA on established liver metastases, pDNA encoding human IL-2 administered with EP was selected as the lead therapeutic candidate for the intralesional treatment of solid tumors, metastatic melanoma in particular.

3.3. Nonclinical Repeated-Dose Study in Rats

The purpose of this IND-enabling study was to assess the safety of VCL-IM01 [human IL-2 pDNA (VCL-1102) in phosphate-buffered saline (PBS)] when administered repeatedly to Sprague-Dawley rats by s.c. injection, followed by EP. Sixty Sprague-Dawley rats were randomly assigned to three groups (10 rats/sex/group) and received 0.5 mg or 5 mg per injection of VCL-IM01 or the control article (PBS, pH 7.2). The dose volume was 1 mL/injection for all rats, including controls.

The test and control articles were administered in the dorsal area by s.c. injection on days 1, 3, 5, 7, and 9. Following test or control article administration, the site of

A. Tumors injected with pDNA + EP

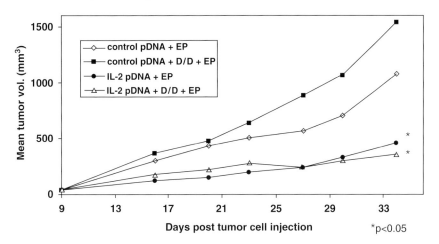

Days post tumor cell injection   *p<0.05

B. Tumors injected with pDNA, no EP

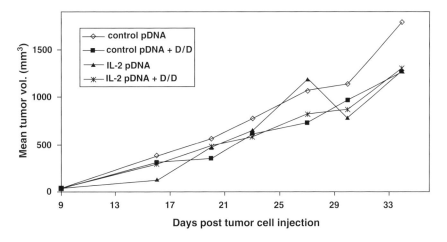

Days post tumor cell injection

Fig. 28.1 Electroporation (EP) enhancement of the antitumor efficacy of IL-2 pDNA in intratumoral (i.t.)-treated mice (days 9, 13, and 17; $n = 10$ mice/group) bearing s.c. Cloudman melanoma tumors (20–80 mm³). **A**. Mice injected i.t. with control lacking the IL-2 gene pDNA administered ± EP have statistically significant ($p < 0.05$; Mann–Whitney U test) higher mean tumor volumes at day 35 (*) compared to animals that received IL-2 pDNA ± DMRIE/DOPE (D/D) cationic lipid + EP. **B**. There was no statistically significant difference in mean tumor volumes in mice injected i.t. with control pDNA or IL-2 pDNA ± D/D when delivered without EP

injection was electroporated using the MedPulser® EPT System. Treatments were administered to four dorsal sites, with each of the first 4 treatments administered to a naïve site each time. The last treatment was administered to the first site treated a week after the first treatment.

During the course of the study, the following were evaluated: mortality, clinical signs of toxicity, injection site reactogenicity, body weight changes, food consumption, ophthalmoscopic changes, antibodies to double-stranded DNA (dsDNA), clinical chemistry, coagulation, hematology, urinalysis, macroscopic observations, organ weights, and histopathological changes. To assess the degree of reactogenicity after treatment, edema and erythema were evaluated separately using a modified Draize Scoring system (17) with a scale of 0–4, where 0 indicates *no signs of edema or erythema*, and 4 indicates *severe edema* (raised more than 1 mm and extending beyond the area of exposure) *or severe erythema* (beet or crimson red) *and/or slight eschar formation* (injuries in depth). On day 11, approximately 48 h after the last injection, 5 rats/sex/group were sacrificed and a necropsy examination was performed. The remaining 5 rats/sex/group were observed for an additional 26 days, then sacrificed (Or euthanized) (day 37).

There were no mortalities and no evidence of drug-related toxicity observed during the course of this study. Additionally, there was no evidence of anti-dsDNA antibody production, an indicator of autoimmune disease. Some dose-related subacute inflammation (neutrophilic infiltration, lymphocytic infiltration, and fibroplasia) in the skin/subcutis and underlying muscle of the injection sites was observed subsequent to VCL-IM01 administration (5 mg of VCL-IM01 per injection). This treatment-related inflammation was found to be subclinical and was absent in most of the animals by day 37 (28 days after the last injection). There was no remarkable difference between injection site 1 (receiving 2 separate injections on days 1 and 9) when compared with the other injection sites that received a single treatment.

After treatment, there was no erythema associated with injection/EP of either PBS or VCL-IM01. Very slight (barely perceptible) to slight (edges of area not definable) edema was associated with injection/EP, that dissipated in most instances within 1–3 days after a single injection to each site irrespective of the material (control or test article) injected. There was very slight (barely perceptible) to moderate (area well-defined) edema in the high-dose VCL-IM01 group only after two injections to a single site, that diminished within 5 days. However, this was not toxicologically relevant as no differences were noted after comparing multiple with single injection sites histologically.

On the basis of these results, the s.c. administration of VCL-IM01 to anesthetized Sprague-Dawley rats followed by EP (using the MedPulser® EPT System) five times at doses of 0.5 and 5 mg per injection was well tolerated.

In another study, to determine serum human IL-2 levels posttreatment, four groups of five female Sprague-Dawley (HSD:SD) rats, each weighing ~160–200 g, were injected s.c. with 1 mL of PBS (group 1), 1 mg VCL-IM01/mL (group 2), or 5 mg VCL-IM01/mL (group 3). Injected sites were electroporated using the MedPulser® EPT system. Each animal received eight single injections administered every other day with EP. Injection/EP was rotated over four administration sites, such that each of the four sites was injected and electroporated twice at eight-day intervals. Group 4 animals were injected s.c. with 1 mL of VR-1012 (a blank plasmid with the same backbone as VCL-1102) a total of five times, once every 7 days at the same injection site with EP. Clinical signs of toxicity were evaluated daily.

Blood samples (0.5 mL/sample) were collected from each animal once prior to treatment, on day 10 (24 h following the 5th treatment on day 9), and on day 16 to evaluate serum levels of human IL-2 posttreatment. Levels were determined using a human IL-2 ELISA (R&D Systems).

The results from this study provided no evidence of toxicity posttreatment and no detectable human IL-2 in serum either at 24 h or 5 days after the last of five repeated injections.

Collectively the results from these nonclinical studies, in addition to previously published safety study results of VCL-1102 pDNA administered intravenously with rapid clearance of the plasmid (18), supported clinical testing of VCL-IM01 in a phase I trial.

3.4. Phase I Clinical Trial

The safety of EP-assisted treatment of metastatic melanoma with human IL-2 pDNA (VCL-IM01) is currently under evaluation in a multi-center, open-label dose-escalation phase I trial (Vical Incorporated). In this trial, metastatic melanoma patients may receive VCL-IM01 at doses of 0.5 mg (1 tumor), 1.5 mg (1 tumor), 5 mg (1 tumor), or up to 15 mg (up to 3 tumors, 5 mg per tumor) administered i.t. by conventional needle and syringe with EP using the Medpulser EPT System. The primary endpoint of the study is safety after EP-assisted treatment with VCL-IM01 in patients with recurrent metastatic melanoma. Secondary outcomes include over-all response rate, duration of response, treated tumor response rate, assessment of injected tumor(s) for induration, inflammation, and erythema, and serum levels of IL-2. The expected total enrollment for the trial is 30 patients.

Patients will be enrolled to receive one course of treatment, comprising two cycles. Each cycle consists of four weekly intralesional VCL-IM01 pDNA injections with EP, followed by an observation period of 2–4 weeks. Tumor burden, response rate, and disease progression will be assessed using Response Evaluation Criteria in Solid Tumor (RECIST) guidelines. These evaluations will be conducted at screening, at the end of cycle 2 (week 14), and at six months (week 26) from the initial drug administration. Partial or complete responses will be confirmed at least 4 weeks after the initial response determination.

The injectable tumor(s) for administration of VCL-IM01 will usually be the largest and most easily measured tumor of <25 cm^2; however, the tumor(s) selected for injection will be at the discretion of the investigator. The selected tumor(s) will be injected at every treatment visit until the tumor(s) has diminished in size to the point at which it is no longer considered injectable by the clinician; at this point another tumor may be selected for injection.

The results from this phase I clinical trial will provide safety information as well as preliminary efficacy data in humans for EP-assisted delivery of VCL-IM01, an important treatment candidate designed to provide effective cancer therapy without the side effects associated with systemic IL-2 therapy.

4. Notes

1. Prior to tumor injection and EP, mice are anesthetized with inhalant isoflurane until negative toe-pinch reflex. Tumors are then shaved and swabbed with 70% isopropyl alcohol prior to the procedure.
2. Since the tissue culture conditions and strain of mice may vary from laboratory to laboratory, a tumor growth kinetics study should be performed initially to determine the optimal dose of cells to achieve tumor take rates of approximately 80–100%.
3. Plasmid preparations should be assayed for levels of endotoxin prior to in vivo injection, using a method such as Limulus Amoebocyte Lysate (LAL) assay (Associates of Cape Cod, MA). Endotoxin levels of <10 EU/mg may be acceptable for in vivo studies.
4. For tumor efficacy studies, an initial dose-response study should be performed in which the dose and timing of intralesionally injected pDNA, followed by EP, are varied to determine the optimal dose and schedule of pDNA treatment with EP for a particular tumor model.

References

1. Hanna, E., Zhang, X., Woodlis, J., Breau, R., Suen, J., and Li, S. (2001) Intramuscular electroporation delivery of IL-12 gene for treatment of squamous cell carcinoma located at distant site. *Cancer Gene Ther.* **8**, 151–157.
2. Li, H., Cao, M.Y., Lee, Y., et al. (2005) Virulizin, a novel immunotherapy agent, activates NK cells through induction of IL-12 expression in macrophages. *Cancer Immunol. Immunother.* **54**, 1115–1126.
3. Lucas, M.L. and Heller, R. (2003) IL-12 gene therapy using an electrically mediated nonviral approach reduces metastatic growth of melanoma. *DNA Cell Biol.* **22**, 755–763.
4. Tomura, T., Nishi, T., Goto, T., et al. (2001) Intratumoral delivery of interleukin 12 expression plasmids with in vivo electroporation is effective for colon and renal cancer. *Hum. Gene Ther.* **12**, 1265–1176.
5. Torrero, M.N., Henk, W.G., and Li, S. (2006) Regression of high-grade malignancy in mice by bleomycin and interleukin-12 electrochemogenetherapy. *Clin. Cancer Res.* **12**, 257–263.
6. Kishida, T., Asada, H., Satoh, E., et al. (2001) *In vivo* electroporation-mediated transfer of interleukin-12 and interleukin-18 genes induces significant antitumor effects against melanoma in mice. *Gene Ther.* **8**, 1234–1240.
7. Yamashita, Y., Shimada, M., Hasegawa, H., et al. (2001) Electroporation-mediacted Interleukin-12 gene therapy for hepatocellular carcinoma in the mice model. *Cancer Res.* **61**, 1005–1012.
8. Li, S., Zhang, X., Xia, X., et al. (2001) Intramuscular electroporation delivery of IFN-alpha gene therapy for inhibition of tumor growth located at a distant site. *Gene Ther.* **8**, 400–407.
9. Lohr, F., Lo, D.Y., Zaharoff, D.A., et al. (2001) Effective tumor therapy with plasmid-encoded cytokines combined with *in vivo* electroporation. *Cancer Res.* **61**, 3281–3284.
10. Matsubara, H., Gunji, Y., Maeda, T., et al. (2001) Electroporation-mediated transfer of cytokine genes into human esophageal tumors produces anti-tumor effects in mice. *Anticancer Res.* **21**, 2501–2504.
11. Chi, C.H., Wang, Y.S., Lai, Y.S., and Chi, K.H. (2002) Anti-tumor effect of *in vivo* IL-2 and GM-CSF electrogene therapy in murine hepatoma model. *Anticancer Res.* **22**, 3153–3322.
12. Lucas, M.L., Heller, L., Coppola, D., and Heller, R. (2002) IL-12 plasmid delivery *in vivo* electroporation for the successful treatment of established subcutaneous B16.F10 melanoma. *Mol. Ther.* **5**, 668–675.
13. Hartikka, J., Sawdey, M., Cornefert-Jensen, F., et al. (1996) An improved plasmid DNA expression vector for direct injection into skeletal muscle. *Hum. Gene Ther.* **7**, 1205–1217.

14. Grossblatt, N. (ed.) (1996) *The guide for the care and use of laboratory animals*, 7th edn. National Academies Press, Washington, DC.
15. Tomayko, M.M. and Reynolds, C.P. (1989) Determination of subcutaneous tumor size in athymic (nude) mice. *Cancer Chemother. Pharmacol.* **24**, 148–154.
16. Horton, H.M., Planchon, R.R., Sawdey, M., et al. (2007) Therapy of primary and metastatic murine tumors by intratumoral administration of interleukin-2 plasmid DNA with electroporation. *Cancer Therapy* **5**, 125–132.
17. Draize, J.H. (1959) Appraisal of the safety of chemicals in food, drugs and cosmetics. The Association of Food and Drug Officials of the United States, Austin, TX, pp. 49.
18. Parker, S.E., Vahlsing, H.L., Lew, D., et al. (1999) Cancer gene therapy using plasmid DNA—pharmacokinetics and safety evaluation of an IL-2 plasmid DNA expression vector in rodents and nonhuman primates. *Biopharm. Appl. Technol. Biopharm. Dev.* **12**, 18–24.

Part V
Treatment of Other Diseases via Electroporation Gene Therapy

Chapter 29
Factor IX Gene Therapy for Hemophilia

Jason G. Fewell

Abstract Using gene therapy to produce systemic levels of human factor IX for the treatment of hemophilia B has been clinically evaluated using viral-based vectors. The efficacy of this approach has been limited because of immune responses against the viral components. An alternative approach is to use physical methods such as in vivo electroporation to deliver plasmid DNA, thereby avoiding some of the complications associated with viral-based delivery systems. A method describing intramuscular injection of plasmid formulated with an anionic polymer and followed by electroporation, which can produce high transfection efficiency and high levels of systemic factor IX protein following a single administration, is provided here.

Keywords: electroporation, muscle, factor IX, plasmid, gene therapy, hemophilia

1. Introduction

Treating monogenic disorders is a long-standing goal for gene therapy technology. In principal, the concept is simple: use the body's own cells to overexpress a functional protein to replace a protein that is mutated or missing. In vitro, this technique is common practice where cells can be induced to transiently or stably produce endogenous or foreign protein in abundance. Transition into an in vivo or therapeutic setting has been much more challenging.

Perhaps the most widely recognized monogenetic disorder is hemophilia (type A or B). In both cases, mutations have caused the loss of transcript or function proteins (factor VIII or factor IX, respectively), which in turn leads to an inability to effectively form blood clots. Although mutations in factor VIII are much more common in the general population, factor IX (F.IX) gene therapy has been the focus of a greater amount of research. This is partly because the F.IX protein is much smaller than factor VIII and technically more manageable for use, with a variety of gene delivery systems.

Arguably, the most successful preclinical gene therapy work for hemophilia type B has come from using adeno-associated virus (AAV) delivery systems. Using this vector to administer F.IX gene intramuscularly and intrahepatically has produced

long-term therapeutic levels of functional F.IX protein in both hemophilic mice and dogs, thereby leading to the correction in blood clotting deficits in these animals (1, 2). It was with much anticipation that clinical trials were started, whereby AAV containing the human F.IX (hF.IX) gene was injected intramuscularly into multiple sites. Results from this trial indicated that significant levels of hF.IX were produced in some patients, but to achieve the required efficacy injection into ~500 sites would be required. Consequently this approach was deemed impractical. A better F.IX protein expression profile was achieved from injecting AAV/F.IX into the portal vein of dogs, resulting in high transfection efficiency of the hepatocytes (2). When this approach was evaluated clinically in humans it was seen that the hF.IX expression levels that were initially achieved gradually declined over ~8 weeks. The decline was accompanied by an elevation in liver transaminases and an apparent cell-mediated immunity that targeted antigens of the AAV capsid (3, 4). Unfortunately, the preclinical animal studies did not predict this response. Many additional studies have been performed using a variety of viral vectors, producing a wide range of results (5). Almost universally, the underlying message is that avoidance of the intact human immune systems will be difficult using viral-based approaches.

Development of nonviral delivery systems for treating hemophilia offer an alternative approach that may avoid activation of the immune response that plagues viral vector delivery. However, nonviral vectors have very low efficiencies relative to many of the viral vectors, making it unlikely that therapeutic systemic levels of hF.IX could be achieved by direct injection of plasmid alone. Use of physical delivery approaches in vivo, such as ultrasound and electroporation, have been shown to increase transgene expression in muscle following delivery of "naked" plasmid (6, 7). Electroporation, in particular, can produce transgene expression levels that are several logs greater than those obtained by injection without electroporation (8). Further, electroporation can be used to enhance gene delivery into many tissue types, making it potentially useful for treating a wide variety of diseases (9). By utilizing electroporation for delivering the hF.IX gene into muscle, we can begin to approach the systemic protein levels necessary to achieve a therapeutic response in hemophilic animals (10, 11).

In this chapter, a method utilizing electroporation for efficient delivery of plasmid encoding for hF.IX and formulated with a polymeric delivery system will be described. From the available literature it is possible to find much work devoted to optimizing the electroporation parameters to maximize gene expression and minimize tissue damage, and there are many considerations that are dependent upon the size of the animal and the target organ. Here, the focus will be on delivery to muscle as this tissue is easily accessible and has been shown in various animal models and in humans that muscle is capable of secreting functional hF.IX protein (12). Also, transfection of muscle offers the possibility of maintaining high levels of transgene expression for long duration because of low cellular turnover rates (1). Use of the anionic polymer poly-L-glutamate has been shown to significantly enhance transgene expression levels following electroporation relative to unformulated plasmid (11, 13). It is believed that the poly-L-glutamate formulation allows for greater plasmid DNA retention following electroporation and, perhaps, increased nuclear trafficking (14). Use of this polymer to formulate hF.IX plasmid is incorporated

into the protocol. Commercially available ELISAs and hF.IX antibodies readily allow for detection of hF.IX protein in plasma and localization in electroporated muscle tissue.

2. Materials

2.1. Plasmid DNA

1. Plasmid DNA encoding for hF.IX. This plasmid DNA can be produced by inserting a synthetic coding sequence of hF.IX into a commercially available plasmid vector. One expression vector that works well for in vivo delivery is the pCI vector from Promega (Madison, WI), where the hF.IX sequence can be inserted behind the β-globin or IgG chimeric intron and is under the control of CMV enhancer or promoter. The coding sequence is then terminated by the late SV40 polyadenlyation signal. The construct should provide high levels of constitutive expression in mammalian cells.
2. EndoFree Plasmid DNA Giga Kit (Qiagen, Valencia, CA).

2.2. Plasmid DNA Formulation

1. Sterilized glass vials and stoppers with crimping caps.
2. Poly-L-glutamic acid, sodium salt (mol. Wt., 15,000–50,000): 25 mg/mL, and store at 4°C (*see* **Note 1**).
3. 5M NaCl.
4. 100 mM Tris-HCl, pH 7.5.

2.3. Animals

1. Female C57BL/6 mice (6–8 weeks old). House animals (5 per cage) in ventilated caging systems and maintain under approved IACUC protocols (*see* **Note 2**). Maintain 12 h light and dark cycles. Acclimatize animals for at least 3 days prior to experimentation.

2.4. Intramuscular Injection and Electroporation

1. Insulin syringes, 0.3 cm³ (Becton Dickinson, Franklin Lakes, NJ).
2. Ethanol, 70%.

3. Power trim clippers or equivalent.
4. IsoFlo (Isoflurane).
5. IMPAC Anesthesia center (VetEquip, Pleasanton, CA) or equivalent.
6. ECM 830 Square Wave Electroporation System (Harvard Apparatus, Holliston, MA).
7. Stainless steel caliper electrode, $1.5 \times 1.5 \, cm^2$, Model 384 L (Harvard Apparatus, Holliston, MA).

2.5. Blood or Tissue Collection and Assay

1. Disposable Pasture pipettes (5.75 in.).
2. EDTA Microtainer tubes (Becton Dickinson, Franklin Lakes, NJ).
3. Asserachrom IX: Ag human F.IX ELISA kit (Diagnositca Stago, Parsippany, NJ).
4. Paraformaldehyde, 3%, in PBS.
5. Methanol.
6. Normal goat serum, 20% (diluted in PBS)
7. Anti–factor IX antibody produced in rabbit, diluted 1:6,000 in PBS containing 1% BSA (wt./vol.).
8. Goat anti–rabbit IgG diluted 1:400 in PBS containing 1% BSA (wt./vol.).
9. Elite ABC reagent (Vector Laboratories, Burlingame, CA) diluted 1:80 in PBS containing 1% BSA (wt./vol.).
10. DAB solution (Vector Laboratories, Burlingame, CA).
11. Mayer's hematoxylin.
12. Permount.

3. Methods

3.1. Plasmid DNA Formulation

1. Purify hF.IX plasmid DNA (~10 mg) using the EndoFree Plasmid Giga Kit according to manufacturer's instructions. Store purified plasmid DNA at a final concentration of 4.0 mg/mL (in sterile water) at −20°C.
2. Prepare formulated plasmid DNA (final DNA concentration of 1.0 mg/mL and final poly-L-glutamate concentration of 6.0 mg/mL) by combining the individual components in a sterile vial (with gentle mixing) in the following order (*see* **Note 3**). Pipette 250 μL of hF.IX DNA followed by 240 μL of the poly-L-glutamate stock solution (6.0 mg/mL). Add an additional 430 μL of water for injection, 30 μL NaCl stock solution (5 M), and finally add 50 μL Tris-HCl stock solution (150 mM).
3. Measure osmolality if possible (should be 290 mOsm ± 10%).
4. Measure pH (should be ~7.4).
5. Seal vials with stoppers and crimping caps.

3.2. Plasmid DNA Delivery (Intramuscular) and Electroporation

1. Fill insulin syringes with appropriate volume of formulated plasmid to be injected into a muscle. Typically for the tibialis cranialis muscle of mice, a 25-μL injection volume works well. If alternative or multiple muscles are targeted, the volume can be adjusted according to the size of the muscle(s).
2. Anesthetize mice in a plexiglas chamber of the anesthesia center. Once the animal is deeply anesthetized, transfer the animal onto a surgical pad and maintain level of anesthesia via a nose cone.
3. Shave the hindlimbs of the mice with the clippers. Remove as much of the fur as possible around the entire hindlimb to provide a good contact surface for the plate electrodes.
4. Wipe the shaved area with 70% alcohol.
5. Grasp the leg of the animals by the foot and extend the leg. Orient the tibialis muscle (or muscle of choice) such that needle is parallel with the muscle fibers. Inject through the skin and into the belly of the muscle at ~30–45°. Slowly inject 25 μL of formulated plasmid into the muscle and remove the needle. Proceed to the contralateral leg and repeat the process.
6. To deliver the electrical pulses, grasp the foot of the animal firmly and extend the leg fully. Place the caliper electrodes around the leg such that the injected muscle is centered between the steel plates (Fig. 29.1). Compress the plates until snug around the muscle, and measure the distance between the plates. Determine the necessary voltage required to deliver 375 V/cm and set the electroporator to

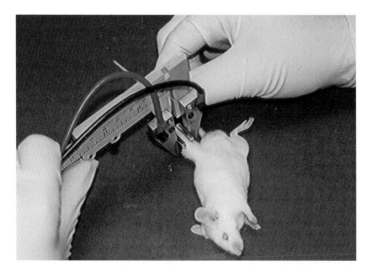

Fig. 29.1 Orientation of the caliper electrodes around the extended mouse leg (a CD-1 mouse is being shown)

deliver 2 pulses at 375 V/cm for each pulse, with a 25 ms pulse duration and a frequency of 1 Hz. Repeat this step on the contralateral leg (*see* **Note 4**). Electoporation should occur no later than 5 min following injection.

7. Place the animal back in cage for recovery.

3.3. Plasma Collection and Analysis

1. Use a disposable pasture pipette to collect ~400 µL of blood from the retro-orbital plexus of the mice (*see* **Note 5**).
2. Immediately transfer blood into EDTA microtainer tube and invert several times to mix.
3. Centrifuge at 1500–2000*g* for 2 min.
4. Carefully transfer plasma into a separate tube. Use immediately for hF.IX ELISA or store at −80°C.
5. Analyze samples using the Asserachrom IX:Ag human hF.IX ELISA according to manufacturer's instruction.

3.4. Immunohistochemistry for Localization of hF.IX Expression

1. Euthanize mice using an approved method.
2. Carefully remove the tibialis muscle.
3. Prepare 10-µm frozen sections from muscle tissues (transverse orientation) using standard methodologies.
4. Fix tissues in 3% paraformaldehyde in PBS (pH 7.4) for 15 min.
5. Rinse in PBS for 5 min.
6. Incubate in methanol for 10 min.
7. Wash three times in PBS (5 min each).
8. Block in 20% normal goat serum for 1 h.
9. Incubate in anti–hF.IX antibody diluted 1:6,000 for 1 h at room temperature (or overnight at 4°C).
10. Rinse sections in PBS 3 times (5 min each).
11. Incubate with biotinylated goat anti–rabbit IgG diluted 1:400 in PBS for 30 min at room temperature.
12. Rinse sections in PBS (5 min).
13. Visualize by incubating sections with Elite ABC reagent (1:80 dilution) for 30 min.
14. Incubate for 5 min in DAB solution followed by a rinse in PBS.
15. Counter stain with Mayer's hematoxylin.
16. Dry slides and cover slips with Permount mounting media.

4. Notes

1. Plasmid formulated with poly-L-glutamate and used with electroporation has been shown to enhance transgene expression levels in muscle from 2 to 10 fold (relative to unformulated plasmid) and produces more consistent in vivo results compared with other anionic polymeric formulations such as poly-L-aspartate and poly-acrylic acid (14). There is data indicating that efficacy is somewhat dependent on molecular weight of the poly-L-glutamate. Loss of efficacy occurs when using poly-L-glutamate with a molecular weight above 50 kDa (14).
2. Other mice strains can be used as well. In immunocompetent mice, systemic expression of hF.IX protein will be transient because of the host animal's humoral immune response and development of hF.IX antibodies. If long-term expression or repeated injections are desired, then SCID Beige mice can be used following the same protocol. Larger animals, such as rats, can also be used; however, this will require a larger set of caliper electrodes ($2.0 \times 2.0\,cm^2$ plates, available from Harvard Apparatus). In general, use of caliper electrodes is not acceptable when the size of the tissue to be electroporated will result in a space greater than 1 cm between plates. This is due to excessive thermal tissue damage that will be produced to maintain the 375 V/cm electroporation parameters. Thus, a transition to other electrode types will be required for large animal models (such as canines). Commonly used configurations are four or six needle electrodes (1 cm diameter) that are inserted intramuscularly to a depth of about 1 cm. Parallel pairs of needles are sequentially pulsed to produce the electrical field (15).
3. A protocol for making 1.0 mL of formulated plasmid is provided. This protocol can be adjusted to any volume based on need. Additionally, 1.0 mg/mL final DNA concentration generally works well for most applications; however, optimized delivery concentrations can be determined empirically by adjusting the DNA concentrations. Use of poly-L-glutamte at 6.0 mg/mL can work well across a wide range of plasmid concentrations (14). It is important to maintain physiological values for osmolality and pH.
4. The diameter of the hindlimb of a 20 g mouse across the location of the tibialis cranialis muscle will normally be 3–4 mm. This will necessitate applying 112–150 V to produce the required voltage of 375 V/cm. The electroporation parameters of 375 V/cm, 25 ms per pulse, at a frequency of 1 Hz have been shown to produce high levels of transgene expression and minimal tissue damage. The electric pulses will cause all muscles within the electric field to strongly contract; therefore, it is important to maintain a firm grasp of the foot during the electroporation process to prevent the leg from pulling out from between the caliper electrodes between the first and second pulse.
5. Peak expression levels of hF.IX in plasma will occur between 3 and 7 days after treatment. In immunocompetent animals, all expression will be gone by ~11 days owing to humoral immune responses. In SCID mice, fairly stable hF.IX plasma expression levels can be maintained for more than 100 days following a single treatment, although there is a slow decay in expression over time.

References

1. Herzog, R.W., Yang, E.Y., Couto, L.B., et al. (1997) Long-term correction of canine hemophilia B by gene transfer of blood coagulation factor IX mediated by adeno-associated viral vector. *Proc. Natl. Acad. Sci. USA*. **94**, 5804–5809.
2. Arruda, V.R., Stedman, H.H., Nichols, T.C., et al. (2005). Regional intravascular delivery of AAV-2-F.IX to skeletal muscle achieves long-term correction of hemophilia B in a large animal model. *Blood* **105**, 3458–3464.

3. Manno, C.S., Chew, A.J., Hutchison, S., et al. (2003) AAV-mediated factor IX gene transfer to skeletal muscle in patients with severe hemophilia B. *Blood.* **101**, 2963–2972.

4. Manno, C.S., Arruda, V.R., Pierce, G.F., et al. (2006). Successful transduction of liver in hemophilia by AAV-Factor IX and limitations imposed by the host immune response. *Nat. Med.* **12**, 342–347.

5. Chuah, M.K.L., Collen, D., and VandenDriessche, T. (2001) Gene therapy for hemophilia. *J. Gene Med.* **3**, 3–20.

6. Miao, C.H., Brayman, A.A., Loeb, K.R., et al. (2005). Ultrasound enhances gene delivery of human factor IX plasmid. *Hum. Gene Ther.* **16**, 893–905.

7. Somiari S., Glasspool-Malone, J., Drabick J.J., et al. (2000). Theory and in vivo application of electroporative gene delivery. *Mol. Ther.* **2**, 178–187.

8. Vicat, J.M., Boisseau, S., Jourdes, P., et al. (2000). Muscle transfection by electroporation with high-voltage and short-pulse currents provides high-level and long-lasting gene expression. *Hum. Gene Ther.* **11**, 909–916.

9. Heller, R. (2003) Delivery of plasmid DNA using in vivo electroporation. *Preclinica.* **1**, 198–208.

10. Bettan, M., Emmanuel, F., Darteil, R., et al. (2000) High-level protein secretion into blood circulation after electric pulse-mediated gene transfer into skeletal muscle. *Mol. Ther.* **2**, 204–210.

11. Fewell, J.G., MacLaughlin, F., Mehta, V., et al. (2001) Gene therapy for the treatment of hemophilia B using PINC-formulated plasmid delivered to muscle with electroporation. *Mol. Ther.* **3**, 574–583.

12. Arruda, V.R., Hagstrom, J.N., Deitch, J., et al. (2001) Posttranslational modifications of recombinant myotube-synthesized human factor IX. *Blood.* **97**, 130–138.

13. Draghia-Akli, R., Khan, A.S., Cummings, K.K., Parghi, D., Carpenter, R.H., Brown, P.A. (2002) Electrical enhancement of formulated plasmid delivery in animals. *Technol. Cancer Res. Treat.* **1**, 365–372.

14. Nicol, F., Wong, M., MacLaughlin, et al. (2002) Poly-L-glutamate, an anionic polymer, enhances transgene expression for plasmids delivered by intramuscular injection with in vivo electroporation. *Gene Ther.* **9**, 1351–1358.

15. Rabussay, D., Dev, N.B., Fewell, J., Smith, L.C., Widera, G., and Zhang, L. (2003) Enhancement of therapeutic drug and DNA delivery into cells by electroporation. *J. Phys. D: Appl. Phys.* **36**, 348–363.

Chapter 30
KGF-1 for Wound Healing in Animal Models

Guy P. Marti, Parsa Mohebi, Lixin Liu, Jiaai Wang, Tomoharu Miyashita, and John W. Harmon

Abstract Keratinocyte growth factor-1 (KGF-1) is a member of the fibroblast growth factor (FGF) family FGF7 and is expressed in normal and wounded skin. KGF-1 is massively produced in the early stages of the wound healing process as well as during the later remodeling process (1, 2). We have studied the effects of the electroporation of a KGF-1 plasmid into excisional wounds of different rodent models mimicking diseases known to impair the normal wound healing process.

We have used a genetically diabetic mouse model and a septic rat model in our experiments, and we have shown improvement of the healing rate (92% of the wounds are healed at day 12 vs. 40% of the control), the quality of epithelialization (histological score of 3.3 vs. 1.5), and the density of new blood vessels (85% more new blood vessels in the superficial layers than that of the control) (3, 4). Considering these results, we believe we can further explore the treatment modalities for using the electroporation-assisted transfection of DNA plasmid expression vectors of growth factors to enhance cutaneous wound healing.

Keywords: KGF, gene therapy, electroporation, cutaneous wound healing

1. Introduction

According to its paracrine cell signaling mechanism, keratinocyte growth factor-1 (KGF-1) is produced by various cell types (fibroblasts, endothelial cells, smooth-muscle cells, and dendritic epidermal T cells) but not by keratinocytes themselves (5, 6). KGF-1 has been shown to induce migration and multiplication of keratinocytes (7–10). Those dermal keratinocytes will produce various cytokines and growth factors during the process, such as IL-1, IL-6, TNF-α, VEGF, GM-CSF, PDGFa, FGF2, TGFβ, and EGF (11, 12). Interestingly, in certain specific locations, keratinocytes will undergo transdifferentiation into different phenotypes during wound healing (e.g. fibroblast, myofribroblast, and macrophage), expressing various genes involved in the wound

S. Li (ed.), *Electroporation Protocols: Preclinical and Clinical Gene Medicine.*
From *Methods in Molecular Biology, Vol. 423.*
© Humana Press 2008

healing process (13–15). Therefore, although KGF-1 specifically targets keratinocytes, we believe that it plays a major role in cutaneous wound healing.

In numerous previous studies, KGF has been tested for injuries of skin, oral mucosa, tympanic membrane, and GI track as well as on different injury models that include incisional, excisional, burn, irradiated, and ischemic. All these trials have been with topical application of the peptide growth factor. The effects have been minimal because topically applying growth factors to wounds has major drawbacks. In another study of exogenously applied fibroblast growth factor-1 (FGF-1), it was necessary to repeatedly administer large amounts of recombinant FGF-1 to achieve a significant improvement in wound healing (16). Large amounts of FGF-1 were required because of the short half-life of the reagent raising the costs and complexity of potential therapeutic use. In addition to this rapid protease degradation, the wound eschar can block access to the wound.

Previous methods of in vivo plasmid transfection faced significant drawbacks. Injection of naked DNA plasmid requires prohibitively high levels of plasmid because of the extremely low cell uptake efficiency. At such high concentrations, the plasmid vectors themselves interfere with wound healing (17). The gene gun suffers the disadvantage of variable results from the same gun, often requiring repeated administration to achieve a desired result (18, 19). Adenovirus has been found to have high transfection efficiency, but, at the same time, it is currently under scrutiny for its potential association with serious medical risk (20). Therefore, we chose electroporation as an in vivo method of plasmid transfection. Here, we illustrate an effective protocol to deliver KGF-1 into the targeted tissue for healing of wounds.

2. Materials

2.1. Plasmid Selection

1. Plasmids: pCDNA3.1/KGF-1 was obtained from Invitrogen (Carlsbad, CA). gWIZ-luc was obtained from Gene Therapy Systems (San Diego, CA). The plasmid purification kit was obtained from Qiagen (Valencia, CA).
2. Cell culture and in vitro transfection: NIH3T3 cells were maintained in DMEM supplemented with 10% fetal bovine serum in a humidified incubator at 37°C with 5% CO_2.
3. ELISA: KGF-1 standards and KGF-1 human antibody were obtained from R&D Research Systems (Minneapolis, MN).

2.2. In Vivo Electroporation

1. Generator and electrode: The electroporator is the ECM 830 BTX model from Inovio. (San Diego, CA). The electrode was a custom designed pin electrode, consisting of two rows of parallel acupuncture needles separated by 5 mm. Each

row contained seven acupuncture 0.12 GA needles (AcuGlide, Helio Medical, San Jose, CA).

2. Bioluminescent images: The camera used was a cooled charged coupled device camera (IVIS, Xenogen, Alameda, CA). The image analysis software was Living Image (Xenogen, Alameda, CA).

2.3. Detection of Human KGF-1 RNA

1. PCR kit.
2. RT-PCR kit and RNA isolation kit.

2.4. Animal Models

1. Diabetic mice: Female 6–8-week old BKS.Cg-m. Lepr^{db-db} mice were obtained from Jackson Laboratories (Bar Harbor, ME). Diabetic mice have to be caged separately. These animals grow fatter and slower with time and cannot fight back easily in a group cage. The group will feed on the weakest animal's back and introduce a bias in the normal wound healing course.
2. For the septic rat model, use female 6–8-week old Sprague-Dawley rats.

2.5. Animal Anesthetics

1. Sterile 10 mL bottle with a rubber stopper.
2. Ketamine, 100 mg/mL.
3. Xylazine, 100 mg/mL.
4. Sterile water for injection.

3. Methods

3.1. Plasmid Selection

1. Use gene-specific PCR primers incorporating restriction enzyme sites for Pst1 and BamH1. PCR amplify KGF-1 from pCDNA3.1-KGF and ligate into the corresponding restriction enzyme sites in gWIZ (Gene Therapy Systems, San Diego, CA). Purify all plasmids following manufacturer's instructions. Store Plasmid DNA at −20°C. Clone Human KGF-1 into three different expression vectors, each utilizing the CMV promoter for transcription with a SV40 polyadenylation signal to terminate the transcription.

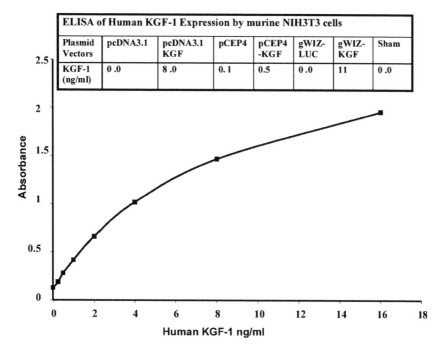

ELISA of Human KGF-1 Expression by murine NIH3T3 cells							
Plasmid Vectors	pcDNA3.1	pcDNA3.1 KGF	pCEP4	pCEP4 -KGF	gWIZ- LUC	gWIZ- KGF	Sham
KGF-1 (ng/ml)	0.0	8.0	0.1	0.5	0.0	11	0.0

Fig. 30.1 ELISA of human KGF-1 in NIH3T3 cells. NIH3T3 cells were transfected with various constructs and assayed for presence of human KGF-1 and compared with known standards

2. Seed NIH3T3 cells in six-well plates at a density of 2×10^5 cells/well. Incubate DNA constructs with lipofectamine at room temperature for 30 min. Prior to transfection, rinse cells in Opti-mem (Invitrogen) before adding the lipid–DNA complex and incubate together at 37°C for 5 h. Harvest cell supernates for 24–48 h after transfection, and assay for the presence of human KGF-1 protein by ELISA.

3. Analyze expression of human KGF-1 protein by an ELISA after transfection into the murine cell line NIH3T3. Apply supernates from NIH3T3 cells expressing KGF-1 to ELISA plates coated with human KGF-1 antibody and compare with known standards. The vector gWIZ-KGF was selected because it was the most productive of the three. pCDNA3.1/GS-KGF also efficiently expressed KGF although not as abundantly as gWIZ-KGF. The larger episomal vector pCEP4-KGF was very inefficient at expressing KGF protein; therefore, its use was discontinued. Results are shown in Fig. 30.1.

3.2. Method for Delivering Anesthesia to the Animals

1. Mouse anesthesia: In a sterile 10 mL bottle with a rubber stopper, mix 1 mL of ketamine (100 mg/mL), 0.1 mL of xylazine (100 mg/mL), and 8.9 mL of sterile water for injection. Shake well before use. Keep away from light and in a cool

place. Inject 0.1 mL/10 g i.p. Repeat half a dose at a time whenever necessary (approximately every 30 min). Prevent heat loss until the animal recovers.
2. Rat anesthesia: Similar to the mouse anesthesia method, but mix 8.75 mL of ketamine (100 mg/mL) and 1.25 mL of xylazine (100 mg/mL). Administer 0.05–0.10 mL/100 g i.p. Repeat as required with 1/3 to 1/2 doses at a time (approximately every 30 min).

3.3. In Vivo Electroporation

1. Manufacturing requirements for the electrode must include a precise parallel arrangement of the needles. Since acupuncture needles are fragile, their tips must not protrude more than 2 mm from the holding base.
2. Voltage delivery depends on the distance between the two electrodes. The voltage settings on the generator control panel have to show 900 V because the distance between the positively and negatively charged needles is 0.5 cm. Like others, our laboratory has conducted numerous studies to improve the transfection across rodent skin (21–24). By increasing the voltage and pulse duration, there are more electrically charged molecules entering the cells. Both parameters cannot be increased at the same time without cell toxicity.
3. The injection of the plasmid has to be done strictly intradermally. A skin bleb will demonstrate the intradermal delivery of the product. Animals have to be electroporated at the site of injection within 2 min of plasmid administration (*see* **Note 1**).
4. The array of needles must cover the skin bleb. The electrode is applied firmly on the skin without exerting excessive pressure (*see* **Note 2**).
5. Electroporation parameters and delivery: Administer six square wave pulses at an amplitude of 1800 V for a duration of 100 μs, and an interval between pulses of 125 ms.

3.4. Bioluminescent Images

After animals are sedated, intraperitoneally inject 140 μL of 15 mg/mL of D-luciferin in PBS. After taking a conventional light photograph, bioluminescent images are acquired using a cooled charged coupled device camera. Take luminescent images 30 min after luciferin administration, during which time the light emission has been shown to be in a plateau phase. Bioluminescent images are overlaid onto the conventional image of each animal, and the light emission, corrected for background luminescence, is calculated for each injection site using image analysis software. The different spectral colors represent a linear scale of the intensity of luminescence, corresponding to the total number of photons emitted per second from a square centimeter of tissue. The color scale is standardized for all output images (minimum, 15,000 photons/s/cm^2; maximum, 1,000,000 photons/s/cm^2). An example is shown in Fig. 30.2.

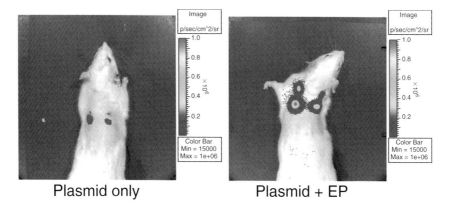

Fig. 30.2 Both animals were injected intradermally with the gWIZ-Luc plasmid containing a CMV promoter and luciferase transgene. The animal on the right side was electroporated (EP), but the animal on the left side was not. The light emission (photons/s) is standardized with a color scale shown on this picture as a difference in grey intensity

3.5. Detection of Human KGF-1 RNA

1. Perform conventional PCR using a standard PCR kit from Invitrogen according to manufacturer's instructions. The primers are as follows:
   OJW007: 5′-AACTGCAGATGCACAAATGGAT-3′ and OJW004: 5′-TTAAG TTATTGCCATAGGAAG-3′.
2. For RT-PCR: In a separate experiment 25 or 100 μg of pCDNA3.1-KGF plasmid DNA are intradermally injected into Balb-C mice. After 17 days, mice are killed and its skin tissue is removed and frozen in liquid nitrogen. The tissue is then homogenized and purified using an RNA isolation kit (Qiagen). RNA is purified according to manufacturer's instructions. Aliquots of RNA are used in an RT-PCR reaction employing an RT-PCR kit from Qiagen. Primers are designed to ensure that only exogenously expressed murine KGF-1 mRNA would be detected, yielding a 600 bp product.
3. Plasmid-specific KGF-1 primers used are as follows:
   5′-CACCATGCACAAATGGATACTGACATG-3′ and
   5′-AGTTATTGCCATAGGAAGAAAGT-3′.

3.6. Rat Model of Sepsis

Our original model of sepsis is the product of a series of experiments aiming to create general severe sepsis without having a prohibitive mortality rate as well as setting the conditions for a delayed wound healing. In initial experiments, we used agar pellets made from BORT II strain of *E. coli* bacteria at 10^2–10^7 concentrations. Excessive bacterial concentration led to 100% mortality, but lowering the concentration

led to a model of sepsis without wound healing impairment. Then we used the cecal ligation model of sepsis. To have an acceptable mortality rate of 15–20%, we partially ligated the cecum (50% of its volume) and pushed a load of feces in the cecum, but without puncture. This model showed a delayed cutaneous wound healing (wounds were 2.5-fold larger at day 7 in infected rats) (25, 26).

3.7. Wound Assessment by Plannimetry

Tracing fresh wounds onto acetate paper could introduce bias in the study. Rodent skin tends to shift on the carcass, and muscles and skin tend to stretch. It is important to quickly immobilize the animal to minimize the examiner's action on the fresh wound. Measurements should be taken on the day of wounding to be compared with a wound template. Later in the study, measurements have to consider the epithelialization front. The eschar has to be removed gently, otherwise new fresh thin epithelium could be stripped off the wound. It is recommended to soak the wound with wet gauze for 5 min before removing the scab. Two lines should be traced on the acetate paper: one outer line tracing the edges of normal skin and one inner line tracing the epithelialized border (*see* **Note 3**).

3.8. Histology of Wounds

Histologic sections from each wound are analyzed and graded by two examiners who are blinded as to the identity of each sample. Histological assessment is done with a grading scale for quality of epithelialization. There is one grading scale for mice and one grading scale for rats. For vascularity assessment, choose two random spots in the superficial layer and two in the deeper layer of the wound and count the number of identified blood vessels (*see* **Notes 4,5**).

3.9. Statistical Analysis

Results must be presented as means ± standard error of the mean. Differences in means between groups are analyzed for significance using Student's t-test or ANOVA as appropriate using SigmaStat (SYSTAT Software Inc., Point Richmond, CA).

4. Notes

1. With the mouse models that have thin skin, this bleb must have the following three criteria: sharp edges from the skin surface, the skin on top of the bleb must whiten, and the bleb must persist at least 5 s.
2. Prior to and during EP delivery, the skin of the animal must be dry; otherwise an electrical arch can occur, resulting in tissue burn.

3. Plannimetry: Even though the calculations are made in pixels, we have found that two different systems will not calculate the same amount of pixels for the same image. The scanner and the calculation software have to be identical within the same study.
4. Histological grading for mouse skin: grade 1, incomplete epithelium with micro-ulcers; grade 2, thin epithelium, unresolved inflammation; grade 3, intact, not reticulated epithelium; and grade 4, reticulated mature epithelium.
5. Histological grading for rat skin: grade 1, incomplete epithelialization with open areas, inflammatory cells, hemorrhage present in subcutaneous tissue; grade 2, thin, but complete epithelium, 3–5 cell thickness, continuing subcutaneous inflammatory reaction; grade 3, thicker epithelial coverings, complete with 5–10 cell thickness, subcutaneous tissue well healed; and grade 4, reticulated epithelium, 10–15 cell thickness, subcutaneous tissue well healed with dense scar.

References

1. Marchese, C. (1995) Modulation of KGF and its receptor in re-epithelializing human skin. *J. Exp. Med.* **182**, 1369–1376.
2. Werner, S. (1992) Large induction of KGF expression in the dermis during wound healing. *Proc. Natl. Acad. Sci. USA*. **89**, 6896–6900.
3. Smola, H., Thiekötter, G., and Fusenig, N.E. (1993) Mutual induction of growth factor gene expression by epidermal–dermal cell interaction. *J. Cell. Biol.* **122**, 417–429.
4. Winkles, J.A., Alberts, G.F., Chedid, M., Taylor, W.G., Demartino, S., and Rubin, J.S. (1997) Differential expression of the keratinocyte growth factor (KGF) and KGF receptor genes in human vascular smooth muscle cells and arteries. *J. Cell. Physiol.* **173**, 380–386.
5. Pierce, G.F., Yanagihara, D., Klopchin, K., et al. (1994) Stimulation of all epithelial elements during skin regeneration by keratinocyte growth factor. *J. Exp. Med.* **179**, 831–840.
6. Finch, P.W., Rubin, J.S., Miki, T., Ron, D., and Aaronson, S.A. (1989) Human KGF is FGF-related with properties of a paracrine effector of epithelial cells. *Science*. **145**, 752–755.
7. Sutherland, J., Denyer, M., and Britland, S. (2005) Motogenic substrata and chemokinetic growth factors for human skin cells. *J. Anat.* **207**, 67–78.
8. Finch, P.W. and Rubin, J.S. (2004) Keratinocyte growth factor fibroblast growth factor 7, a homeostatic factor with therapeutic potential for epithelial protection and repair. *Adv. Cancer Res.* **91**, 69–136.
9. Goldman, R. (2004) Growth factors and chronic wound healing, Past, Present and future. *Adv. Skin Wound Care*. **17**, 24–35.
10. Galkowska, H., Olszewski, W.L., and Wojewodzka, U. (2005) Keratinocyte and dermal vascular endothelial cell capacities remain unimpaired in the margin of chronic venous ulcer. *Arch. Dermatol. Res.* **296**, 286–295.
11. Chakravarti, S., Wu, F., Vij, N., Roberts, L., and Joyce, S. (2004) Microarray studies reveal macrophage-like function of stromal keratocytes in the cornea. *Invest. Ophthalmol. Vis. Sci.* **45**, 3475–3484.
12. Berryhill, B.L., Kader, R., Kane, B., Birk, D.E., Feng, J., and Hassell, J.R. (2002) Partial restoration of the keratocyte phenotype to bovine keratocytes made fibroblastic by serum. *Invest. Ophthalmol. Vis. Sci.* **43**, 3416–3421.
13. Seo, S.K., Gebhardt, B.M., Lim, H.Y., et al. (2001) Murine keratocytes function as antigen-presenting cells. *Eur. J. Immunol.* **31**, 3318–3328.
14. Byrnes, C.K., Khan, F.H., Nass, P.H., Hatoum, C., Duncan, M.D., and Harmon, J.W. (2001) Success and limitations of a naked plasmid transfection protocol for keratinocyte growth factor-1 to enhance cutaneous wound healing. *Wound Repair Regen.* **9**, 341–346.
15. Andree, C., Swain, W.F., Page, C.P., et al. (1994) In vivo transfer and expression of a human epidermal growth factor gene accelerates wound repair. *Proc. Natl. Acad. Sci. USA*. **91**, 12188–12192.

16. Eming, S.A., Whitsitt, J.S., He, L., Krieg, T., Morgan, J.R., and Davidson, J.M. (1999) Particle-mediated gene transfer of PDGF isoforms promotes wound repair. *J. Invest. Dermatol.* **112**, 297–302.

17. Felgner, P. and Rhodes, G. (1991) Gene therapeutics. *Nature.* **349**, 351.

18. Bohnen, J.M., Matlow, A.G., Mustard, R.A., Christie, N.A., and Kavouris, B. (1988) Antibiotic efficacy in intraabdominal sepsis: a clinically relevant model. *Can. J. Microbiol.* **34**, 323–326.

19. Wichterman, K.A. (1980) Sepsis and septic shock—a review of laboratory models and a proposal. *J. Surg. Res.* **29**, 189–201.

20. Glasspool-Malone, J., Somiari, S., Drabick, J.J., and Malone, R.W. (2000) Efficient nonviral cutaneous transfection. *Mol. Ther.* **2**, 140–146.

21. Faurie, C., Golzio, M., Moller, P., Teissie, J., and Rols, M.P. (2003) Cell and animal imaging of electrically mediated gene transfer. *DNA Cell. Biol.* **22**, 777–783.

22. Faurie, C., Phez, E., Golzio, M., et al. (2004) Effect of electric field vectoriality on electrically mediated gene delivery in mammalian cells. *Biochim. Biophys. Acta.* **1665**, 92–100.

23. Pavselj, N. and Preat, V. (2005) DNA electrotransfer into the skin using a combination of one high- and one low-voltage pulse. *J. Control. Release.* **106**, 407–415.

24. Ferguson, M., Byrnes, C., Sun, L., et al. (2005) Wound healing enhancement: electroporation to address a classic problem of military medicine. *World J. Surg.* **29**, S55–S59.

25. Marti, G., Ferguson, M., Wang, J., et al. (2004) Electroporative transfection with KGF-1 DNA improves wound healing in a diabetic mouse model. *Gene Ther.* **11**, 1780–1785.

26. Sun, L., Xu, L., Chang, H., et al. (1997) Transfection with aFGF cDNA improves wound healing. *J. Invest. Dermatol.* **108**, 313–318.

Chapter 31
Hepatocyte Growth Factor Gene Therapy for Hypertension

Kazuo Komamura, Jun-ichi Miyazaki, Enyu Imai, Kunio Matsumoto, Toshikazu Nakamura, and Masatsugu Hori

Abstract Hepatocyte growth factor (HGF) has mitogenic, motogenic, and morphogenic biological activities as well as helps in regenerating various tissues. In cardiovascular organs, HGF was reported to have anti-apoptotic, anti-fibrotic, and vasodilating effects. HGF has close relationships with hypertension, arteriosclerosis, and heart failure. HGF enhances renal regeneration and suppresses the progression of hypertension. Intramuscular electroporation of the therapeutic gene is a simple, economic, and low toxic method compared with systemic administration of the purified proteins or peptides. We outline the technique of intramuscular electroporation of HGF gene as a remedy for hypertension.

Keywords: electroporation, gene transfer, gene therapy, hepatocyte growth factor, hypertension, kidney, muscle

1. Introduction

Hepatocyte growth factor (HGF), a heterodimeric molecule composed of a 69 kD α-chain and a 34 kD β-chain, was found to be a potent mitogen for fully differentiated hepatocytes (Fig. 31.1A) (1, 2). HGF has multiple biological activities in a wide variety of cells, including mitogenic, motogenic (enhancement of cell movement), morphogenic, and anti-apoptotic activities (3, 4). The motogenic action of HGF was deduced from the unexpected finding that characterization of scatter factor showed it to be identical to HGF (5, 6). Scatter factor was originally identified and purified as a fibroblast-derived epithelial cell motility factor (7). HGF affects cell–cell interaction and cell–extracellular matrix (ECM) interaction and stimulates or activates proteolytic networks involved in the breakdown of ECM proteins (8, 9). Thus, typical biological activities of HGF are involved in construction, remodeling, and protection of tissue structures during development and regeneration. The receptor for HGF was identified in 1991 to be a c-met protooncogene product (10). Intracellular signaling pathways of HGF and c-Met are involved in the development of epithelial tissues of the liver, placenta, kidney, lung, mammary

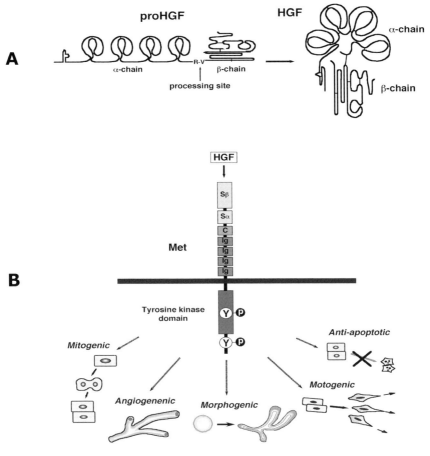

Fig. 31.1 Schematic structures of hepatocyte growth factor and its receptor. Prohepatocyte growth factor (HGF) and mature HGF (**A**). Typical biological activities of HGF mediated by c-Met/HGF receptor and intracellular signal transducers, which associate with tyrosine-phospho-rylated c-Met (**B**)

gland, and teeth, as a mediator in epithelial–mesenchymal interactions for organo-genesis (11). c-Met is also expressed in the heart, blood vessels, and nervous sys-tem, and it is involved in various functions, including angiogenesis, morphogenesis, motogenesis, and anti-apoptosis (Fig. 31.1B).

In clinical settings, serum HGF levels are elevated not only in patients with hepatic disease (12), malignancies (13), or end-stage renal disease (14) but also in patients with atherosclerosis and proliferative diabetic retinopathy (15, 16). Nakamura et al. (17) reported elevation of serum HGF concentrations in hypertensive patients, which was correlated with the severity of hypertensive target organ damage. Komai et al. (18) reported that serum HGF concentration was

correlated with human arterial stiffness and forearm vasoreactivity. Thus, HGF has close relationships with hypertension and related diseases. Recent studies suggest that HGF enhances renal regeneration and suppresses the onset of acute renal failure caused by renal toxins, renal ischemia, or unilateral nephrectomy (19–21). In the following section, we outline the technique of electroporation for HGF gene transfection aimed at hypertension therapy.

1.1. Application of In Vivo Electroporation for Cardiovascular Disease

Electroporation-mediated gene transfer has been used effectively in the muscles of mouse, rat, rabbit, and monkey models (22), and it has been applied to gene transfer into cardiac muscle (23). Thus, this method should have broad applications in physiologic and pharmacologic studies using experimental animals. It is likely that further improvement of this method will provide a new approach to efficient DNA vaccination and gene therapy for human diseases. Among the potentially treatable human diseases are various autoimmune diseases, chronic inflammatory disorders, infections, malignancies, and acquired or inherited serum protein deficiencies. This method may also be applied to the constitutive overexpression of vascular endothelial growth factor or HGF to induce therapeutic angiogenesis in patients with critical limb ischemia.

1.2. Therapeutic Use of HGF for Hypertension

Suppression of renal HGF expression by angiotensin-II and transforming growth factor-β (TGF-β) might accelerate renal injury, such as glomerulosclerosis and tubular degeneration, because HGF prevents apoptosis of endothelial and epithelial cells mediated by several conditions, including hypertension (24, 25). HGF exerts mitogenic responses in renal epithelial cells derived from distinct regions and species, including rabbit and rat proximal tubular cells (26, 27) and rat glomerular epithelial cells (28). HGF stimulates the proliferation of renal epithelial cell lines, including a rat visceral glomerular cell line (29), proximal tubular cell lines (30), and a murine medullary collecting duct epithelial cell line (31). Likewise, HGF exhibits mitogenic action on renal endothelial cells (32). HGF has no apparent effect on rat and human mesangial cell proliferation (32, 33) and is weakly mitogenic (33, 34), although mesangial cells do express the c-Met receptor (32, 34).

In addition to biological activities of HGF on renal cells, involvement of HGF in renal regeneration has been proposed based on findings that renal and plasma HGF levels and c-Met receptor expression are regulated in response to renal injuries. An induction of HGF mRNA in the kidney following renal injuries indicates that the kidney is one source of HGF. In the kidney, stromal cells, such

as macrophages and endothelial and mesangial cells, express HGF (35); thus, HGF seems to act through paracrine- and autocrine-related pathways. On the other hand, HGF mRNA expression is up-regulated in distant intact organs, such as the lung, liver, and spleen, as well as the injured kidney following acute renal injury, including unilateral nephrectomy. It is noteworthy that the induction of HGF mRNA expression in distant noninjured organs was seen in cases of hepatic and heart injuries (36, 37). These observations suggest that HGF increases in the blood circulation may be derived from noninjured organs as well as injured organs. Therefore, renotropic systems supported by HGF may involve two distinct pathways: HGF locally produced in the kidney acting through paracrine and autocrine mechanisms and HGF produced in distant organs acting through an endocrine mechanism.

In ICR strain-derived glomerulonephritis, mice not given HGF, molecular and cellular events leading to end-stage chronic renal disease progressed during this period. Renal TGF-β levels, the number of α-SMA-positive myofibroblasts, type I collagen and fibronectin accumulation, and the number of tubular apoptosis increased, whereas the number of proliferating tubular cells decreased. In contrast, in mice treated with HGF, TGF-β and PDGF expression, ECM accumulation, and the number of myofibroblasts and tubular apoptosis decreased, whereas the number of regenerating tubular cells increased. Consistent with these changes, the beneficial effects of HGF on the clinical outcome were apparent as decreases in levels of serum creatinine, blood urea nitrogen (BUN), and urine albumin, and as a diminution in histologic renal injury (38). It should be emphasized that serum creatinine, BUN, and urine albumin levels and histologic renal injury in mice treated with HGF were less than such events seen at the start of HGF administration. These results indicate that treatment with HGF has therapeutic effects in cases of chronic renal disease, rather than the HGF supplements merely retarding or inhibiting the progression of chronic renal disease.

2. Materials

2.1. Experimental Animal

1. Rats: Male Sprague-Dawley rats weighing 150 g.
2. Anesthetic: Pentobarbital sodium (Nembutal).

2.2. Plasmid DNA

1. pCAGGS plasmid (provided by Prof. Jun-ichi Miyazaki's Laboratory, http://www.med.osaka-u.ac.jp/pub/nutri/www/index.html).
2. *E. coli* HB101.

3. Microcentrifuge.
4. Spectrophotometer.
5. PBS: 137 mM NaCl, 2.68 mM KCI, 8.1 mM Na$_2$HPO$_4$, 1.47 mM KH$_2$PO$_4$, pH 7.4.
6. pKSCX-HGF or therapeutic HGF plasmid (provided by Prof. Miyazaki's Laboratory).

2.3. Intrazmuscular DNA Injection and Electroporation (39)

1. Pentobarbital sodium.
2. Bupivacaine hydrochloride.
3. Insulin syringe: 27 GA needle.
4. TE buffer: 10 mM Tris-HCl, pH 7.5, 1 mM ethylene diamine tetraacetate.
5. Electric pulse generator (T820, BTX, San Diego, CA) with a switch box (MBX-4, BTX) and a graphic pulse analyzer (BTX400, BTX).

2.4. Assessment of the Efficiency of Gene Transfer

1. CsCl-purified preparations of pCAGGS-IL-5 and pCAGGS-lacZ plasmid DNA at a concentration of 1.5 μg/μL in PBS.
2. Murine IL-5 ELISA kit (Endogen, Woburm, MA).
3. Paraformaldehyde (4%) in PBS.
4. X-gal (5-bromo-4-chloro-3-indolyl-β-D-galactopyranoside): 40mM in dimethyl sulfoxide.
5. O.C.T. compound.
6. Dry ice-acetone.
7. Cryostat.
8. Slide glasses coated with 3-aminopropyltriethoxysilane (Sigma, St. Louis, MO).
9. 1.5% glutaraldehyde in PBS.
10. Eosin.

3. Methods

In the following section, we describe the preparation of plasmid genes and the method of HGF gene transfer into tibialis anterior muscles of adult rats by in vivo electroporation. Modification of the described method may be necessary when other types of muscles or species are used.

3.1. *Prepare Plasmid DNA*

1. Transfom the pCAGGS plasmid DNA (Fig. 31.2A) into *E. coli* HB101 or another bacteria strain.
2. Lyse bacteria pellets using the alkaline lysing method and purify the plasmid DNA by two cycles of ethidium bromide-CsCl equilibrium density gradient ultracentrifugation.
3. Purify plasmid DNA further using isopropanol precipitation, phenol and phenol–chloroform extraction, and ethanol precipitation.
4. Dissolve plasmid DNA in pure water, and determine the quantity and quality using spectrophotometer at 260 and 280 nm.
5. Prior to injection, dilute plasmid DNA to its final concentration, 1–1.5 µg/µL in PBS. Because the salt concentration seems to affect the efficiency of gene transfer, the final DNA solution is made by adding 1 part of 10× PBS to 9 parts of DNA solution.
6. Construct the therapeutic HGF encoding plasmid DNA, pKSCX-HGF, by inserting rat HGF cDNA (2.2 kb) into the unique EcoRI site between a chicken β-actin promoter and a-3′-flanking sequence of the rabbi&tbdot; β-globin gene of the pKSCX vector (Fig. 31.2B).

3.2. *Intramuscular Injection of plasmid DNA*

1. Perform subtotal nephrectomy by right subcapsular nephrectomy combined with infarction of approximately two-thirds of the left kidney by selective ligation of two of three to four extra-renal branches of the left renal artery. Two weeks after

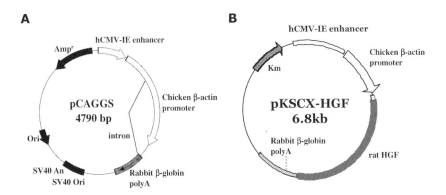

Fig. 31.2 Structure of expression plasmids pCAGGS (**A**) and pKSCX-HGF (**B**). Plasmid pCAGGS has the cytomegalovirus immediate early (CMV-IE) enhancer-chicken β-actin hybrid (CAG) promoter and a 3′-flanking sequence and a polyadenylation signal of the rabbit β-globin gene. This promoter has extremely high activity in the muscle cell. The expression cassette of pKSCX-HGF contains chicken β-actin promoter, rat HGF, and rabbit β-globin poly A

this renal ablation procedure, divide the nephrectomized rats into two groups. Treat one group with HGF electroporation gene therapy and treat the other group with empty plasmid DNA, pCAGGS.

2. Anesthetize rats by intraperitoneal injection of 50 mg/kg body weight of pentobarbital sodium.

3. Inject 200 μL of 0.5% bupivacaine into the right and left tibialis anterior muscles using a disposable insulin syringe with a 27 GA needle.

4. Anesthetize rats with intraperitoneal pentobarbital 3 days later. Inject a 100 μL solution containing 200 μg of pKSCX-HGF into the bupivacaine-treated muscles.

3.3. Electroporation In Vivo

1. Insert a pair of electrode needles into the muscle to a depth of 5 mm to encompass the DNA injection sites (*see* **Note 1**) (Fig. 31.3A). Electrodes contain a pair of stainless steel needles, with 5 mm length, 0.4 mm diameter, and a fixed gap of

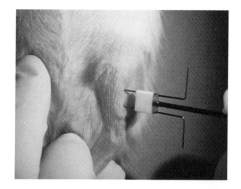

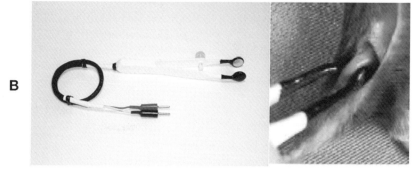

Fig. 31.3 Delivering electric pulses. Electric pulses were delivered using a needle electrode connected with electric pulse generator (**A**). A tweezer-type electrode for electroporation (**B**). It can be used for electroporation for skeletal muscle or whole kidney.

5 mm between the two needles. Connect the electrodes to the electric pulse generator via a switch box, which can produce square waves, i.e., the voltage remains constant during the pulse duration. Monitor electric pulses using a graphic pulse analyzer. Push on the chamber resistance switch and monitor the resistance value. When the monitor reads $1–2\,k\Omega$, it indicates that the electrodes are correctly inserted into the muscle. Otherwise, change the insertion site of the electrodes.

2. Deliver three 50 ms electric pulses using an electric pulse generator followed by three more pulses of the opposite polarity to each injection site at a rate of 1 pulse/s.

3.3. Assess the Efficiency of Gene Transfer

Before introducing the gene of interest by electroporation, it is important to test the effectiveness of the experimental procedures using a positive control. Perform the gene transfer efficiency test using a reporter gene, such as IL-5 or β-galactosidase.

3.3.1. Determine IL-5 Expression

1. Inject the bilateral tibialis anterior muscles of anesthetized mice with 50 μg each of pCAGGS-IL-5 plasmid DNA at a concentration of 1.5 μg/μL in PBS, and deliver electric pulses at 100 V, as described in sect. 3.2.
2. Five days after injection, obtain serum samples from the tail vein of the mice.
3. Assay the serum samples for IL-5 using an ELISA kit (Endogen), according to supplier's instructions.

3.3.2. β-Galactosidase Expression

1. Inject the bilateral tibialis anterior muscles of anesthetized mice with 50 μg each of pCAGGS-lacZ plasmid DNA at a concentration of 1.5 μg/μL in PBS, and deliver electric pulses at 100 V as described in sect. 3.2.
2. Four to five days after injection, sacrifice the mice by cervical dislocation.
3. Fix the tibialis anterior muscles in cold 4% paraformaldehyde in PBS for 3 h, and then wash in PBS for 1 h.
4. Stain the muscle sample at 37°C for 18 h in the presence of 1 mM X-gal in order to detect E.coli β-galactosidase activity in whole muscle.
5. For transverse sections, embed the muscle in O.C.T. compound and freeze in dry ice-acetone.
6. Slice serial sections (15 μm thick) with a cryostat and place on slide glasses coated with 3-aminopropyltriethoxysilane.

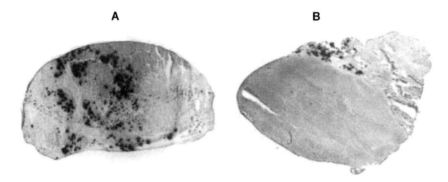

Fig. 31.4 Histochemical staining for β-galactosidase activity in the muscle after gene transfer of pCAGGS-lacZ DNA with or without electropulsation. The bupivacaine-treated portions of the bilateral tibialis anterior muscles were injected with pCAGGS-lacZ plasmid DNA and treated with (**A**) or without (**B**) electric pulses. Five days later, the muscle was excised and stained for β-galactosidase activity. Transverse sections of the muscle samples were also stained for β-ned with eosin

7. Fix the slices in 1.5% glutaraldehyde for 10 min at room temperature, and then wash three times in PBS.
8. Incubate the samples at 37°C for 3 h in the presence of 1 mM X-gal.
9. Counterstain the muscle sections with eosin.
10. Observe the sections with a microscope for X-gal staining (Fig. 31.4).

3.3.3. Determine the Levels of HGF in Serum

1. Inject, using an insulin syringe with a 27 GA needle, a 100 μL solution containing 200 μg of purified closed circular plasmid pKSCX-HGF into the tibialis anterior muscles. in a volume of 100 μL
2. Five days after injection, obtain serum samples from the tail vein of the rats.
3. Assay the level of HGF in serum samples using a rat HGF EIA kit (Institute of Immunology Co. Ltd., Tokyo, Japan), according to the supplier's instructions (39) (*see* **Note 2**).

Acknowledgments This work was supported by grant-in-aid for Scientific Research 11670729 and 14570709 from the Ministry of Education, Culture, Sport, Science, and Technology of Japan and by the program for promotion of Fundamental Studies in Health Sciences of the Pharmaceuticals and Medical Devices Agency (PMDA).

4. Notes

1. Other types of electrodes may be used: Tweezer-type electrodes have been successfully used in our laboratory and in other laboratories (40) (Fig. 31.3B).

2. In an experiment of nephrectomy hypertension, male Sprague-Dawley rats weighing 150 g were used for subtotal nephrectomy. pKSCX -HGF (400 μg/animal) or empty pCAGGS plasmid was transfected on 3 separate days (day 0, day 7, and day 14). At day 21, mean arterial pressure of HGF treated nephrectomized rats was 122 ± 13 mm Hg, which was lower than the pressure 147 ± 14 mm Hg of the control rats that received empty pCAGGS plasmid. Using the same model, Tanaka et al. reported that glomerulosclerosis and interstitial fibrosis following 5/6 nephrectomy were ameliorated by electroporation of pKSCX-HGF plasmid into skeletal muscle (41).

References

1. Nakamura, T., Nawa, K., Ichihara, A., Kaise, N., and Nishino, T. (1987) Purification and subunit structure of hepatocyte growth factor from rat platelets. *FEBS Lett.* **244**, 311–316.
2. Nakamura, T., Nishizawa, T., Hagiya, M., et al. (1989) Molecular cloning and expression of human hepatocyte growth factor. *Nature.* **342**, 440–443.
3. Matsumoto, K. and Nakamura, T. (1996) Emerging multipotent aspects of hepatocyte growth factor. *J. Biochem. (Tokyo)* **119**, 591–600.
4. Matsumoto, K. and Nakamura, T. (1997) Hepatocyte growth factor (HGF) as a tissue organizer for organogenesis and regeneration. *Biochem. Biophys. Res. Commun.* **239**, 639–644.
5. Konishi, T., Takehara, T., Tsuji, T., Ohsato, K., Matsumoto, K., and Nakamura, T. (1991) Scatter factor from human embryonic lung fibroblasts is probably identical to hepatocyte growth factor. *Biochem. Biophys. Res. Commun.* **180**, 765–773.
6. Weidner, K.M., Arakaki, N., Hartmann, G., et al. (1991) Evidence for the identity of human scatter factor and human hepatocyte growth factor. *Proc. Natl. Acad. Sci. USA.* **88**, 7001–7005.
7. Weidner, K.M., Behrens, J., Vandekerckhove, J., and Birchmeier, W. (1990) Scatter factor: molecular characteristics and effect on the invasiveness of epithelial cells. *J. Cell Biol.* **111**, 2097–2108.
8. Pepper, M.S., Matsumoto, K., Nakamura, T., Orci, L., and Montesano, R. (1992) Hepatocyte growth factor increases urokinase-type plasminogen activator (u-PA) and u-PA receptor expression in Madin-Darby canine kidney epithelial cells. *J. Biol. Chem.* **267**, 20493–20496.
9. Jiang, W.G., Martin, T.A., Parr, C., Davies, G., Matsumoto, K., and Nakamura, T. (2005) Hepatocyte growth factor, its receptor, and their potential value in cancer therapies. *Crit. Rev. Oncol. Hematol.* **53**, 35–69.
10. Bottaro, D.P., Rubin, J.S., Faletto, D.L., et al. (1991) Identification of the hepatocyte growth factor receptor as the c-met proto-oncogene product. *Science.* **251**, 802–804.
11. Birchmeier, C. and Gherardi, E.(1998) Developmental roles of HGF/SF and its receptor, the c-Met tyrosine kinase. *Trends Cell Biol.* **8**, 404–410.
12. Shiota, G., Umeki, K., Okano, J., and Kawasaki, H. (1995) Hepatocyte growth factor and acute phase proteins in patients with chronic liver diseases. *J. Med.* **26**, 295–308.
13. Seidel, C., Borset, M., Turesson, I., Abildgaard, N., Sundan, A., and Waage, A. (1998) Elevated serum concentrations of hepatocyte growth factor in patients with multiple myeloma. *Blood.* **91**, 806–812.
14. Rampino, T., Libetta, C., De Simone, W., et al. (1998) Hemodialysis stimulates hepatocyte growth factor release. *Kidney Int.* **53**, 1382–1388.
15. Nishimura, M., Nakano, K., Ushiyama, M., et al. (1998) Increased serum concentrations of human hepatocyte growth factor in proliferative diabetic retinopathy. *J. Clin. Endocrinol. Metab.* **83**, 195–198.
16. Tateishi, J., Waku, S., Masutani, M., Ohyanagi, M., and Iwasaki, T. (2002) Hepatocyte growth factor as a potential predictor of the presence of atherosclerotic aorto-iliac artery disease. *Am. Heart J.* **143**, 272–276.

17. Nakamura, S., Moriguchi, A., Morishita, R., et al. (1998) A novel vascular modulator, hepatocyte growth factor (HGF), as a potential index of the severity of hypertension. *Biochem. Biophys. Res. Commun.* **242**, 238–243.

18. Komai, N., Ohishi, M., Morishita, R., et al. (2002) Serum hepatocyte growth factor concentration is correlated with the forearm vasodilator response in hypertensive patients. *Am. J. Hypertens.* **15**, 499–506.

19. Kawaida, K., Matsumoto, K., Shimazu, H., and Nakamura, T. (1994) Hepatocyte growth factor prevents acute renal failure and accelerates renal regeneration in mice. *Proc. Natl. Acad. Sci. USA.* **91**, 4357–4361.

20. Miller, S.B., Martin, D.R., Kissane, J., and Hammerman, M.R. (1994) Hepatocyte growth factor accelerates recovery from acute ischemic renal injury in rats. *Am. J. Physiol.* **266**, F129–F134.

21. Mizuno, S., Kurosawa, T., Matsumoto, K., Mizuno-Horikawa, Y., Okamoto, M., and Nakamura, T. (1998) Hepatocyte growth factor prevents renal fibrosis and dysfunction in a mouse model of chronic renal disease. *J. Clin. Invest.* **101**, 1827–1834.

22. Mir, L.M., Bureau, M.F., Gehl, J., et al. (1999) High-efficiency gene transfer into skeletal muscle mediated by electric pulses. *Proc. Natl. Acad. Sci. USA.* **96**, 4262–4267.

23. Harrison, R.L., Byrne, B.J., and Tung, L. (1998) Electroporation-mediated gene transfer in cardiac tissue. *FEBS Lett.* **435**, 1–5.

24. Nakano, N., Morishita, R., Moriguchi, A., et al. (1998) Negative regulation of local hepatocyte growth factor expression by angiotensin II and transforming growth factor-β in blood vessels: potential role of HGF in cardiovascular disease. *Hypertension.* **32**, 444–451.

25. Matsumoto, K., Morishita, R., Moriguchi, A., et al. (1999) Prevention of renal damage by angiotensin II blockade, accompanied by increased renal hepatocyte growth factor in experimental hypertensive rats. *Hypertension.* **34**, 279–284.

26. Igawa, T., Kanda, S., Kanetake, H., et al. (1991) Hepatocyte growth factor is a potent mitogen for cultured rabbit renal tubular epithelial cells. *Biochem. Biophys. Res. Commun.* **174**, 831–838.

27. Kan, M., Zhang, G.H., Zarnegar, R., et al. (1991) Hepatocyte growth factor/hepatopoietin A stimulates the growth of rat kidney proximal tubule epithelial cells (RPTE), rat nonparenchymal liver cells, human melanoma cells, mouse keratinocytes and stimulates anchorage-independent growth of SV-40 transformed RPTE. *Biochem. Biophys. Res. Commun.* **174**, 331–337.

28. Harris, R.C., Burns, K.D., Alattar, M., Homma, T., and Nakamura, T. (1993) Hepatocyte growth factor stimulates phosphoinositide hydrolysis and mitogenesis in cultured renal epithelial cells. *Life Sci.* **52**, 1091–1100.

29. Kawaguchi, M., Kawashima, F., Ohshima, K., Kawaguchi, S., and Wada, H. (1994) Hepatocyte growth factor is a potent promoter of mitogenesis in cultured rat visceral glomerular epithelial cells. *Cell. Mol. Biol. (Noisy-le-grand)* **40**, 1103–1111.

30. Ishibashi, K., Sasaki, S., Sakamoto, H., et al. (1992) Hepatocyte growth factor is a paracrine factor for renal epithelial cells: stimulation of DNA synthesis and NA,K-ATPase activity. *Biochem. Biophys. Res. Commun.* **182**, 960–965.

31. Cantley, L.G., Barros, E.J., Gandhi, M., Rauchman, M., and Nigam, S.K. (1994) Regulation of mitogenesis, motogenesis, and tubulogenesis by hepatocyte growth factor in renal collecting duct cells. *Am. J. Physiol.* **267**, F271–280.

32. Yo, Y., Morishita, R., Nakamura, S., et al. (1998) Potential role of hepatocyte growth factor in the maintenance of renal structure: anti-apoptotic action of HGF on epithelial cells. *Kidney Int.* **54**, 1128–1138.

33. Kolatsi-Joannou, M., Woolf, A.S., Hardman, P., White, S.J., Gordge, M., and Henderson, R.M. (1995) The hepatocyte growth factor/scatter factor (HGF/SF) receptor, met, transduces a morphogenetic signal in renal glomerular fibromuscular mesangial cells. *J Cell. Sci.* **108**, 3703–3714.

34. Kallincos, N.C., Pollard, A.N., and Couper, J.J. (1998) Evidence for a functional hepatocyte growth factor receptor in human mesangial cells. *Regul. Pept.* **74**, 137–142.

35. Igawa, T., Matsumoto, K., Kanda, S., Saito, Y., and Nakamura, T. (1993) Hepatocyte growth factor may function as a renotropic factor for regeneration in rats with acute renal injury. *Am. J. Physiol.* **265**, F61–F69.

36. Miyazawa, K., Shimomura, T., Kitamura, A., Kondo, J., Morimoto, Y., and Kitamura, N. (1993) Molecular cloning and sequence analysis of the cDNA for a human serine protease reponsible for activation of hepatocyte growth factor. Structural similarity of the protease precursor to blood coagulation factor XII. *J. Biol. Chem.* **268**, 10024–10028.

37. Ono, K., Matsumori, A., Shioi, T., Furukawa, Y., and Sasayama, S. (1997) Enhanced expression of hepatocyte growth factor/c-Met by myocardial ischemia and reperfusion in a rat model. *Circulation.* **95**, 2552–2558.

38. Mizuno, S., Kurosawa, T., Matsumoto, K., Mizuno-Horikawa, Y., Okamoto, M., and Nakamura, T. (1998) Hepatocyte growth factor prevents renal fibrosis and dysfunction in a mouse model of chronic renal disease. *J. Clin. Invest.* **101**, 1827–1834.

39. Komamura, K., Tatsumi, R., Miyazaki, J., et al. (2004) Treatment of dilated cardiomyopathy with electroporation of hepatocyte growth factor gene into skeletal muscle. *Hypertension.* **44**, 365–371.

40. Umeda, Y., Marui, T., Matsuno, Y., et al. (2004) Skeletal muscle targeting in vivo electroporation-mediated HGF gene therapy of bleomycin-induced pulmonary fibrosis in mice. *Lab. Invest.* **84**, 836–844.

41. Tanaka, T., Ichimaru, N., Takahara, S., et al. (2002) In vivo gene transfer of hepatocyte growth factor to skeletal muscle prevents changes in rat kidneys after 5/6 nephrectomy. *Am. J. Transplant.* **2**, 828–836.

Chapter 32
Electroporation of Corrective Nucleic Acids (CNA) In Vivo to Promote Gene Correction in Dystrophic Muscle

Robert M.I. Kapsa, Sharon H.A. Wong, and Anita F. Quigley

Abstract Non-viral gene transfer into skeletal muscle in vivo is enhanced by electroporation (EP) to efficiencies far beyond any other (non-EP) method reported to date. Electroporation consistently delivers high levels of transgene to muscle and has been used extensively for the delivery of therapeutic transgenes to dystrophic mouse muscle such as the *mdx* mouse model of human Duchenne muscular dystrophy (DMD). Since the earliest applications, electroporation has consistently and reproducibly achieved highly efficient DNA delivery to a high proportion (greater than 70%) of fibres in treated muscles. This manuscript describes a methodology for introduction of corrective nucleic acids (CNAs) for the purpose of correcting the dystrophin gene ($\underline{DMD}^{mdx}$) mutation responsible for muscular dystrophy in the *mdx* mouse model of human DMD by targeted corrective gene conversion (TCGC).

Keywords: targeted corrective gene conversion, TCGC, mutation correction, in vivo, small fragment homologous replacement, SFHR, *mdx* mouse, electroporation, dystrophin

1. Introduction

Duchenne muscular dystrophy (DMD) is an X-linked disorder that affects 1 in 3,500 liveborn males with progressive loss of muscle mass and rapid exhaustion of regenerative capacity resulting from sarcolemmal instability mediated by a functional lack of the cytoskeletal protein, dystrophin (1–4). In DMD, mutations in the 2.4 million bp dystrophin gene (*DMD*) cause codon reading frameshift or nonsense mutation and consequential premature termination of dystrophin protein sequence expression (5, 6). The *mdx* mouse model of human DMD has a C-to-T nonsense transition at mRNA position bp 3185 in exon 23 of the X chromosome, causing a truncation of the dystrophin protein expressed from the *mdx* locus (*DMD*^{mdx}) (7, 8).

Targeted correction of mutations stands to deliver a genuine "cure" for disorders by total removal of the disease-causing mutation from the cell's genome. Targeted corrective

gene conversion (TCGC) promotes phenotypic correction for the entire remaining life-time of the cell, with the corrected gene's expression regulated by endogenous systems in a context relevant to the cell's normal biological patterning. We have applied TCGC via small fragment homologous replacement (SFHR) to the DMD^{mdx} both in vitro and in vivo, with a key finding that one of the major hurdles to this technology is the delivery of CNA to myonuclei containing the mutation, most particularly in vivo (9, 10). Electroporation presents an alternative method for delivery to myonuclei, with poten-tially less adverse effects on subsequent cell function (11–13).

This study describes the methodology that we have used for in vivo correction of the murine DMD^{mdx} gene by EP of CNA in dystrophic *mdx* muscle, with particular emphasis on the means by which the process is optimised using reporter plasmid and by which identification of corrected loci are optimised in the *mdx* mouse system. By analogy, this protocol applies to other situations in which TCGC presents a potential therapy for a hereditary muscle condition.

2. Materials

2.1. Animals

Male and female C57Bl/10 *mdx* mice, 8 weeks of age, were purchased from the Animal Resource Centre (ARC, Western Australia) (*see* **Notes 1,2**).

2.2. Nucleic Acids

1. Reporter Plasmid: Kanamycin resistance-bearing expression vector pEGFP-N2 (BD Biosciences Clontech, CA), encoding a red-shifted variant of the *Aequorea victoria* enhanced green fluorescent protein (EGFP) under control of the CMV promoter.
2. Competent *Escherichia coli* strain JM109.
3. Luria-Bertani medium: 10 g/L Bacto Tryptone, 5 g/L w/v Yeast Extract, 10 g/L NaCl, and 15 g/L Bacto agar for plates.
4. Kanamycin.
5. Miniprep DNA purification system.
6. Mouse isotonicity phosphate-buffered saline.

2.3. Corrective Nucleic Acids

1. QIAamp tissue DNA extraction protocol according to specification.
2. QIAquick PCR purification kit.

3. Bio-X-Act 3′–5′ proof-reading DNA polymerase.
4. Bio-X-Act buffer system.
5. 15 mM MgCl$_2$.
6. 10 mM dATP, 10 mM dTTP, 10 mM dGTP, 10 mM dCTP.
7. 15 mM Tris-HCl, pH 8.8.
8. MasterCycler gradient PCR machine.
9. DNA oligonucleotide primers with and without 5′ end-labelling with Texas Red (see Fig. 32.1, Table 32.1)

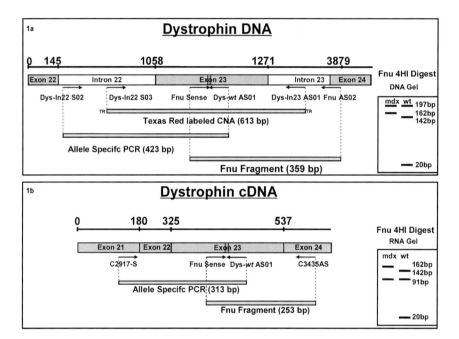

Fig. 32.1 PCR amplification of CNA and detection of converted mdx loci. Texas red (TR) labeled primers (Dys In22 S-03 Dys In23 AS-01) are used to PCR amplify a 613 bp DNA fragment from total DNA extracted from a wild type mouse (**1a**). The labeled PCR fragment (corrective nucleic acid) is then purified and concentrated prior to injection into the TA of mice. Wt (gene) loci can be detected by either allele specific PCR or by RFLP. Allele specific primers are used to generate a PCR product in the presence of the wild type locus only, at a specific annealing temperature. The presence of wt locus can also be detected using a modified primer (Fnu sense) to amplify a 359 bp band. The PCR product is then digested with *Fnu*4H I to produce a 162 and 197 bp band in the presence of the *mdx* loci. The presence of the wt loci introduces a second *Fnu*4H I restriction site resulting in further digestion of the 162 bp band into 20 bp and 142 bp, which can be resolved by PAGE. The strategy for detection of corrected transcript arising from corrected (i.e. *wt*) *mdx* dystrophin gene loci is shown in Fig. 32.1b

Table 32.1 Primer Sequences for Texas Red (TR) Labeled CNA PCR, Allele Specific and *Fnu* 4HI RFLP Detection of Both *wt* and *mdx* Dystrophin Loci from DNA and RNA

Primer name	Label	Primer sequence										Direction	Purpose
		5'									3'		
		3	6	9	12	15	18	21	24	27	30		
Dys In22 S-03	*TR*	GTT	TCA	CTG	TAG	GTA	AGT	AAA	TGT	ATC	AC	Sense	TR-CNA and CNA
Dys In23 AS-01	*TR*	GGC	TTT	TGA	TAT	CAT	CAA	TAT	CTT	TGA	AGG	Antisense	TR-CNA and CNA
Dys In22 S-02		GTT	GAT	TCT	AAA	AAT	CCC	ATG	TTG			Sense	Allele specific
Dys wt AS-01		GTC	ACT	CAG	ATA	GTT	GAA	GCC	ATT	TTG		Antisense	Allele specific
c2917-S		GAA	AGA	AAA	GGG	ACA	GGG					Sense	Allele specific
Fnu S		CAA	AGT	TCT	TTG	AAA	GAG	CAG				Sense	*Fnu* 4HI PCR-RFLP
Fnu AS-02		AAA	TAG	GCA	AGT	TGC	AAT	CC				Antisense	*Fnu* 4HI PCR-RFLP
c3435-AS		AAA	ACA	TCA	ACT	TCA	GCC	ATC	C			Antisense	*Fnu* 4HI RT-PCR-RFLP

2.4. Surgical and In Vivo Procedures

1. Ibuprofen (Herron, Brisbane, Australia), or other non-steroidal anti-inflammatory propionic acid derivative analgesic suspended in 0.9% saline (50 mg/mL).
2. Isoflurane/1.5% oxygen mix.
3. Midget anaesthetic/oxygen mixer.
4. Insulin syringes.
5. Hamilton syringe: 10 or 20 µL.
6. BTX ECM830 electropulsator (Fig. 32.2, Inovio, San Diego, CA).
7. Tweezer electroporator electrodes (Fig. 32.3), manufactured by silver-soldered tin plates on stainless steel surgical tweezers.
8. Michel clips (Becton Dickinson, MD).
9. Betadine (Faulding Pharmaceuticals, SA, Australia).
10. Dispase I.
11. Type II Collagenase.

2.5. Immunohistology and Microscopic Analyses

1. Antibodies (Ab): dystrophin Ab; Alexa Fluor 488 FITC donkey anti-Rabbit secondary Ab.
2. Histochemicals: hematoxylin; eosin.
3. Olympus IX70 inverted microscope.
4. Donkey serum.
5. PBS, pH 7.4.
6. Fluorescent mounting media.
7. IsoPentane (BDH).
8. Cryostat.
9. Glass slides and coverslips.

2.6. Molecular Detection Analyses

1. 40 mM Tris acetate, 1 mM EDTA buffer (TAE, 1x), pH 9.0
2. G-Tip high molecular weight (Genome)-specific DNA extraction system.
3. *Fnu* 4HI restriction enzyme (*see* **Note 3**).
4. Taq polymerase.
5. Agarose gels: 1% w:v in TAE.
6. Hybond N + nylon membrane.
7. TriPure™ DNA, RNA and protein isolation reagent.
8. RNAase-free DNAase I and buffer system.
9. Omniscript™ reverse transcriptase system.
10. NaOH: 0.4 M.

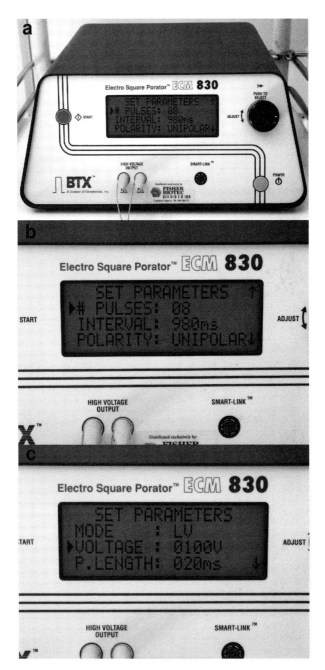

Fig. 32.2 Electroporation parameters. A BTX 830 Electro Square Porator was used for electroporation of TA muscles (**a**) Each TA was subjected to one series of pulses, consisting of eight unipolar 100 V pulses of 20 ms in length with 980 ms intervals (**b** and **c**)

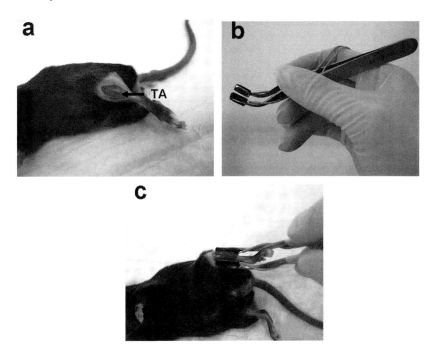

Fig. 32.3 The electroporation procedure. The mouse is anesthetised and hind limb shaved before surgical exposure of the TA by gently cutting the skin (**a**). The TA is held either side with electroporation tweezers (**b**) during the procedure, while the mouse is under anesthetic (**c**). After electroporation, the skin is closed over and fixed in place with Michel clips before the mouse is allowed to recover

11. Church buffer: 100 mM Na_2HPO_4, 0.685% (v/v) H_3PO_4, pH 7.2, 0.5% BSA, 7% SDS, 1 μM EDTA.
12. Imager.
13. 10× standard saline citrate (10× SSC).
14. Sodium dodecyl sulfate (SDS).

3. Methods

3.1. Animals

Male and female C57Bl/10 *mdx* mice, 8 weeks of age, are housed in numbers no greater than five per cage and need to be closely monitored thoughout the experimental stages to ensure there is no adverse reaction to the treatments. Animals are allowed free access to food and water. All procedures should be conducted with approval from the institutional Animal Experimental Ethics

Committee and conform to the National guidelines for the care and use of experimental animals set by the national animal ethics body.

3.2. Generation of Nucleic Acids

3.2.1. Propagation of Reporter Plasmid

1. Transform competent JM109 bacteria with pEGFP-N2 vector by heat shock (30 min on ice followed by 42°C for 2 min).
2. Incubate the transformed bacteria for 1–1.5 h at 37°C (with shaking) in non-selective Luria Bertani media.
3. Transfer and spread the transformed bacteria on selective agar media plates containing 50 μg/mL kanamycin.
4. Incubate the plated bacteria overnight at 37°C.
5. Pick several Kn-resistant colonies and incubate them O/N at 37°C in 30 mL of fresh LB media containing 50 μg/mL kanamycin to propagate pEGFP-N2 vector.
6. Extract plasmid DNA by using the Wizard® Miniprep DNA Purification System according to the manufacturer's specifications.
7. Suspend the purified plasmid in mouse isotonicity PBS to a final concentration of 1.5 μg/μL for injection into *mdx* muscle.

3.2.2. Corrective Nucleic Acid

1. Extract and column-purify DNA from the liver of the female *wt* mouse (*see* **Note 2**) using the QIAamp Tissue DNA extraction protocol according to manufacturer's specification (Qiagen).
2. Suspend the DNA in 15 mM tris HCl, pH 9.0, and store at −20°C until required.
3. Make up 100 μL PCR reactions containing 50 ng of total DNA, 0.4 μM of each primer, 200 μM dNTPs, 1.5 mM $MgCl_2$, 0.5 units of Bio-X-Act 3′–5′ proof-reading DNA polymerase in buffer supplied by the manufacturer.
4. Amplify a 613 bp amplicon (Amplicon C) using oligonucleotide primers Dys In22 S-03 and Dys In23 AS-01 (Fig. 32.1a, Table 32.1) by 29 cycles of 92°C/45 s (denaturation), 65°C/2.5 min (annealing/extension) with an initial cycle of 92°C/2 min and 65°C/2.5 min (total 30 cycles) using an Eppendorf gradient Mastercycler.
5. Visualise and then resolve Amplicon C PCR product from reagent components by electrophoresis (1% agarose, TAE), followed by ion exchange chromatography according to manufacturer's specification (QIAquick PCR Purification System, Qiagen).
6. Suspend the gel-purified Amplicon C in mouse isotonicity PBS at a final concentration of 0.1 μg/μL (0.27 pmol/μL).

3.3. Preparation of Mice for Treating

1. Administer a non-steroidal anti-inflammatory analgesic (Ibuprofen, 5 mg/kg).
2. Suspend in 0.9% saline and inject by neck skin-fold subcutaneous injection.
3. Anaesthetise each 8-week old male *mdx* mouse by nose-cone inhalation of isoflurane/1.5–2% oxygen mix administered.
4. Maintain controlled isoFluorane/1.5–2% oxygen anaesthesia throughout the surgical procedure.

3.4. Injection of Nucleic Acids into Dystrophic TA Muscle

1. Shave both TA muscles of each mouse.
2. Swab the shaved area with betadine to disinfect the surgical area.
3. Make a single skin-deep incision longitudinally up the muscle's length to expose the TA muscle (Fig. 32.3).
4. The exposed right (RHS) TA muscle is injected (single injection) with 15 μg of pEGFP-N2 or 1 μg CNA as follows: A 10 μL Hamilton syringe is charged with 1.5 μg/μL pEGFP-N2 or 0.1 μg/μL CNA and inserted into the right TA near the ankle up to thee-quarters of the way up the TA muscle (visible). Nucleic acid is discharged slowly with gradual withdrawal of the needle towards the entry site.
5. Inject the left (LHS) TA muscle of each mouse as above, except with 10 μL of saline (no CNA or plasmid) to provide a relevant vehicle control for subsequent analysis of gene correction (TCGC) effects.

3.5. Electroporation of Dystrophic Muscle

1. Grip the exposed TA muscle firmly between the electrode plates on the tweezers, taking care not to include the Tibia between the plates.
2. Apply eight square wave electric pulses each of 20 ms duration, and 1 Hz frequency with output voltage of 200 V/cm (distance between the plates) (Fig. 32.2) to the muscles with (RHS) and without (LHS) the nucleic acid. Saline-injected TAs are used as negative controls.
3. Seal the wound with Michel clips, and sterilise (swab) with 1% w:v Betadine, and allow the mice to recover for 7–14 days.

3.6. Evaluation and Optimisation of Electroporation Efficiency

Wait 7–14 days after EP to allow sufficient gene expression, and then sacrifice the mice by cervical dislocation and harvest the electroporated TAs to evaluate efficiency of electroporation (pEGFP-N2 vector) and TCGC (Amplicon C CNA).

1. Cut the TA muscles from the legs of the mice and immediately snap freeze them in a liquid nitrogen-cooled isopentane bath and store at –80°C until required.

2. Cut serial transverse sections ($8\,\mu m$) along the entire muscles' lengths, collecting every 26th section on a slide series with the 25 sections immediately following collection in marked 1.5 mL Eppendorf tubes to allow nucleic (DNA and RNA) acid analyses of regions/sections showing the highest level of electroporation.

3. Evaluate the viability and efficiency of electroporation in the electroporated muscle(s) by visualising pEGFP-N2 expression (i.e. EGFP) in unstained sections using a filter (WIBA, Olympus) with excitation/emission wavelengths appropriate for EGFP of 488/508 nm (Fig. 32.4a). The sections containing the most EGFP-positive (EGFP $^{+\,ve}$) fibres are then recorded and expressed as a percentage of the total number of fibres in the TA. We have observed up to 89% of fibres expressing the EGFP transgene in such applications. Single fibre preparations can be used to evaluate the penetration of differentiated fibres as opposed to the mononucleated (precursor) cells in the muscle (Fig. 32.4a) (*see* **Note 4**). To establish distribution/targeting of CNA by the EP protocol, inject and electroporate anesthetised normal (non-*mdx*) C57Bl/10 mice with CNA 5′ end-labelled with Texas Red (TR). Sacrifice the mice without allowing recovery from the anesthetic by cervical dislocation 2 h after injection and remove injected muscles as described in the preceding sections. Cut serial sections and stain every second one with Hematoxylin and Eosin to identify nuclei. Visualise the TR-CNA by fluorescence using a filter with excitation/emission ranges of 545–580 nm/610–700 nm (Fig. 32.4b, MIY, Olympus).

4. Evaluate Penetration of Muscle Components by in vivo electroporation.

5. After electroporation with pEGFP, extract single fibres from treated and untreated muscles by incubation for at least 2 h in media containing a dispase–collagenase enzymatic mixture. Allow the muscles to digest; checking regularly by fluorescence microscopy to detect sufficient disaggregation of fibres, and as possible, remove EGFP positive fibres for individual evaluation by microscopy (9).

3.7. Molecular Analysis of Corrected Locus (Dystrophin Locus)

In the first instance, the sections harvested from the treated (RHS) and untreated (LHS) TA muscles are assessed for the relative frequencies of dystrophin expression by anti dystrophin Ab:

3.7.1. Evaluation of Dystrophin Restored by In Vivo Electroporation of CNA

1. After being cut, sections can be stored frozen at −80°C prior to staining for dystrophin expression.

2. Incubate the sections at room temperature in 10% donkey serum in 1× PBS

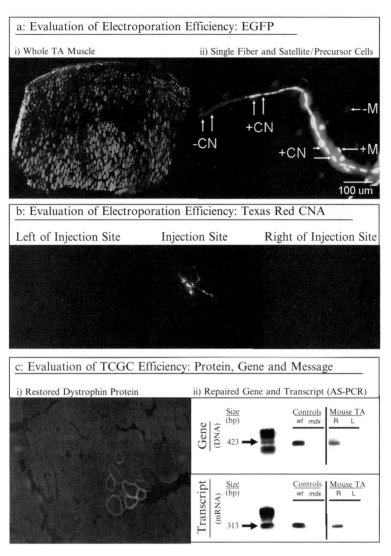

Fig. 32.4 Evaluation of in vivo electroporation into mdx mouse muscle. (**a**). Transgenic EGFP, electroporated into TA muscle shows up as bright green under fluorescence. In this section, 89% of fibres were seen to express the transgene. Single fibre analysis allows evaluation of electroporation-mediated penetration of cell nuclei in fibres (CN+) or muscle precursor cells (M+) and comparison to fibres (CN-) or precursor cells (M-) that have not been penetrated by the transgene. (**b**). CNA (double stranded) labelled with Texas Red at each 5′ end can be traced after injection and electroporation into muscle. The injection site retains much of the CNA injected, but in concordance with the great percentage of cells penetrated by transgene, the CNA spreads throughout the entire section. (**c**). Correction of the *mdx* dystrophin locus can be evaluated at the gene, transcript, and protein levels. Fibres expressing dystrophin are detected using anti-dystrophin immunohistochemistry, and corroborated by allele-specific analyses of dystrophin gene and transcript. All methods shown here indicate a very low amount of *mdx* dystrophin locus correction (i.e. 1% of loci). In conjunction with **a** and **b**, this suggests that a small amount of TCGC can occur in the absence of DNA replication in the nuclei of mature muscle fibres

3. Incubate the sections with 50 μL H-300 anti dystrophin primary Ab diluted 1:200 in 10% donkey serum (37°C/2 h).
4. Wash the section at room temperature twice in 1× PBS for 5 min each time.
5. Overlay the sections with secondary Ab (Alexa Fluor) and incubate at 37°C for 1 h.
6. Wash the section twice in 1× PBS (5 min each wash).
7. Mount the sections in fluorescent mounting media.
8. Dystrophin expressed from corrected nuclei will appear as fluorescent green at the periphery of fibres in areas of transverse sections otherwise devoid of dystrophin expression in the muscle (Fig. 32.4c). With this protocol, we have detected between 0.5 and 1% of total fibres corrected in treated muscles, with minimal detection of revertant fibres which arise in *mdx* muscle as a result of aberrant transcriptional processing of primary dystrophin transcript (9).

3.7.2. AS-PCR for Detection of *wt* Loci (*see* Note 3)

From the treated TA muscles, isolate DNA from the 25 sections immediately flanking (following) the sections showing the greatest expression of restored dystrophin protein by immunohistochemistry. Using the G-Tip High Molecular Weight DNA extraction system according to manufacturer's instruction excludes residual CNA carry over (by size) into subsequent PCR methods used to detect corrected loci (*see* **Note 3**). Perform Allele-specific PCR (AS-PCR) using a gradient PCR machine (MasterCycler, Eppendorf) as follows:

1. Antisense oligonucleotide primer Dys-*wt* AS-01 (3′ mismatch for the *mdx* nucleotide) and sense primer Dys In22 S-02 (Table 32.1) are used to facilitate selective detection of *wt* dystrophin locus.
2. Make PCR reactions (50 μL) consisting of 50 ng gDNA, 0.1 μM of each primer, 0.2 μM of each dNTP, 1.5 mM $MgCl_2$ and 2.5 units of Taq polymerase.
3. Set the thermal cycler to a regime that consists of a first cycle of 92°C for 2 min and 65.3°C for 2 min, followed by 29 further cycles of 92°C for 30 s and 65.3°C for 2 min (total 30 cycles). Using this protocol, a 423-bp analytical PCR product is generated only in the presence of *wt* DNA template (*see* **Note 5**).

3.7.3. Detection of AS-PCR Product by Southern Hybridisation

1. Electrophorese the AS-PCR products on a 1% TAE agarose gel/buffer system.
2. Transfer the PCR products to a Hybond N$^+$ membrane by standard alkaline capillary transfer overnight (tissue/sponge stack) utilising 0.4 M NaOH.
3. Generate a probe by PCR using primers Dys-*wt* AS-01 and Dys In22 S-02 (Table 32.1) and *wt* gDNA template. Reverse-phase column purify the product (as before) and label it by incorporation of [αP^{32}] dATP (500 μCi to 2 mCi) using Klenow DNA polymerase (5–10 U/mL) under standard conditions.

4. Hybridise the labelled probe to the DNA on the membrane in Church Buffer. Before addition of the labelled probe to the hybridisation buffer (Church buffer), incubate the membrane in the buffer for 2 h at 65°C.
5. Add the probe to the buffer and leave it to hybridise overnight at 65 C with gentle agitation.
6. Wash the hybridised membrane in 2× SSC/0.1% SDS three times before washing for 30 min at 65°C in 1× SSC/0.1% SDS, 30 min in 0.5× SSC/0.1% SDS and 45 min in 0.25× SSC/0.1% SDS (*see* **Note 6**), checking after each wash to see if the probe can be visualised.
7. Wash the membrane briefly in 0.12× SSC before sealing in plastic wrap and exposure to phosphor plates (Molecular Dynamics).

3.8. Molecular Analysis of Corrected Locus Expression (Transcript) for Determination of wt Transcript in the TAs

1. Isolate total RNA from the 25 TAs sections immediately adjacent to (preceding) the section with the highest dystrophin Ab signal using TriPure™ Isolation Reagent according to manufacturer instructions (Roche).
2. Remove DNA (and therefore CNA carryover) from the RNA by incubation of the RNA in DNAase I/Buffer (Promega) for at least an hour at 37°C (*see* **Note 7**).
3. Perform first-strand reverse transcription (rt) on 250 ng of RNA, using Omniscript reverse transcriptase under manufacturer's conditions.
4. Amplify a secondary 803 bp *DMD* rtPCR product from the cDNA, using c3606-AS and c2801-S primers as previously described (10).
5. Perform an allele-specific RT-PCR (AS-rtPCR) off the 803 bp RT-PCR product generated from the treated *mdx* mouse TA cDNA using antisense primer Dys-*wt* AS-01 and sense primer c2917-S which falls on exon 22 (Fig. 32.1b and Table 32.1). Use AS-rtPCR conditions essentially as for the AS-PCR described above, with the exception of an annealing temperature of 69.9°C (*see* **Note 8**). Make sure that the amount of template used per PCR reaction is kept constant at 200 pg. Using this protocol, a 313-bp analytical PCR product is amplified only in the presence of *wt* DNA.
6. Resolve the AS-rtPCR products by agarose gel and determine the presence of *wt* transcript by Southern analysis (as for DNA) using a labelled 313-bp AS-rtPCR product obtained from the *wt* template as a probe (Fig. 32.4c-Transcript).

3.9. Statistical Analyses

Statistical analysis of dystrophin positive fibres is required to establish statistical significance between positive fibres arising from corrected dystrophin gene loci as

opposed to those arising as revertant *mdx* fibres via aberrant primary transcript processing. The student's T test facilitates this adequately.

4. Notes

1. Male *mdx* mice are used for correction studies to emulate the human condition and to provide just a single mutant locus in the cell to be corrected.

2. Female *mdx* mice are used to generate templates, but not for correction as they have two mutant loci per cell.

3. AS-PCR techniques are necessary when less than 1–2% of loci have been corrected. In cases where corrected loci are present at greater than these low frequencies, detection by restriction analysis using *Fnu* 4HI RFLP can be used and percentage corrected loci or transcripts can be quantified by densitometry. While the specific strategy is not described in detail here, the general principles for preparation of DNA and transcript from treated tissues are as described for the AS-PCR methodologies. Figure 32.1 shows the topographical schematics of the strategy at the *mdx* dystrophin gene locus, and the primers used are included in Table 32.1. The detailed methodology has been published elsewhere (14).

4. If more specificity is required for identification of muscle components penetrated by in vivo electroporation of pEGFP vector, satellite cells can be identified by Pax7 immunohistochemistry (Pax7 Ab, R&D Systems, MN) (9).

5. The location of primer Dys In22 S-02 is in an intronic region upstream of exon 23 and the *mdx* mutation locus, and in a gene region not present on Amplicon C confers *wt* chromosomal locus specificity to the AS-PCR assay. This AS-PCR protocol was tested for the likelihood of artefactual positive result by CNA/gDNA(*mdx*) titration using more than 40 times the CNA: gDNA(*mdx*) ratios used in real experiments (manuscript in preparation). This is an important control which must be performed on all correction protocols to ensure the limits of the respective system with regards to its ability (or not) to generate artefactually positive results.

6. The radioactivity of the membrane should be checked at intervals between washes to ensure that the wash has not been too stringent. If the entire probe washes off the filter, the hybridisation will need to be repeated.

7. The DNAase step is crucial to avoid false positive results arising from the carryover of CNA at the transcript stage. Carrying out RNA analyses without a DNAase step significantly compromises the integrity and reliability of positive results and should *always* be included.

8. Allele-specific PCR needs to be done in a gradient thermal cycler to allow absolute determination of the optimal annealing conditions to differentiate the alleles being tested.

References

1. Mastaglia, F.L. and Kakulas, B.A. (1969) Regeneration in Duchenne muscular dystrophy: a histological and histochemical study. *Brain.* **92**, 809–818.

2. Hoffman, E.P., Brown, R.H., Jr., and Kunkel, L.M. (1987) Dystrophin: the protein product of the Duchenne muscular dystrophy locus. *Cell.* **51**, 919–928.

3. Monaco, A.P., Bertelson, C.J., Liechti-Gallati, S., Moser, H., and Kunkel, L.M. (1988) An explanation for the phenotypic differences between patients bearing partial deletions of the DMD locus. *Genomics.* **2**, 90–95.

4. Hawke, T.J. and Garry, D.J. (2001) Myogenic satellite cells: physiology to molecular biology. *J. Appl. Physiol.* **91**, 534–551.

5. Malhotra, S.B., Hart, K.A., Klamut, H.J., et al. (1988) Frame-shift deletions in patients with Duchenne and Becker muscular dystrophy. *Science.* **242**, 755–759.
6. Koenig, M., Beggs, A.H., Moyer, M., et al. (1989) The molecular basis for Duchenne versus Becker muscular dystrophy: correlation of severity with type of deletion. *Am. J. Hum. Genet.* **45**, 498–506.
7. Ryder-Cook, A.S., Sicinski, P., Thomas, K., et al. (1988) Localization of the mdx mutation within the mouse dystrophin gene. *EMBO J.* **7**, 3017–3021.
8. Sicinski P., Geng Y., Ryder-Cook A.S., Barnard E.A., Darlison M.G., and Barnard P.J. (1989) The molecular basis of muscular dystrophy in the mdx mouse: a point mutation. *Science.* **244**, 1578–1580.
9. Wong, S.H.A., Lowes, K.N., Quigley, A.F., et al. (2005) DNA electroporation in vivo targets mature fibres in dystrophic *mdx* muscle. *Neuromus Disorders.* **15**, 630–641.
10. Kapsa, R., Quigley, A., Lynch, G.S., et al. (2001) In vivo and in vitro correction of the mdx dystrophin gene nonsense mutation by short-fragment homologous replacement. *Hum. Gene Ther.,* **12**, 629–642.
11. Wolff, J.A., Malone, R.W., Williams, P., et al. (1990) Direct gene transfer into mouse muscle in vivo. *Science.* **247**, 1465–1468.
12. Mir, L.M., Bureau, M.F., Rangara, R., Schwartz, B., and Scherman, D. (1998) Long-term, high level in vivo gene expression after electric pulse-mediated gene transfer into skeletal muscle. *C. R. Acad. Sci. III.* **321**, 893–899.
13. Vilquin, J.T., Kennel, P.F., Paturneau-Jouas, M., et al. (2001) Electrotransfer of naked DNA in the skeletal muscles of animal models of muscular dystrophies. *Gene Ther.* **8**, 1097–1107.
14. Todaro, M., Quigley, A., Kita, M., Chin, J., Lowes, K., Kornberg, A.J., Cook, M.J., and Kapsa, R. (2007) Effective detection of corrected dystrophin loci in *mdx* mouse myogenic precursors. *Hum Mutat.* **28**(8), 816–823.

Chapter 33
Gene Delivery to Dystrophic Muscle

Kim E. Wells, Jill McMahon, Helen Foster, Aurora Ferrer, and Dominic J. Wells

Abstract Electroporation is a powerful method for gene delivery to dystrophic muscle in the *mdx* mouse model of Duchenne muscular dystrophy. Successful transfer of reporter and therapeutic plasmids and antisense oligonucleotides has been demonstrated. However, the efficiency falls with increasing plasmid size. Although it is unlikely that the electrotransfer approach will be useful clinically, it is an important experimental tool, particularly in testing potential immune responses to gene transfer in the absence of vector proteins.

Keywords: electrotransfer, dystrophin, plasmid, mdx, DMD

1. Introduction

Duchenne muscular dystrophy (DMD) is an X-linked recessive, progressive, and lethal muscle wasting disorder caused by the absence of a critical protein, dystrophin (1). A variety of viral and plasmid based nonviral vectors have been used for gene delivery to dystrophic muscle (2). Indeed, the injection of naked plasmid DNA was used to perform the first successful example of somatic dystrophin gene transfer into the *mdx* mouse (3), a mouse model of DMD. However, naked plasmid is an inefficient gene transfer system and the intramuscular administration does not benefit from complexation with charged moieties such as lipids. A major breakthrough in nonviral vector based gene delivery to skeletal muscle using in vivo electroporation was reported by three independent groups in the late 1990s (4–6). Subsequently, a number of groups have used this physical method of gene delivery to introduce dystrophin cDNAs into *mdx* dystrophic muscle (7–12). In vivo electroporation has also been used to deliver antisense oligonucleotides to induce skipping of exon 23 and thus remove the premature stop codon from the murine dystrophin mRNA (13).

    The protocol used in the studies from our group (8, 12, 13) were based on a method that we had developed previously that involved pretreatment of the muscle with a dilute solution of hyaluronidase 2 h prior to the injection of plasmid DNA

and the application of the electrical pulses (14). This method was very successful when applied to dystrophic muscle.

2. Materials

All chemicals obtained from Sigma (Poole, Dorset, UK) unless otherwise noted.

2.1. Plasmid DNA

1. Four reporter plasmid DNA constructs, CMVβ, 9.7CB, 16CB, and 21.7CB. pCMVβ (7.2 kb), which contains the β-galactosidase gene under the control of the CMV immediate-early promoter and enhancer, were obtained from Clontech, Palo Alto, CA 9.7CB, 16CB, and 21.7CB plasmids were constructed by inserting the β-galactosidase expression cassette from pCMVβ into stuffer plasmid DNA backbone as described previously (15), producing plasmids of sizes 9.7, 16, and 21.7 kb, respectively.
2. The full-length and minidystrophin plasmids containing the mouse or human cDNAs for full-length or minidystrophin (deleted for exons 17–48) under the control of the CMV immediate-early promoter and enhancer have been previously described (12, 16).
3. ENDOfree plasmid Giga kit (Qiagen, Crawley, UK).
4. Saline: 8.5% sodium chloride in water for injection (10× stock).
5. Water for injection.

2.2. In Vivo Experiments

1. Wild type and *mdx* mice from in-house animal breeding colonies. All mice maintained under Home Office Animals (Scientific Procedures) Act 1986 guidelines. All experiments conducted under Project Licence PPL 70/5812.
2. Insulin syringes.
3. Fentanyl and fluanisone (Hypnorm®, VetPharma Ltd, Sherburn in Elmer, Leeds, UK).
4. Midazolam (Hypnovel®, Roche, Welwyn Garden City, UK).
5. AErrane—isoflurane.
6. Anesthetic machine with oxygen cylinder and isoflurane vaporizer.
7. Electroporator BTX EC830.
8. BTX Tweezertrodes™.
9. Hyaluronidase.
10. Animal shaver (*see* **Note 1**).

2.3. Collection and Processing of Mouse Samples

1. Euthasate (National Veterinary Supplies, UK).
2. Aluminium foil squares.
3. Cork discs.
4. Liquid nitrogen dewar.
5. Isopentane.
6. Cryo-M-Bed (Bright Instruments, Huntingdon, UK).
7. IMS (industrial methylated spirits) (*see* **Note 2**).
8. Aminopropyltriethoxysilane (APES) for coating microscope slides.

2.4. β-Galactosidase Histochemistry

1. β-Galactosidase staining solution mix for 25 mL total volume: 22.5 mL (phosphate-buffered saline, PBS), 1.25 mL potassium ferricyanide (100 mM stock, 0.329 g in 10 mL water), 1.25 mL potassium ferrocyanide (100 mM stock, 0.422 g in 10 mL water), 0.025 mL deoxycholate (10% stock), 0.025 mL NP40 (20% stock), 0.25 mL MgCl$_2$ (100 mM stock), 0.5 mL X-gal (2% (50 mM) X-gal in DMSO). Store X-gal stock solution at −20°C wrapped in silver foil.
2. Alcoholic eosin: approximately 2% w/v in 95% IMS.
3. Paraformaldehyde.

2.5. Dystrophin Immunostaining

1. Immuno pen (ImmEdge, Vector Laboratories, Peterborough, UK).
2. Dulbeco's Ca^{2+}Mg^{2+}free PBS.
3. Tween20.
4. Avidin/Biotin block kit (Vector).
5. Dried nonfat milk powder.
6. Fetal calf serum (FCS).
7. Primary antibody: DysC3750 (8), rabbit polyclonal raised against last 17 amino acids of dystrophin using a peptide conjugated to keyhole limpet hemocyanin (Cymbus Biotech, Chandlers Ford, UK).
8. Secondary antibody: biotinylated swine anti rabbit (Dako Cytomation, Ely, UK).
9. ABC-HRP complex.
10. Enzyme substrate: Diaminobenzidine (DAB) at 1 mg/mL in 50 mM Tris-HCl, pH 7.4.
11. Nickel solution: 50 mM ammonium nickel sulphate (2.5 g ammonium nickel sulphate in 100 mL 50 mM Tris-HCl, pH 7.4).
12. 100%, 95%, and 75% IMS (*see* **Note 2**).
13. Xylene.
14. DPX mountant.

2.6. β-Galactosidase *ELISA*

1. Dulbeco's Ca^{2+}/Mg^{2+} free PBS.
2. *β-Galactosidase* ELISA kit (Roche Diagnostics, Burgess Hill, UK).
2. Protease inhibitor cocktail (50 μg/mL antipain, 10 μg/mL aprotinin, and 0.5 μg/mL leupeptin).

3. Methods

3.1. Intramuscular Injection and Electroporation for Treating Muscle.

1. Anesthetize mice by intraperitoneal injection of 5 μL/g of a Hypnorm, Hypnovel, and H_2O mix (1:1:2 by volume).
2. Shave the hind legs to visualize the tibialis anterior (TA) muscles (*see* **Note 1**).
3. Inject each TA muscle percutaneously with 10 units (25 μL of 0.4 units/μL) of bovine hyaluronidase. Injections were carried out using a 29 GA needle insulin syringe inserted just below the tibial plateau and then extended longitudinally to the muscle fibers proximal to distal direction. The injection is made in the mid-belly of the muscle.
4. Two hours later, anesthetize the mice using isofluorane either in an induction chamber or, if still sedated, using a facemask.
5. Inject 25 μL of plasmid solution in 0.85% saline into the TA as above. β-galactosidase plasmids are injected at 1 mg/mL unless otherwise stated and dystrophin plasmids at 2 mg/mL.
6. Hold the injected leg steady and apply the 7 mm circular electrodes (BTX Tweezertrodes™) to the medial and lateral sides of the lower hind limb with reasonable pressure to maintain contact with the skin surface. Electrode jelly is used on the electrode plates to ensure good electrical contact. A voltage of 175 V/cm is applied in ten 20 ms square wave pulses at 1 Hz using a BTX electroporator as shown in Fig. 33.1 (*see* **Notes 3,4**).
7. Collect samples 7 days after the injection of plasmid DNA unless otherwise stated (*see* **Note 5**).

3.2. Collection of Treated Muscles

1. Euthanize mice by cervical dislocation or overdose of euthasate.
2. Excise TA muscles post mortem, place in Eppendorf tubes and snap-freeze in liquid nitrogen for analysis by ELISA or western blot.
3. Alternatively, mount TA muscles on a cork block, lightly covering the muscles with Cryo-M-Bed and snap-freeze in liquid nitrogen-cooled isopentane for histological analysis. Frozen blocks are wrapped in aluminum foil for storage at −70°C.

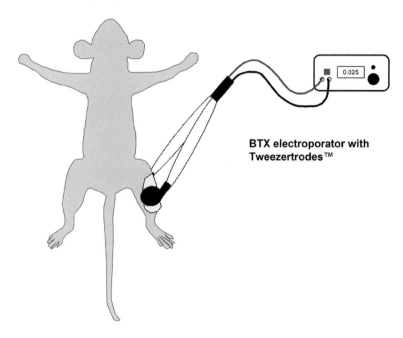

Fig. 33.1 Schematic of the process of electroporation of the tibialis anterior. Diagram shows the placement of the Tweezertrodes on the lower hind limb of the mouse after pretreatment of the tibialis anterior (shaved region) with hyaluronidase and 2 h later with a solution of plasmid DNA

3.3. Preparation of Slides and Cutting Muscle Sections

3.3.1. Coating Slides with APES

1. Use clean slides. Commercially supplied washed slides are clean enough. If not, soak overnight in decon and wash well with hot running water followed by two washes in good quality H_2O. Dry in oven, 37°C or 60°C. Rack slides into as many slide racks as you can find so that you treat a couple of hundred slides in one session.
2. Immerse slides in a freshly prepared solution of 2% APES in 100% IMS for 1 min. Use square glass troughs which take about 400–500 mL.
3. Wash slides in 100% IMS.
4. Wash slides twice in good quality water.
5. Dry overnight in oven. Store in dust-free environment at room temperature.

3.3.2. Cutting Muscle Sections

1. For analysis of transfection efficiency, cut 10 μm cryostat sections at a minimum of ten evenly spaced levels throughout each muscle and lift onto APES-coated

slides. Make a set of 12 semi-serial slides, each slide having a section from each of the ten evenly spaced levels throughout the muscle (Fig. 33.2).

2. Up to 4 TAs can be mounted on the same cork block, e.g. treated and control muscles from two mice, if small slices of liver and kidney are used to mark the different muscle pairs (Fig. 33.2).

3. Store slides at −70°C after air-drying.

3.4. β-Galactosidase Histochemistry Protocol

β-galactosidase is a useful reporter gene for assessing the efficiency of plasmid electroporation in skeletal muscle as it can be easily detected in both muscle sections and muscle homogenates to assess both the number of transfected fibers and the total level of expression. Dystrophin plasmids for treatment of dystrophic muscle appear less efficient than the β-galactosidase reporter gene and experiments using the same β-galactosidase expression cassette in plasmids with increasing amounts

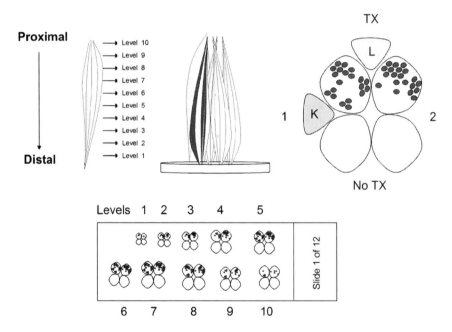

Fig. 33.2 Schematic for the sectioning of skeletal muscle. Sections are collected from 10–12 evenly spaced levels along the length of the muscle. Up to four tibialis anterior muscles can be placed on a cork block with liver and kidney slivers to mark the muscles. Sections are collected onto 12 slides such that each slide contains a section from each of the 10–12 levels and slides are semi-serial. Thus staining for one feature can be matched with staining for another feature on the adjacent slide. The section with the highest number of transfected fibers will vary between muscles depending on the precise needle placement and so the optimal section may vary between the four muscles in the block

of "stuffer" DNA show that this decreased efficiency is in part due to plasmid size (Fig. 33.3).

1. Air dry (for at least 20 min at room temperature).
2. Fix in 2% paraformaldehyde in 1× PBS for 2–5 min only.
3. Wash 3 times in PBS.
4. Incubate in staining solution for 1–24 h at 37°C.
5. Wash briefly in PBS or water (*see* **Note 6**).
6. Rinse in 95% IMS.
7. Counterstain in a saturated solution of alcoholic eosin for ~2 min (*see* **Note 7**).
8. Transfer slides straight into 100% IMS and change solution three times to dehydrate completely.
9. Clear in xylene and mount in DPX.
10. View slides under a light microscope. Using a video camera, images can be imported to computer using SnapperTool version 2.04 software (Datacell,

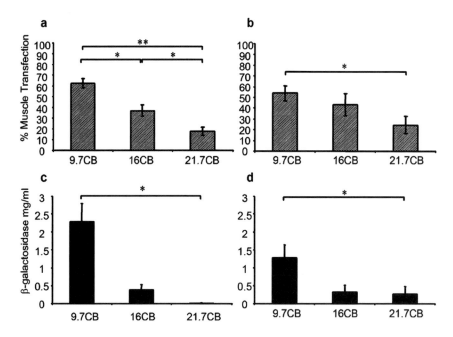

Fig. 33.3 Effects of size on the efficiency of plasmid electrotransfer. Male 11-week-old CD1 wild-type mice were injected with plasmids carrying the same expression cassette but with different amounts of plasmid "stuffer" (CMVβ, 9.7CB, 16CB, and 21.7CB) and subject to in vivo electroporation as described in the text. One group was treated at 1 mg/mL for all three plasmids (constant dose) and another group at 1 mg/mL, 1.5 mg/mL, and 2 mg/mL for plasmids 9.7CB, 16CB, and 21.7CB, respectively (constant copy). Percentage transfection and total expression determined as described in the text. Both transfection and total expression decrease with increasing plasmid size when a constant dose of plasmid is administered and this effect is still significant even after adjustment of the quantity of plasmid delivered to ensure an equal copy number of plasmids. $* = p < 0.05$, $** = p < 0.01$

Maidenhead, UK). Transfection efficiencies can be calculated by measuring the cross-sectional area of blue fibers and expressing this as a percentage of the total cross-sectional area of the muscle. The results for each muscle are taken from the section showing the highest percentage transfection using SigmaScan Pro version 5.0.0 image analysis software (SPSS, Birmingham, UK.)

3.5. β-Galactosidase ELISA

1. For analysis of total β-galactosidase expression, a TA muscle is homogenized in a microfuge tube with a micropestle in 200 μL of lysis buffer from the β-galactosidase ELISA kit together with a cocktail of protease inhibitors. The volume is increased to a total of 400 μL and the tissue subjected to three rounds of freeze-thaw cycles.
2. After mixing thoroughly, the solution is spun at RCF 15,000g for 10 min to pellet all insoluble material.
3. Store supernatants at –70°C until assayed, using the β-galactosidase ELISA kit as per the manufacturer's instructions.

3.6. Immunostaining Analysis for Dystrophin Expression

1. Set up a staining tray with damp tissue in the central space between the rods. Leave tray with lid on to develop a damp environment for staining slides. Do not dilute antibodies until they are needed.
2. Air dry slides for at least 30 min. Label all slides with pencil indicating antibody used, dilution of primary antibody, and date.
3. Ring sections with immuno pen.
4. Wet sections briefly in PBS/Tween 0.05%.
5. Block for 10 min in avidin solution from Vector block set.
6. Wash briefly in PBS/T (2–5 min).
7. Block for 10 min in biotin solution from Vector block set.
8. Rinse briefly in PBS/T.
9. Block for 30 min in 5% dried milk in PBS/T.
10. Dilute an aliquot of DysC3750 out to 1:500 in PBS/T plus 1% FCS (diluted aliquots can be stored frozen).
11. Rinse slides briefly in PBS/T and then incubate in diluted DysC3750 for 1 h (*see* **Note 8**).
12. Wash 3 times for 5–10 min each in PBS/T.
13. Make Vector ABC reagent (Vector stain Elite kit) in 50 mM tris HCl, pH 7.5, using 2 drops of A and 2 drops of B in 5 mL. Leave at room temperature for 30 min then keep on ice.
14. Incubate in Dako biotinylated swine anti rabbit (E0353) at 1:500 in PBS/T for 30 min.

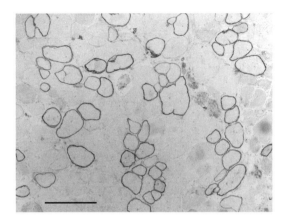

Fig. 33.4 Photomicrograph of a typical dystrophin stain after plasmid electrotransfer. Section from *mdx* mouse with full-length dystrophin delivered via in vivo electroporation, 50 μg in 25 μL. The mouse was killed 7 days after treatment and muscle section immunostained with DysC3750 as described. Scale bar indicates 200 μm

15. Wash three times in PBS/T.
16. Incubate for 30 min in ABC reagent.
17. Wash three times in PBS/T.
18. Develop the reaction using Nickel-enhanced DAB. Mix 4 mL of nickel solution plus 1 mL of DAB and 1 μL of 30% hydrogen peroxide to activate the DAB. Mix thoroughly and apply to the sections.
19. Leave for 2 min or check under the microscope for color developing (*see* **Note 9**). Wash slides in water twice and leave in water while finishing rest of slides.
20. Dehydrate through graded alcohols and mount in DPX. An example of a dystrophin immunostain is shown in Fig. 33.4.

3.7. Statistical Analysis

Two-sided Student's t tests are used where only two groups are being analyzed. For analysis of multiple groups, a one-way ANOVA is routinely used followed by an appropriate multiple paired comparison. P values less than 0.05 are considered statistically significant. Statistical tests are conducted using Sigma-Stat from SPSS (Birmingham, UK).

4. Notes

1. Any beard trimmer will act as a suitable animal shaver for the mouse lower limb.
2. IMS is equally good as ethanol for histology and is a lot cheaper.

3. Electrotransfer should not be performed under injectable anesthesia as the repeated pulses tend to waken the mice. Isoflurane inhalation is safe and mice can be put under deep anesthesia just prior to the period of application of the electrical pulses and then rapidly recovered. During the period of deep anesthesia, the breathing of the mice should be carefully monitored. If it becomes deep, infrequent, and jerky then the anesthetic mask should be removed to allow the anesthesia to lighten.

4. The distance between electrodes placed on the lower hind limbs is 5 or 6 mm, thus the electroporator is set to deliver 87.5 or 105 V.

5. β-galactosidase is immunogenic and transfected fibers are likely to be removed by a cell-mediated immune response. The same is true for human dystrophin in the *mdx* (14). Hence, samples should be collected at 6–7 days to avoid immune rejection.

6. Standard washes may result in loss of the blue stain. A brief rinse to remove the staining solution is all that is required.

7. Standard water-soluble eosin removes the blue stain so alcoholic eosin must be used.

8. We use 300–400 μL per slide as the stain evenness and background are better with a good volume and it limits the chances of scraping the sections when trying to spread out a smaller volume.

9. Incubate for no more than 2 min without checking; if slides are giving a clean background then leave for 3 min maximum; choose one time for all the slides in the set.

References

1. Hoffman, E.P., Brown, R.H., Jr., and Kunkel, L.M. (1987) Dystrophin: the protein product of the Duchenne muscular dystrophy locus. *Cell.* **51**, 919–928.
2. Wells, D.J. and Wells, K.E. (2002) Gene transfer studies in animals: what do they really tell us about the prospects for gene therapy in DMD? *Neuromuscul Disord.* **12**, S11–22.
3. Acsadi, G., Dickson, G., Love, D.R., et al. (1991) Human dystrophin expression in *mdx* mice after intramuscular injection of DNA constructs. *Nature.* **352**, 815–818.
4. Aihara, H. and Miyazaki, J. (1998) Gene transfer into muscle by electroporation in vivo. *Nat. Biotechnol.* **16**, 867–870.
5. Mathiesen, I. (1999) Electropermeabilization of skeletal muscle enhances gene transfer in vivo. *Gene Ther.* **6**, 508–514.
6. Mir, L.M., Bureau, M.F., Gehl, J., et al. (1999) High-efficiency gene transfer into skeletal muscle mediated by electric pulses. *Proc. Natl. Acad. Sci. U.S.A.* **96**, 4262–4267.
7. Vilquin, J.T., Kennel, P.F., Paturneau-Jouas, M., et al. (2001) Electrotransfer of naked DNA in the skeletal muscles of animal models of muscular dystrophies. *Gene Ther.* **8**, 1097–1107.
8. Gollins, H., McMahon, J., Wells, K.E., and Wells, D.J. (2003) High efficiency plasmid gene transfer into dystrophic muscle. *Gene Ther.* **10**, 504–512.
9. Murakami, T., Nishi, T., Kimura, E., et al. (2003) Full-length dystrophin cDNA transfer into skeletal muscle of adult *mdx* mice by electroporation. *Muscle Nerve.* **27**, 237–241.
10. Molnar, M.J., Gilbert, R., Lu, Y., et al. (2004) Factors influencing the efficacy, longevity, and safety of electroporation-assisted plasmid-based gene transfer into mouse muscles. *Mol. Ther.* **10**, 447–455.
11. Wong, S.H., Lowes, K.N., Quigley, A.F., et al. (2005) DNA electroporation in vivo targets mature fibres in dystrophic *mdx* muscle. *Neuromuscul. Disord.* **15**, 630–641.
12. Ferrer, A., Gollins, H., Wells, K.E., Dickson, G., and Wells, D.J. (2004) Long term expression of full-length human dystrophin in transgenic *mdx* mice expressing truncated human dystrophins. *Gene Ther.* **11**, 884–893.

13. Wells, K.E., Fletcher, S., Mann, C.J., Wilton, S.D., and Wells, D.J. (2003) Enhanced in vivo delivery of antisense oligonucleotide to restore dystrophin expression in adult *mdx* mouse muscle. *FEBS Lett.* **552**, 145–149.

14. McMahon, J.M., Signori, E., Wells, K.E., Fazio, V.M., and Wells, D.J. (2001) Optimisation of electrotransfer of plasmid into skeletal muscle by pretreatment with hyaluronidase—increased expression with reduced muscle damage. *Gene Ther.* **8**, 1264–1270.

15. Wells, D.J., McMahon, J., Maule, J., et al. (1998) Evaluation of plasmid DNA for in vivo gene therapy: factors affecting the number of transfected fibers. *J. Pharm. Sci.* **87**, 763–768.

16. Ferrer, A., Wells, K.E., and Wells, D.J. (2000). Immune responses to dystrophin: implications for gene therapy of Duchenne muscular dystrophy. *Gene Ther.* **7**, 1439–1446.

Chapter 34
Effect of Electroporation on Cardiac Electrophysiology

Vadim V. Fedorov, Vladimir P. Nikolski, and Igor R. Efimov

Abstract Defibrillation shocks are commonly used to terminate life-threatening arrhythmias. According to the excitation theory of defibrillation, such shocks are aimed at depolarizing the membranes of most cardiac cells, resulting in resynchronization of electrical activity in the heart. If shock-induced transmembrane potentials are large enough, they can cause transient tissue damage due to electroporation. In this review, evidence is presented that electroporation of the heart tissue can occur during clinically relevant intensities of the external electrical field and that electroporation can affect the outcome of defibrillation therapy, being both pro- and antiarrhythmic.

Here, we present experimental evidence for electroporation in cardiac tissue, which occurs above a threshold of 25 V/cm as evident from propidium iodide uptake, transient diastolic depolarization, and reductions of action potential amplitude and its derivative. These electrophysiological changes can induce tachyarrhythmia, due to conduction block and possibly triggered activity; however, our findings provide the foundation for future design of effective methods to deliver genes and drugs to cardiac tissues, while avoiding possible side effects such as arrhythmia and mechanical stunning.

Keywords: electroporation, heart, defibrillation, optical mapping, propidium iodide

1. Introduction

Electroporation of cardiac muscle is a new approach for gene and drug transfection (1). It can be used for creating a "biological" pacemaker and for treating different heart diseases (2, 3). On the other hand, high-intensity electrical shocks are commonly used in clinical practice to terminate atrial and ventricular fibrillation. If shock-induced changes in transmembrane potential are excessively large, they can cause transient cell membrane damage due to electroporation, resulting in transient or permanent electrical and mechanical dysfunction of the heart (4–7). There is a substantial body of experimental evidence suggesting that electroporation

S. Li (ed.), *Electroporation Protocols: Preclinical and Clinical Gene Medicine.*
From *Methods in Molecular Biology, Vol. 423.*
© Humana Press 2008

of the heart cell membrane can occur during clinically relevant intensities of electrical shocks. While high-intensity shocks are used routinely, the tissue and cellular responses to large currents are not fully understood. In particular, it remains a subject of heated debate whether shock-induced electroporation is pro- or antiarrhythmic in clinical settings (8).

An improved understanding of electroporation may not only reduce the side effects associated with defibrillation and gene transfection therapy, but can help in designing more effective ways to deliver genes and drugs to target cells. In this chapter, we review experimental studies on electroporation in the heart and relate the findings to several recent experimental studies performed in our laboratory.

2. Cellular Responses to Strong Electrical Fields

The phenomenon of electrically modifying the cell membrane conductivity has been known since the 1940s. Goldman (9) measured the voltage-current (V-I) characteristics of the membrane of *Chara australia* and found a phenomenon similar to the dielectric breakdown of cell membrane, i.e., an abrupt increase in the membrane conductance (electroporation) when the membrane was hyperpolarized beyond a certain potential. Electroporation has been most extensively studied in bilayer membrane systems (10–13). Bilayer membrane systems allow for precise control of transmembrane voltage as well as adequate dynamic range and temporal resolution of recordings of electrical conduction through the pore induced by electrical stimuli. Such experimental studies provided a detailed description of the process of electroporation and resealing of the pores. Rapid-freezing electron microscopy of electropermeabilized cells provided direct evidence of the formation of volcano-shaped pores in cell membranes (14). The pore structures rapidly expand to 20–120 nm in diameter during the first 20 ms of electroporation, and begin to shrink and reseal after several seconds.

Electroporation has also been characterized quantitatively in isolated cardiac cells (15–18). Experimental observations allowed construction of mathematical models of the behavior of a single pore (19–21) and their participation in the electrical activity of cardiac myocytes (17, 22, 23) and of the tissue (24, 25) during strong electrical field stimulation.

In whole cell patch clamp experiments, a breakdown in membrane conductance at transmembrane potential thresholds of 0.6–1.1 V was shown in response to 0.1–1.0 kV/s voltage ramps (15, 26), which is unaffected by Na, K, and Ca channel blockers (27). This result is consistent with the formation of ion-nonspecific membrane pores. Application of this technique for detecting electroporation in the tissue is difficult because the increase in cell membrane conductance translates to only a small decrease in total tissue resistance. Additionally, previous modeling work has shown that electroporation occurs only in a very small region of the tissue, perhaps only in a one-cell layer adjacent to the electrode (24, 25). Yet, direct

real-time recording or visualization of electroporation in in vivo or in vitro cardiac tissue remains to be developed.

Information about electroporation can be indirectly inferred from staining of the tissue with fluorescent dyes such as propidium iodide (PI) (1) or ethidium bromide (EB) (18), which penetrate the cells only through the pores, and subsequent histological imaging of intracellular space. Electroporation-induced electrophysiological changes include depression of excitability, resulting in depolarization of the cellular membrane during diastolic interval (8, 16, 28, 29); reduction of the amplitude of action potentials (APs) and the rate of upstroke rise (dV/dt), and elevation of intracellular calcium concentration (16).

3. Electroporation Assessment via Shock-Induced Changes of Transmembrane Potential Morphology

The outcome of defibrillation shocks is determined by the nonlinear transmembrane potential response (ΔV_m) induced by a strong external electrical field in cardiac cells. Recently, we investigated the contribution of electroporation to ΔV_m transients during high-intensity shocks by using optical mapping (7). Rectangular and ramp stimuli (10–20 ms) of different polarities and intensities were applied to the rabbit epicardium during plateau phase of the action potential. Epicardial activation was optically recorded under the custom 6-mm-diameter electrode using voltage-sensitive dye di-4-ANEPS (Fig. 34.1).

Shock-induced electroporation was localized to the circular area of the electrode and was evident from transient diastolic depolarization and reduction of AP amplitude and its derivative. The electroporation threshold in the ventricle, estimated by diastolic potential (DP) elevation reaching 10% of action potential amplitude (APA), was 320 ± 40 mA/cm² for both polarities. Electroporation was both voltage- and polarity-dependent and significantly more pronounced in the atria than in the ventricles (Fig. 34.2). Conduction slowing and block was observed in the electroporated region (8, 30).

Double-barrel microelectrode recordings and optical mapping techniques have shown that weak stimuli produce monotonic transmembrane potential changes (ΔV_m) in single cell (31, 32), cell culture strands (33), and heart tissue (34, 35), as predicted by the cable theory and generalized activating function theory. However, reports on strong shocks of defibrillation strength sharply disagree on the morphology and amplitude of shock-induced responses ΔV_m. When a stimulus is applied to a single cell during the early plateau phase of the action potential, the optical recordings show depolarization of the cathodal end and hyperpolarization of the anodal end of the cell (36, 37).

We found that during the strong shock-induced hyperpolarization (or more accurately negative polarization) ΔV_m first gradually increases in amplitude but soon starts to decay, causing elevation of the average potential of the cell (Fig. 34.3). Previously, similar effects were observed in a single guinea pig ventricular cell (32)

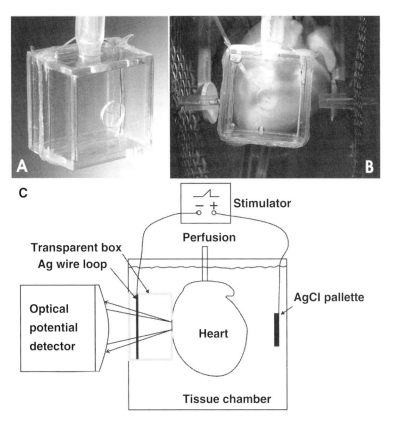

Fig. 34.1 Experimental setup and design of stimulation electrode producing homogeneous field with optically accessible field of view. The 20 × 20-mm silver wire loop was positioned inside the box 10 mm from the heart surface around the 6-mm-diameter opening. (Reproduced from (7), with permission)

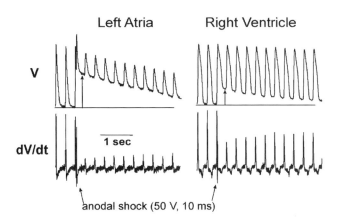

Fig. 34.2 Evidence of shock-induced electroporation in both the atria and the ventricles. Optical recording of transmembrane potentials, *V* (*upper trace*), shows time-dependent postshock reduction of resting potential and action potential amplitude. Maximal upstroke rate of rise (d*V*/d*t*) is also reduced and slowly recovers after shocks (*lower traces*)

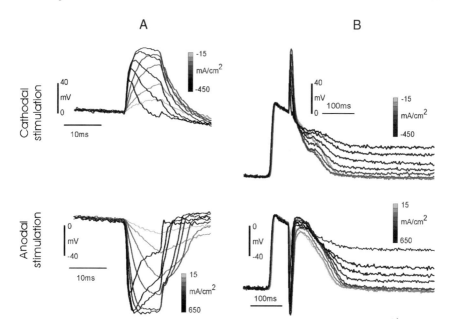

Fig. 34.3 Optical recording of transmembrane potential transients under the electrode during stimulation with different current densities. (**a**) small and (**b**) large timescale. Current strength is grayscale coded. Electroporation is evident from saturation of ΔV_m and elevation of the diastolic potential. (Reproduced from (7), with permission)

and narrow strands of cultured rat myocytes (33). It was concluded that during application of shocks to cell strands within the AP plateau, passive changes of V_m were followed by two voltage- and time-dependent shifts of V_m, possibly reflecting membrane electroporation (33, 38). We also concluded that the appearance of the second phase in hyperpolarizing transients in our whole heart experiments (Fig. 34.3a) is considered a signature of membrane electroporation (7).

Neunlist and Tung (29) presented measurements of frog epicardial ventricular cellular responses recorded from the 150-μm-diameter area of stimulus application, showing that electroporation reduction of AP amplitude is more deleterious for anodal polarities. Moreover, the ΔV_m during diastolic shock (across all intensity levels) was −200% AP amplitude for anodal pulses and +125% AP amplitude for cathodal ones. Electroporation-induced decrease in AP amplitude was shifted toward lower current densities for anodal stimulus than for cathodal stimulus (half-maximal values, 185 and 238 mA/cm², respectively).

Whole heart studies revealed different types of asymmetry for the positive and negative polarizations during strong shocks (see Fig. 34.3) (7). We also found that these effects are accompanied by epicardial postshock elevation of diastolic potential (see Fig. 34.3b). In our study, we determined epicardial ΔV_m responses during high-density electrical current stimuli of both polarities applied at a 6-mm-diameter area of left ventricle (see Fig. 34.1). We detected saturation and subsequent decay

of epicardial polarizations during strong cathodal and anodal shocks applied at the area with a size of several space constants (0.8–1.5 mm at the epicardium) (39). Our optical recordings of negative ΔV_m responses to high-intensity stimuli were in agreement with the results of studies on strands of cultured myocytes (33, 38), single cell (17), and frog ventricle (29) for hyperpolarizing stimuli. We did not observe a plateau or an increase in depolarization transients during cathodal stimuli of the same stimulus strengths that caused a decayed hyperpolarization response.

Fast et al. (35) showed that the initial positive polarization in virtual cathode areas of a wedge preparation changes in response to hyperpolarizing responses as the stimuli strength increases to 30 V/cm and above, similar to the behavior of the middle of a single myocyte in studies by Sharma and Tung (32). Such observations were reported previously by Cheng et al. (40) and Zhou et al. (41), who detected hyperpolarization transients near the cathodal shock electrode. We observed the same phenomena in optical recordings from the epicardial surface (7). This can explain why depolarization saturation, rather than hyperpolarization saturation, is observed at lower shock current densities. Neunlist and Tung (29) stimulated a small area near the electrode, which could affect their measurements because of a virtual electrode effect (42), leading to the development of positive and negative polarizations at nearby locations. In contrast, our experiments were designed to overcome this limitation by stimulating a larger area relative to the field of view (see Fig. 34.1). Also, we did not use transmural sections of tissue, which interrupt fibers and thus could affect the results in a slab preparation (42); however, strong cathodal stimuli resulted in hyperpolarizing responses, which were partially reversible. Thus, we concluded that electroporation was the most plausible explanation for these effects; however, voltage-dependent resealing of the pores (formed during the shock application) would be required to explain the restoration of membrane resistance.

The correlation between anodal (negative) ΔV_m and diastolic ΔV_m elevation was recently reported by Fast and Cheek (43) in myocyte cultures. A similar result was reported earlier by Neunlist and Tung (29) and Cheng et al. (40). Interestingly, Fast and Cheek (43) did not observe nonmonotonic ΔV_m at the cathodal end of the cell strand even at the highest shock strengths. They later demonstrated that the absence of positive ΔV_m decay in the optical recordings of transmembrane potential at the edge of the preparation facing the cathodal electrode during strong shocks results from the spatial averaging of polarizations in the neighboring areas (44, 45).

We observed that a rather small optical hyperpolarization response could be sufficient for electroporation, which is in agreement with the results from earlier reports (26, 29). Among possible reasons for this were a "dog-bone" virtual polarization near the pacing electrode (29), which could attenuate the response in optical recordings because of optical averaging over areas of opposite polarizations, or an insufficient temporal resolution of the optical mapping system (29), which could underestimate the true instantaneous transmembrane voltage produced by a square pulse. Our results for polarization transients recorded during 10-ms ramp waveform stimulation (no temporal resolution limitations) over a 6-mm-diameter area of epicardium (opposite polarization is located 3 mm away from the center recording point) reject such explanations.

It was shown that Ca channel blockers did not influence the electroporation threshold in whole cell patch studies (27) and that they only increased positive ΔV_m in cell culture studies (46). Application of a Ca channel blocker resulted in an increase in the saturation level of depolarizing responses and did not affect hyperpolarizing responses in cell culture (46). We also observed that nifedipine increased the saturation levels for positive, but not negative, ΔV_m during epicardial stimulation. Yet, nifedipine had a clear impact on the saturation level of the depolarization signal in our study, which means that other factors (i.e., space averaging in optical recordings) could also be responsible for saturation and reversal of the depolarizing responses with the increase in stimulus strength.

4. Electroporation Assessment via Membrane-Impermeable Fluorescent Dyes

Another way to demonstrate electroporation is to characterize the permeation of markers (fluorescent dyes) through cell membrane pores. Uptake depends on dye concentration differences inside and outside the cells and on the net electrical charge of the dye. Electroporation in chick embryo myocardial cells caused by field stimulation at 50–200 V/cm enables incorporation of fluorescein-isothiocyanate-labeled dextrans (FITC-dextrans) with molecular masses of 4–10 kDa (47). To investigate electroporation as a delivery method in cardiac tissue, Harrison et al. (1) used three different indicators of electroporation—PI (668 Da), and expression vectors for green fluorescent protein (GFP) and luciferase. Song and Ochi (18) used ethidium bromide (EB, ethidium+, 314 Da) as a fluorescent marker that could pass through small pores produced by mild electroporation in isolated rabbit ventricular myocytes. The internal Ca^{2+} concentration ($[Ca^{2+}]$) is also a useful indicator of electroporation, though Ca^{2+} influx through voltage-gated channels would be expected to contaminate influx through passive pores (45).

Although the method of cell permeabilization is already a routine technique, the complete understanding of its mechanisms remains to be formulated. The most fundamental questions remain unknown: What is the size and density of the pores created by the shock? Do pores grow after the shock? and What is the time of their resealing? For example, because lucifer yellow dye uptake was not observed at 50-V/cm shock strength, which was above the 30-V/cm threshold of nonmonotonic negative ΔV_m, electroporation was not definitive, and an activation of an unknown hyperpolarization-activated channel(s) was proposed as an alternative explanation by Fast and Cheek (43). However, recently this group detected such an uptake by using another fluorescent dye: PI (38). They have shown that the application of a series of shocks with a strength of about 23 V/cm resulted in the uptake of the membrane-impermeable dye, PI. Dye uptake was restricted to the anodal side of strands with the largest negative ΔV_m, indicating the occurrence of membrane electroporation at these locations. The lack of detectable dye uptake in some of the previous studies could be related to the shorter exposure time and to

the lower sensitivity of the lucifer yellow technique. So, PI is a more relevant dye for estimating electroporation, with a 20–30-fold increase in fluorescence after binding to nucleic acids.

In our experiments with PI, we did not detect an immediate fluorescence increase during the shock (7), which suggests that the amount of PI molecules penetrated through the electroporation holes during the 20-ms stimulus was undetectable in our protocol. This also explains why we did not observe a difference in PI uptake for shocks of different polarities despite the positive charge of the PI molecule. Slow diffusion of PI into the cells took place even after the external electrical field was turned off, and thus fluorescence was continuously rising after the shock during dye perfusion in our experiments (7), as it did in cell culture studies (48). These data suggest that in our experiments electroporated cells were repaired within minutes rather than seconds. In our study, we observed the PI dye uptake at a shock strength of $700\,\text{mA/cm}^2$ ($35\,\text{V/cm}$) when we detected nonmonotonic ΔV_m. No PI uptake was observed at $300\,\text{mA/cm}^2$ ($15\,\text{V/cm}$) when ΔV_m was monotonic (see Fig. 34.4). We suggest that DP elevation might be a more sensitive indicator of electroporation than is PI uptake because the maximum DP elevation can be detected within 1 s after shock application (see Fig. 34.2); however, this method cannot be used in the depths of three-dimensional tissues.

Although PI is widely used in electroporation research, these studies are usually conducted on cell suspensions. There were concerns that this molecule may not be well suited for studies on tissues with interconnected cells because of its relatively smaller molecular weight (668 Da) with a radius of about 0.6 nm, which is smaller than the pore of gap junction channel (about 0.8 nm). Thus, it was suggested that PI might diffuse to neighboring cells, creating an appearance of electroporation in intact cells.

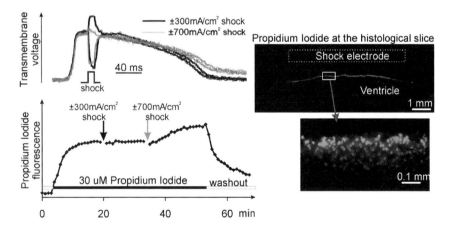

Fig. 34.4 Manifestation of electroporation changes in optical potential recordings is associated with an increase in propidium iodide fluorescence under the stimulation electrode. No increase was observed at sites not under the electrode. Histological images showed typical pattern of nuclear stain in the thin layer of epicardium at the areas where optical potentials had signs of electroporation. (Reproduced from (7), with permission)

Our data show significant differences in the depth of staining with PI, including staining of some interior regions of myocardium, which are isolated from other stained regions. The area of electroporation, identified by PI staining, which occurred in the middle of the papillary muscle confirms that in our experiments diffusion extends less than 0.1 mm. Diffusion of PI molecules is theoretically possible through gap junctions, but perhaps it does not occur over large distances because of rapid binding of PI to the nuclei, which prevents its diffusion in the intracellular space (Fig. 34.5).

A single cell study showed that during 2-kV/cm, 20-μs shocks, the cells with irreversible membrane electroporation accumulate a 5-fold larger amount of PI than do cells that restored their membrane within 10 min after field exposure (48). Also, 1.8-A/cm² stimuli cause irreversible cell damage (5). This indicates that PI accumulation during the strong shock could be related to other factors (barotrauma, hyperthermia) leading to cell death. If such factors are less dependant on proximity

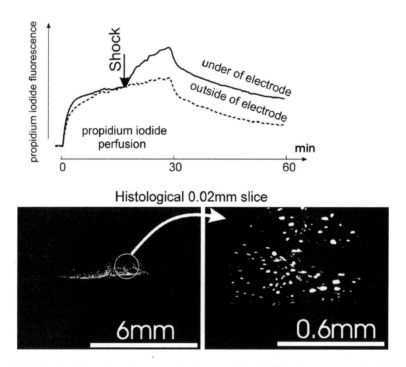

Fig. 34.5 Uptake of membrane-impermeable dye propidium iodide after a strong shock. *Upper panel* shows the initial increase in propidium iodide fluorescence, after beginning perfusion, recorded inside the stimulated area and 3 mm outside the stimulated area. After shock application (1,600 mA/cm², 20 ms), there was an accelerated accumulation of fluorophore in the region current was applied. *Lower panel:* The fluorescent images (made with 4× and 40× lenses) of 20-μm slice sectioned throughout the stimulated area. Electroporated region is clearly demarcated by the propidium-iodide-stained cell nuclei. (Reproduced from (7), with permission)

to the tissue boundaries than is electroporation (24, 25), it can explain the much larger depth of affected tissue after the 1.6-A/cm^2 shock in comparison with the 0.7-A/cm^2 shock (see Figs. 34.4 and 34.5).

5. Is Shock-Induced Electroporation Pro- or Antiarrhythmic?

The main interest in studying the effects of electrical shocks on the heart was driven by the ability of strong shocks to terminate arrhythmias. The mechanism of ventricular defibrillation has been extensively investigated for many decades; however, the definite mechanism of defibrillation is still unclear. Although many hypotheses have been proposed, two major competing theories on ventricular defibrillation are highlighted in the present review: the incapacitation theory (49) and the stimulation theory (50).

We believe that the two theories of defibrillation are extreme in their assessment of the role of electroporation. The incapacitation theory places electroporation as the foundation of the mechanisms of defibrillation, while the other ignores its role in defibrillation. The truth is likely to be in the middle. Indeed, despite significant improvement in defibrillation efficacy, which resulted in a radical reduction of defibrillation thresholds and myocardial damage, there are still extensive clinical and basic electrophysiology data indicating that defibrillation shocks are accompanied by adverse effects. These effects are likely to be associated with electroporation and may include the following effects: (1) transient ectopy, tachycardia, or induction of ventricular fibrillation (28, 43, 51, 52); (2) depression of electrical and mechanical functions (6, 8, 29, 53); (3) bradycardia, complete heart block, and increased pacing thresholds (52, 54, 55); (4) atrial and ventricular mechanical dysfunction (stunning), which is directly related to the strength of shocks (4, 56–59); (5) significant elevation of serum Troponin I in patients after spontaneous cardioverter defibrillator shocks (60); and (6) decrease in the myocardial lactate extraction rate by mitochondria (61).

Both clinical and basic experimental evidence suggest that electroporation is induced (47) by defibrillation shocks and plays a role in defibrillation (62) and postshock metabolism and electromechanical function of the heart. However, the antifibrillatory role of electroporation is not clear, since conflicting data suggest both pro- and antiarrhythmic effects of electroporation (8, 28).

The effects of electroporation on the cardiac tissue or whole heart levels, during application of an external electrical field, are much less understood than those on a single cell. One of the problems of such experiments is related to the strong spatial heterogeneity of electroporation effects, which complicates their interpretation (8, 26, 28, 29).

Several of our own studies were aimed at exploring spatiotemporal patterns and voltage dependence of electroporation in several types of preparations, such as the whole, intact Langendorff-perfused heart (8, 40), the isolated endocardium and septum preparations (8), and 6-mm-diameter areas of epicardium under the stimulating electrode (7, 30). In these studies, we found that electroporation is dependent

upon shock intensity, tissue structure, and electrode configuration. For example, we found that the endocardium is significantly more susceptible to electroporation than is the epicardium (8).

Electroporation can selectively affect small trabeculated structures and bundles of the conduction system of the heart. This could result in transient suppression of excitability and conduction block (6). We found that conduction in a small papillary muscle in the rabbit heart can be transiently inhibited by a strong electrical shock (see Fig. 34.6). We found that such transient blocks can last from one beat to many seconds, depending on the shock strength (8). Potentially, such transient inhibition of conduction in bundles of the heart can lead to initiation of a reentrant or focal arrhythmia, or on the other hand, can have an antiarrhythmic effect via isolation of ectopic foci and the reduction of tissue mass available for arrhythmia maintenance. Recently, we found that rabbit atria are more vulnerable to shock-induced electroporation than are its ventricles (see Fig. 34.2). Moreover, electroporation induces reentrant tachyarrhythmia in atria, due to conduction block and possibly triggered activity (30); however, the same strength shock which induced atrial tachyarrhythmias could also terminate them.

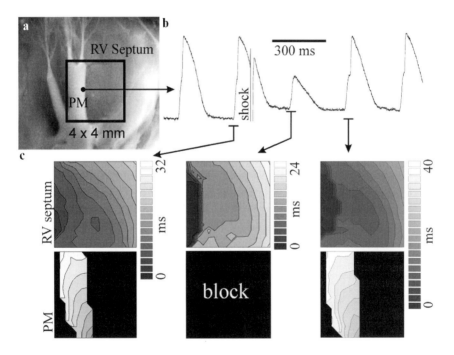

Fig. 34.6 Shock-induced block of conduction in the papillary muscle. (**a**) A 4×4-mm^2 field of view and surrounding structures of the right ventricular septum. *Point* indicates the position of recording site illustrated in panel b. (**b**) Optical recording at the papillary muscle during application of a 300-V, 8-ms monophasic shock. Notice two components in the upstrokes of all responses except the first postshock action potential. (**c**) Activation isochronal maps (1 ms) reconstructed from the two components of optical recordings (see text for detail). (Reproduced from (8), with permission)

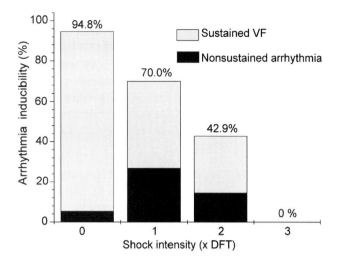

Fig. 34.7 Inducibility of ventricular fibrillation by T wave shock versus preconditioning shock applied 1,200 or 1,500 ms prior to shock. Preconditioning shock intensity is expressed as 0×, 1×, 2×, and 3× defibrillation thresholds. (Reproduced from (8), with permission)

Profibrillatory effects of electroporation and tissue damage are supported by clinical observations of postshock ectopy and arrhythmia, as reviewed here, as well as by basic studies, including the classical studies of Gurvich (63) and several recent studies (28, 43, 64). Antifibrillatory effects of electroporation have been known since the seminal work of Prevost and Battelli (49). We also obtained data which support antifibrillatory effects of electroporation (8). Our results (8) show that a strong electroporating shock applied during a period of electroporation reduces vulnerability to shock-induced arrhythmia.

A test shock applied during the T wave induces arrhythmias in 94.8% of cases without a preconditioning shock (see Fig. 34.7), most of which are sustained (gray bar). Application of the preconditioning shock with intensity equal to the defibrillation threshold (DFT) reduces arrhythmia induction to 70%. Application of the electroporating preconditioning shock with intensity equal to 3× DFT completely eliminates vulnerability.

6. Electroporation-Mediated Gene Transfer in Cardiac Tissue

Although electronic pacemakers are currently the main stay of therapy for sino-atrial node (SAN) dysfunction, heart block, and other electrophysiological abnormalities, they are not optimal. Among their shortcomings are limited battery life, need for permanent catheter implantation into the heart, and lack of response to autonomic neuromediators (65). Recently, several alternatives such as viral

(66) or nonviral (67) gene transfer have been proposed to develop a so-called biological pacemaker.

Neumann et al. (68) first demonstrated that in vitro electroporation of cells in the presence of plasmid DNA resulted in DNA transfer and expression. Since then, this method has been widely used for gene delivery to increase gene transfer and expression in vitro and, more recently, in vivo. (69) Electroporation has been shown to markedly enhance the efficiency of naked DNA transfer (70).

Several studies have been reported using electroporation to transfer DNA to the whole heart. In one study, hearts were removed from stage 18 embryonic chickens, placed in a bath of DNA, and electroporated at 200 V/cm with 10-ms square-wave pulses delivered with a Grass stimulator (1). When no electrical field was applied, no gene expression was detected in any of the hearts. However, when either six or eight pulses were delivered to the hearts, significant levels of GFP or luciferase expression could be detected. Indeed, up to 30% of the cells in the heart showed GFP expression, almost all of which was in cardiomyocytes. Another group performed very similar studies on excised and transplanted mouse hearts (71). They found good expression by using naked DNA coupled with electroporation induced by 20 square-wave 25-ms pulses with strength of 200 V/cm.

Yet, this method of electroporation-mediated gene transfer in cardiac tissue is not widely used. One of the limitations to transfecting cardiac muscle is that the cardiomyocytes are surrounded by an endothelial layer on the endocardium, where it would be most easy to deliver DNA via the blood. Moreover, it is not clear what parameters of electrical stimulus and electrode configurations are optimal to create direct gene transfer to intramural layers of cardiac tissue without significant damage of surface cell layers.

7. Conclusion

Application of electrical shocks is a routine technique to treat cardiac arrhythmias. High-intensity fields (greater than 25 V/cm) generated inside the cardiac tissue cause transient tissue damage due to electroporation. Electroporation can be monitored by changes in the morphology of the transmembrane polarization transients during anodal and cathodal shocks from monotonic to nonmonotonic response, elevation of the resting potential, and postshock reduction of AP amplitude and its derivative. Membrane-impermeable dye (i.e., propidium iodide) uptake signifies that recovery of membrane integrity in cardiac cells can take minutes after shock termination. Such long-lasting effects can have both anti- and proarrhythmic effects in the heart. Further improvement of defibrillation therapy may be able to direct electroporation power precisely to the reentry substrate in order to minimize the adverse effects on contractile heart properties or for delivering gene therapy to the specialized cardiac tissue.

Acknowledgments This work was supported by the National Heart, Lung, and Blood Institute Grant R01HL-074283.

References

1. Harrison, R.L., Byrne, B.J., and Tung, L. (1998) Electroporation-mediated gene transfer in cardiac tissue. FEBS Lett. 435, 1–5.
2. Rosen, M.R., Brink, P.R., Cohen, I.S., and Robinson, R.B. (2004) Genes, stem cells and biological pacemakers. Cardiovasc. Res. 64, 12–23.
3. Kim, J.M., Lim, B.K., Ho, S.H., et al. (2006) TNFR-Fc fusion protein expressed by in vivo electroporation improves survival rates and myocardial injury in coxsackievirus induced murine myocarditis. Biochem. Biophys. Res. Commun. 344, 765–771.
4. Babbs, C.F., Tacker, W.A., VanVleet, J.F., Bourland, J.D., and Geddes, L.A. (1980) Therapeutic indices for transchest defibrillator shocks: effective, damaging, and lethal electrical doses. Am. Heart J. 99, 734–738.
5. Koning, G., Veefkind, A.H., and Schneider, H. (1980) Cardiac damage caused by direct application of defibrillator shocks to isolated Langendorff-perfused rabbit heart. Am. Heart J. 100, 473–482.
6. Yabe, S., Smith, W.M., Daubert, J.P., Wolf, P.D., Rollins, D.L., and Ideker, R.E. (1990) Conduction disturbances caused by high current density electric fields. Circ. Res. 66, 1190–1203.
7. Nikolski, V.P., Sambelashvili, A.T., Krinsky, V.I., and Efimov, I.R. (2004) Effects of electroporation on optically recorded transmembrane potential responses to high-intensity electrical shocks. Am. J. Physiol. Heart Circ. Physiol. 286, H412–H418.
8. Al-Khadra, A.S., Nikolski, V., and Efimov, I.R. (2000) The role of electroporation in defibrillation. Circ. Res. 87, 797–804.
9. Goldman, D.E. (1943) Potential, impedance, and rectification in membranes. J. Gen. Physiol. 27, 37–50.
10. Abidor, I.G., Arakelyen, V.B., Chernomordik, L.V., Chizmadzhev, Y.A., Pastushenko, V.F., and Tarasevich, R. (1979) Electric breakdown of bilayer lipid membranes. I. The main experimental facts and their qualitative discussion. Bioelectrochem. Bioenerg. 6, 37–52.
11. Tung, L., Troiano, G.C., Sharma, V., Raphael, R.M., and Stebe, K.J. (1999) Changes in electroporation thresholds of lipid membranes by surfactants and peptides. Ann. N.Y. Acad. Sci. 888, 249–265.
12. Melikov, K.C., Frolov, V.A., Shcherbakov, A., Samsonov, A.V., Chizmadzhev, Y.A., and Chernomordik, L.V. (2001) Voltage-induced nonconductive pre-pores and metastable single pores in unmodified planar lipid bilayer. Biophys. J. 80, 1829–1836.
13. Tsong, T.Y. (1991) Electroporation of cell membranes. Biophys. J. 60, 297–306.
14. Chang, D.C. and Reese, T.S. (1990) Changes in membrane structure induced by electroporation as revealed by rapid-freezing electron microscopy. Biophys. J. 58, 1–12.
15. O'Neill, R.J. and Tung, L. (1991) Cell-attached patch clamp study of the electropermeabilization of amphibian cardiac cells. Biophys. J. 59, 1028–1039.
16. Krauthamer, V. and Jones, J.L. (1997) Calcium dynamics in cultured heart cells exposed to defibrillator-type electric shocks. Life Sci. 60, 1977–1985.
17. Cheng, D.K., Tung, L., and Sobie, E.A. (1999) Nonuniform responses of transmembrane potential during electric field stimulation of single cardiac cells. Am. J. Physiol. 277, H351–H362.
18. Song, Y.M. and Ochi, R. (2002) Hyperpolarization and lysophosphatidylcholine induce inward currents and ethidium fluorescence in rabbit ventricular myocytes. J. Physiol. 545, 463–473.
19. Krassowska, W. and Neu, J.C. (1994) Response of a single cell to an external electric field. Biophys. J. 66, 1768–1776.
20. Chizmadzhev, Y.A., Kuzmin, P.I., Kumenko, D.A., Zimmerberg, J., and Cohen, F.S. (2000) Dynamics of fusion pores connecting membranes of different tensions. Biophys. J. 78, 2241–2256.
21. Smith, K.C., Neu, J.C., and Krassowska, W. (2004) Model of creation and evolution of stable electropores for DNA delivery. Biophys. J. 86, 2813–2826.
22. DeBruin, K.A. and Krassowska, W. (1999) Modeling electroporation in a single cell. II. Effects of ionic concentrations. Biophys. J. 77, 1225–1233.
23. DeBruin, K.A. and Krassowska, W. (1999) Modeling electroporation in a single cell. I. Effects of field strength and rest potential. Biophys. J. 77, 1213–1224.

24. Aguel, F., DeBruin, K.A., Krassowska, W., and Trayanova, N.A. (1999) Effects of electroporation on the transmembrane potential distribution in a two-dimensional bidomain model of cardiac tissue. *J. Cardiovasc. Electrophysiol.* **10**, 701–714.

25. DeBruin, K.A. and Krassowska, W. (1998) Electroporation and shock-induced transmembrane potential in a cardiac fiber during defibrillation strength shocks. *Ann. Biomed. Eng.* **26**, 584–596.

26. Tung, L., Tovar, O., Neunlist, M., Jain, S.K., and O'Neill, R.J. (1994) Effects of strong electrical shock on cardiac muscle tissue. *Ann. N.Y. Acad. Sci.* **720**, 160–175.

27. Tovar, O. and Tung, L. (1992) Electroporation and recovery of cardiac cell membrane with rectangular voltage pulses. *Am. J. Physiol.* **263**, H1128–H1136.

28. Kodama, I., Shibata, N., Sakuma, I., et al. (1994) Aftereffects of high-intensity DC stimulation on the electromechanical performance of ventricular muscle. *Am. J. Physiol.* **267**, H248–H258.

29. Neunlist, M. and Tung, L. (1997) Dose-dependent reduction of cardiac transmembrane potential by high-intensity electrical shocks. *Am. J. Physiol.* **273**, H2817–H2825.

30. Fedorov, V.V., Hemphill, M., Kostecki, G., Li, L., and Efimov, I.R. (2007) Low electroporation threshold, conduction block, focal activity and reentrant arrhythmia in the rabbit atria: possible mechanisms of stunning and defibrillation failure. Circulation (in press).

31. Gray, R.A., Huelsing, D.J., Aguel, F., and Trayanova, N.A. (2001) Effect of strength and timing of transmembrane current pulses on isolated ventricular myocytes. *J. Cardiovasc. Electrophysiol.* **12**, 1129–1137.

32. Sharma, V. and Tung, L. (2002) Spatial heterogeneity of transmembrane potential responses of single guinea-pig cardiac cells during electric field stimulation. *J. Physiol.* **542**, 477–492.

33. Fast, V.G., Rohr, S., and Ideker, R.E. (2000) Nonlinear changes of transmembrane potential caused by defibrillation shocks in strands of cultured myocytes. *Am. J. Physiol. Heart Circ. Physiol.* **278**, H688–H697.

34. Efimov, I.R., Cheng, Y.N., Biermann, M., Van Wagoner, D.R., Mazgalev, T., and Tchou, P.J. (1997) Transmembrane voltage changes produced by real and virtual electrodes during monophasic defibrillation shock delivered by an implantable electrode. *J. Cardiovasc. Electrophysiol.* **8**, 1031–1045.

35. Fast, V.G., Sharifov, O.F., Cheek, E.R., Newton, J.C., and Ideker, R.E. (2002) Intramural virtual electrodes during defibrillation shocks in left ventricular wall assessed by optical mapping of membrane potential. *Circulation.* **106**, 1007–1014.

36. Knisley, S.B., Blitchington, T.F., Hill, B.C., et al. (1993) Optical measurements of transmembrane potential changes during electric field stimulation of ventricular cells. *Circ. Res.* **72**, 255–270.

37. Windisch, H., Ahammer, H., Schaffer, P., Muller, W., and Platzer, D. (1995) Optical multisite monitoring of cell excitation phenomena in isolated cardiomyocytes. *Pflugers Arch.* **430**, 508–518.

38. Cheek, E.R. and Fast, V.G. (2004) Nonlinear changes of transmembrane potential during electrical shocks: role of membrane electroporation. *Circ. Res.* **94**, 208–214.

39. Akar, F.G., Roth, B.J., and Rosenbaum, D.S. (2001) Optical measurement of cell-to-cell coupling in intact heart using subthreshold electrical stimulation. *Am. J. Physiol. Heart Circ. Physiol.* **281**, H533–H542.

40. Cheng, Y., Tchou, P.J., and Efimov, I.R. (1999) Spatio-temporal characterization of electroporation during defibrillation. *Biophys. J.* **76**, A85.

41. Zhou, X., Ideker, R.E., Blitchington, T.F., Smith, W.M., and Knisley, S.B. (1995). Optical transmembrane potential measurements during defibrillation-strength shocks in perfused rabbit hearts. *Circ. Res.* **77**, 593–602.

42. Neunlist, M. and Tung, L. (1995) Spatial distribution of cardiac transmembrane potentials around an extracellular electrode: dependence on fiber orientation. *Biophys. J.* **68**, 2310–2322.

43. Fast, V.G. and Cheek, E.R. (2002) Optical mapping of arrhythmias induced by strong electrical shocks in myocyte cultures. *Circ. Res.* **90**, 664–670.

44. Sharifov, O.F., Ideker, R.E., and Fast, V.G. (2004) High-resolution optical mapping of intramural virtual electrodes in porcine left ventricular wall. *Cardiovasc. Res.* **64**, 448–456.

45. Fast, V.G., Cheek, E.R., Pollard, A.E., and Ideker, R.E. (2004) Effects of electrical shocks on Ca_i^{2+} and V_m in myocyte cultures. *Circ. Res.* **94**, 1589–1597.

46. Cheek, E.R., Ideker, R.E., and Fast, V.G. (2000) Nonlinear changes of transmembrane potential during defibrillation shocks: role of Ca(2+) current. *Circ. Res.* **87**, 453–459.

47. Jones, J.L., Jones, R.E., and Balasky, G. (1987) Microlesion formation in myocardial cells by high-intensity electric field stimulation. *Am. J. Physiol.* **253**, 480–486.

48. Shirakashi, R., Kostner, C.M., Muller, K.J., Kurschner, M., Zimmermann, U., and Sukhorukov, V.L. (2002) Intracellular delivery of trehalose into mammalian cells by electropermeabilization. *J. Membr. Biol.* **189**, 45–54.

49. Prevost, J.L. and Battelli, F. (1899) Sur quel ques effets des dechanges electriques sur le coer mammifres. *Comptes Rendus Seances Acad. Sci.* **129**, 1267.

50. Gurvich, N.L. and Yuniev, G.S. (1939) Restoration of regular rhythm in the mammalian fibrillating heart. *Byulletin Eksper. Biol. Med.* **8**, 55–58.

51. Donoso, E., Cohn, L.J., and Friedberg, C.K. (1967) Ventricular arrhythmias after precordial electric shock. *Am. Heart J.* **73**, 595–601.

52. Waldecker, B., Brugada, P., Zehender, M., Stevenson, W., and Welens, H.J. (1986) Ventricular arrhythmias after precordial electric shock. *Am. J. Cardiol.* **57**, 120–123.

53. Tovar, O. and Tung, L. (1991) Electroporation of cardiac cell membranes with monophasic or biphasic rectangular pulses. *Pacing Clin. Electrophysiol.* **14**, 1887–1892.

54. Eysmann, S.B., Marchlinski, F.E., Buxton, A.E., and Josephson, M.E. (1986) Electrocardiographic changes after cardioversion of ventricular arrhythmias. *Circulation.* **73**, 73–81.

55. Stickney, R.E., Doherty, A., Kudenchuk, P.J., et al. (1999) Survival and postshock ECG rhythms for out-of-hospital defibrillation. *PACE.* **22**, 740.

56. Sparks, P.B., Kulkarni, R., Vohra, J.K., et al. (1998) Effect of direct current shocks on left atrial mechanical function in patients with structural heart disease. *J. Am. Coll. Cardiol.* **31**, 1395–1399.

57. Sparks, P.B., Jayaprakash, S., Mond, H.G., Vohra, J.K., Grigg, L.E., and Kalman, J.M. (1999) Left atrial mechanical function after brief duration atrial fibrillation. *J. Am. Coll. Cardiol.* **33**, 342–349.

58. Grimm, R.A., Stewart, W.J., Arheart, K., Thomas, J.D., and Klein, A.L. (1997) Left atrial appendage "stunning" after electrical cardioversion of atrial flutter: an attenuated response compared with atrial fibrillation as the mechanism for lower susceptibility to thromboembolic events. *J. Am. Coll. Cardiol.* **29**, 582–589.

59. Kam, R.M., Garan, H., McGovern, B.A., Ruskin, J.N., and Harthorne, J.W. (1997) Transient right bundle branch block causing R wave attenuation postdefibrillation. *Pacing Clin. Electrophysiol.* **20**, 130–131.

60. Hasdemir, C., Shah, N., Rao, A.P., et al. (2002) Analysis of troponin I levels after spontaneous implantable cardioverter defibrillator shocks. *J. Cardiovasc. Electrophysiol.* **13**, 144–150.

61. Osswald, S., Trouton, T.G., O'Nunain, S.S., Holden, H.B., Ruskin, J.N., and Garan, H. (1994) Relation between shock-related myocardial injury and defibrillation efficacy of monophasic and biphasic shocks in a canine model. *Circulation.* **90**, 2501–2509.

62. Peleska, B. (1965) Problems of defibrillation and stimulation of the myocardium. *Zentralbl. Chir.* **90**, 1174–1188.

63. Gurvich, N.L. (1975) *The main principles of cardiac defibrillation.* Medicine, Moscow.

64. Ohuchi, K., Fukui, Y., Sakuma, I., Shibata, N., Honjo, H., and Kodama, I. (2002) A dynamic action potential model analysis of shock-induced aftereffects in ventricular muscle by reversible breakdown of cell membrane. *IEEE Trans. Biomed. Eng.* **49**, 18–30.

65. Pyatt, J.R., Somauroo, J.D., Jackson, M., et al. (2002) Long-term survival after permanent pacemaker implantation: analysis of predictors for increased mortality. *Europace.* **4**, 113–119.

66. Miake, J., Marban, E., and Nuss, H.B. (2002) Biological pacemaker created by gene transfer. *Nature.* **419**, 132–133.

67. Potapova, I., Plotnikov, A., Lu, Z., et al. (2004) Human mesenchymal stem cells as a gene delivery system to create cardiac pacemakers. *Circ. Res.* **94**, 952–959.

68. Neumann, E., Schaefer-Ridder, M., Wang, Y., and Hofschneider, P.H. (1982) Gene transfer into mouse lyoma cells by electroporation in high electric fields. *EMBO J.* **1**, 841–845.

69. Dean, D.A. (2005) Nonviral gene transfer to skeletal, smooth, and cardiac muscle in living animals. *Am. J. Physiol. Cell Physiol.* **289**, C233–C245.

70. Aihara, H. and Miyazaki, J. (1998) Gene transfer into muscle by electroporation in vivo. *Nat. Biotechnol.* **16**, 867–870.

71. Wang, Y., Bai, Y., Price, C., et al. (2001) Combination of electroporation and DNA/dendrimer complexes enhances gene transfer into murine cardiac transplants. *Am. J. Transplant.* **1**, 334–338.

Chapter 35
Muscle and Fat Mass Modulation in Different Clinical Models

Ruxandra Draghia-Akli and Amir S. Khan

Abstract Studies described in the recent literature support the idea that gene therapy can lead to genuine clinical benefits when mediated by plasmid delivery in conjunction with electroporation. Plasmid-mediated muscle-targeted gene transfer offers the potential of a cost-effective pharmaceutical-grade therapy delivered by simple intramuscular injection. This approach is particularly appropriate for modulating muscle and fat mass and their intrinsic properties, from treatment of conditions such as cachexia associated with chronic diseases, autoimmune diseases, e.g., myasthenia gravis, to stimulation or suppression of appetite, and further to *in vivo* manipulation of glucose metabolism and fat deposition in patients with diabetes, or to basic studies of muscle-specific transcription factors and their impact in development. Recent innovations, including in situ electroporation, enabling sustained systemic protein delivery within the therapeutic range, are reviewed. Translation of these advances to human clinical trials will enable muscle- and fat-targeted gene therapy to become a viable therapeutic alternative.

Keywords: electroporation, plasmid, muscle, fat, body composition, gene therapy

1. Introduction

The success of gene therapy depends on the efficient expression of genes into appropriate target cells without causing cell injury, oncogenic mutation, or inflammation. An ideal therapeutic approach would allow for readministration of the vector several times, especially if targeting chronic diseases. Few available technologies meet all of these requirements. Most of the gene therapy studies have been performed with viral vectors, and serious limitations in terms of immunogenicity and pathogenicity have become apparent (1). Plasmid-based gene therapy has fewer safety concerns, including lower vector immunogenicity, permitting readministration. Historically, the simple injection of naked plasmid has been sufficient to produce detectable levels of reporter transgenes and persistent

S. Li (ed.), *Electroporation Protocols: Preclinical and Clinical Gene Medicine.*
From *Methods in Molecular Biology, Vol. 423.*
© Humana Press 2008

levels of anti-inflammatory agents and other mediators, although levels of gene expression were generally much lower than with viral vectors (2, 3). So, a major limitation of classical nonviral gene therapy has been low transfection efficiency (4). One of the most versatile and efficient methods of enhancing gene transfer involves the application of electric field pulses after the injection of nucleic acids (DNA, RNA, and/or oligonucleotides) into tissues. The method is safe, provided appropriate electrical parameters are chosen. The transfection efficiency of electroporation (EP) is many times greater than that of naked DNA injection alone, with markedly reduced interanimal variability (5).

There is now conclusive evidence that plasmid-mediated gene therapy in conjunction with EP can have significant and long-lived clinical benefits (6). This approach is particularly appropriate for long-term circulating therapeutic protein replacement, supplementation, or modulation currently requiring repeated injection therapy (6, 7). As an approach to protein drug delivery, plasmid-based EP transfer has been proven safe and effective in preclinical models of immunological, endocrine, neoplastic, and other diseases, in particular conditions that affect muscle and fat mass, and is highly promising in the treatment of a wide variety of diseases or as a basic research tool.

2. Electroporation in Basic and Clinical Models of Muscle and Fat Mass Accretion and Regulation

2.1. Leptin and Appetite Regulation

Leptin is a hormone predominantly produced by adipocytes and, functionally, a key regulator of body weight. Loss-of-function mutations of the leptin gene or its receptor in mice result in syndromes of obesity and type 2 diabetes (*ob/ob* and *db/db* mice, respectively). Although human obesity is only rarely caused by these mutations, many studies suggest that the administration of leptin may ameliorate obesity from other causes. Thus, leptin gene therapy for the control of obesity was developed. Recent publications have described the transfer of the leptin gene into muscle, using either a hydrodynamic method (8, 9) or an EP method (8–11). In mice treated using EP, elevated serum leptin concentrations up to 90 ng/mL were recorded (a 200-fold increase, compared with that in control mice). Seven weeks after gene transfer, the body weight of both young and adult leptin-cDNA-treated mice was about 20% lighter than the control. The levels of retroperitoneal fats and serum triglycerides of leptin-cDNA-treated mice were markedly lower than those of the control, but a lesser effect was found on food intake (10). Furthermore, the production of insulin was lowered in treated mice, but blood glucose remained normal.

Conversely, to investigate whether *in vivo* gene transfer causes leptin-antagonistic effects on food intake, animal body weight, and fat tissue weight, a R128Q mutated-leptin gene was used (9), and EP was compared with the

hydrodynamic method of gene delivery. Mutated-leptin gene transfer by EP caused significant increases in body weight starting the fifth day (5.4% increase, compared with that of control; $p < 0.05$). This resulted in significant increases in tissue weight of epididymal white fat and neuropeptide Y mRNA expression in the hypothalamus, compared with those of the control group up to 3 weeks after gene transfer ($p < 0.05$), suggesting that mutated-leptin gene transfer successfully produced leptin-antagonistic effects by modulating the central regulator of energy homeostasis. Also, the extent of leptin-antagonistic effects by EP was much higher than that by hydrodynamics-based gene delivery, with at least single gene transfer.

2.2. Muscle and Fat Tissue Metabolism

Skeletal muscle is a major peripheral tissue that accounts for approximately 40% of total body weight and 50% of energy expenditure and is a primary site of glucose metabolism and fatty acid oxidation (12). Consequently, muscle has a significant role in insulin sensitivity, obesity, body composition (relative proportion of muscle and fat mass), the blood-lipid profile, and modulation of muscle function with respect to intracellular fat deposition is widely studied. Electroporation has been used to deliver plasmids encoding for a multitude of molecules implicated in glucose transport in muscle cells (such as GLUT-1, GLUT-4 (13)), in β-adrenergic receptor (β-AR)-mediated adaptive thermogenesis (12), as well as to deliver plasmids encoding the enzyme 1,2-acyl CoA:diacylglyceroltransferase-1 (DGAT1), which plays an important role in triglyceride storage (14), and signaling involved in the regulation of the phosphorylation state of several kinases in skeletal muscle, including c-Jun NH(2)-terminal kinase (JNK) (15).

For instance, β-AR null mice develop severe obesity on a high-fat diet. When Nur77-specific small interfering RNAs (siNur77) were delivered into skeletal muscle by injection and EP, the endogenous Nur77 mRNA expression was repressed. As a result, gene and protein expression of AMP-activated protein kinase γ3, UCP3, CD36, adiponectin receptor 2, GLUT4, and caveolin-3 which are associated with the regulation of energy expenditure and lipid homeostasis were attenuated. The resulting reduced expression of these genes consequently decreased lipolysis. The authors speculate that Nur77 agonists would stimulate lipolysis and increase energy expenditure in skeletal muscle and suggest that selective activators of Nur77 may have therapeutic utility in the treatment of obesity.

In vivo EP in rodents was used to alter local gene expression in skeletal muscle as a means of investigating the role of specific proteins in glucose metabolism and their correlation to phenomena associated with impaired metabolism at the peripheral level in diabetes (13). Importantly, this and other studies showed that the EP procedure itself did not impact the expression of stress proteins (16). EP of the GLUT-1 gene resulted in a 57% increase in GLUT-1 protein, accompanied by a proportionate increase in basal 2-deoxyglucose tracer uptake into muscles of starved rats. EP of

plasmids expressing two short hairpin RNAs (shRNAs) for GLUT-4 demonstrated to reduce specific protein expression and 2-deoxyglucose tracer uptake in 3T3-L1 adipocytes into mouse muscle caused a 51% reduction in GLUT-4 protein, associated with attenuated clearance of tracer to muscle after a glucose load.

In adipose tissue, the microsomal enzyme DGAT1 plays an important role in triglyceride storage. DGAT1 is also expressed in skeletal muscle. When a DGAT1-expressing plasmid formulated in saline was injected in the left tibialis anterior (TA) muscle of rats (14), followed by the application of eight EP pulses, using the contralateral leg as sham-electroporated control, a significant overexpression of the DGAT1 protein was observed. The functionality of DGAT1 overexpression was underscored by the pronounced diet-responsive increase in intramyocellular lipid (IMCL) storage. In chow-fed rats, DGAT1-positive myocytes showed significantly higher IMCL content when compared with those in the control leg, which was almost devoid of IMCL ($1.99\% \pm 1.13\%$ vs. $0.017\% \pm 0.014\%$ of total area fraction; $p < 0.05$). High-fat feeding increased IMCL levels in both DGAT1-positive and control myocytes, resulting in very high IMCL levels in DGAT1-overexpressing myocytes ($4.96\% \pm 1.47\%$ vs. $0.80\% \pm 0.14\%$; $p < 0.05$).

These results highlight the utility of *in vivo* EP for the acute manipulation of muscle gene expression in the study of the role of specific proteins involved in muscle cell metabolism.

2.3. Cachexia

Cachexia is a classic clinical phenomenon, which simply means "poor condition" in Greek, and is marked by weight loss and increased mortality, affecting more than 5 million people in the United States alone. The major cause of cachexia appears to be excess cytokines. Other potential mediators include deficiencies of testosterone and insulin-like growth factor I (IGF-I) deficiency and excess of myostatin and glucocorticoids. Numerous diseases can result in cachexia, each by a slightly different mechanism (17). Cachexia can occur in the setting of cancer (18) as well as in other chronic conditions, such as HIV/AIDS (19), chronic pulmonary obstructive disease, or cardiac failure (20). Weight loss is the most obvious manifestation of cachexia. It is present in more than half of cancer patients. Other clinical manifestations include anorexia, muscle wasting, loss of adipose tissue, and fatigue, resulting in poor performance status. In addition to poor prognosis and impaired response to therapy, cachexia may be a direct cause of death (21).

A few molecules delivered as plasmid injection, followed by EP, have addressed this important medical problem. They are mainly growth hormone releasing hormone (GHRH) (22, 23) and IGF-I (24). For instance, a myogenic plasmid expressing GHRH was administered by injection, followed by EP, to severely debilitated dogs with naturally occurring tumors (22) or mice with implanted tumors (23, 25). In dogs, serum concentrations of IGF-I, the downstream effector of GHRH, were significantly increased in 12 of 16 dogs at a median of 16 days post-treatment (serum IGF-I, 37.9 $\pm$ 13.1 to 55.4 $\pm$ 20.4 ng/mL, baseline to first follow-up evaluation, respectively, $p < 0.01$). These increases ranged from 21 to 120% of the pre-treatment values and were

generally sustained or higher at the end of the study. Treated dogs maintained their weights over the 56-day study and did not show any adverse effects from the GHRH gene transfer. A long-term evaluation (26) of the dogs receiving plasmid GHRH by electrotransfer revealed biochemical and hematological improvements (serum IGF-I,

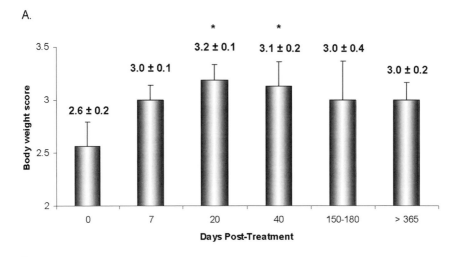

Fig. 35.1 Body weight score (**A**) and activity level score (**B**) as rated by the owners and attending veterinarians of dogs after a single treatment with GHRH-expressing plasmid. The animals were followed to one year to elucidate the long-term effects of the plasmid-GHRH administration. ($^*p < 0.05$, compared with baseline scores at day 0). Scores are rated using a modified Karnofsky scale modified for veterinary use (27): 5 = *significantly increased* (able to perform better than immediately predisease level), 4 = *increased* (able to perform at predisease level), 3 = *no change* (decreased activity from predisease level, but able to function as an acceptable pet), 2 = *decreased* (ambulatory only to the point of eating and consistently defecating and urinating in acceptable areas), 1 = *significantly decreased* (must be force-fed; unable to confine urinations and defecations to acceptable areas)

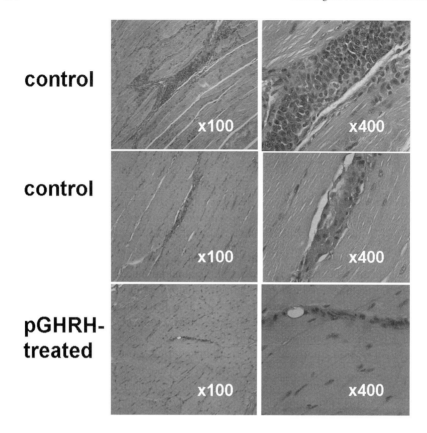

Fig. 35.2 Muscle biopsies are performed from tumor-bearing mice treated with either GHRH-expressing plasmid or saline. Half of all control animals exhibit muscle atrophy associated with focal areas of interstitial mononuclear cell infiltration of lymphoid origin. Histology images are captured at both 100× and 400× to show detail

43.2 ± 1.8 to 70.3 ± 13.6 ng/mL, baseline to 180 days post-treatment, $p < 0.003$) and increased weight scores of up to 15% up to 265 days ($p < 0.001$) (Fig. 35.1a). Muscle mass was improved and maintained in the treated dogs, and, as a consequence, their exercise and activity level increased during the 440-day study (26–38%, $p < 0.0005$, and 30–47%, $p < 0.001$, compared with pre-treatment levels) (Fig. 35.1b). In mice with transplanted tumors (23, 25), muscle mass was maintained during the study in treated animals while 50% of controls exhibited histological evidence of muscle atrophy (Fig. 35.2). In a glucocorticoid-induced form of cachexia, localized overexpression of IGF-I in tibialis anterior muscle was induced by injection of IGF-I cDNA, followed by EP, 3 days before starting dexamethasone injections (0.1 mg/kg/day, s.c.) (24). Dexamethasone induced atrophy of the TA muscle, as illustrated by reduction in muscle mass (403 ± 11 vs. 461 ± 19 mg, $p < 0.05$) and fiber cross-sectional area (1759 ± 131 vs. 2517 ± 93 μm², $p < 0.05$), paralleled by a decrease in the muscle IGF-I content (7.2 ± 0.9 vs. 15.7 ± 1.4 ng/g of muscle, $p < 0.001$). As the result of

IGF-I gene transfer, the muscle IGF-I content increased 2-fold (15.8 ± 1.2 vs. 7.2 ± 0.9ng/g of muscle, $p < 0.001$). In addition, the muscle mass (437 ± 8 vs. 403 ± 11mg, $p < 0.01$) and the fiber cross-sectional area (2269 ± 129 vs. $1759 \pm 131 \mu m^2, p < 0.05$) increased in the TA muscle electroporated with IGF-I DNA than in the contralateral muscle electroporated with a control plasmid. These therapeutic options appeared to be safe and capable of stimulating the release of therapeutic levels of hormones in both small and large animals and might be beneficial in patients with cachexia.

Conversely, molecules implicated in the pathogenesis of cachexia, such as tumor necrosis factor alpha (TNF-α), could be studied using this tool (28). A mechanistic understanding of muscle wasting was hampered by limited *in vivo* cytokine models (29). The EP procedure overcame this limitation and resulted in the production of elevated levels of circulating TNF-α. This was followed by body weight loss, upregulation of atrogin-1, and muscle atrophy, including muscles distant from the site of gene transfer, and a significant inhibition of regeneration following muscle injury. New therapies to inhibit/decrease cytokine levels or modulate their action at the cytokine receptor level can now be studied.

2.4. Myasthenia Gravis

Muscle loss can nevertheless have different causes. Myasthenia gravis (MG) is an autoimmune disorder usually caused by autoantibodies to the acetylcholine receptor (AChR) (30). The AChR is clustered and anchored in the postsynaptic membrane of the neuromuscular junction (NMJ) by rapsyn, a cytoplasmic protein. Endogenous rapsyn expression may be an important determinant of AChR loss and neuromuscular transmission failure in the human disease, and upregulation of rapsyn expression could be used therapeutically. To examine a potential therapeutic application of rapsyn upregulation, young rats were induced to develop MG by passive transfer of AChR antibody, mAb 35. *In vivo* EP was then used to overexpress rapsyn unilaterally in one tibialis anterior. In control rats, transfected muscle fibers had extrasynaptic rapsyn aggregates as well as slightly increased rapsyn and AChR concentrations at the NMJ. In MG rats, despite deposits of the membrane attack complex, the rapsyn-overexpressing muscles showed no decrement in the compound muscle action potentials (CMAPs), no loss of AChR, and the majority had normal postsynaptic folds, whereas endplates of untreated muscles showed typical AChR loss and morphological damage. These data suggest that increasing rapsyn expression could be a potential treatment for selected muscles of patients with myasthenia gravis.

2.5. Transcription Factors

Many transcription factors are involved in the regulation of muscle and fat tissue metabolism, differentiation, and repartition. Numerous recent studies are focusing

on using the EP technique to study the role of transcription factors in embryonic and adult life. For instance, TEF-1 transcription factors regulate gene expression in skeletal muscle but are not muscle-specific (31). Instead, they rely on the muscle-specific cofactor vestigial-like 2 (Vgl-2), a protein related to *Drosophila* vestigial; however, the mechanism whereby Vgl-2 regulates TEF-1 factors and the requirement for Vgl-2 for muscle-specific gene expression were not known. Using an antisense morpholino, the expression of Vgl-2 was blocked in chick embryos by direct injection and EP. The results demonstrated that Vgl-2 is required for muscle gene expression, in part by switching DNA binding of TEF-1 factors during muscle differentiation.

Similar experiments were developed (32) to study the proper patterning of somites to give rise to sclerotome, dermomyotome, and myotome that involves the coordination of many different cellular processes, including lineage specification, cell proliferation, cell death, and differentiation, by intercellular signals mediated through Wnt family. To test the proliferative effect of Wnt-3a *in vivo*, studies were done in which Wnt-3a was ectopically expressed in chick neural tubes via EP. Ectopic expression of Wnt-3a *in vivo* resulted in a mediolateral expansion of the dermomyotome and myotome. These results demonstrate that small changes in proliferation, such as those achieved using EP, can dramatically influence patterning and morphogenesis.

2.6. Veterinary Applications—Modulation of Muscle and Fat Mass in Farm Animals

In many studies, we have tested a GHRH-expressing plasmid expressed in skeletal muscle following intramuscular injection enhanced by EP. The expressed GHRH is released in the systemic circulation, stimulates the animal's pituitary to produce and release growth hormone (GH), and, as a result, augments serum IGF-I levels. Young pigs directly injected with as little as 0.1 mg of a GHRH-expressing plasmid had significantly greater weight gain, significant increase in lean body mass, and a decrease in fat mass when compared with controls (33).

We have also demonstrated that when pregnant sows (with body weights between 250 and 400 kg) are injected intramuscularly and electroporated at day 85 of gestation with 1–5 mg of a GHRH-expressing plasmid, the offspring have optimal health and growth characteristics (34) due to both improved intrauterine weight gain and enhanced maternal lactation performance. Thus, the piglets from treated animals were 12% larger at birth and 28% larger at weaning, compared with those from controls ($p < 0.05$), and exhibited a significantly reduced morbidity and mortality (35). Lean body mass was improved at 175 days postnatal and at slaughter: percent lean body mass increased from 53.6 ± 0.66 in controls to 56.1 ± 0.88 in the offspring of plasmid GHRH-treated animals, $p < 0.02$ and fat mass/total body weight was decreased from 10.32 ± 0.37 in controls, to 9.39 ± 0.29 in the offspring of plasmid-GHRH-treated animals, $p < 0.02$. An important finding is that the analysis

of more than 300 plasmid-GHRH-treated animals revealed that expression was maintained for at least 1 year, and the beneficial effects on the offspring occurred for three consecutive pregnancies in the treated animals after a single plasmid administration (expression driven by a synthetic muscle-specific promoter (36), plasmid delivered by intramuscular injection + constant current EP) without redosing (Draghia-Akli, manuscript in preparation).

In a recent study, 32 Holstein heifers (average weights, 547 ± 43 kg) were treated with 2.5 mg of GHRH-expressing plasmid delivered as a single injection, followed by EP, in the third trimester of pregnancy, and 20 heifers were used as controls (37). Animals treated with GHRH-expressing plasmid demonstrated improved immune function and health status, significantly increased body weights at 100 days of milk production, and improved body condition score (BCS). BCS is a reflection of the body fat reserves carried by the animal (38, 39) and the relative proportion between fat and muscle mass. The fat reserves can be used by the cow during early lactation when the animals tend to be in a negative energy balance, exhibiting decreases in body weight (40, 41). The BCS of heifers differed between groups at the time of stress and negative energy balance. Heifers treated with the GHRH-expressing plasmid showed an improvement ($p < 0.0001$) in BCS (Fig. 35.3). During the first 100 days in milk, treated animals lost an average of 3.5 kg (0.06% of total body weight) ($p < 0.02$) while control cows lost on average 26.4 kg (4.6% of body weight). The better BCS correlated with an increase in the serum IGF-I levels: day 100 – day 60 = 22.4 ± 4 ng/mL for GHRH-treated heifers (119.7 ± 6.9 ng/mL at day 100 vs. 97.3 ± 6.6 ng/mL at day 60) versus 8 ± 7.4 ng/mL for controls (99.8 ± 3.9 ng/mL at day 100 vs. 91.8 ± 6.8 ng/mL at day 60) ($p < 0.04$).

In a separate study, young pigs (33) injected with GHRH-expressing plasmid gained proportionally less fat than did controls and were leaner at the end of the study (4.34 ± 0.04 g of fat gained/kg of fat free mass gained per day for injection at

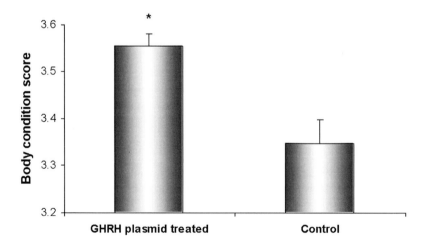

Fig. 35.3 Body condition scores in cows treated with plasmid-expressing GHRH or controls. Body condition scores were significantly improved in plasmid-GHRH-treated animals (*$p < 0.0001$)

birth, 4.4 ± 0.04 g for injection at 7 days, vs. controls 5.63 ± 0.34 g, $p < 0.05$). Body composition studies by DEXA (measuring total body fat, nonbone lean tissue mass, and bone mineral content), K40 potassium (measuring lean body mass), and carcass neutron activation analysis (measuring nitrogen) showed a proportional increase of all internal organs (e.g., heart, lung, liver, spleen, brain, adrenals, stomach, kidney, pancreas, intestine) in plasmid-GHRH-treated animals. Bone mineral density was significantly higher in treated animals.

The beneficial results obtained with plasmid-based GHRH in companion and farm animals demonstrate that by combining adequate plasmid design with EP, one can obtain physiological levels of a transgene product and beneficial physiological effects even in very large mammal models (weighing 500 kg or more), giving hope that soon this and other applications may be translated to a number of human applications.

3. Conclusions

Nonviral gene therapy holds great promise for the treatment of many diseases. Unlike protein or viral vector therapy, it allows long-term and relatively constant delivery of many protein drugs. These vectors are nonimmunogenic and can be expressed in muscle for months; however, nonviral vector expression has been much superior in rodents than in large animals. This problem of scalability has been addressed in various ways, but the application of EP remains one of the simplest and most effective methods of gene transfer. This approach has been successfully employed in preclinical studies to deliver hormones, cytokines, transcription factors, or other molecules. One of the most promising applications, however, is in the systemic delivery of protein drugs such as endocrine hormones for the modulation of muscle and fat mass, including a possible treatment for cachexia. The feasibility of these approaches in large animals, including dogs, pigs, and cows, is a clear indication that these could also be applied to human disease.

References

1. Tan, P.H., Chan, C.L., and George, A.J. (2006) Strategies to improve non-viral vectors—potential applications in clinical transplantation. *Expert. Opin. Biol. Ther.* **6**, 619–630.
2. Wolff, J.A., Ludtke, J.J., Acsadi, G., Williams, P., and Jani, A. (1992) Long-term persistence of plasmid DNA and foreign gene expression in mouse muscle. *Hum. Mol. Genet.* **1**, 363–369.
3. Wolff, J.A., Malone, R.W., Williams, P., et al. (1990) Direct gene transfer into mouse muscle in vivo. *Science.* **247**, 1465–1468.
4. Acsadi, G., Dickson, G., Love, D.R., et al. (1991) Human dystrophin expression in *mdx* mice after intramuscular injection of DNA constructs. *Nature.* **352**, 815–818.
5. Andre, F. and Mir, L.M. (2004) DNA electrotransfer: its principles and an updated review of its therapeutic applications. *Gene Ther.* **11**, S33–S42.

6. Ratanamart, J. and Shaw, J.A. (2006) Plasmid-mediated muscle-targeted gene therapy for circulating therapeutic protein replacement: a tale of the tortoise and the hare? *Curr. Gene Ther.* **6**, 93–110.

7. Prud'homme, G.J., Glinka, Y., Khan, A.S., and Draghia-Akli, R. (2006) Electroporation-enhanced nonviral gene transfer for the prevention or treatment of immunological, endocrine and neoplastic diseases. *Curr. Gene Ther.* **6**, 243–273.

8. Xiang, L., Murai, A., Sugahara, K., Yasui, A., and Muramatsu, T. (2003) Effects of leptin gene expression in mice in vivo by electroporation and hydrodynamics-based gene delivery. *Biochem. Biophys. Res. Commun.* **307**, 440–445.

9. Xiang, L., Murai, A., and Muramatsu, T. (2005) Mutated-leptin gene transfer induces increases in body weight by electroporation and hydrodynamics-based gene delivery in mice. *Int. J. Mol. Med.* **16**, 1015–1020.

10. Wang, X.D., Liu, J., Yang, J.C., Chen, W.Q., and Tang, J.G. (2003) Mice body weight gain is prevented after naked human leptin cDNA transfer into skeletal muscle by electroporation. *J. Gene Med.* **5**, 966–976.

11. Wang, X.D., Tang, J.G., Xie, X.L., et al. (2005) A comprehensive study of optimal conditions for naked plasmid DNA transfer into skeletal muscle by electroporation. *J. Gene Med.* **7**, 1235–1245.

12. Maxwell, M.A., Cleasby, M.E., Harding, A., Stark, A., Cooney, G.J., and Muscat, G.E. (2005) Nur77 regulates lipolysis in skeletal muscle cells. Evidence for cross-talk between the β-adrenergic and an orphan nuclear hormone receptor pathway. *J. Biol. Chem.* **280**, 12573–12584.

13. Cleasby, M.E., Davey, J.R., Reinten, T.A., et al. (2005) Acute bidirectional manipulation of muscle glucose uptake by in vivo electrotransfer of constructs targeting glucose transporter genes. *Diabetes.* **54**, 2702–2711.

14. Roorda, B.D., Hesselink, M.K., Schaart, G., et al. (2005) DGAT1 overexpression in muscle by in vivo DNA electroporation increases intramyocellular lipid content. *J. Lipid Res.* **46**, 230–236.

15. Fujii, N., Boppart, M.D., Dufresne, S.D., Crowley, P.F., Jozsi, A.C., et al. (2004) Overexpression or ablation of JNK in skeletal muscle has no effect on glycogen synthase activity. *Am. J. Physiol. Cell Physiol.* **287**, C200–C208.

16. Rubenstrunk, A., Mahfoudi, A., and Scherman, D. (2004) Delivery of electric pulses for DNA electrotransfer to mouse muscle does not induce the expression of stress related genes. *Cell Biol. Toxicol.* **20**, 25–31.

17. Morley, J.E., Thomas, D.R., and Wilson, M.M. (2006) Cachexia: pathophysiology and clinical relevance. *Am. J. Clin. Nutr.* **83**, 735–743.

18. Argiles, J.M., Busquets, S., Felipe, A., and Lopez-Soriano, F.J. (2006) Muscle wasting in cancer and ageing: cachexia versus sarcopenia. *Adv. Gerontol.* **18**, 39–54.

19. Wanke, C. (2004) Pathogenesis and consequences of HIV-associated wasting. *J. Acquir. Immune. Defic. Syndr.* **37** (Suppl. 5), S277–S279.

20. Springer, J., Filippatos, G., Akashi, Y.J., and Anker, S.D. (2006) Prognosis and therapy approaches of cardiac cachexia. *Curr. Opin. Cardiol.* **21**, 229–233.

21. Nelson, K.A. (2000) The cancer anorexia-cachexia syndrome. *Semin. Oncol.* **27**, 64–68.

22. Draghia-Akli, R., Hahn, K.A., King, G.K., Cummings, K., and Carpenter, R.H. (2002) Effects of plasmid mediated growth hormone releasing hormone in severely debilitated dogs with cancer. *Mol. Ther.* **6**, 830–836.

23. Khan, A.S., Anscombe, I.W., Cummings, K.K., Pope, M.A., Smith, L.C., and Draghia-Akli, R. (2005) Growth hormone releasing hormone plasmid supplementation, a potential treatment for cancer cachexia, does not increase tumor growth in nude mice. *Cancer Gene Ther.* **12**, 54–60.

24. Schakman, O., Gilson, H., de Coninck, V., et al. (2005) Insulin-like growth factor-I gene transfer by electroporation prevents skeletal muscle atrophy in glucocorticoid-treated rats. *Endocrinology.* **146**, 1789–1797.

25. Khan, A.S., Anscombe, I.W., Cummings, K.K., Pope, M.A., Smith, L.C., and Draghia-Akli, R. (2003) Effects of plasmid-mediated growth hormone releasing hormone supplementation on LL-2 adenocarcinoma in mice. *Mol. Ther.* **8**, 459–466.

26. Tone, C.M., Cardoza, D.M., Carpenter, R.H., and Draghia-Akli, R. (2004) Long-term effects of plasmid-mediated growth hormone releasing hormone in dogs. *Cancer Gene Ther.* **11**, 389–396.

27. Morrison, W. (2002) Clinical evaluation of cancer patients. In: Morrison, W. (ed.). *Cancer in dogs and cats: medical and surgical management*, 2nd edn. Teton New Media, West Lafayette, IN, pp. 59–67.

28. Coletti, D., Moresi, V., Adamo, S., Molinaro, M., and Sassoon, D. (2005) Tumor necrosis factor-alpha gene transfer induces cachexia and inhibits muscle regeneration. *Genesis.* **43**, 120–128.

29. Figueras, M., Busquets, S., Carbo, N., Almendro, V., Argiles, J.M., and Lopez-Soriano, F.J. (2005) Cancer cachexia results in an increase in TNF-α receptor gene expression in both skeletal muscle and adipose tissue. *Int. J. Oncol.* **27**, 855–860.

30. Losen, M., Stassen, M.H., Martinez-Martinez, P., et al. (2005) Increased expression of rapsyn in muscles prevents acetylcholine receptor loss in experimental autoimmune myasthenia gravis. *Brain.* **128**, 2327–2337.

31. Chen, H.H., Maeda, T., Mullett, S.J., and Stewart, A.F. (2004) Transcription cofactor Vgl-2 is required for skeletal muscle differentiation. *Genesis.* **39**, 273–279.

32. Galli, L.M., Willert, K., Nusse, R., et al. (2004) A proliferative role for Wnt-3a in chick somites. *Dev. Biol.* **269**, 489–504.

33. Draghia-Akli, R., Ellis, K.M., Hill, L.A., Malone, P.B., and Fiorotto, M.L. (2003) High-efficiency growth hormone releasing hormone plasmid vector administration into skeletal muscle mediated by electroporation in pigs. *FASEB J.* **17**, 526–528.

34. Khan, A.S., Fiorotto, M.L., Cummings, K.K., Pope, M.A., Brown, P.A., and Draghia-Akli, R. (2003) Maternal GHRH plasmid administration changes pituitary cell lineage and improves progeny growth of pigs. *Am. J. Physiol. Endocrinol. Metab.* **285**, E224–E231.

35. Draghia-Akli, R. and Fiorotto, M.L. (2004) A new plasmid-mediated approach to supplement somatotropin production in pigs. *J. Anim. Sci.* **82**, E264–E269.

36. Li, X., Eastman, E.M., Schwartz, R.J., and Draghia-Akli, R. (1999) Synthetic muscle promoters: activities exceeding naturally occurring regulatory sequences. *Nat. Biotechnol.* **17**, 241–245.

37. Brown, P.A., Davis, W.C., and Draghia-Akli, R. (2004) Immune enhancing effects of growth hormone releasing hormone delivered by plasmid injection and electroporation. *Mol. Ther.* **10**, 644–651.

38. Apple, J.K., Davis, J.C., Stephenson, J., Hankins, J.E., Davis, J.R., and Beaty, S.L. (1999) Influence of body condition score on carcass characteristics and subprimal yield from cull beef cows. *J. Anim. Sci.* **77**, 2660–2669.

39. Komaragiri, M.V., Casper, D.P., and Erdman, R.A. (1998) Factors affecting body tissue mobilization in early lactation dairy cows. 2. Effect of dietary fat on mobilization of body fat and protein. *J. Dairy Sci.* **81**, 169–175.

40. Dechow, C.D., Rogers, G.W., and Clay, J.S. (2002) Heritability and correlations among body condition score loss, body condition score, production and reproductive performance. *J. Dairy Sci.* **85**, 3062–3070.

41. DeRouen, S.M., Franke, D.E., Morrison, D.G., et al. (1994) Prepartum body condition and weight influences on reproductive performance of first-calf beef cows. *J. Anim. Sci.* **72**, 1119–1125.

Part VI
Applications of Electroporation for DNA Vaccination

Chapter 36
DNA Vaccination for Prostate Cancer

Anna-Karin Roos, Alan King, and Pavel Pisa

Abstract DNA-based cancer vaccines have been used successfully in mice to induce cytotoxic T lymphocytes (CTLs) specific for prostate antigens. Translation of a prostate-specific antigen (PSA) DNA vaccine into a phase I clinical trial demonstrated that PSA-specific immune responses could be induced but at a significantly lower level compared with those in mice. To enhance the efficacy of DNA vaccination against prostate cancer, we have explored and optimized intradermal electroporation as an effective way of delivering a PSA DNA vaccine. The results demonstrated that intradermal DNA vaccination using low amounts of DNA, followed by two sets of electrical pulses of different length and voltage, effectively induced PSA-specific T cells. Here we describe in detail how to perform intradermal DNA electroporation to induce high gene expression in skin and, more important, how to induce and analyze PSA-specific T cell responses.

Keywords: prostate cancer, electroporation, skin, DNA vaccine, interferon-γ, cytotoxic T lymphocyte, immunotherapy

1. Introduction

Prostate cancer is the most common cancer and the fourth leading cause of cancer-related death in men in the developed countries worldwide (1). No treatment options are available once the tumor becomes hormone refractory, and, therefore, new treatments are urgently needed. We have previously shown that DNA vaccination with prostate-specific antigen (PSA) in mice (2, 3) and humans (4) can evoke PSA-specific immune responses; however, DNA vaccination in humans has not been as efficient as in mice. One reason for the lower efficacy of DNA vaccines in large animals could be low uptake of DNA, but with modern in vivo electroporation techniques, transgene products can be expressed at levels that have only been achieved with viral vectors (5). Moreover, in vivo electroporation has been shown to induce low levels of inflammation, which might further enhance the immune response. Most important, electroporative delivery has been shown to increase the efficacy of DNA vaccines in large animals (6, 7).

S. Li (ed.), *Electroporation Protocols: Preclinical and Clinical Gene Medicine.*   463
From *Methods in Molecular Biology, Vol. 423.*
© Humana Press 2008

Table 36.1 Features of electroporation-enhanced skin and muscle DNA immunizations

	Skin	Muscle
Antibody induction	Yes	Yes
CTL induction	Yes	Yes
Access to tissue	Easy	Difficult
Length of antigen expression	Moderate	Long
Local APC density	High	Low
Adverse effects	Low to moderate	High

In vivo electroporation can be performed in a wide variety of tissues (8), but, for DNA vaccination against infectious diseases and cancer, muscle is the most commonly used target tissue. In skin and in muscle, the predominant production of antigen responsible for induction of the immune response is by nonantigen presenting cells, as demonstrated using cell specific promoters (9, 10). However, in the skin, there is a higher density of antigen presenting cells (APCs), and direct transfection of APCs is more likely in the skin than in muscle. Additionally, more APCs are available for cross-priming in the skin, which makes skin an even more desirable target tissue for DNA vaccination than muscle. The APCs in the skin, Langerhans cells and dermal dendritic cells, can, after antigen uptake, migrate to lymph nodes, where efficient antigen presentation to T cells occurs (11). Furthermore, skin is easily accessible to electrodes and electroporative DNA delivery, and electroporation is less painful in skin than in muscle. A comparison of electroporation-enhanced skin and muscle DNA immunization is outlined in Table 36.1.

We recently investigated a panel of different in vivo electroporation conditions for the delivery of a PSA DNA vaccine to skin (12). The best pulsing condition (2 pulses of 1,125 V/cm for 50 μs and 8 pulses of 275 V/cm for 10 ms) increased gene expression 1,000 fold compared with intradermal DNA delivery without electroporation. The same electroporation condition also induced higher levels of PSA-specific CD8$^+$ T cells than did nonelectroporative delivery of the same amount of DNA into skin and muscle (12). These results demonstrate that DNA transfer by intradermal electroporation is an effective vaccination strategy for the induction of CD8$^+$ T cells.

In this chapter we provide the details and technical material needed to perform intradermal PSA DNA electroporation. We describe how to analyze transgene expression and the induction/function of antigen-specific T cells in spleen after vaccination. We have also outlined how the kinetics of the antigen-specific T cell response can be monitored in peripheral blood.

2. Materials

2.1. Plasmid DNA

1. Plasmids: The pVax1 vector (Invitrogen, Carlsbad, CA) was used as the backbone for both plasmids in the study. For immunological studies, the plasmid pVax-PSA (3,977 bp) was constructed by inserting the gene coding

for the full-length human PSA protein (obtained from Dr. Tim Ratliff, Washington University, St. Louis, MO). For gene expression, pVax-luc (4,663 bp) was constructed by inserting the complementary DNA (cDNA) for firefly luciferase from the pGL2-basic vector (Promega, Madison, WI) into vector pVax1.

2. Plasmid production and purification: Endotoxin-Free Plasmid Purification Kit.
3. Plasmid solvent: phosphate-buffered saline (PBS) without Mg^{2+} or Ca^{2+}.

2.2. *Intradermal DNA Injection and Electroporation*

1. Animals: female C57Bl/6 (H-2^b) or Balb/c (H-2^d) mice (6–8 week old).
2. Anesthesia equipment: oxygen cylinder, isoflurane vaporizer, and breathing mask (Univentor 400, Zejtun, Malta).
3. Anesthesia reagent: isoflurane.
4. For DNA injections: 29-GA insulin-grade syringes.
5. Electroporation device: PA-4000S – Advanced PulseAgile® Rectangular Wave Electroporation System (Cyto Pulse Sciences, Glen Burnie, MD).
6. Electrodes: needle-array electrodes (NE-4-4) with two parallel rows of four 2-mm pins (1.5 × 4-mm gaps) (Cyto Pulse Sciences) (see Fig. 36.1.).
7. Oscilloscope: PCS64i digital oscilloscope (Velleman Components N.V., Belgium).

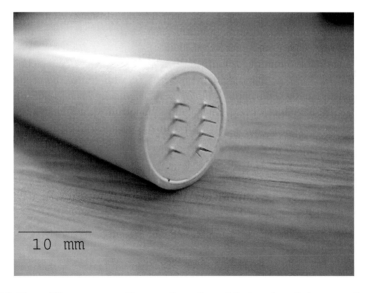

10 mm

Fig. 36.1 Four-millimeter-gap needle-array electrode used for intradermal electroporation

2.3. Luciferase Assay

1. Homogenization: mortar, pestle, and scalpels.
2. Luciferase detection: Enhanced Luciferase Assay kit, including lysis buffer (BD Biosciences, PharMingen, San Jose, CA).
3. Luminometer: Wallac Victor Multilabel Counter (PerkinElmer, Life Sciences, Upplands Väsby, Sweden). White enhanced 96-well microplates (BD Biosciences).
4. Purified firefly luciferase protein (BD Biosciences).

2.4. Cell Preparation from Murine Spleen and Peripheral Blood

1. Citrate-phosphate-dextrose solution with adenine (CPD-A) (Sigma-Aldrich, Stockholm, Sweden) anticoagulant for collection of peripheral blood.
2. Ammonium chloride lysing reagent for erythrocyte lysis.
3. 70-μm cell strainers to obtain single cell suspensions and remove fat.
4. Handling media: Dulbecco's modified Eagle's medium (DMEM) supplemented with 10 mM HEPES, 5×10^{-5} M 2-mercaptoethanol, 25 μg gentamycin/mL, and 1% fetal calf serum (FCS).

2.5. In Vitro T Cell Restimulation Cultures

1. Complete medium: DMEM handling medium +2 mM L-glutamine, 1% nonessential amino acids, and 5% FCS.
2. Human recombinant IL-2 (Proleukin, Chiron Corporation, Emeryville, CA).
3. For restimulation of PSA-specific T cells: synthetic peptide psa65-73 (HCIRNKSVI).

2.6. Intracellular Cytokine Staining Assay

1. Synthetic peptides: psa65-73 (HCIRNKSVI) and LCMV-derived peptide GP33 (KAVYNFATC$_{33-41}$).
2. Positive control for stainings: PMA and ionomycin.
3. Intracellular staining kit: CytoFix/CytoPerm Plus kit containing GolgiPlug reagent and PermWash solution (BD Biosciences).
4. Antibodies: rat IgG2a-FITC-labeled anti-mouse CD8a (0.5 mg/mL) and rat IgG1-PE-labeled anti-mouse IFN-γ (0.2 mg/mL). Purified rat IgG for blocking nonspecific binding.
5. Staining buffer: 0.5% BSA and 0.05% sodium azide in PBS (*see* **Note 1**).

2.7. Cr-Release Assay

1. Cell lines: mouse EL4 (H-2^b) cell line and EL4/PSA (clonally derived cell line expressing PSA) (*see* **Note 2**).
2. Culture medium for EL4 and EL4/PSA: DMEM supplemented with 2 mM L-glutamine, 10 mM HEPES, 25 µg gentamycin/mL, and 10% FCS. Selective antibiotic G418 (500 µg/mL; Invitrogen) should be added to cultures with EL4/PSA.
3. Na$_2$[^{51}Cr]O$_4$ to label target cells.
4. Triton X-100 for maximum release of ^{51}Cr from target cells.
5. Automatic gamma counter.

3. Methods

Here we describe a detailed vaccination protocol using a previously optimized electroporation condition (12) for induction of PSA-specific CD8$^+$ T cells. Evaluation of gene expression after vaccination as well as evaluation of the presence and function of PSA-specific T cells are described. Balb/c mice were used for gene expression experiments, because their light skin makes marking of the DNA injection site easy and precise. C57Bl/6 mice were used for induction of immune responses, since an immunodominant H-2D^b-restricted epitope of human PSA was recently identified in this strain of mice (13).

3.1. Intradermal DNA Vaccination and Electroporation

1. Dilute the plasmid DNA to 0.5 µg/µL (*see* **Note 3**) in PBS (*see* **Note 4**).
2. Remove all the hair on the lower back/flanks of mice with an electric razor.
3. Anesthetize one mouse at a time in an induction chamber filled with 4% isoflurane and 96% oxygen until the mouse has lost its frightening reflex. Move the mouse to a face mask and continue to deliver 2.5% isoflurane in oxygen (*see* **Note 5**).
4. Use an insulin-grade syringe and slowly inject 20 µL DNA into the skin on both sides of the spine on the lower back of the mouse (*see* **Note 6**).
5. Immediately insert a 4-mm gap needle-array electrode (*see* **Note 7**) over one of the intradermal injection sites (*see* **Note 8**).
6. Administer 2 pulses of 450 V for 50 µs and 8 pulses of 110 V for 10 ms. Set the interval between the two 450-V pulses to 300 ms, the interval between the two 450-V pulses and the eight 110-V pulses to 500 ms, and the intervals between the eight 110-V pulses to 300 ms (*see* **Note 9**).
7. Remove the electrode carefully and immediately insert it over the second DNA injection site.

8. Repeat administration of pulses.
9. If the electroporation was performed to evaluate gene expression, mark the margins of the electroporated skin area with a permanent-marking pen.

3.2. Collection of Skin, Peripheral Blood, and Spleen after Vaccination

1. For measurement of gene expression after electroporation, kill the mouse by cervical dislocation 24 h after electroporation (*see* **Note 10**). Use sterile scissors and forceps to remove the electroporated skin area. Keep skin in a tube on ice and transfer as fast as possible to −80°C.
2. To follow the kinetics of the PSA-specific CD8$^+$ T cell response, collect peripheral blood between days 9–16 after vaccination. Restrain the mouse and place it under a heating lamp. Carefully make a cut across the tail vein with a razor blade. Let the blood drip into an eppendorf tube with 100 µL CPD-A anticoagulant until the total volume is around 200 µL. Tap the tube to mix anticoagulant and blood and keep at room temperature (RT).
3. To measure the PSA-specific CD8$^+$ T cell response in the spleen, kill the mouse by cervical dislocation (days 12–17) and spray the whole mouse with 70% ethanol. Use sterile scissors and forceps to remove the spleen and transfer it to a tube with 2 mL handling media. Keep spleens at RT if processed within 30 min; otherwise keep them at 4°C.

3.3. Analysis of Skin for Reporter Gene Expression

1. Keep all materials on ice and pre-chill a tabletop centrifuge to 4°C.
2. Place the skin in a mortar on ice with 100 µL of ice-cold lysis buffer and cut into small pieces with scalpels (*see* **Note 11**). Use a pestle to thoroughly homogenize the sample.
3. Transfer the homogenized skin to a tube and rinse the mortar with 400 µL ice-cold lysis buffer. Pool into the same tube. Store the sample on ice until all samples are processed.
4. Vortex all samples at 14,000 rpm for 20 min at RT. Centrifuge (3,000 × g, 4 min, and 4°C) and collect supernatants. Centrifuge supernatants again (10,000 × g, 3 min, and 4°C).
5. Transfer supernatants to new tubes and measure or store at −80°C (*see* **Note 12**).
6. Determine the bioluminescence by adding a 50-µL aliquot of each sample to a 96-well luminescence plate together with the substrates from a luciferase assay kit and measure for 10 s on a luminometer.
7. Make a standard curve using recombinant firefly luciferase protein to determine the luciferase content (relative light units) in the skin samples. Determine background luminescence (skin injected intradermally with empty vector, pVax, and electroporated) and subtract from all samples.

3.4. Cell Preparation from Murine Spleen and Peripheral Blood

1. To prepare lymphocytes from peripheral blood, transfer blood to a 15 mL tube and add 1 mL ammonium chloride lysing reagent. Incubate at RT for 5 min. Carefully add 10 mL handling media and centrifuge (all centrifugation steps are at 200 × g for 6 min).
2. Resuspend cells in 10 mL complete media and centrifuge.
3. Resuspend cell pellet in 200 µL complete media. Use 100 µL cell suspension per peptide stimulation.
4. For preparation of splenocytes, obtain single-cell suspensions by mincing spleens using the back of a 1-mL syringe and pass cells through a 70-µm cell strainer.
5. Add 10 mL handling media, resuspend, and centrifuge.
6. Loosen the pellet and carefully add 2 mL ammonium chloride lysing reagent.
7. Incubate for 3 min at RT, add 12 mL handling media, and centrifuge.
8. Resuspend pellet in 10 mL handling media, and pass cell suspension through a cell strainer again to remove clumps, and centrifuge.
9. Resuspend cells to 10×10^6 cells/mL in complete media and use 100 µL cell suspension per peptide stimulation.

3.5. In Vitro Restimulation of PSA-Specific T Cells

1. Per well in a 12-well plate add 10×10^6 splenocytes, 1 nM of the synthetic peptide psa65–73, and 20 IU/mL human recombinant IL-2 in a total of 5 mL complete media.
2. To do an IFN-γ secretion assay, set up 1 well of restimulation, or, to do a Cr-release assay, set up 3 wells of restimulation per spleen.
3. Incubate in a humidified incubator at 37°C for 5 days.

3.6. Analysis of IFN-γ Secretion by PSA-Specific T Cells

1. Collect the cells from one restimulation well and centrifuge ($200 \times g$ for 5 min). Remove supernatant completely and add 300 µL complete media.
2. Add 100 µL cell suspension ($\sim 1 \times 10^6$ cells) from restimulation cultures, fresh splenocytes, or peripheral blood to each of two wells in a 96-well U-bottom plate.
3. Add 100 µL complete media with 200 nM peptide psa65–73 to one well containing splenocytes and 100 µL complete media with 200 nM peptide GP33 (negative control) to the other well containing splenocytes from the same mouse/restimulation culture.
4. Prepare a positive control by adding 100 µL complete media containing PMA (50 ng/mL) and ionomycin (500 ng/mL) to 100 µL cell suspension from any mouse/restimulation culture.

5. Incubate the plate at 37°C for 2.5 h. Carefully and without disturbing cells, add 50 μL complete media containing 0.25 μL GolgiPlug reagent and continue incubating cells at 37°C for another 2.5 h.

6. Centrifuge the plate and remove supernatant by turning the plate upside down on a tissue. All plate centrifugation steps are at RCF $787 \times g$ for 1 min.

7. For staining of the cells with anti-CD8a and anti-IFN-γ antibodies, use the CytoFix/CytoPerm Plus kit and follow the manufacturer's instructions.

8. For preparing antibody mastermixes, add 1/6 μL anti-CD8a antibody +100 μg/mL rat IgG in total 50 μL staining buffer per sample and 1/5 μL anti-IFN-γ antibody +100 μg/mL rat IgG in total 50 μL 1× PermWash buffer per sample.

9. After IFN-γ staining, resuspend the cells in 200 μL staining buffer and transfer cell suspension to 5 mL Fluorescence Activated Cell Sorting (FACS) tubes containing 800 μL staining buffer (*see* **Note 13**).

10. Do FACS analysis and set compensation using the PMA/ionomycin sample.

11. Determine the number of PSA-specific CD8$^+$ T cells by subtracting the number of IFN-γ-producing CD8$^+$ T cells in the well stimulated with the GP33 peptide from the number of IFN-γ-producing CD8$^+$ T cells in the well stimulated with peptide psa65–73.

3.7. Determination of CTL Activity by Cr-Release Assay

1. Collect target cells (EL4 and EL4/PSA) and centrifuge at RCF $200 \times g$ for 6 min (same for all centrifugation steps). Count cells, wash in culture media, and centrifuge. Resuspend 1.5×10^6 cells in culture media and centrifuge.

2. Remove media completely and resuspend cell pellet in 90 μL culture media.

3. Add 10 μL Cr^{51}, carefully tap tube to mix, and incubate at 37°C for 1.5 h.

4. Wash target cells thrice in culture media and adjust the cells to 1×10^5/mL in complete media for splenocytes.

5. Collect splenocytes (effectors) from restimulation cultures (pool 3 wells from same spleen) and centrifuge. Adjust the cells to 1×10^7/mL in complete media.

6. Mix effector and target cells to a total volume of 200 μL in a 96-well U-bottom plate at effector to target ratios 50:1, 25:1, and 12.5:1. Set up each ratio with both targets in triplicate. Keep target numbers constant at 1×10^4 cells per well.

7. Prepare 3 wells with 1×10^4 targets + 100 μL complete media (spontaneous release) and 3 wells with 1×10^4 targets + 100 μL 5% Triton X-100 (maximum release), for both EL4 and EL4/PSA. Centrifuge plate at $200 \times g$ for 3 min and incubate at 37°C for 4.5 h.

8. Centrifuge plate at $200 \times g$ for 3 min and carefully collect 100 μL supernatant without touching the cell pellet. Transfer supernatants to cytotox tubes and measure radioactivity with an automatic gamma counter.

9. Calculate specific lysis for both targets with the following formula: (experimental release – spontaneous release)/(maximum release – spontaneous release).

Spontaneous release should be less than 10% of maximum release. To determine the percent of PSA-specific lysis, the percent lysis of EL4 cells should be subtracted from the percent lysis of EL4/PSA cells.

4. Notes

1. Filter the staining buffer (0.2-μm pore membrane) and store at 4°C up to 6 months.
2. EL4/PSA was constructed by stable transfection of EL4 with pcDNA3-PSA. The culture media from the transfected cell line EL4/PSA should be routinely checked for PSA.
3. If higher DNA doses are required for other plasmids/antigens/animal species, DNA concentrations up to 9 μg/μL can be delivered intradermally in combination with in vivo electroporation. For pVax-PSA, this increase does not cause adverse effects or lower the immune responses in mice (unpublished results).
4. Plasmids are most often diluted in PBS, sterile saline, or water. Water should be avoided since it is more painful than other solvents when injected into the skin and also results in a lower transfection of DNA into cells (12).
5. For humans, local anesthesia such as 1% lidocaine has been used for the delivery of high-voltage pulses to cutaneous and subcutaneous tumors (14).
6. It is crucial that the intradermal injection is correctly administered in the skin and not subcutaneously. Intradermal injections are given most easily with two people, one holding and stretching the skin of the mouse and the other injecting the DNA. The skin should rise and a clear bubble should be visible.
7. Different designs of electrodes have been used for intradermal DNA delivery, but the most common types are needle (invasive) or caliper/plate (noninvasive) variants. In humans, it was reported that high-voltage delivery using a needle electrode caused less discomfort than a plate electrode, as it did not cause burning of the skin (15).
8. Place the electrode so that the needles are on the outside of the DNA injection site (not penetrating the bubble). Wiggle the electrode carefully to make sure that all the needles of the electrode are inserted into the skin.
9. These specific voltages are to be used with a 4-mm-gap needle-array electrode. Before turning on the voltage, an oscilloscope can be activated to record the current during electroporation, and later be used to calculate skin resistance.
10. Gene expression in skin is usually measured between 24 and 48 h after DNA injection site when transgene expression is maximal.
11. Work quickly, as skin is easiest to cut in thin slices/pieces when frozen. It is important to keep materials and samples cold to avoid protein degradation.
12. The supernatants can be frozen at −80°C and analyzed later, but samples slowly lose activity every time they are thawed and refrozen.
13. Cells can be left overnight at 4°C in staining buffer at this stage.

References

1. Ferlay, J., Bray, F., Pisani, P., and Parkin, D.M. (2004) *GLOBOCAN 2002: cancer incidence, mortality and prevalence worldwide*. IARC CancerBase. IARC Press, Lyon.
2. Pavlenko, M., Roos, A.K., Leder, C., et al. (2004) Comparison of PSA-specific CD8+ CTL responses and antitumor immunity generated by plasmid DNA vaccines encoding PSA-HSP chimeric proteins. *Cancer Immunol. Immunother.* **53**, 1085–1092.

3. Roos, A.K., Pavlenko, M., Charo, J., Egevad, L., and Pisa, P. (2005) Induction of PSA-specific CTLs and anti-tumor immunity by a genetic prostate cancer vaccine. *Prostate.* **62**, 217–223.

4. Pavlenko, M., Roos, A.K., Lundqvist, A., et al. (2004) A phase I trial of DNA vaccination with a plasmid expressing prostate-specific antigen in patients with hormone-refractory prostate cancer. *Br. J. Cancer.* **91**, 688–694.

5. Draghia-Akli, R., Khan, A.S., Pope, M.A., and Brown, P.A. (2005) Innovative electroporation for therapeutic and vaccination applications. *Gene Ther. Mol. Biol.* **9**, 329–338.

6. Otten, G., Schaefer, M., Doe, B., et al. (2004) Enhancement of DNA vaccine potency in rhesus macaques by electroporation. *Vaccine.* **22**, 2489–2493.

7. Babiuk, S., Baca-Estrada, M.E., Foldvari, M., et al. (2002) Electroporation improves the efficacy of DNA vaccines in large animals. *Vaccine.* **20**, 3399–3408.

8. Prud'homme, G.J., Glinka, Y., Khan, A.S., and Draghia-Akli, R. (2006) Electroporation-enhanced nonviral gene transfer for the prevention or treatment of immunological, endocrine and neoplastic diseases. *Curr. Gene Ther.* **6**, 243–273.

9. Corr, M., von Damm, A., Lee, D.J., and Tighe, H. (1999) In vivo priming by DNA injection occurs predominantly by antigen transfer. *J. Immunol.* **163**, 4721–4727.

10. Cho, J.H., Youn, J.W., and Sung, Y.C. (2001) Cross-priming as a predominant mechanism for inducing CD8(+) T cell responses in gene gun DNA immunization. *J. Immunol.* **167**, 5549–5557.

11. Peachman, K.K., Rao, M., and Alving, C.R. (2003) Immunization with DNA through the skin. *Methods.* **31**, 232–242.

12. Roos, A.K., Moreno, S., Leder, C., Pavlenko, M., King, A., and Pisa, P. (2006) Enhancement of cellular immune response to a prostate cancer DNA vaccine by intradermal electroporation. *Mol. Ther.* **13**, 320–327.

13. Pavlenko, M., Leder, C., Roos, A.K., Levitsky, V., and Pisa, P. (2005) Identification of an immunodominant H-2D(b)-restricted CTL epitope of human PSA. *Prostate.* **64**, 50–59.

14. Heller, R., Jaroszeski, M., Atkin, A., et al. (1996) In vivo gene electroinjection and expression in rat liver. *FEBS Lett.* **389**, 225–228.

15. Heller, R., Jaroszeski, M.J., Reintgen, D.S., et al. (1998) Treatment of cutaneous and subcutaneous tumors with electrochemotherapy using intralesional bleomycin. *Cancer.* **83**, 148–157.

Chapter 37
HER2/neu DNA Vaccination for Breast Tumors

Arianna Smorlesi, Francesca Papalini, Sara Pierpaoli, and Mauro Provinciali

Abstract Several studies of DNA vaccination against HER2/neu showed the effectiveness of immunization protocols in models of transplantable or spontaneous tumors. The DNA delivery system plays a crucial role in the success of DNA vaccination. In particular, our studies of DNA vaccination against HER2/neu tumor antigen showed that intramuscular injection of the vaccine followed by electroporation elicits an optimal protection against the development of spontaneous HER2/*neu*-tumors occurring in transgenic mice.

Keywords: DNA vaccination, electroporation, HER2/neu, breast tumor, immune response

1. Introduction

Experimental evidence of the potential of genetic vaccination as an immunological strategy against cancer has been provided in recent years. Direct injection of plasmid DNA encoding a selected antigen into mouse muscle or skin results in the expression of the gene product and can elicit a specific immune response (1). This DNA vaccine model represents a promising, practical, and effective way to elicit immune responses against HER2/neu [2–4], although genetic vaccines, when translated from murine models to large animal models and clinical human use, often lose their potency (5). Studies on genetic vaccination against HER2/*neu* are focused on improving DNA vaccines to induce strong effector and memory responses to this oncoprotein through the use of adjuvant molecules (6–9).

Methods that can significantly improve plasmid DNA transduction will greatly enhance the efficiency of DNA vaccines (10). Electroporation, which has been widely used to transport DNA and other molecules into living cells in vitro (11), has been demonstrated to be safe and efficacious in vivo by enhancing the local efficacy of chemotherapeutic agents and by improving the local transfection efficiency of DNA plasmid over plasmid injection alone.

S. Li (ed.), *Electroporation Protocols: Preclinical and Clinical Gene Medicine.*   473
From *Methods in Molecular Biology, Vol. 423.*
© Humana Press 2008

In a recent study (12), we compared the efficacy of different procedures of DNA delivery showing that intramuscular injection of the vaccine followed by electroporation results in the best outcome of the vaccine against a spontaneous HER2/neu-expressing tumor. Others demonstrated in a different transgenic model the efficacy of electroporation compared with intramuscular injection alone in the administration of a therapeutic DNA vaccine against HER2/neu antigen (13).

Here we describe a study where immunization with a plasmid DNA encoding HER2/neu by intramuscular injection followed by electroporation to prevent spontaneous carcinogenesis in HER2/neu transgenic mice was compared with immunization by intramuscular injection alone. Tumor appearance and growth were then monitored and humoral and cellular immunity induced by the vaccine were analyzed.

2. Materials

2.1. Plasmid DNA

1. pCMV-ECDTM plasmid, encoding extracellular and transmembrane region of HER2/neu antigen under the control of the CMV early promoter/enhancer was used to immunize mice.
2. pCMV-βgal plasmid encoding unrelated antigen (β-galactosidase) under the control of the CMV early promoter/enhancer was used to treat control mice.

2.2. Animals

FVB/*neu*-T female transgenic mice (14), expressing the activated rat HER2/*neu* oncogene (HER2/*neu* oncomice with H-2q FVB/n background) were purchased from Charles River (Hollister, CA) and maintained under specific pathogen-free conditions in our animal facility. Mice were housed in plastic non-galvanised cages (Four to six mice per cage) maintained at a constant temperature (20°C ± 1°C) and humidity (50% ± 5%) on a 12 h light/12 h dark cycle and fed with standard pellet food and tap water.

2.3. Immunization

1. Two months-old FVB/neu-T mice.
2. Insulin syringes.
3. Isoflurane.

4. Anaesthetic apparatus with oxygen cylinder and isoflurane vaporizer.
5. Electroporation apparatus (ECM 830 field generator, Genetronix, San Diego, CA) and caliper electrode (Genetronix).

2.4. Cell Cultures

1. N202.1A tumor cells: a cloned cell line overexpressing rat HER2/*neu* oncogene that was established in vitro from a lobular carcinoma spontaneously arising in FVB*neu*-N mice (15).
2. Medium for N202.1A cells culture: DMEM supplemented with penicillin (100 U/mL) and streptomycin (100 μg/mL) and supplemented with 20% FBS.
3. Trypsin-EDTA solution for harvesting adherent cells from cell culture flasks.
4. Phosphate-buffered saline (PBS).
5. Mitomycin C.
6. Biosafety cabinet.

2.5. Analysis of Antibody Response

1. N202.1A cells.
2. PBS with 2% BSA and 0.05% sodium azide.
3. Antibodies: fluorescein (FITC)-conjugated rabbit anti-mouse Ig, FITC-anti-mouse IgG1, FITC-anti-mouse IgG3, FITC-anti-mouse IgG2a, and FITC-anti-mouse IgG2b.
4. Isoton II solution (Coulter, Hialeah, FL).
5. Centrifuge.
6. 4°C refrigerator.
7. Coulter XL flow cytometer (Coulter, Hialeah, FL).

2.6. Analysis of Cellular Response

2.6.1. In Vitro Stimulation of Spleen Cells

1. Ca^{2+} and Mg^{2+}-free PBS solution.
2. CO_2 camera, sterile scissors, and tweezers.
3. Sterile cell culture plates, 60 mesh sieves, tweezers, and 10 mL syringes for teasing spleen from experimental mice.
4. RPMI 1640 medium supplemented with penicillin (100 U/mL), streptomycin (100 μg/mL), 10% FBS, and interleukin-2 (IL-2, 50 U/mL, Chiron Corporation, Emeryville, CA).
5. Mitomycin-treated N2021.A cells.

6. Cell culture plates or flasks for co-culture of splenocytes with N2021.A cells.
7. Optical microscope and Burker camera for counting cells.
8. Centrifuge.
9. Biosafety sterile cabinet.
10. Humidified, 5% CO_2, 37°C incubator.

2.6.2. Analysis of Cellular Response (Cytokine Release)

1. Golgi Plug (BD Pharmingen).
2. Buffers for intracellular staining of cytokines: 2% formalin; PBS containing 5% FCS and 0.01% NaN_3; PBS containing 5% FCS and 0.05% saponin.
3. Antibodies: FITC-conjugated anti-CD4 and anti-CD8 monoclonal antibodies; PE-conjugated anti-IL4, anti-IL10, and anti-IFNγ monoclonal antibodies.
4. Cytometer tubes.
5. 4°C refrigerator.
6. Centrifuge.
7. Coulter XL flow cytometer.

2.6.3. Analysis of Cellular Response (Cytotoxic Activity)

1. PBS and PBS with 1% BSA for washing cells.
2. Burker camera, Turk solution, and optical microscope for counting cells.
3. N2021.A cells labelled with carboxyfluorescein diacetate (75 μg/mL c'FDA prepared by diluting a stock solution of 20 mg/mL c'FDA in acetone in PBS, stored at −20°C).
4. RPMI supplemented with penicillin (100 U/mL) and streptomycin (100 μg/mL), 5% FBS.
5. Cell lysis buffer: 1% Triton in 0.05 M borate buffer, pH 9.2.
6. 96-well round microtiter plates.
7. Humidified, 5% CO_2, 37°C incubator.
8. Centrifuge.
9. 1420 VICTOR[2] multilabel counter (Wallac, Turku, Finland).

3. Methods

To evaluate whether electroporation, compared with intramuscular injection alone, improves the efficiency of a HER2/neu DNA vaccine against the development of spontaneous HER2/neu-positive mammary carcinoma, HER2/neu transgenic mice were immunized with pCMV-ECDTM plasmid given by intramuscular injection

followed or not by electroporation; mice were then monitored to evaluate the effect of the vaccine on the spontaneous appearance and growth of tumors and to analyze the immune response elicited against HER2/neu antigen.

3.1. Preparation of DNA Vaccine

1. Use Giga plasmid DNA purification kit (Qiagen, MI, Italy) to prepare large amount of pCMV-ECDTM and pCMV-βgal plasmids according to the manufacturer's instructions.
2. Resuspend DNA in 150 μM NaCl at the concentration of 1 μg/μl; 6 mg/mL poly-L-glutamic acid was added to DNA solution for administration by electroporation (*see* **Note 1**).

3.2. Immunization of Mice

Two-month old HER2/neu transgenic mice were vaccinated with pCMV-ECDTM (experimental group) or with pCMV vector (control group) through intramuscular injection with or without electroporation. Three administrations of the vaccine were carried out at 8, 12, and 16 weeks of age in accordance with previously optimized protocols.

3.2.1. Intramuscular Injection Without Electroporation

1. Anaesthetize mouse by isoflurane inhalation (*see* **Note 2**).
2. Wet mouse leg with 70% alcohol to visualize quadricep muscles.
3. Make a cut on the skin to expose the muscle (*see* **Note 3**).
4. While holding the leg with pliers, inject 100 μL DNA dissolved in saline solution (0.1 μg/μL) using an insulin syringe (*see* **Note 4**).
5. Sew up the cut with surgical thread or surgical clamp.

3.2.2. Intramuscular Injection with Electroporation

1. Anaesthetize mouse by isoflurane inhalation.
2. Wet mouse legs with 70% alcohol to visualize tibialis muscles.
3. Inject each tibialis muscle with 25 μL DNA dissolved in saline solution (150 mM NaCl) containing 6 mg/mL poly-L-glutamate (1 μg/μL) with an insulin syringe (*see* **Note 4**).
4. Hold the muscle with a caliper covering the injected area and apply a field strength of 200 V/cm with a pulse length of 25 ms and three pulses using ECM 830 field generator.

3.3. Monitoring of Tumor Growth

After immunization, treated and control mice were inspected twice per week to evaluate incidence and growth of tumors (*see* **Note 5**). Mice with no evidence of tumors at the end of the observation period were classified as tumor-free, whereas mice with a tumor of at least 3 mm mean diameter were classified as tumor bearers. Number of tumor masses/animal was also registered. All mice bearing neoplastic masses exceeding 10 mm mean diameter were killed for humane reasons (see Fig. 37.1).

3.4. Analysis of Humoral Response

Two weeks after the end of immunization protocols, sera of treated and control mice were harvested and stored at −80°C until used (*see* **Note 6**).

In order to assess the presence of $p185^{neu}$ specific antibody, the ability of sera to bind $p185^{neu}$ was evaluated by flow cytometry according to the following procedure.

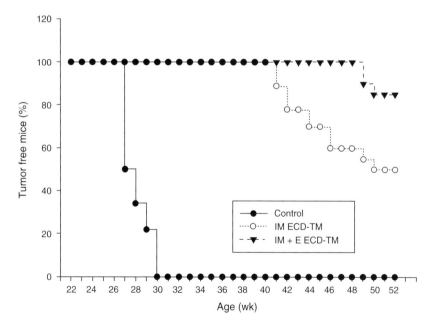

Fig. 37.1 Inhibition of mammary carcinogenesis in HER-2/neu transgenic mice vaccinated with pCMV-ECD-TM plasmid using IM or IM electroporation (IM + E) DNA delivery. Percentages of tumor-free mice were calculated as cumulative number of tumor bearer and tumor free mice. Difference in tumor incidence, as assessed by the Mantel-Haenszel log-rank test, was significant between control vs. IM ECDTM or IM + E ECDTM ($p < 0.0001$), and IM ECDTM + E vs. IM ECDTM ($p < 0.05$)

1. Harvest N202.1A cells when approaching confluence and wash them with cold PBS supplemented with 2% BSA and 0.05% sodium azide.
2. Dispense cells in cytometer tubes (2×10^5 cells/tube): one tube for each serum and control sample (*see* **Note 7**).
3. Incubate cells (2×10^5 cells/sample) with 50 µL of 1:10 dilution in PBS-azide-BSA of control or immune sera (30 min at 4°C).
4. Wash cells with PBS-azide-BSA.
5. Incubate cells with fluorescein-conjugated rabbit anti-mouse Igs or, in case of IgG idiotype analysis, anti-mouse IgG1, anti-mouse IgG3, anti-mouse IgG2a, and anti-mouse IgG2b as second-step Abs (30 min at 4°C).
6. Wash cells with PBS-azide-BSA.
7. Resuspend cells in Isoton II (Coulter, Hialeah, FL) and evaluate through a Coulter XL flow cytometer.
8. Calculate the specific 202/1A binding potential (Sbp) of the sera with the following equation: [(% positive cells with test serum) (fluorescence mean) × serum dilution − (% positive cells with control serum) (fluorescence mean) × serum dilution] (Fig. 37.2).

3.5. In Vitro Stimulation of Spleen Cells

Spleen cells from treated and control mice were prepared and incubated in vitro in the presence of N202.1A tumor cells in order to analyze the production of cytokines and the activation of cytotoxicity in response to antigen stimulation.

Two weeks after the end of immunization protocol, at least four mice of each experimental group were sacrificed, and their spleens were harvested for in vitro stimulation assay.

3.5.1. Preparation of Stimulators Cells

1. Harvest N202.1A (*see* **Note 8**) at 80% confluence with trypsin/EDTA.
2. Wash cells in PBS, count, and resuspend the appropriate amount of cells in RPMI 1640 with penicillin (100 U/mL) and streptomycin (100 µg/mL).
3. Add 60 µg/mL of mitomycin C and incubate cells for 30 min at 37°C and 5% CO_2.
4. Wash with PBS and centrifuge cells.
5. Resuspend in RPMI 1640 with penicillin (100 U/mL), streptomycin (100 µg/mL), 10% FBS, and 50 U/mL interleukin-2 at a concentration 1×10^5 cells/mL.
6. Place on appropriate cell culture plates or flasks and incubate at 37°C and 5% CO_2 until effectors will be prepared.

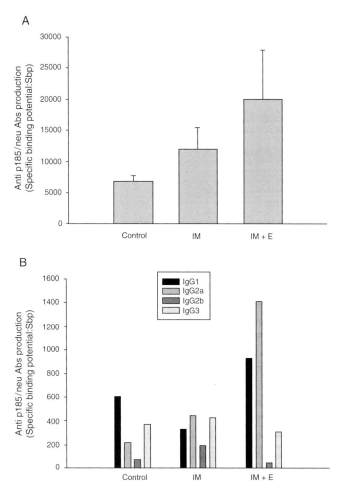

Fig. 37.2 Humoral immunity in FVB*neu*-T transgenic mice vaccinated with pCMV-ECDTM plasmid using IM or IM electroporation (IM + E) DNA delivery. Specific binding potential of sera from treated and control animals to p185 protein (**A**) and anti-p185^{neu} IgGs isotytpes (**B**) are shown. Sbp was evaluated by flow cytometry after indirect immunofluorescence

3.5.2. Preparation of Effector Cells

1. Sacrifice mice by CO_2 inhalation.
2. Dip the mouse in 70% EtOH and remove the spleen using sterile surgery instruments and place it in a tube containing sterile PBS.
3. Under a safety hood, transfer the spleen in a 60 mm tissue culture dish and mechanically disrupt the spleen by teasing it through a 60 mesh sieve in Ca^{2+} and Mg^{2+} PBS to obtain a single cell suspension.

4. Pipet to resuspend splenocytes and transfer into centrifuge tubes to wash three times in PBS by centrifuging 10 min at RCF 215g.
5. Resuspend cells in RPMI 1640 with penicillin (100 U/mL), streptomycin (100 µg/mL), 10% FBS, and 50 U/mL interleukin-2, count cells, and adjust to a concentration of 2×10^6 cells/mL.
6. Plate effector cells on stimulator cells already plated by adding the same volume of splenocyte suspension to achieve the final concentration of 1×10^6 splenocytes/mL and a ratio stimulators:lymphocytes of 20:1.
7. Incubate at 37°C and 5% CO_2 for appropriate time: 2 days for cytokine assay or 5 days for cytotoxicity assay.

3.6. Analysis of Cytokine Expression

After 2 days of in vitro stimulation with N202.1A cells, spleen cells were treated with Golgi Plug, (*see* **Note 9**) then harvested and stained for intracellular cytokines according to the following procedure.

1. Harvest spleen cells, centrifuge, and wash them with PBS containing 5% FCS and 0.01% NaN_3.
2. Dispense cells in cytometer tubes ($2.5–5 \times 10^5$ cells/sample), centrifuge, aspirate supernatant, and resuspend pellet in 50 µL PBS/FCS/NaN_3.
3. Add anti-CD4 or anti-CD8 monoclonal antibodies (5 µg/10^6 cells) to the cells and incubate at 4°C for 30 min.
4. Wash cells with PBS/NaN_3, centrifuge, and resuspend pellet in 200 µL 2% formalin and incubate at 4°C for 15 min to fix cells.
5. Wash cells in PBS containing 5% FCS and 0.05% saponin, centrifuge, and resuspend pellet in 50 µL of the same buffer.
6. Appropriately (*see* **Note 10**) add monoclonal antibodies to cytokines (5 µg/10^6 cells of anti-IL10, anti-IL4, or anti-IFNγ (CALTAG) PE-conjugated) and incubate for 15 min at room temperature.
7. Wash cells in PBS containing 5% FCS and 0.01% saponin, centrifuge, and resuspend pellet in 450 µL of PBS.
8. Analyze labelled cells through a Coulter XL flow cytometer (see Fig. 37.3).

3.7. Analysis of CTL Activity

After 5 days of in vitro stimulation with N202.1A cells, spleen cells were harvested to analyze their cytotoxic activity. Cytotoxic assay was performed using a fluorimetric method according to the following procedure.

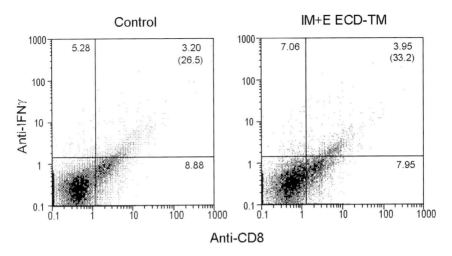

Fig. 37.3 Intracellular cytokine staining in HER-2/neu transgenic mice vaccinated with pCMV-ECDTM plasmid using intramuscualr (IM) or IM electroporation (IM + E) DNA delivery. The percentage of CD8 T cells containing IFNγ was determined through double-staining flow cytometry after in vitro lymphocyte incubation with mitomycin-treated N202/1A tumor cells. Data are expressed as percentage of positive cells. Numbers in brackets indicate the percentage of IFN–γ producing cells among total CD8 T lymphocytes

3.7.1. Preparation of Effector Cells

1. Harvest spleen cells from in vitro stimulation plates, wash in PBS, centrifuge, and resuspend cells in PBS.
2. Count viable cells and resuspend in RPMI with 10% FCS adjusting to a concentration of 5×10^6 cells/mL.

3.7.2. Preparation of Target Cells

1. Harvest N202.1A cells with trypsin/EDTA, wash twice in PBS.
2. Resuspend pellet in 1 mL working solution of PBS/c'FDA (*see* **Note 11**) and incubate at 37°C, 5% CO_2 incubator for 30 min.
3. Wash labelled cells three times in PBS with 1% BSA.
4. Resuspend cells at a concentration of 0.5×10^5/mL.
5. Plate 0.5×10^4 target cells onto effectors in 200 mL total volume per well.

3.7.3. Cytotoxicity Assay

1. Plate effector and target cells in 200 μL total volume in 96-well round microtiter plates. Test effector:target cell ratios from 100:1 to 12.5:1. Test each ratio in triplicate.

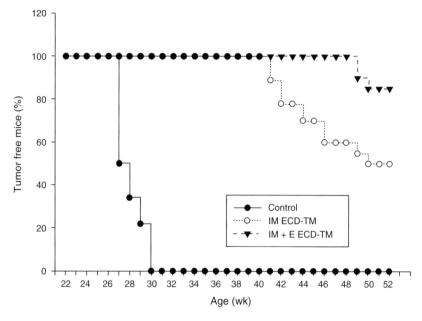

Fig. 37.4 Cell mediated cytotoxic activity in HER-2/neu transgenic mice vaccinated with pCMV-ECDTM plasmid using IM or IM electroporation (IM + E) DNA delivery. Cytotoxicity was performed through a fluorimetric assay. Data are reported as percent of lysis obtained at different E:T ratios. Difference in cytotoxicity was significant between IM or IM + E vs. control group (*, $p < 0.05$), and between IM + E vs IM group (**, $p < 0.05$)

2. Incubate plates at 37°C and 5% CO_2 incubator for 3 h and then centrifuge at 700g for 5 min.
3. Separate the supernatant from the cellular part by rapidly inverting the plate and flicking the supernatant out.
4. Add 100 μL of 1% Triton 100 in 0.05 M borate buffer, pH 9.2, to each well.
5. Keep the plate at 4°C for 20 h.
6. Read plate for fluorescence with a 1420 VICTOR^2 multilabel counter.
7. Calculate the percentage of specific lysis as follows (see Fig. 37.4): % specific lysis: [(Fmed-Fexp)/Fmed] × 100 where F = fluorescence of lysed cells after the supernatant has been removed (Fmed = F from target incubated in medium alone; Fexp = F from target incubated with effector).

3.8. Statistical Analysis

Differences in tumor incidence were evaluated by the Mantel-Haenszel log-rank test; differences in tumor multiplicity were evaluated by Student's t test. Differences in immune parameters were evaluated by ANOVA followed by Student-Newman-Keuls post hoc test when appropriate. Differences were considered statistically significant when $p < 0.05$.

4. Notes

1. The addiction of poly-L-glutamate to DNA suspension at the concentration of 6 mg/mL improves the efficiency of DNA transfection in vivo.
2. Isoflurane inhalation is safe for mice and they recover very fast and can be subjected to it several times. Initially subjected to a dosage of 4% of isoflurane, the mouse can be kept sleeping until needed by reducing the dosage to 0.4% isoflurane.
3. Pilot experiments that we performed previously in studies of DNA vaccination against HER2/neu showed us that the exposition of the muscle improves the quality of injection by avoiding dispersions of DNA suspension and results in a better outcome of the vaccine (unpublished observations). A further trick to avoid DNA dispersion is to inject the DNA very slowly; a bubble appears in the area of injection following a correct administration of DNA.
4. The different amounts of DNA injected in the two protocols used for (intramuscular) i.m. injection and i.m. plus electroporation have been established following pilot experiments taking into account for the diverse efficacy of in vivo transfection of DNA in the two techniques.
5. Monitoring of tumor growth should be performed by one person from the beginning to the end of experiment to avoid variations in the identification of newly appeared masses.
6. To collect sera of mice, harvest blood from each mouse, let serum separate from cellular parts, centrifuge (15 min at 14,000 rpm), recover serum by micropipetting, and store at −80°C.
7. Positive control: N202.1A cells to label with FITC-conjugated anti-HER2/neu antibody (Ab-4, Oncogene Science, Cambridge, MA); negative controls: N202.1A to incubate with control sera and N202.1A to incubate without serum.
8. Expression level of HER2/neu on N202.1A cells is correlated with quality of cells and their confluence status; it is advisable to check HER2/neu expression level by cytofluorimetry (stain with Ab-4, Oncogene Science, Cambridge, MA) prior to using the cells for in vitro stimulation.
9. Golgi Plug, avoiding the release of cytokines, is toxic for cells; it is added directly in the medium (1 μl/mL) 4–5 h before the end of in vitro stimulation.
10. We usually stain for CD4-IL4, CD4-IL10, or CD8-IFNγ.
11. A stock solution of carboxyfluorescein diacetate (20 mg/mL acetone, stored at −20°C) is diluted in PBS to give a final concentration of 75 μg/mL.

References

1. Tang, D., De Vit, M., and Johnston, S.A. (1992) Genetic immunization is a simple method for eliciting an immune response. *Nature*. **356**,152–154.
2. Hynes, N.E. and Stern, D.F. (1994) The biology of erbB2/neu/HER2 and its role in cancer, *BBA*. **1198**, 165–184.
3. Amici, A., Smorlesi, G., Noce, G., et al. (2000) DNA vaccination with full-length or truncated Neu induced protective immunity against the development of spontaneous mammary tumors in HER-2/neu transgenic mice. *Gene Ther*. **7**, 703–706.
4. Quaglino, E., Rolla, S., Iezzi, M., et al. (2004) Concordant morphologic and gene expression data show that a vaccine halts HER2/neu preneoplastic lesions. *J. Clin. Invest*. **113**, 709–717.
5. Conry, R.M., Curiel, D.T., Strong, T.V., et al. (2002) Safety and immunogenicity of DNA vaccine encoding carcinoembryonic antigen and hepatitis B surface antigen in colorectal carcinoma patients. *Clin. Cancer Res*. **8**, 2782–2787.
6. Cappello, P., Triebel, F., Iezzi, M., et al. (2003) LAG-3 enables DNA vaccination to persistently prevent mammary carcinogenesis in HER-2/neu transgenic BALB/c mice. *Cancer Res*. **63**, 2518–2525.

7. Disis, M.L., Scholler, N., Dahlin, A., et al. (2003) Plasmid-based vaccines encoding rat *neu* and immune stimulatory molecules can elicit rat *neu*-specific immunity. *Mol. Cancer Ther.* **2**, 995–1002.

8. Renard, V., Sonderbye, L., Ebbehoj, K., et al. (2003) HER-2 DNA and protein vaccines containing potent Th cell epitopes induce distinct protective and therapeutic antitumor responses in HER-2 transgenic mice. *J. Immunol.* **171**,1588–1595.

9. Rovero, S., Boggio, K., Carlo, E.D., et al. (2001) Insertion of the DNA for the 163–171 peptide of IL1beta enables a DNA vaccine encoding p185(neu) to inhibit mammary carcinogenesis in Her-2/neu transgenic BALB/c mice. *Gene Ther.* **8**, 447–452.

10. Kirman, J.R. and Seder, R.A. (2003) DNA vaccination: the answer to stable, protective T-cell memory? *Curr. Opin. Immunol.* **15**, 471–476.

11. Neumann, E., Schaefer-Ridder, M., Wang, Y., and Hofshneider, P.H. (1982) Gene transfer into mouse lymphoma cells by electroporation in high electric fields. *EMBO J.* **1**, 841–845.

12. Smorlesi, A., Papalini, F., Amici, A., et al. (2006) Evaluation of different plasmid DNA delivery systems for immunization against HER2/neu in a transgenic murine model of mammary carcinoma *Vaccine.* **24**, 1766–1775.

13. Quaglino, E., Iezzi, M., Mastini, C., et al. (2004) Electroporated DNA vaccine clears away multifocal mammary carcinomas in her-2/neu transgenic mice. *Cancer Res.* **64**, 2858–2864.

14. Guy, C.T., Cardiff, R.D., and Muller, W.J. (1996) Activated neu induces rapid tumour progression. *J. Biol. Chem.* **271**, 7673–7678.

15. Lollini, P.L., Nicoletti, G., Landuzzi, L., et al. (1998) Down regulation of major histocompatibility complex class I expression in mammary carcinoma of HER2/*neu* transgenic mice. *Int. J. Cancer.* **77**, 937–941.

Chapter 38
Electroporation-Mediated HBV DNA Vaccination in Primate Models

Yong-Gang Zhao and Yuhong Xu

Abstract Electroporation has been shown to be an effective method to improve the efficiency of gene expression and the immunogenicity of DNA vaccines. To optimize the procedure and test for its efficacy in more clinically relevant large animal models, we studied the effects of electroporation-mediated DNA vaccination with different electro-pulse parameters in rhesus macaques. Plasmid DNA encoding the HBV preS2-S and an adjuvant plasmid encoding a fused gene of IL-2 and IFN-gamma were injected intramuscularly followed by electroporation once a month for several months. The humoral as well as cellular immune responses were closely followed for more than a year. The different electro-pulse parameters resulted in considerably different intensities in immune responses, suggesting that optimization of electroporation parameters is important in developing clinical application of DNA vaccination.

Keywords: electroporation, DNA vaccine, primates, HBV, immune response

1. Introduction

DNA-based vaccination has been shown to elicit significant cell mediated immune responses in several animal models (1, 2); therefore, lot of efforts are being made to develop therapeutic DNA vaccines for the treatment of chronic Hepatitis B virus (HBV) infection (3, 4). However, recent studies suggested that additional immune enhancement strategies are probably needed and could be very important for efficient vaccination or immunotherapy, especially when used in large animals including humans (5–7). Electroporation-mediated intramuscular delivery is considered one of the most promising methods (8–12).

The procedure of electroporation-mediated DNA vaccination usually involves the application of a series of electric pulses following the DNA vaccine injection. The detailed parameters of the pulse sequence have been shown to be important (13–15), but most earlier studies focused on maximizing transgene expressions and were done using mouse models, which cannot be easily extrapolated to large

animals when the immune responses are the main concern (10, 11). The studies that had used primate models only tested very limited parameters (16) and some showed results from electroporation combined with virus vector deliveries; therefore, for the better designing of electroporation protocols for potential applications in humans, we carried out a detailed study examining the effects of electroporation-mediated DNA vaccination with different parameters, using more relevant primate models.

2. Materials

2.1. Plasmids and Recombinant Proteins

1. The HBV antigen encoding plasmid was constructed using the pcDNA3.1 vector with inserted HBV preS2-S sequence.
2. The adjuvant plasmid contained the *h*IL-2 signal peptide sequence, the *h*IL-2 sequence, a 24mer linker (encoding GGGGGSSC), and the *h*IFN-γ sequence, cloned sequentially in a pCR-Blunt vector (*see* **Note 1**).
3. The antigen plasmid and the adjuvant plasmid were mixed with each other at 1:1 weight ratio for all vaccination studies.
4. Working solutions were prepared by diluting 1 mg DNA in 1 mL 0.5× saline (75 mM sodium chloride).
5. Recombinant HBs antigen (adw subtype) was obtained from SmithKline Beecham (Rixensart, Belgium).

2.2. Electropulse Generator and Electrodes

1. The electropulse generator was custom made and can generate square wave signals within the voltage range of 0–300 V and pulse duration range of 10 μs to 100 ms in any combination sequences (*see* **Note 2**).
2. The seven-needle electrode array was made of silver Chinese acupuncture needles (0.3 mm diameter, 12 mm length) (*see* **Note 3**). Six needles were arranged in a circular pattern ($r = 6$ mm) as the anodes, while the seventh needle was placed in the center as the cathode.

2.3. Animal Experiments

1. Healthy male/female rhesus macaques, 3–5 years old, 3–5 kg in weight, were provided by Suzhou Xishan C.A.S. Laboratory Animals Co. LTD. (Xishan, Suzhou, China), where they were housed in compliance with regulations of the

Chinese Council for Animal Care. Their use in this experiment and the experimental procedures were approved by the Jiangsu Province Department of Experimental Animals Management.
2. Anesthetic ketamine was provided by Suzhou Xishan C.A.S. Laboratory Animals Co. Ltd. (Xishan, Suzhou, China).
3. Veterinary syringes and needles were provided by Suzhou Xishan C.A.S. Laboratory Animals Co. Ltd. (Xishan, Suzhou, China).

2.4. Antibody ELISA

1. Anti-hepatitis B surface protein antibody titers were measured using the Anti-HBs ELISA kit (Sino-American Biotechnology Co., China) (see **Note 4**). The Anti-HBs ELISA kit includes antigen precoated plates, enzyme-labeled buffer, substrate buffer A, substrate buffer B, and stop solution.
2. Anti-HBs antibody protein standard (Sino-American Biotechnology Co., China).

2.5. IFN-γ ELISpot

1. ELISPOT assay plates: multi-screen 96-well plates.
2. Monoclonal antibody against human IFN-γ (BMS 107) from Bender Med Systems (Vienna, Austria).
3. Fetal bovine serum.
4. RPMI medium.
5. Lymphocytes separation medium.
6. Heparin sodium.
7. Concanavalin A.
8. Biotinylated anti-IFN-γ antibody (clone 7B6–1) from MABTECH (Sweden).
9. Streptavidin–alkaline phosphatase conjugates.
10. 5-Bromo, 4-chloro, 3-indolyl phosphatase (BCIP).
11. The HBs epitope containing peptide pool was designed with the help of an epitope prediction software TEPITOPE (courtesy of Juergen hammer, Section of Bioinformatics, Genetics, and Genomics, Hoffmann–La Roche Inc., Nutley). The specific sequences are as follows: HBsP9–30: LGPLLVLQAGFFLL TKILTIPQ; HBsP21–40: LLTKILTIPQSLDSWWTSLN; HBsP39–49: LNFLGGTPVCL; HBsP72–88: YRWMCLRRFIIFLCILL; HBsP82–100: IFL CILLLCLIFLLVLLDY; HBsP103–117: MLPVCPLIPGSSTTS; HBsP163–176: WEWASVRFSWLSLL; HBsP176–192: LVPFVQWFVGLSPTVWL; HBsP197– 213: MMWFWGPSLYNILSPFI; HBsP206–224: YNILSPFIPLLPIFFCLWA.

3. Methods

3.1. DNA Injection and Electroporation

1. Anesthetize the macaque by ketamine injection at a dose of 2 mg/kg weight.
2. Put the anesthetized macaque on the experiment bed with neck, wrists, and ankle straps.
3. Shave the inner side of both legs to expose the areas on top of the gastrocnemius muscles.
4. Select the needle injection point and mark it as a red dot. Then inject 250 μg of antigen encoding plasmid (1 mg/mL) in 0.5× saline (*see* **Note 5**).
5. Within 1 min after the injection, insert the electrodes into the muscle with the center cathode aligning with the red dot. Apply the specific series of electroporation pulses.
6. Repeat the injection and the electroporation treatment on the other leg.

3.2. Electrical Pulse Sequences

1. Four different electric pulse sequences were selected for testing. Figure 38.1 provides a schematic drawing of the waveforms. Type I consisted of six 36 V unipolar square pulses with a 50 ms duration and 1 s interval. Type II had six 90 V unipolar square pulses with a 50 ms duration and 1 s interval. Type III included one set of 90 V bipolar pulses with a 50 ms duration, and, after 8 s, followed by six 90 V unipolar pulses with 50 ms duration and 1 s interval. Finally, Type IV consisted of one set of 240 V bipolar pulses with a 100 μs duration, and, after 8 s, followed by six 24 V unipolar pulses with a 50 ms duration and 1 s interval (*see* **Note 6**).

3.3. Vaccination Schedules

1. Prime all animals at week 0 using the specific electroporation sequences.
2. At week 4 and week 8, boost the animals with the same electroporation procedure.
3. At week 16 and week 28, give two more boost injections (*see* **Note 7**).
4. At week 44, inject a commercial subunit vaccine containing 10 μg of recombinant HBs antigen (*see* **Note 8**).

3.4. Measurement of Antibody Titers

1. Antibody titers were measured every 2 weeks throughout the study. Figure 38.2 shows a plot of the numbers for each individual animal (*see* **Note 9**).

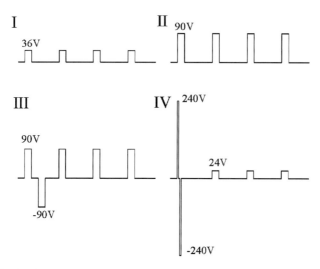

Fig. 38.1 Electrical pulse waveforms applied in the experiments. Type I: six 36 V unipolar square pulses with a 50 ms duration and 1 s interval. Type II: six 90 V unipolar square pulses with a 50 ms duration and 1 s interval. Type III: one pair of 90 V bipolar pulses with 100 μs duration, and, after 8 s, followed by six 90 V unipolar pulses with a 50 ms duration and 1 s interval. Type IV: one pair of 240 V bipolar pulses with 100 μs duration, and, after 8 s, followed by six 24 V unipolar pulses with a 50 ms duration and 1 s interval. Reprinted from Zhao, Y.G. et al. (2006), Anti-HBV Immune Responses in Rhesus Macaques Elicited by Electroporation mediated DNA vaccination, *Vaccine*. 24, 898, with permission from Elsevier

2. Draw 1 mL blood samples at various time points before and during the vaccination experiment.
3. Isolate the sera and immediately freeze at −20°C.
4. When assaying, draw 50 μl of the sera sample and add to each well of a precoated ELISA plate.
5. Add 50 μl of enzyme labeled antibody solution to each well and then incubate at 37°C for 30 min.
6. Wash the wells 6 times with PBS.
7. Add 50 μl substrate solution A to each well, and then add 50 μl substrate solution B to wells.
8. Incubate the plate at 37°C for 15 min.
9. Add 50 μl stop solution to each well.
10. Measure the O.D. value at 450 nm wavelength using an ELISA plate reader and plot the figure (*see* **Note 9**).

3.5. Peripheral Blood Mononuclear Cells Isolation

1. Two weeks after vaccination, draw 15 mL blood samples into 100 U/mL heparin sodium solution.

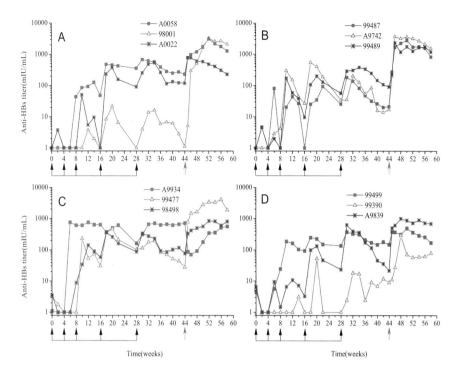

Fig. 38.2 Antibody titers in rhesus macaques after DNA vaccination and electroporation with different electroparameters. Each animal received two 250 μL intramuscular injections in the gastrocnemius muscles, one on each leg, containing a total of 0.5 mg each of the HBs plasmid and the adjuvant plasmid. The vaccine administrations with electroporation were indicated by black arrows pointing at the *X* axis. S protein boost was indicated by the hollow triangles. The dotted line in the bottom of the figure indicates the threshold of 10 U/mL of anti-HBs antibody. A: Group I; B: Group II; C: Group III; D: Group IV. Reprinted from Zhao YG et al. (2006), Anti-HBV Immune Responses in Rhesus Macaques Elicited by Electroporation mediated DNA vaccination, *Vaccine*. 24, 902, with permission from Elsevier

2. Dilute the heparinized blood samples at the ratio of 1:3 to 1× PBS. Layer over the lymphocytes separation medium at a ratio of 1 part lymphocytes separation medium to 2 parts diluted blood.

3. Spin for 30 min at room temperature with no brake at a speed of about 1,800 rpm.

4. Harvest the white band of peripheral blood mononuclear cells (PBMC) and suspend them in 1× PBS.

5. Wash and spin 3 times in RPMI medium at about 200× *g*.

6. Resuspend the pellet in 1× PBS.

7. Count the cells.

3.6. ELISPOT Assays

1. ELISPOT assay plates were coated with monoclonal antibody against human IFN-γ by incubating in 5 mg/mL antibody solution in PBS over night at 4°C (*see* **Note 10**).
2. Wash the plates 6 times with PBS containing 0.05% Tween 20, and block them with 10% FBS in RPMI medium.
3. Plate the isolated PBMC into the ELISPOT plate at the densities of 4×10^5, 2×10^5, or 1×10^5 cells per well.
4. Add the epitope-containing peptide mixture to the culture plates and incubate for 24–36 h.
5. Add concanavalin A to a few wells as the positive control.
6. After the incubation, empty the wells, wash with PBS–Tween 20, and then react with biotinylated anti-IFN-γ antibody at room temperature for 2 h.
7. Wash the wells again with PBS-Tween and react with streptavidin–alkaline phosphatase conjugates for 1 h first and then 5-bromo, 4-chloro, 3-indolyl phosphatase for 15–20 min.
8. Count the blue spots in each well manually under a dissection microscope.

3.7. Statistical Analysis

Antibody titer data were reported as mean ± SEM. We tested the significance of the differences between the individual groups using the one-tailed Student's t-test for unpaired values. For statistical comparison of several groups, we used one-way ANOVA.

Acknowledgements This work was supported by the National 863 High Technology Research and Development Program of China (2002AA2Z3317) and Shanghai QiMingXing Program (04QMH1404).

4. Notes

1. We proved that the fusion protein expressed by the adjuvant plasmid transfected cells had both IL-2 and IFN-γ bioactivities. When administered together with the antigen plasmid in mice, it not only boosted the maximal antibody titer by about 4-fold but also extended the duration of positive antibody response from 4 months to over 6 months.
2. The electro-pulse generator is now made commercially available from Shanghai Teressa Health Science Inc. (Shanghai, China).
3. Commercial acupuncture needles are available in various sizes and lengths. Such a relatively thin needle type was chosen because they seem to cause less pain in mice experiments.
4. Anti-hepatitis B surface protein antibody titers were measured using the ELISA Kit. The kit is approved for human diagnostic uses; however, because it is based on the sandwich method, it can also be used for animal samples.

5. In an earlier study exploring the effective dose response for electroporation-mediated DNA vaccination in rhesus macaques, we tested three different doses (0.1 mg, 0.5 mg, and 1 mg) with three animals in each group. In the 0.1 mg dose group, one animal did not respond, and the other two only responded after the third treatment. In the 0.5 and 1 mg dose groups, all animals responded after the three electroporation-mediated DNA injections. Therefore, we picked the 0.5 mg dose for this study.

6. The high-voltage and low-voltage combination sequences were designed based on previous studies that examined the effects of pulse parameters to transgene expression. The high voltage, short-duration pulse (HV) was considered important for cell electropermeabilization, and the low-voltage, long-duration pulses (LV) were used for the electrophoretic movement of DNA (13, 17). However, high transgene expression may not always translate to high immune responses. Therefore, it is important to test the various parameters for immune response.

7. Since not all the animals in the experiment groups showed definitive responses after three vaccinations, we added two more boosts at later time points, for better comparison between different experiment groups.

8. The final injection of recombinant antigen was designed to examine the anamnestic responses in different experiment subjects.

9. Antibody titer values over 10 U/mL were considered positive.

10. There was no anti-rhesus IFN-γ antibody commercially available. But it seems anti-human IFN-γ antibodies can be used for rhesus samples.

References

1. Donnelly, J.J., Ulmer, J.B., Shiver, J.W., and Liu, M.A. (1997) DNA vaccines. *Annu. Rev. Immunol.* **15**, 617–648.

2. Liu, M.A. (2003) DNA vaccines: a review. *J. Intern. Med.* **253**, 402–410.

3. Davis, H.L. (1999). DNA vaccines for prophylactic or therapeutic immunization against hepatitis B virus. *Mt. Sinai J. Med.* **66**, 84–90.

4. Thermet, A., Rollier, C., Zoulim, F., Trepo, C., and Cova, L. (2003) Progress in DNA vaccine for prophylaxis and therapy of hepatitis B. *Vaccine.* **21**, 659–662.

5. MacGregor, R.R., Boyer, J.D., Ugen, K.E., et al. (1998) First human trial of a DNA-based vaccine for treatment of human immunodeficiency virus type 1 infection: safety and host response. *J. Infect. Dis.* **178**, 92–100.

6. Turnes, C.G., Aleixo, J.A., Monteiro, A.V., and Dellagostin, O.A. (1999) DNA inoculation with a plasmid vector carrying the faeG adhesin gene of *Escherichia coli* K88ab induced immune responses in mice and pigs. *Vaccine.* **17**, 2089–2095.

7. Ugen, K.E., Nyland, S.B., Boyer, J.D., et al. (1998) DNA vaccination with HIV-1 expressing constructs elicits immune responses in humans. *Vaccine.* **16**, 1818–1821.

8. Selby, M., Goldbeck, C., Pertile, T., Walsh, R., and Ulmer, J. (2000) Enhancement of DNA vaccine potency by electroporation in vivo. *J. Biotechnol.* **83**, 147–152.

9. Bachy, M., Boudet, F., Bureau, M., et al. (2001) Electric pulses increase the immunogenicity of an influenza DNA vaccine injected intramuscularly in the mouse. *Vaccine* **19**, 1688–1693.

10. Babiuk, L.A., Pontarollo, R., Babiuk, S., Loehr, B. and van Drunen Littel-van den Hurk, S. (2003) Induction of immune responses by DNA vaccines in large animals. *Vaccine.* **21**, 649–658.

11. Scheerlinck, J.P., Karlis, J., Tjelle, T.E., Presidente, P.J., Mathiesen, I., and Newton, S.E. (2004) In vivo electroporation improves immune responses to DNA vaccination in sheep. *Vaccine.* **22**, 1820–1825.

12. Widera, G., Austin, M., Rabussay, D., et al. (2000) Increased DNA vaccine delivery and immunogenicity by electroporation in vivo. *J. Immunol.* **164**, 4635–4640.

13. Bureau, M.F., Gehl, J., Deleuze, V., Mir, L.M., and Scherman, D. (2000) Importance of association between permeabilization and electrophoretic forces for intramuscular DNA electrotransfer. *Biochim. Biophys. Acta.* **1474**, 353–359.
14. Gehl, J. and Mir, L.M. (1999) Determination of optimal parameters for in vivo gene transfer by electroporation, using a rapid in vivo test for cell permeabilization. *Biochem. Biophys. Res. Commun.* **261**, 377–380.
15. Mir, L.M., Bureau, M.F., Gehl, J., et al. (1999) High-efficiency gene transfer into skeletal muscle mediated by electric pulses. *Proc. Natl. Acad. Sci. U.S.A.* **96**, 4262–4267.
16. Otten, G., Schaefer, M., Doe, B., et al. (2004) Enhancement of DNA vaccine potency in rhesus macaques by electroporation. *Vaccine.* **22**, 2489–2493.
17. Satkauskas, S., Bureau, M.F., Puc, M., et al. (2002) Mechanisms of in vivo DNA electrotransfer: respective contributions of cell electropermeabilization and DNA electrophoresis. *Mol. Ther.* **5**, 133–140.

Chapter 39
Taking Electroporation-Based Delivery of DNA Vaccination into Humans: A Generic Clinical Protocol

Torunn Elisabeth Tjelle, Dietmar Rabussay, Christian Ottensmeier, Iacob Mathiesen, and Rune Kjeken

Abstract We are presently aware of two early-phase DNA vaccine clinical trials in humans using electroporation-enhanced vaccine delivery. Moreover, two phase I immunogenetherapy studies are in progress and several tolerability studies have been performed on healthy volunteers. We have used knowledge from these studies to compose a template for clinical protocols involving electroporation-mediated gene delivery. In this template the emphasis will be on aspects related to electroporation. In addition, we will discuss general topics concerning electroporation-augmented DNA vaccination in human subjects.

Keywords: clinical trial, electroporation, electroporation device, DNA vaccination

1. Introduction

Human DNA vaccines are based on the premise that it may be preferable and more effective to express antigens from DNA constructs delivered into cells of the body rather than to inject the same antigens, in various forms, directly into humans (1). DNA vaccines offer certain advantages over conventional vaccines, including relatively high thermal stability, presentation of unique epitopes only accessible in nascent antigens, and the triggering of both humoral and cellular responses (2). DNA vaccines delivered either as plasmid DNA by needle and syringe, needle-free injection, or gene gun, or delivered via viral or lipid vectors have been effective in numerous cases in small and sometimes in large animals (2–8). Two animal DNA vaccines have recently been approved in the USA and Canada for the vaccination of horses against West Nile virus and a second vaccine protecting salmon against infectious hematopoietic necrosis virus, respectively (9, 10). However, despite a large number of clinical studies (74 active and 102 inactive studies in the USA as of June 2006, see www. clinicaltrials.gov/ct/) no DNA vaccine for humans, although safe, has yet met applicable efficacy requirements.

Delivery of plasmid DNA augmented by electroporation has shown encouraging results in small and large animals (2, 11–13) and is now being tested in humans.

S. Li (ed.), *Electroporation Protocols: Preclinical and Clinical Gene Medicine.*
From *Methods in Molecular Biology, Vol. 423.*
© Humana Press 2008

DNA encoding the antigen(s) of choice is injected into the target tissue (muscle, skin, or tumor tissue), followed by electroporation to enhance transfer of the DNA into the target cells across the barrier of the outer cell membrane. Electroporation generally results in a 100-fold or greater increase in antigen expression as well as a local inflammatory effect. These two factors are thought to be primarily responsible for the enhanced immune response observed after electroporation: The antibody response is faster, reaches higher titers, and can be achieved with lower DNA doses than in the absence of electroporation; the cellular response is also enhanced significantly (2, 12, 14). Compared with other delivery methods (e.g., viral vectors, lipid formulations, and the gene gun), electroporation-assisted delivery provides advantages in terms of safety, nontoxicity, and the ability to scale up from small animals to nonhuman primates and possibly humans (3, 11, 15–18). Previous questions as to the tolerability of electroporation, the potential enhancement of plasmid integration into chromosomes by electroporation, and the potential induction of antibodies against plasmid DNA have now been answered favorably. Tolerability has been demonstrated in studies involving healthy human volunteers (19, 20). Animal safety, toxicology, and biodistribution studies have been required by the FDA prior to clinical trials, including studies of plasmid integration into chromosomal DNA following electroporation-mediated delivery. The data obtained by different investigators were sufficient for regulatory approval for several clinical trials (15, 18, 21–26). To date, anti-DNA antibodies have not been detected in study subjects subjected to plasmid delivery via electroporation, but patients continue to be monitored.

Two different electroporation procedures are being used in ongoing clinical trials. In the first procedure, DNA is injected (by needle and syringe) followed by insertion of a four-needle electrode array at the site of injection to deliver electrical pulses (27), in the second procedure two standard syringes with injection needles are mounted on a movable sledge. As the needles are advanced into the muscle tissue, DNA is injected at a predetermined rate. When DNA injection is completed, electrical pulses are delivered via the two injection needles now serving as electrodes (28).

Ongoing clinical studies using electroporation-mediated DNA delivery include two early-phase safety and/or efficacy studies involving intramuscular injection of a vaccine followed by electroporation. One of the studies, the first DNA vaccine delivered with electroporation in humans, is sponsored by Southampton University Hospitals, UK (29), and another by Merck (26). Moreover, two immune therapy studies employing IL-12 or IL-2 genes are active in the USA (23–25). The latter studies are designed to treat malignant melanoma, and the vaccine is injected intratumorally, followed by electroporation. The IL-12 study was the first clinical study ever to deliver DNA to humans via electroporation.

The information provided in this chapter is intended to be helpful in designing protocols for performing electroporation in the context of human clinical studies. It is based on experience gathered from tolerability studies in human volunteers as well as preclinical and ongoing clinical studies. Given the scope of this chapter, we have omitted much of the nonelectroporation-related details usually

provided in clinical protocols, such as information related to the specific agent to be delivered and the disease to be treated or prevented. Instead we have emphasized information relevant to the use of electroporation, which may be applicable in future studies.

2. Materials

2.1. Anesthesia and/or Pain Relief

1. If necessary, offer patients a short acting intravenous benzodiazepine or other sedative drugs, either with the first or subsequent vaccinations.
2. If necessary, offer patients painkillers (eg. paracetamol) before treatment (*see* **Note 1**).

2.2. Electroporation Devices

Use a qualified electroporation device. Three electroporation devices are already in clinical trials; however, none of these devices are presently commercially available. When selecting a device, it is recommended to use a supplier that has a track record of maintaining a quality system and specific device documentation sufficient to obtain regulatory approval for using the device in the intended clinical trial. The following three devices are presently in clinical trials:

1. The Elgen system consists of a square wave pulse generator, interfacing with the user via a host PC, and a combined injection/electrode device, which injects the DNA during needle insertion and uses of-the-shelf syringes and needles (Fig. 39.1). The output pulses used in human studies so far were set at a constant current of 250 mA, corresponding to about 60–70 V. The injection/electrode device makes use of two disposable standard syringes with needles (21 G, 0.8 × 40 mm) that are placed in a movable carriage inside the injection/electrode device housing. The needles are electrically connected to the pulse generator. The system is used for both injecting the DNA and delivering the electrical pulses. The two parallel needles, 8 mm apart, are manually inserted into the muscle tissue to a depth of 1.5 cm. A gearing mechanism automatically injects the DNA (200 μL/syringe) during needle insertion, thus ensuring distribution of the injection volume along the entire needle paths. This experimental device has been approved by the MHRA for the use in a clinical trial in the UK. The device is supplied by Inovio AS, Norway, a subsidiary of Inovio Biomedical Corporation, USA. This prototype has been extensively modified and named Elgen1000 (CE marked), consisting of a control unit (the pulse generator) and an injector unit (a fully automatic version of the injection/electrode device).

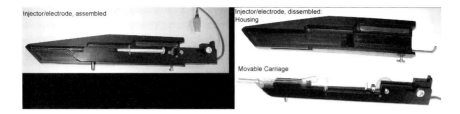

Fig. 39.1 The Elgen system injector/electrode device

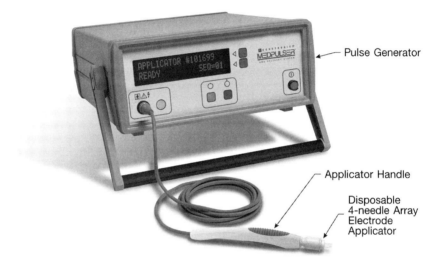

Fig. 39.2 The MedPulser DNA Delivery System

2. The MedPulser DNA Delivery System (DDS) (Fig. 39.2) supplied by Inovio Biomedical Corporation consists of a pulse generator and a reusable applicator with a disposable tip containing a four-needle array electrode (needle length, 15 mm; electrode distance, 4.3 mm). Typically, DNA vaccine is injected intramuscularly, followed by insertion of the electrode array encompassing the injection site and subsequent pulse delivery. The MedPulser DDS delivers two unipolar pulses of 60 ms at 106 V, with a frequency of 4 Hz. This device has been approved for use in clinical trials in the USA and Italy. The device can also be used for the electroporation of skin or tumor tissue.

3. The MedPulser DNA Electroporation Therapy System, also supplied by Inovio Biomedical Corporation, is similar to the MedPulser DDS. However, it uses a six-needle electrode array, with the needles either integrated into the applicator or contained in a disposable tip (needle length up to 3 cm; electrode distance, 8.6 mm). This system delivers six bipolar, rotating pulses of 100 µs each at 1,130 V, with a frequency of 4 Hz. It is presently used in two Phase I clinical trials in the USA for intratumoral DNA delivery.

2.3. DNA Vaccine

Use DNA vaccines according to European regulatory guidelines on DNA vaccines (EMEA website: http://www.emea.eu.int/index/indexh1.htm, see "Note for Guidance on the Quality, Preclinical and Clinical Aspects of Gene Transfer Medicinal Products (Adopted April 2001)", http://www.emea.eu.int/pdfs/human/bwp/308899en.pdf) or according to US guidelines, which can be found at the FDA website (http://www.fda.gov/cber/guidelines.htm) (*see* **Note 2**).

3. Methods

3.1. Study Design

1. Select a patient group. For therapeutic vaccines, a clinical phase I/II study patient group will probably include patients who have gone through conventional treatments without success; however, after careful risk benefit analysis, it may be more appropriate to include study subjects whose tumor is either not measurable (adjuvant treatment) or who have a minimal tumor burden. Alternatively, the vaccination can be administered in parallel with other treatments if the established treatment is not expected to interfere (e.g. it will be important to avoid situations where immunosuppressive agents are used, such as corticosteroids and purine analogues). While currently most protocols will avoid the concurrent use of chemotherapeutic agents, it may be appropriate to test such combinations in the future as there is a growing body of data suggesting that chemotherapy may be beneficial by affecting the number and function of, for example, regulatory T cells. For prophylactic vaccines, a clinical phase I study patient group will include healthy volunteers, but may also be offered to patients at high risk of developing malignancies.
2. Determine objectives and endpoints for the trial. The most frequent primary endpoint for phase I studies is toxicity. Secondary endpoints may be induction of humoral and/or cellular immune responses as well as changes in surrogate disease markers in blood and disease regression for therapeutic vaccines.
3. Set up a two-arm study comparing escalating doses of plasmid DNA given by intramuscular injection with or without subsequent electroporation.
4. Allow a time interval for observing and documenting safety issues, if any, after each dose escalation (without electroporation) before administering the lowest dose with electroporation.
5. Choose doses according to preclinical safety and efficacy studies, generally in the milligram range for the plasmid-DNA-only study arm. For the electroporation group, lower doses are recommended. To allow direct comparison of the effectiveness of both strategies, design the dose-escalation with 1 or more identical dose levels in each arm.

6. Vaccinate the patient at day 0 and repeat the vaccination 2–3 times at intervals of 3–4 weeks, followed by additional booster injections at later time points, typically after 6 and 12 months.
7. Allow crossover between study arms at the times of booster administration.

3.2. Inclusion and Exclusion Criteria

1. Determine patient inclusion criteria. Inclusion criteria for the study could be based on general health status and specific immunological criteria as well as carefully defining the study population in terms of the underlying condition. This step is particularly important for early phase studies, which will usually enroll small cohorts of patients to reduce interpatient variability.
2. Determine patient exclusion criteria specific for electroporation treatment.

   1. Patients having metal implants in the vicinity of the intended treatment sites are recommended to be excluded from electroporation treatment because of the potential of electrical contact between the electrodes and the implant, which may cause a short circuit.
   2. Patients at risk of congestive heart failure are recommended to be excluded since electroporation causes involuntarily muscle contractions, which might cause stress to a weak heart (*see* **Note 3**).
   3. Until more safety data is available, avoid patients who have a pacemaker in situ and ensure that this device is programmable if patients with pacemakers are to be included.
   4. Depending on the electrode configuration one might consider obesity as an exclusion criterion. Thick adipose tissue at the site of injection could make it difficult to reach the underlying muscle tissue and therefore increase the risk of inadequate electroporation (see sect. 3.4, no. 7) (*see* **Note 4**).

3.3. Information for the Patient

Prepare an information folder for enrolled patients, containing the following information.

1. General information about the trial.
2. A short introduction on electroporation as a delivery method for DNA vaccines.
3. A notice about the expected involuntarily muscle contraction, which might feel unpleasant or distressing.
4. Information about short-term pain: Electroporation results in short but intense pain. The best comparison may be to the sensation one experiences when touching a live wire or electric fence; however, the sensation is limited to the electroporated muscle

or muscles, such as in the thigh or the arm. This discomfort or pain will only last for a very short period (only during the pulsing, less than 1 s).

5. Information about the expected duration of posttreatment pain or discomfort: One might feel a mild ache for a few days, similar to what can be experienced following exercise or a strenuous walk.

6. Explanation of the effect on muscle tissue: Some local and transient damage is introduced in the muscle at the injection or electroporation site, and this damage is expected to abate within a few weeks.

7. Explanation of the effect the enhanced immune response might have on the treated area: A desired outcome of the treatment is to generate an immune response against the vaccine, which can be felt as soreness at the site of treatment. Since the site is small and muscle tissue will regenerate and repair, this is not expected to cause serious complications (*see* **Note 5**).

8. Information about optional sedation or pain relief.

9. Information on a brief questionnaire, which should be answered immediately after the injection and/or electroporation and 2 days after the treatment, to learn more about the experience persons have with electroporation.

10. Information about possible allergic reactions to the treatment. Depending on the device used, participants in the study should be informed about potential reactions to heavy metals dissolving from electrodes during treatment (*see* **Note 6**).

3.4. Vaccine Administration and Electroporation

1. Inform the patient that no specific preparations prior to electroporation are necessary.

2. Brief the patient again about the procedure, contraction, short pain, and questionnaire.

3. Let the patient find a comfortable position, either lying down on the back or sitting in a chair.

4. Give the patient a short acting intravenous benzodiazepine if this is to be offered and if the patient wishes to take up this offer.

5. Hold the muscle and inject the vaccine intramuscularly into the deltoid or the thigh. Use an injection volume of 0.2–0.5 mL delivered either by the Elgen injector/electrode system or a normal intramuscular injection (*see* **Note 7**).

6. Apply electrical pulses using invasive two- or four-needle array, with needle electrodes of 1.5 cm length and 4–10 mm apart.

7. Ensure that both the injection and electroporation target the muscle tissue and not only the adipose tissue (*see* **Note 4**).

8. Use a pulse pattern recommended by the provider of the electroporation device. Both devices used for muscle electroporation in the clinic apply short (20–60 ms), low voltage (40–106 V), unipolar pulses (2–6 pulses) at 4–10 Hz. Therefore, the procedure is very short.

9. Observe the patients closely for a period of at least 1 h after treatment.

3.5. Assessment of Pain and Distress

During, immediately after, and 48 h after the treatment, collect relevant information to determine the level of pain or discomfort. A modified visual analogue scale or a standard pain questionnaire may be helpful (*see* **Note 8**).

3.6. Examination of Patients or Follow Up

Follow up the patients at short intervals the first time after treatment to assess unexpected side effects and toxicities. The precise frequency will need to be decided according to the particulars of the study and, in particular, the schedule chosen for vaccine delivery and will also be influenced by the time-points, at which samples for immunological monitoring are to be collected. One possible schedule is to review the patient every second week for the first 8 weeks after treatment. After that, the examinations should be as frequent as the disease requires and according to standard clinical practice. For an early phase study it is expected that intervals of 1–2 months will be considered adequate.

Examinations should include the following.

1. Perform a general physical evaluation: WHO performance status and various immunological parameters.
2. Carefully examine the injection site for clinical evidence of a local reaction, including measurements of the circumference of the extremity, where appropriate.
3. Monitor for anti-DNA antibodies, rheumatoid factors, and evidence of muscle destruction (LDH and CK). If these tests become significantly positive after being normal, or other signs of autoimmunity appear, vaccinations should be terminated and rheumatology consultation should be sought.

4. Notes

1. So far, both of these options have been declined by all patients treated with intramuscular electroporations.
2. For a more general review of safety requirements for DNA vaccines, see, e.g., Glenting and Wessels (30).
3. New York Heart Association scale is as follows:
    Class III: patients with cardiac disease resulting in marked limitations of physical activity, who are comfortable at rest, but less than ordinary physical activity causes dyspnea (or fatigue, palpitation, or anginal pain),
    Class IV: patients with cardiac disease resulting in inability to carry out physical activity without discomfort. Symptoms of dyspnea (or angina) may be present even at rest, and, if any physical activity is undertaken, discomfort is increased.

4. It is imperative that DNA injection and electrodes reach into the muscle tissue. Thickness of adipose tissue in humans varies according to gender, size, and anatomical site and may exceed 1–2 cm (31). So, DNA injection and electrode insertion depth may need to be adjusted accordingly. Typically, electroporation delivered only into adipose tissue will be less painful because of reduced electrical current caused by the higher electrical resistance of the adipose tissue. Electroporation of adipose tissue will most likely not elicit an efficient immune response against the vaccine (experimental evidence is presently not available). To reduce the risk of inadequate treatment, proper protocol design, including planning for contingencies involving technical or procedural shortcomings, must be ensured. In addition, proper choice of electroporation equipment and procedure, and adequate training of study personnel need to be emphasized.

5. If the treatment (immunization) is successful, the immune system might attack the muscle cells producing the vaccine and eliminate them. The duration of the soreness due to local inflammation at the site of expression will therefore depend on the presence of antigens at the treatment site and the effectiveness of the immune stimulation by the vaccine.

6. The Elgen system uses regular stainless steel electrodes and trace amounts of nickel and chromium are released during electroporation. Although we consider the possibility of an allergic reaction due to electroporation unlikely, patients with metal (nickel or chromium) allergies should be informed about this risk. Inovio's MedPulser® (DDS) uses gold-plated electrodes to avoid toxic metal complications.

7. Most electrode configurations will not affect more than 0.5 mL of muscle volume, depending on the injection procedure; therefore, increasing the injection volume may not increase the vaccine effect since the reach of the electrical field might become a limiting factor. If a larger volume is needed to administer a high DNA dose, one may consider dividing the dose between two or more injection sites. However, increasing the number of treatment sites will also increase discomfort and pain. One might also consider administering a second injection to a different muscle group, preferably one draining to a different lymph node since it may be advantageous to stimulate several lymph nodes during vaccination (32).

8. Electroporation-induced pain has not been a limiting factor. Experience from all our studies confirms that patients subjected to electroporation do come back for repeated treatments. In addition, three different feasibility studies in healthy volunteers subjected to up to four sequential treatments using various electrical conditions and different electrode configurations have shown that electroporation without anesthesia is tolerable. When reporting pain during electroporation, patients typically score from 3–8 on a 1–10 visual analogue pain scale ((27) and Tjelle 2003). While this may seem high, one should keep in mind the extremely short duration of the procedure. Nevertheless, when planning a study involving human subjects it may be prudent to choose equipment and pulse conditions, which minimize invasiveness and pain while still eliciting efficient immune responses against the plasmid encoded antigen(s). From the growing body of data accumulated, it seems clear that a wide range of electroporation conditions (2–10 pulses, 10–60 ms duration, and 60–250 V/cm) are efficient for DNA vaccination in large animals (including primates) and should, therefore, presumably work well in humans. Electroporation-induced pain will be lower using milder electrical conditions (lower voltage, pulse number, and pulse length) and smaller electrodes and electrode arrays ((27) and Tjelle 2003).

References

1. Weiner, D.B. (1995) New vaccine strategies. *Mol. Med. Today.* **1**, 108–109.
2. Widera, G., Austin, M., Rabussay, D., et al. (2000) Increased DNA vaccine delivery and immunogenicity by electroporation in vivo. *J. Immunol.* **164**, 4635–4640.
3. Rabussay, D., Dev, N.B., Fewell, J., Smith, L.C., Widera, G., and Zhang, L. (2003) Enhancement of therapeutic drug and DNA delivery into cells by electroporation. *J. Phys. D.* **36**, 348–363.

4. Babiuk, S., Baca-Estrada, M.E., Foldvari, M., et al. (2003) Needle-free topical electroporation improves gene expression from plasmids administered in porcine skin. *Mol. Ther.* **8**, 992–998.

5. Vangasseri, D.P., Han, S.J., and Huang, L. (2005) Lipid-protamine-DNA-mediated antigen delivery. *Curr. Drug Deliv.* **2**, 401–406.

6. Spik, K., Shurtleff, A., McElroy, A.K., Guttieri, M.C., Hooper, J.W., and Schmaljohn, C. (2006) Immunogenicity of combination DNA vaccines for Rift Valley fever virus, tick-borne encephalitis virus, Hantaan virus, and Crimean Congo hemorrhagic fever virus. *Vaccine.* **24**, 4657–4666.

7. Rompato, G., Ling, E., Chen, Z., Van, K.H., and Garmendia, A.E. (2006) Positive inductive effect of IL-2 on virus-specific cellular responses elicited by a PRRSV-ORF7 DNA vaccine in swine. *Vet. Immunol. Immunopathol.* **109**, 151–160.

8. Ulmer, J. B., Wahren, B., and Liu, M.A. (2006) Gene-based vaccines: recent technical and clinical advances. *Trends Mol. Med.* **12**, 216–222.

9. Minke, J.M., Siger, L., Karaca, K., et al. (2004) Recombinant canarypoxvirus vaccine carrying the prM/E genes of West Nile virus protects horses against a West Nile virus-mosquito challenge. *Arch. Virol. Suppl.* 221–230.

10. Garver, K.A., LaPatra, S.E., and Kurath, G. (2005) Efficacy of an infectious hematopoietic necrosis (IHN) virus DNA vaccine in Chinook Oncorhynchus tshawytscha and sockeye O. nerka salmon. *Dis. Aquat. Organ.* **64**, 13–22.

11. Babiuk, S., Baca-Estrada, M., Foldvari, M., et al. (2002) Electroporation improves the efficacy of DNA vaccines in large animals. *Vaccine.* **20**, 3399.

12. Otten, G.R., Schaefer, M., Doe, B., et al. (2006) Potent immunogenicity of an HIV-1 gag-pol fusion DNA vaccine delivered by in vivo electroporation. *Vaccine.* **24**, 4503–4509.

13. Li, Z., Zhang, H., Fan, X., et al. (2006) DNA electroporation prime and protein boost strategy enhances humoral immunity of tuberculosis DNA vaccines in mice and non-human primates. *Vaccine.* **24**, 4565–4568.

14. Babiuk, S., Baca-Estrada, M.E., Foldvari, M., et al. (2004) Increased gene expression and inflammatory cell infiltration caused by electroporation are both important for improving the efficacy of DNA vaccines. *J. Biotechnol.* **110**, 1–10.

15. Heller, L., Merkler, K., Westover, J., et al. (2006) Evaluation of toxicity following electrically mediated interleukin-12 gene delivery in a B16 mouse melanoma model. *Clin. Cancer Res.* **12**, 3177–3183.

16. Otten, G., Schaefer, M., Doe, B., et al. (2004) Enhancement of DNA vaccine potency in rhesus macaques by electroporation. *Vaccine.* **22**, 2489–2493.

17. Draghia-Akli, R., Ellis, K.M., Hill, L.A., Malone, P.B., and Fiorotto, M.L. (2003) High-efficiency growth hormone-releasing hormone plasmid vector administration into skeletal muscle mediated by electroporation in pigs. *FASEB J.* **17**, 526–528.

18. Wloch, M. K., Hartikka, J., Bozoukova, V., et al. (2006) Electroporation-assisted intramuscular delivery of plasmid DNA vaccines: evaluation of immunogenicity, plasmid DNA persistence and integration. The Secondary International Conference on Modern Vaccines Adjuvants and Delivery Systems, The Royal Society of Medicine, London, UK.

19. Kjeken, R., Tjelle, T.E., Kvale, D., and Mathiesen, I. (2004) Electroporation of skeletal muscle in humans. Executive Summaries. 7th Annual Meeting of the American Society of Gene Therapy.

20. Zhang, L. and Rabussay, D. (2005) Progress toward the development of electroporation for muscle-targeted DNA vaccines. International Symposium on Bioelectrochemistry ad Bioenergetics, Portugal, Coimbra.

21. Wang, Z., Troilo, P.J., Wang, X., et al. (2004) Detection of integration of plasmid DNA into host genomic DNA following intramuscular injection and electroporation. *Gene Ther.* **11**, 711–721.

22. Fons, M. (2006) Clinical Electroporation. Executive Summaries. 9th Annual Meeting of the Americal Society of Gene Therapy, 24.

23. Heller, R. (2006) Electroporation-based gene therapy in humans. Executive Summaries. 9th Annual Meeting of the Americal Society of Gene Therapy, 36.
24. H.Lee Moffitt Cancer and Research Institute. (2006) Phase I Trial of intratumoral pIL-12 electroporation in malignant melanoma. www. clinicaltrials. gov/ct/, NCT00323206.
25. Vical. (2006) A Phase I trial to evaluate the safety of intra-tumoral VCL-IM01 followed by electroporation in meatstatic melanoma. www. clinicaltrials. gov/ct/, NCT00223899.
26. Merck. (2006) V930 First Man (FIM) Study. www. clinicaltrials. gov/ct/, NCT00250419.
27. Rabussay, D., Widera, G., Zhang, L., et al. (2004) Toward the development of electroporation for delivery of DNA vaccines to humans. Executive Summaries. 7th Annual Meeting of the American Society of Gene Therapy.
28. Tjelle, T. E., Salte, R., Mathiesen, I., and Kjeken, R. (2006) A novel electroporation device for gene delivery in large animals and humans. *Vaccine*. **24**, 4667–4670.
29. Southampton General Hospital. (2006). Vaccine treatment for prostate cancer that has come back (GTAC No 089). http://www. cancerhelp. org. uk/trials/.
30. Glenting, J. and Wessels, S. (2005) Ensuring safety of DNA vaccines. *Microb. Cell Fact.* **4**, 26.
31. Poland, G.A., Borrud, A., Jacobson, R.M., et al. (1997) Determination of deltoid fat pad thickness. Implications for needle length in adult immunization. *JAMA*. **277**, 1709–1711.
32. Estcourt, M.J., Letourneau, S., McMichael, A.J., and Hanke, T. (2005) Vaccine route, dose and type of delivery vector determine patterns of primary CD8 + T cell responses. *Eur. J. Immunol.* **35**, 2532–2540.

Chapter 40
Production of Monoclonal Antibody by DNA Immunization with Electroporation

Kaw Yan Chua, John D.A. Ramos, and Nge Cheong

Abstract DNA immunization with in vivo electroporation is an efficient alternative protocol for the production of monoclonal antibodies (mAb). Generation of mAb by DNA immunization is a novel approach to circumvent the following technical hurdles associated with problematic antigens: low abundance and protein instability and use of recombinant proteins that lack posttranslational modifications. This chapter describes the use of a DNA-based immunization protocol for the production of mAb against a house dust mite allergen, designated as Blo t 11, which is a para-myosin homologue found in *Blomia tropicalis* mites. The Blo t 11 cDNA fused at the N terminus to the sequence of a signal peptide was cloned into the pCI mammalian expression vector. The DNA construct was injected intramuscularly with in vivo electroporation into mice, and the specific antibody production in mice was analyzed by enzyme-linked immunosorbent assay (ELISA). Hybridomas were generated by fusing mouse splenocytes with myeloma cells using the ClonaCell™-HY Hybridoma Cloning Kit. Six hybridoma clones secreting Blo t 11 mAb were successfully generated, and these mAb are useful reagents for immunoaffinity purification and immunoassays.

Keywords: monoclonal antibody, DNA immunization, electroporation, hybridoma, dust mite allergen

1. Introduction

Monoclonal antibodies (mAb) are powerful immunological and biochemical tools in basic and applied research. In recent years, mAb have been fully exploited for development of invaluable diagnostic and therapeutic reagents. Conventional hybridoma techniques for the generation of mAb require the use of exogenously produced protein immunogens for immunization in laboratory animals (1). In the case of low abundance proteins or recombinant proteins that lack posttranslational modifications, mAb production may be problematic and challenging. The DNA immunization approach is a simple, cheap, and efficient strategy to circumvent

S. Li (ed.), *Electroporation Protocols: Preclinical and Clinical Gene Medicine.*
From *Methods in Molecular Biology, Vol. 423.*
© Humana Press 2008

laborious protein production and purification and, yet, can effectively induce antigen-specific immune response.

Wolff et al. first showed that intramuscular (i.m.) immunization with naked plasmid DNA resulted in direct gene transfer into mouse muscles (2). Subsequently, several reports revealed that DNA immunization is an effective strategy for the delivery of DNA constructs encoding specific immunogens into host cells, inducing both antigen-specific humoral and cellular immune responses (3–6). The method of DNA immunization requires the gene of interest to be cloned in a suitable mammalian expression vector with a strong promoter (e.g., CMV enhancer/ promoter region) and a polyadenylation/transcriptional termination sequence. The two commonly used routes of DNA immunization include the intramuscular (i.m.) injection and the use of "gene gun" to propel gold beads with DNA into the epidermis (6).

Intramuscular immunization of naked DNA constructs in mice resulted in the in situ expression of the encoded proteins (2), thereby generating antibody responses in the immunized organism (3, 7). In recent years, the use of electroporation in conjunction with intramuscular immunization has attracted considerable attention (8, 9). Muscle electroporation using high-frequency, low-voltage electric pulses has been reported to increase the production and secretion of recombinant erythropoietin in mouse skeletal muscles by more than 100-fold (8). Electroporation was shown to increase the expression of the recombinant protein encoded by the immunized gene and to elevate the immune responses of animals against the expressed protein (9, 10). Electroporation enhances uptake of genes by permeabilization of the muscle fibers, resulting in a very high frequency of transfected bundles as well as higher intensity of expression. The enhancement of immunity apparently follows the increased availability of antigen.

Production of specific antibodies by DNA immunization is initiated by the expression of the protein encoded by the immunized gene (6). The encoded protein is then targeted as an internal immunogen to the host immune system. Antibodies produced by DNA immunization have been reported to recognize both linear and conformational antigens (11). DNA immunization coding for a defined portion of the *Helicobacter pylori* vacuolating cytotoxin in mice resulted in the generation of highly specific mAb recognizing the natural protein (12). Polyclonal antibodies against apobec-1 (13), an editing enzyme that is difficult to purify, and tobacco mosaic virus coat protein (TMV-CP) (14) have been generated by DNA immunization in rabbits. Our laboratory was the first to show that allergen-specific antibody and cellular responses, induced by allergen-gene immunization, were protective against mite allergy (15). Subsequently, mAb for a number of mite allergens have been generated in our laboratory by the DNA immunization approach (16–19). In addition, the DNA immunization approach has been successfully used to generate mAb for native human thyrotropin receptor (20), the phosphoprotein (21), and blood group antigens (22). These studies demonstrate the advantages of DNA immunization, whereby a full spectrum of native epitopes of an antigen can be expressed in situ in the host's mammalian system, and thus a better conformation or protein presentation can be achieved,

compared with administration of recombinant protein expressed and purified in vitro. In addition, complex mixtures of antigens can be screened for immunogenicity or protective efficacy using DNA immunization far more rapidly than by purifying the individual protein components.

Here, we describe the generation of mAb for a house dust mite allergen, designated as Blo t 11, which is a paramyosin homologue found in *Blomia tropicalis* mites. The molecular weight of the Blo t 11 protein is ~102 kDa, and the production of intact protein was problematic because of its susceptibility to degradation (23); therefore, it was not feasible to immunize animals with the conventional protein immunization protocol. We have resolved the problem by adopting the DNA immunization approach. The Blo t 11 cDNA fused at the N terminus to the sequence of a signal peptide was cloned into the pCI mammalian expression vector (Promega). The signal peptide sequence used was derived from the cDNA (complementary DNA) encoding for Der p 5 dust mite allergen (15). The DNA construct was

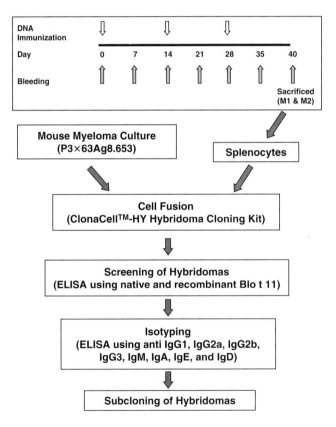

Fig. 40.1 Schematic representation of the DNA immunization regimen in BALB/cJ mice and the stepwise protocols used to generate and screen hybridomas secreting Blo t 11 mAb, followed by production and characterization of the various Blo t 11 mAb

injected intramuscularly with in vivo electroporation into mice and the specific antibody production in mice was analyzed by enzyme-linked immunosorbent assay (ELISA). The generation, screening, and subcloning of hybridomas were performed using the ClonaCell™-HY Hybridoma Cloning Kit (Fig. 40.1).

2. Materials

2.1. DNA Construct

1. Mammalian expression vector pCI (Promega) containing Der p 5 signal peptide sequence (Dp5LS) (*see* **Note 1**).
2. PCR-amplified Blot 11 gene (2,625-bp open reading frame) using forward primer 5′ CGTCCGGAATGGCGGCTCGATCAGCA 3′ containing *Bsp*E1 restriction enzyme site and reverse primer 5′ AGAGAGCTCGAAC TTTTAA GCGGCCGCCG 3′ containing *Not*1 restriction enzyme site.
3. Pfu polymerase: Store at −20 °C.
4. Restriction enzymes *Bsp*E1 and *Not*1: Store at −20 °C.
5. T4 DNA ligase: Store at −20 °C.
6. Competent *Escherichia coli* DH5α: Thaw in ice immediately before use. Store in 200 μL aliquots at −80 °C.
7. Luria-Bertani agar plates with ampicillin: Dissolve 25 g of Luria-Bertani, Miller (Becton Dickinson, Sparks, MD), and 20 g of Bacto Agar™ (Becton Dickinson) in distilled water to a final volume of 1 L. Mix properly using a magnetic stirrer, autoclave for 15 min at 121 psi, cool to ~55 °C, add 1 mL of ampicillin (100 μg/mL), shake, and pour into plates. Store Luria-Bertani agar-ampicillin plates at 4–8 °C.
8. Luria-Bertani broth with ampicillin: Dissolve 25 g of Luria-Bertani in distilled water to a final volume of 1 L. Mix properly using a magnetic stirrer, autoclave for 15 min at 121 psi, cool to ~55 °C, add 1 mL of ampicillin (100 μg/mL), shake, and store at 4–8 °C.
9. Wizard® Plus SV Miniprep DNA Purification System (Promega, Madison, WI): Store at room temperature.
10. QIAGEN Plasmid Giga Kit (Qiagen, Hilden, Germany). Store at room temperature.
11. Agarose gel (1% w/v): Dissolve 1 g of Seakem® LE agarose (BioWhittaker Molecular Applications, Rockland, ME) in 100 mL of 1× TAE buffer, microwave mixture, add ethidium bromide (final concentration of 0.5 μg/mL), pour on casting tray, and allow to polymerize.
12. TAE buffer (50× stock solution): Dissolve 242 g of Tris base in 800 mL of distilled water. Then 57.1 mL of glacial acetic acid and 100 mL of 0.5 M EDTA are added. Adjust pH to 8.0 and add distilled water to a final volume of 1 L. Store solution at room temperature.

13. ABI Prism 377 Automated DNA Sequencer (PE Applied Biosystems, Foster City, CA).
14. GenElute™ Agarose Spin Column (Sigma-Aldrich, St. Louis, MO).

2.2. DNA Immunization

1. Five 6-week-old female BALB/cJ mice. All mice are housed under conventional conditions and fed with mice food pellets and water ad libitum (see **Note 2**).
2. Electro Square Porator (ECM 830) attached to a two-needle array electrode with cable (model no. 532) (BTX, CA).
3. U-100 insulin syringe with 29-G needle.
4. Phosphate-buffered saline (PBS): Dissolve 8 g of NaCl, 0.2 g of KCl, 1.44 g of Na_2HPO_4, and 0.24 g KH_2PO_4 in distilled water to a final volume of 1 L. Adjust pH to 7.4 with HCl.

2.3. Enzyme-Linked Immunosorbent Assay

1. Antigens: Recombinant and native Blo t 11 allergens were used (see **Note 3**). Store in aliquots at −20 °C.
2. ELISA plates (Costar Flat bottom high binding plate 3590).
3. ELISA coating buffer: 0.1 M $NaHCO_3$, pH 8.3.
4. Washing buffer: 1× phosphate-buffered saline, pH 7.4 (sect. 2.2, item 4), containing 0.05% (v/v) Tween 20.
5. ELISA plate washer (Tecan, Austria).
6. Blocking buffer: 1% (w/v) bovine serum albumin in washing buffer.
7. Primary antibody (antisera) is diluted in blocking buffer (see **Note 4**).
8. Secondary antibody: Biotinylated anti-mouse immunoglobulin is diluted 5,000× in blocking buffer.
9. Alkaline phosphatase-conjugated ExtrAvidin is diluted 5,000× in blocking buffer.
10. Substrate solution: *p*-nitrophenyl phosphate (pNPP) tablets are dissolved and diluted in distilled water (see **Note 5**).
11. ELISA plate reader.

2.4. Cell Fusion and Cloning of Hybridomas

1. ClonaCell™-HY Hybridoma Cloning Kit (StemCell Technologies Inc., Vancouver, Canada).
2. Myeloma cells, a sub clone of P3X63Ag8.653 (ATCC, Manassas, VA).
3. 96-well Falcon® tissue culture plates.

4. Biotinylated anti-mouse IgG1, IgG2a, IgG2b, IgG3, IgM, IgA, IgE and IgD.
5. Trypan blue.

2.5. Ascites Production and Purification

1. Dulbecco's modified Eagle's medium (DMEM).
2. 10% fetal Calf Serum.
3. Trypan blue.
4. Pristane (2,6,10,14-tetra-methyl-pentadecane).
5. 8-week-old female BALB/c mice.
6. Protein A- or G-agarose column.

3. Methods

3.1. Preparation of DNA Construct

1. PCR-amplify the 2,625-bp open reading frame of the Blo t 11 gene by using a 25 µL PCR reaction consisting of 1 µL of plasmid DNA, (*see* **Note 6**) 1 µL each of 10 µM forward and reverse primers, 0.5 µL of 10 mM dNTPs, 2.5 µL of 10× PCR buffer, 0.4 µL of *Pfu* polymerase enzyme, and 18.6 µL of deionized distilled water. Amplification is done with 30 cycles: each cycle consists of 94 °C for 30 s, 55 °C for 30 s, and 72 °C for 3 min with a final elongation step of 72 °C for 10 min, using the Programmable Thermalcycler.
2. Electrophorese the 25 µL PCR product in 1% agarose gel for 20 min at 110 V using a BioRad Agarose electrophoresis unit. Run a 1-kb DNA marker to determine the molecular weight of the PCR product. Visualize and document the DNA band using a gel documentation system.
3. Cut the DNA band from the agarose gel examined under a UV transilluminator and purify the PCR product using a GenElute™ Agarose Spin Column.
4. Separately digest the purified Blo t 11 gene PCR product and the pCI Dp5LS plasmid by using *Bsp*E1 and *Not*1 for 1 h at 37 °C. Electrophorese and purify the digested samples as in step 3.
5. Ligate the *Bsp*E1- and *Not*1-digested Blo t 11 gene into the *Bsp*E1- and *Not*1-linearized pCI-Dp5LS plasmid for 14–16 h at 16 °C.
6. Transform the ligated samples into *E. coli* DH5α by heat shock method. A transformation reaction mixture of 200 µL of thawed competent cells with 10 ng of the ligation mixture is incubated on ice for 30 min, followed by heat shock at 42 °C for 2 min and 30 s in a water bath, and finally incubated on ice for 10 min.
7. Resuspend the transformation reaction (200 µL) in 800 µL of Luria-Bertani broth and incubate at 37 °C for 1 h with constant shaking at 250 rpm. Plate cells onto selective Luria-Bertani ampicillin plates and incubate these plates overnight at 37 °C (*see* **Note 7**).

8. Pick 3–5 colonies and culture them separately in 5 mL Luria-Bertani broth with ampicillin. The broth cultures are incubated overnight at 37 °C with constant shaking at 250 rpm. Each of the selected colonies is also plated onto a Luria-Bertani ampicillin plate, which is incubated overnight at 37 °C.

9. Pellet bacterial cells from the overnight cultures by centrifugation for the isolation of plasmid DNA using the Wizard® Plus SV Miniprep DNA purification system. Presence of insert is analyzed by PCR, RE digest, and DNA sequencing.

10. A clone containing a plasmid with the inserted gene in its proper orientation is cultured in 1 L of Luria-Bertani broth for the large-scale isolation of plasmid DNA. Luria-Bertani broth cultures are set up following similar protocols as in step 8.

11. Plasmids are isolated and purified using QIAGEN Plasmid Giga Kit, and a sample is sequenced using the ABI Prism™ 377 Automated DNA Sequencer.

12. Resuspend the purified plasmid DNA in sterile water, analyze the purified DNA in 1% agarose gel, and quantitate the DNA by spectroscopy. Aliquots are stored at −20 °C until further use.

13. Glycerol stocks of the positive transformants consisting of a 1:1 mixture of overnight bacterial culture and sterile 70% glycerol are prepared, mixed properly, and stored at −80 °C for future use.

3.2. Immunization of DNA Constructs

1. Acclimatize five 6-week-old female BALB/cJ for at least a week.

2. Immunize mice intramuscularly with 50 μg of the pCI-Dp5LS-Blo t 11 plasmid DNA (suspended in 50 μL of PBS) on the quadriceps muscle of the hind leg by using U-100 insulin syringe with 29-G needle.

3. Immediately after immunization, each mouse is pulsed at the site of the immunization using the Electro Square Porator (ECM 830) attached to a two-needle array electrode with cable (model no. 532) (BTX, Genetronix, CA), using the following parameters: 4 pulses of 82 V with a pulse length of 20 ms and an interval of 200 ms.

4. Boost each mouse twice with electroporation at days 14 and 28 using the same parameters mentioned earlier.

5. Collect blood via infra-orbital every week to monitor the titer of antibody production. Isolate sera from the blood samples and store at −180 °C until further use.

3.3. Determination of Antibody Titer Production

1. Suspend 50 μL of antigens (5 μg/mL) in ELISA coating buffer and coat the suspension onto 96-well ELISA plates in duplicates by incubation for 14–16 h at 4 °C with gentle shaking.

2. Wash ELISA plates thrice with 200 μL of washing buffer per well using an automated ELISA plate washer.

3. Incubate each well with 200 μL blocking buffer for 1 h at 22 °C. Use a multichannel pipette to dispense reagents. Wash plates as in step 2 after the incubation period.

4. Dilute antisera in blocking buffer. Add 50 μL of different dilutions (*see* **Note 4**) of the primary antibody (antisera) extracted from the different immunized mice at different time points to corresponding wells. Incubate the primary antibody for 1 h at room temperature with constant gentle shaking using a horizontal shaker. Wash plates as in step 2 after the incubation period.

5. Dispense 50 μL of diluted secondary antibody into each well. Incubate for 1 h at room temperature with constant gentle shaking using a horizontal shaker. Wash plates as in step 2 after the incubation period.

6. Incubate each well with 50 μL of diluted alkaline phosphatase-conjugated ExtrAvidin solution. Incubate for 1 h at room temperature with constant gentle shaking using a horizontal shaker. Wash plates as in step 2 after the incubation period.

7. Develop reaction by incubating each well with 50 μL of substrate solution. Incubate for 30 min up to 1 h at room temperature in the dark.

8. Read absorbance at 405 nm using an ELISA plate reader. Plot the absorbance and determine the kinetics of antibody titer production of each mouse. An example of antibody response of 2 mice immunized with a DNA construct is shown in Fig. 40.2.

3.4. Cell Fusion and Cloning of Hybridomas

1. A week prior to cell fusion, culture parental myeloma cells to confluence in medium A of the ClonaCell™-HY Hybridoma Cloning Kit. Subculture cells into several flasks reaching the mid-log phase growth on the day of fusion.

2. Determine the viability of the myeloma cells by the trypan blue exclusion assay and count the number using a hemocytometer.

3. Resuspend 2×10^7 viable myeloma cells in 30 mL of medium A of the ClonaCell™-HY Hybridoma Cloning Kit.

4. Euthanize two mice with the highest antibody titer by cervical dislocation and dissect the spleens under sterile conditions. Splenocytes are separated by grinding the spleen tissue using the frosted/coarse end of a glass slide. These cells are resuspended in medium B of the ClonaCell™-HY Hybridoma Cloning Kit.

5. Fuse a total of 1×10^8 viable cells with previously prepared myeloma cells using the supplied PEG solution and medium B in the ClonaCell™-HY Hybridoma Cloning Kit.

6. Incubate the resulting hybridomas overnight at 37 °C in medium C. Then plate them using medium D (methylcellulose-based medium) and incubate them for 14 days at 37 °C until isolated colonies appear.

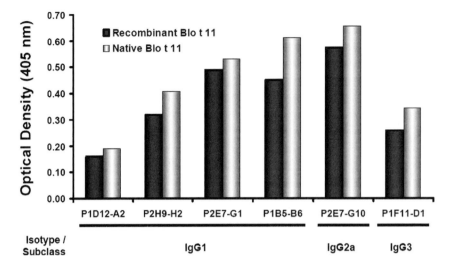

Fig. 40.2 Kinetics of native Blo t 11 specific antibody productions of the two high responder mice (M1 and M2) immunized intramuscularly with pCI-Dp5LS-Blo t 11 construct via electroporation according to methods described in sect. 3.2. Immunization regimen used is shown in Fig. 40.1. Mice were bled weekly for antibody analysis by ELISA, as described in sect. 2.3. Mice were euthanized on day 40, and suspension cells prepared from the spleens were used for hybridoma production according to protocols summarized in Fig. 40.1

7. Pick isolated colonies using a pipette tip and culture hybridomas in 96-well Falcon® tissue culture plates.
8. Collect hybridoma clone supernatants and assay for recombinant and native Blo t-11-specific antibody by ELISA, as described earlier.
9. Subculture antigen-specific antibody-producing clones and expand into 6-well Falcon® culture plates using DMEM medium supplemented with 10% fetal calf serum.
10. Isotype each positive clone by ELISA (as described earlier) using biotinylated anti-mouse IgG1, IgG2a, IgG2b, IgG3, IgM, IgA, IgE and IgD.

3.5. Ascites Production and Purification

1. One week prior to inoculation of hybridoma cells, inject intraperitoneally 1 mL pristane into each mouse by using a 22-G needle.
2. Culture high antibody titer producing hybridoma clones in DMEM medium supplemented with 10% fetal calf serum and expand for inoculation into mice for ascites production.
3. Inject 5×10^6 viable cells in 1 mL PBS intraperitoneally into each mouse by using a 22-G needle.

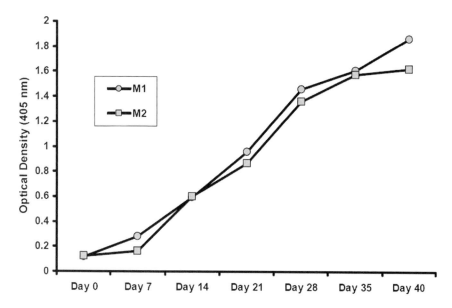

Fig. 40.3 The IgG reactivity of six representative mAb (C1 – C6) to native and recombinant Blo t 11 proteins. The antibody titers were determined by ELISA and presented as optical densities measured at a wavelength of 405 nm. The indicated subclasses of these mAb were determined by ELISA, as described in sect. 3.5

4. About 1–2 weeks after cell inoculation, collect ascites fluid from each mouse by gravity flow into 15-mL conical centrifuge tubes using an 18-G needle inserted 1–2 cm into the abdominal cavity (*see* **Note 8**).

5. Assay ascites fluid for antibody titer and specificity by ELISA, as described earlier, using recombinant and native antigens (*see* **Note 9**). A panel showing the IgG reactivity of 6 different mAb to Blo t 11 allergen is shown in Fig. 40.3.

6. Purifiy the mAb from the pooled ascites fluid by affinity chromatography using protein A or G agarose column (*see* **Note 10**).

Acknowledgments This study was supported by grants from the Agency for Science, Technology, and Research (A*STAR), Singapore (R-178-000-010-303, R-178-000-101-305), and Academic Research Fund, National University of Singapore, Singapore (R-178-000-010-112).

4. Notes

1. The induction of protein-specific antibodies can be enhanced by increasing the levels of protein expression in vivo following DNA immunization. Inclusion of a signal peptide sequence in a DNA construct will help to enhance the in vivo expression of the secretory protein. Our previous studies showed that up to a 10-fold increase in the titers of antibody production in immunized mice were observed with constructs containing the signal peptide sequence

derived from Der p 5 mite allergen cDNA (24). Der p 5 derived signal peptide sequence can be replaced with signal peptide sequences from other genes. We had shown that the use of pSecTag2 vectors (Invitrogen Life Technologies), which are expression vectors designed for high-level protein expression in mammalian hosts, could significantly increase the magnitude of antibody responses in DNA-immunized mice. Proteins expressed from pSecTag2 are fused at the N terminus with Ig κ-chain signal peptide sequence for protein secretion.

2. All animals should be treated humanely and all animal experiments were performed according to Institutional Guidelines for Animal Care and Handling of National University of Singapore.

3. Recombinant Blo t 11 allergen was expressed in *E. coli* as a GST-fusion protein (25) while the native Blo t 11 was immunoaffinity purified from a crude *B. tropicalis* extract (23).

4. We recommend determining the best dilution at which the least noise to background ratio can be observed by diluting the antisera at 10×, 100×, 1,000×, and 5,000×.

5. *p*-nitrophenyl phosphate solution is light-sensitive. Turn off lights or close windows when pipetting the solution. Use the recommended volume of distilled water for dilution of each tablet according to the manufacturer's instructions.

6. The full-length Blo t 11 gene was cloned by screening a *B. tropicalis* (Bt) cDNA library by plaque lift method, 5′ rapid amplification of cDNA ends (RACE), and by long-distance PCR (25). The Bt cDNA library was constructed according to standard protocol by inserting cDNA into *Eco*RI and *Xho*I site of Uni-ZAP® expression vector (Stratagene, La Jolla, CA).

7. It is a common practice to plate 200 and 800 μL of the transformation reaction into two separate Luria-Bertani plates, as the efficiency of the ligation and transformation procedures vary from one experiment to another. The goal is to produce well-isolated bacterial colonies after the overnight incubation.

8. Mice were euthanized after two collections of ascites fluid.

9. The mAb assays could also be performed by Western blot analysis and dot blot analysis.

10. It is also a good idea to assay the purified mAb for specificity using different immunoassays with both recombinant and native antigens.

References

1. Kohler, G. and Milstein, C. (1975) Continuous cultures of fused cells secreting antibody of predefined specificity. *Nature*. **256**, 495–497.

2. Wolff, J.A., Malone, R.W., Williams, P., et al. (1990) Direct gene transfer into mouse muscle *in vivo*. *Science*. **247**, 1465–1468.

3. Tang, D.C., DeVit, M., and Johnston, S.A. (1992) Genetic immunization is a simple method for eliciting an immune response. *Nature*. **356**, 152–154.

4. Srivastava, I.K. and Liu, M.A. (2003) Gene vaccines. *Ann. Intern. Med.* **138**, 550–559.

5. Davis, H.L. (1997) Plasmid DNA expression systems for the purpose of immunization. *Curr. Opin. Biotechnol.* **8**, 635–640.

6. Donnelly, J.J., Ulmer, J.B., Shiver, J.W., and Liu, M.A. (1997) DNA vaccines. *Annu. Rev. Immunol.* **15**, 617–648.

7. Barry, M.A., Barry, M.E., and Johnston, S.A. (1994) Production of monoclonal antibodies by genetic immunization. *Biotechniques*. **16**, 616–620.

8. Rizzuto, G., Cappelletti, M., Maione, D., et al. (1999) Efficient and regulated erythropoietin production by naked DNA injection and muscle electroporation. *Proc. Natl. Acad. Sci. U.S.A.* **96**, 6417–6422.

9. Widera, G., Austin, M., Rabussay, D., et al. (2000) Increased DNA vaccine delivery and immunogenicity by electroporation *in vivo*. *J. Immunol.* **164**, 4635–4640.

10. Mir, L.M., Bureau, M.F., Gehl, J., et al. (1999) High efficiency gene transfer into skeletal muscle mediated by electric pulses. *Proc. Natl. Acad. Sci. U.S.A.* **96**, 4262–4267.

11. Wang, B., Boyer, J., Srikantan, V., et al. (1995) Induction of humoral and cellular immune responses to the human immunodeficiency type 1 virus in nonhuman primates by *in vivo* DNA inoculation. *Virology.* **211**, 102–112.

12. Ulivieri, C., Burroni, D., Telford, J.L., and Baldari, C.T. (1996) Generation of a monoclonal antibody to a defined portion of the *Helicobacter pylori* vacuolating cytotoxin by DNA immunization. *J. Biotechnol.* **51**, 191–194.

13. Yeung, S.C., Anderson, J., Kobayashi, K., Oka, K., and Chan, L. (1997) Production of rabbit polyclonal antibody against apobec-1 by genetic immunization. *J. Lipid Res.* **38**, 2627–2632.

14. Hinrichs, J., Berger, S., and Shaw, J.G. (1997) Induction of antibodies to plant viral proteins by DNA-based immunization. *J. Virol. Methods.* **66**, 195–202.

15. Hsu, C.H., Chua, K.Y., Tao, M.H., et al. (1996) Immunoprophylaxis of allergen-induced immunoglobulin E synthesis and airway hyperresponsiveness *in vivo* by genetic immunization. *Nat. Med.* **2**, 540–544.

16. Yang, L., Cheong, N., Wang, D.Y., et al. (2003) Generation of monoclonal antibodies against Blo t 3 using DNA immunization with *in vivo* electroporation. *Clin. Exp. Allergy.* **33**, 663–668.

17. Ramos, J.D., Cheong, N., Teo, A.S., Kuo, I.C., Lee, B.W., and Chua, K.Y. (2004) Production of monoclonal antibodies for immunoaffinity purification and quantitation of Blo t 1 allergen in mite and dust extracts. *Clin. Exp. Allergy.* **34**, 604–610.

18. Ramos, J.D., Teo, A.S., Lee, B.W., Cheong, N., and Chua, K.Y. (2004) DNA immunization for the production of monoclonal antibodies to Blo t 11, a paramyosin homolog from *Blomia tropicalis. Allergy.* **59**, 539–547.

19. Yi, F.C., Lee, B.W., Cheong, N., and Chua, K.Y. (2005) Quantification of Blo t 5 in mite and dust extracts by two-site ELISA. *Allergy.* **60**, 108–112.

20. Costagliola, S., Rodien, P., Many, M.C., Ludgate, M., and Vassart, G. (1998) Genetic immunization against the human thyrotropin receptor causes thyroiditis and allows production of monoclonal antibodies recognizing the native receptor. *J. Immunol.* **160**, 1458–1465.

21. Kilpatrick, K.E., Danger, D.P., Hull-Ryde, E.A., and Dallas, W. (2000) High-affinity monoclonal antibodies to PED/PEA-15 generated using 5 microgram of DNA. *Hybridoma.* **19**, 297–302.

22. Tearina Chu, T.H., Halverson, G.R., Yazdanbakhsh, K., Oyen, R., and Reid, M.E. (2001) A DNA-based immunization protocol to produce monoclonal antibodies to blood group antigens. *Br. J. Haematol.* **113**, 32–36.

23. Ramos, J.D.A., Teo, S.M.A., Ou, K.L., et al. (2003) Comparative allergenicity studies of native and recombinant *Blomia tropicalis* Paramyosin (Blo t 11). *Allergy.* **58**, 412–419.

24. Wolfowicz, C.B., HuangFu, T.Q., and Chua, K.Y. (2003) Expression and immunogenecity of the major house dust mite allergen Der p 1 following DNA immunization. *Vaccine.* **21**,1195–1204.

25. Ramos, J.D.A., Cheong, N., Lee, B.W., and Chua, K.Y. (2001) cDNA cloning and expression of Blo t 11, the *Blomia tropicalis* allergen homologous to paramyosin. *Int. Arch. Allergy Immunol.* **126**, 286–293.

Index

Printed In The United States Of America